Karst Land System and Sustainable Development

Karst Land System and Sustainable Development

Editor

Xiaoyong Bai

MDPI

Basel • Beijing • Wuhan • Barcelona • Belgrade • Novi Sad • Cluj • Manchester

Editor
Xiaoyong Bai
Chinese Academy of Sciences
Guiyang
China

Editorial Office
MDPI
St. Alban-Anlage 66
4052 Basel, Switzerland

This is a reprint of articles from the Special Issue published online in the open access journal *Land* (ISSN 2073-445X) (available at: https://www.mdpi.com/journal/land/special_issues/karst_land).

For citation purposes, cite each article independently as indicated on the article page online and as indicated below:

Lastname, A.A.; Lastname, B.B. Article Title. *Journal Name* **Year**, *Volume Number*, Page Range.

ISBN 978-3-0365-9220-6 (Hbk)
ISBN 978-3-0365-9221-3 (PDF)
doi.org/10.3390/books978-3-0365-9221-3

Contents

land MDPI

Article

Accounting for Value Changes in Cultivated Land Resources within the Karst Mountain Area of Southwest China, 2001–2020

Lu Zhang [1,2,†], Zhongfa Zhou [3,*], Quan Chen [1,†], Lan Wu [4], Qing Feng [3], Dan Luo [4] and Tangyin Wu [4]

1 School of Karst Science, Guizhou Normal University, Guiyang 550001, China; sophiazl@gznu.edu.cn (L.Z.); 201407075@gznu.edu.cn (Q.C.)
2 Real Estate Registration Center of Guizhou Province, Guiyang 550001, China
3 National Engineering Research Center for Karst Rocky Desertification Control, Guiyang 550001, China; 20030170040@gznu.edu.cn
4 The State Key Laboratory Incubation Base for Karst Mountain Ecology Environment of Guizhou Province, Guiyang 550001, China; 21010170543@gznu.edu.cn (L.W.); 21010170535@gznu.edu.cn (D.L.); 21010170545@gznu.edu.cn (T.W.)
* Correspondence: fa6897@gznu.edu.cn
† These authors contributed equally to this work and should be considered co-first authors.

Abstract: Cultivated land resources are important natural resource assets that are related to food security and sustainable development. Due to the many restrictive factors of the karst landform on agricultural production, the quantity and quality of cultivated land in the karst mountainous areas in Southwest China are poor. Reclaiming cultivated land to develop economy or to avoid transitional reclamation to protect ecology is an important proposition in this area. Analyzing changes in the physical and monetary value of cropland resources can help us to formulate more reasonable policies for the development and utilization of cultivated land resources, and to achieve a win-win scenario for economic development and ecological protection. Using multi-source remote sensing data and 20-year landcover data obtained by the GEE platform, this paper evaluated the cropland resources of the karst mountain areas of China at the pixel level. It was found that under the apparent outflow of the physical account of the cultivated land resources, the monetary value still maintained growth, proving that the current cultivated land-use policy in Guizhou Province has significantly improved the value of local cultivated land resources.

Keywords: cultivated land resource; value changes; karst mountain area; remote sensing; land use policy

Citation: Zhang, L.; Zhou, Z.; Chen, Q.; Wu, L.; Feng, Q.; Luo, D.; Wu, T. Accounting for Value Changes in Cultivated Land Resources within the Karst Mountain Area of Southwest China, 2001–2020. *Land* **2022**, *11*, 765. https://doi.org/10.3390/land11060765

Academic Editor: Le Yu

Received: 28 April 2022
Accepted: 22 May 2022
Published: 24 May 2022

1. Introduction

Natural resource assets are important means of production that are derived from nature, and that play a decisive role in economic and social development. The coordinated relationship between resource consumption, environmental protection, and economic growth has become a subject that affects human destiny [1]. Therefore, we need to find a method for tracking changes in nature, and for determining how changes are linked to economic and other human activities, to reflect the interactions between man and nature. Considering the increasing demand for statistics on natural capital within analytical policy frameworks on environmental sustainability, human well-being, and economic growth and development, advancing this emerging statistical field has become increasingly urgent [2].

Many scholars have performed statistical accounting for various natural resources, such as land resources and forests. Natural resource asset accounting uses the theories of statistics, accounting, resource science, and other disciplines to make a reasonable valuation of natural resources within certain periods of space and time, reflecting quantitative and structural changes to their physical quantity and value [3,4]. The purpose for this is to understand the current situation of natural resources, and the reasonable occupation, use, benefits, and disposal of natural resource assets, and finally, to solve the contradiction between resource utilization and environmental protection.

In 1993, the United Nations and the World Bank incorporated natural resources and the environment into a system of national economic accounting (SNA), and successively issued SEEA-1993 and SEEA-2003, in which physical value is used to describe interactions between the economy and the environment in various fields [5,6]. In March 2012, the 2012 System of Environmental Economic Accounting—Central Framework (SEEA-CF) was adopted as the international general guide, making it the first international statistical standard for environmental economic accounting, and it was supplemented by SEEA Experimental Ecosystem Accounting (SEEA-EEA) and SEEA Applications and Expansion. SEEA applies the accounting concepts, structures, rules, and principles of environmental information that are included in the System of National Accounts (SNA), and it uses a single framework to integrate environmental information (often measured in physical quantity) and economic information (often measured in value) [5–9]. It mainly covers the measurement of three areas: the physical flow of material and energy within and between the economy and the environment, and stocks of environmental assets and changes in these stocks, as well as environmentally related economic activities and transactions [10].

SNA, SEEA, and SEEA-EEA account research provides a good theoretical basis for the accounting of natural resource assets, but traditional SNA and SEEA accounting takes the natural environment as a kind of production material and adopts methods for which it can be presented to reflect the stock of the means of production and the flow in economic activities. Experience exists in related areas of assessment, such as land-cover and land-use statistics, but the integration of different areas of expertise into an accounting framework is new. In the latest SEEA-EEA specification, the principle of using surveying and mapping results has also been emphasized. At present, many studies also focus on how to use remote sensing data to support natural capital accounting [11]. Since natural resources have inherent location attributes, natural resources of the same quantity or quality will show great geographical differentiation in different locations; that is, simple presentation and accounting methods will not include the important spatial characteristics of natural resources. As a result, using multi-remote sensing data to conduct natural resource value not only allows the quantity and quality indicators of accounting objects to be obtained quickly, reducing the workload of manual investigation, but it can also evaluate the accounting results in the spatial dimension, so that the accounting results can better serve the decision-making processes.

In October 2016, UNSD, UNEP, CBD, and EU initiated NCAVES. The project lasted 3 years and was implemented in China, Brazil, India, Mexico, and South Africa. This project aimed to assist China in advancing the country's knowledge agenda for environmental and ecosystem accounting, and to initiate the pilot testing of SEEA Experimental Ecosystem Accounting (SEEA-EEA), as well as ecosystem valuation and macro-economic analysis, with a view toward improving the management of natural biotic resources, ecosystems, and their services at the national level, and mainstreaming biodiversity and ecosystems in national level policy planning and implementation [12]. Guizhou Province is one of the pilots in China, and many scholars have conducted much research into the natural resource balance sheet, GEP, ESV, and other fields, but the subject, object, and method of accounting need to be unified [13–15].

Cropland accounts for 10.20% of the global land surface area, which is the most important resource for agricultural production, and it plays an important role in ensuring food security, ecological security, and sustainable development [16,17]. The cultivated land resource is a natural resource that has been domesticated by human beings. Its growth and decline are not only restricted by natural laws, but are significantly affected by human activities. Compared to other kinds of natural resource assets, cultivated land resources can not only provide necessary food for survival, but they also participate in the energy transformation and material cycle of nature as an ecosystem, which is closely related to human society. This thus establishes how cropland value contributes to physical and monetary changes in long time series, which can assist with the analysis of the change range, flow characteristics, and reasons for change.

The formation of a karst landform is the result of the long-term dissolution of limestone and other soluble rocks by groundwater or surface water. The surface water is dissolved and eroded along the joints and fissures of soluble rocks, forming an uneven and broken surface shape. As one of the three karst-concentrated distribution areas in the world, the karst area in southern China has many factors that are not conducive to agricultural production. These factors, such as bedrock exposure, small soil stock, and discontinuous distribution [18,19], make agricultural planting difficult, and the cost of cultivated land management is very high. Additionally, due to the development of karst, the surface water is difficult to maintain, which means there is a serious water shortage in this region, but at the same time, the discharge of surface water in the rainy season is too late, causing water accumulation in some karst depressions. Therefore, karst areas in southern China are often accompanied by poverty; because both the quality and quantity of cultivated land are poor, the more cultivated the land is, the poorer the people, and the contradiction between man and land is very prominent. As the core area of karst in southern China, Guizhou has serious rocky desertification and a large area of rock exposure. By exploring the impact of human activities on cultivated land, we can determine the positive policies that can improve the value of cultivated land resources, something that is of great significance for ameliorating the current situation of poverty in China's poor areas within the karst [20–22].

Landcover data provide the most direct feedback when accounting for cultivated land resource physical quantity, but cropland resource assessment methods will inevitably require more detailed spatial data. As the development of remote sensing and big-data technology have already brought a new approach towards accounting, we can obtain multi-source remote sensing data more quickly to assist with the accounting work, improve the accuracy of the accounting, and reduce the cost. This research aimed to realize the dynamic monitoring of the spatial pattern evolution of cropland resources via physical accounting, using multi-remote sensing data [23]. It can make up for the defects in the SEEA-CF accounting framework, which only presents data rather than spatial information. Meanwhile, in order to quantify the change rules of the cultivated land resource value, and to observe whether effective land management policies have been adopted, this paper evaluated the changes of cultivated land resource value in Guizhou Province from 2001 to 2020. By analyzing the impact of the economy and other human activities on cropland, it proved that the current cultivated land use policy in Guizhou has significantly improved the value of local cultivated land resources. This provides a reference for the rational utilization of cultivated land resources.

2. Materials and Methods

2.1. Study Area

Guizhou province is located in the inland area of Southwest China, to the east of Yunnan Guizhou Plateau, and is located between 24°37′–29°13′ N and 103°36′–190°35′ E, which is an important ecological barrier in the upper reaches of the Yangtze River and the Pearl River [24,25]. Meanwhile, as the junction of the Eurasian plate and the Indian Ocean plate, its terrain is high in the west and low in the east, tilting from the middle to the north, and from the east and to the south. The landform of the whole province can be divided into four basic types: plateau, mountain, hill, and basin. Moreover, Guizhou province is one of the three karst-concentrated distribution areas in the world, the core area of East Asia, which is also the largest distribution area and the strongest conical karst development in China. With high mountains, deep valleys, and steep terrain, 92.5% of the area of the province is mountainous and hilly, and 109,100 square km comprises exposed karst landform, which means the surface is extremely fragmented and lacks the cropland resources for agriculture [26–28]. In addition, due to the increasing population, the cultivated land area continues to reduce, meaning that the percapita cultivated land area is less than 300 square meters, which is far lower than the average level in China [29].

Moreover, the proportion of cultivated land with a thick soil layer, high fertility, and good conditions of water conservation is low (Figure 1).

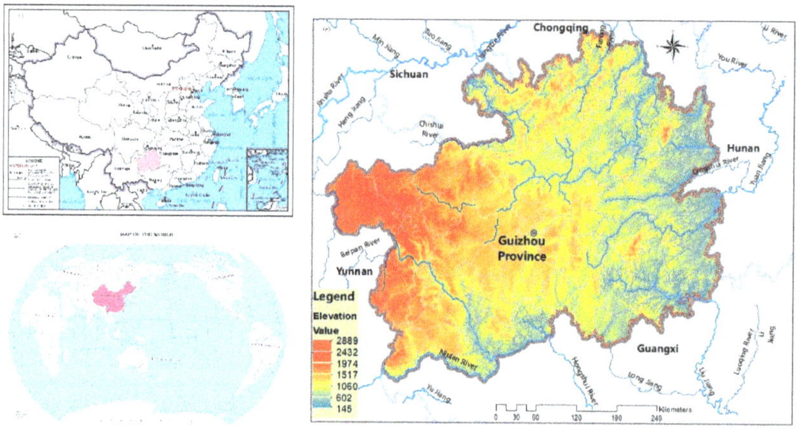

Figure 1. Location of the study area.

2.2. Dataset

2.2.1. Spatial Data

We used Google Earth Engine (GEE) to gather and to calculate the spatial data for this analysis. GEE is an interactive platform that provides geospatial processing services that are powered by the Google Cloud Platform [30]. With Earth Engine, we can perform geospatial processing at a scale that is free of charge, and we can carry out high-impact, data-driven scientific research involving large geospatial datasets [31,32]. In this research, we adopted multi-remote sensing time series data from 2000 to 2020, to detect the impact of land use changes on the value of cropland resources. Landcover data were derived from images collected by the MODIS sensor (the MCD12Q1 V6 product), which provides global land cover types at yearly intervals (250 m × 250 m). The digital elevation models (DEMs) used Shuttle Radar Topography Mission (SRTM) data at a 30 m resolution. Additionally, we estimated the Landsat net primary production (NPP) using Landsat Surface Reflectance for CONUS (Landsat net primary production CONUS) [33]. Beyond these, we selected the GPM data (Monthly Global Precipitation Measurement v6) to revise the existing results of ecological value. Global Precipitation Measurement (GPM) is an international satellite mission that provides next-generation observations of rain and snow worldwide, every three hours. The Integrated Multi-Satellite Retrievals for GPM (IMERG) is a unified algorithm that provides rainfall estimates by combining data from all passive-microwave instruments in the GPM Constellation.

2.2.2. Socioeconomic Data

Socioeconomic data, including the yields of major farm crops (YMFC), the gross output value of farming (GOVF), the gross domestic product (GDP), the permanent resident population (PRP) and the employment in agriculture were obtained from the Guizhou statistical yearbook (2001–2021) (http://stjj.guizhou.gov.cn/ accessed on 4 April 2022). In addition, we derived the grain prices from 2001 to 2020 from the "The National Compilation of Cost-benefit data of Agricultural Products" as a reference (Table 1).

Table 1. Data sources for assessing cultivated land resources value.

	Resource Type	Data Sources
Spatial Data	Land cover (MCD12Q1 V6) Digital elevation models (DEMs) Landsat net primary production (NPP) Global precipitation measurement (GPM)	Google Earth Engine Plateform (https://developers.google.cn/earth-engine/datasets) accessed on 4 April 2022
Socioeconomic Data	Yields of major farm crops (YMFC) Gross output value of farming (GOVF) Gross domestic product (GDP) Permanent resident population (PRP) Financial expenditure Employments in agriculture	Guizhou statistical yearbook (2001–2021) (http://stjj.guizhou.gov.cn/) accessed on 4 April 2022
	Grain prices	The National Compilation of Cost-benefit data of Agricultural Products

2.3. Methods

2.3.1. Cropland Resources Value Accounting Framework

To make a scientific evaluation of the value of cropland resources, we established three accounting accounts [34]: the physical quantity account, the conditional account, and the monetary account [35–38]. Among them, the physical quantity account was used to reflect the changes in the number and scope of cultivated land in the study area from 2000 to 2020, and to provide necessary data for value accounting, while the quality account was used to record the quality status of cultivated land in the study area. Since it is obvious that the value of cropland varies along the quality status, there will be significant differences in crop yield and ecological function. Finally, the monetary account includes two parts. One is the direct value, also called the use value or the commodity value, which is the value that is formed by people's direct harvesting, which is the output value of agricultural products provided by cropland resources. This part can be calculated by the market price method, because agricultural products can directly enter circulation as commodities [39]. The other part is the indirect value, which refers to the ecological service ability of cropland resources as a part of the natural environment when they exist in a natural way, as well as the value of natural resource assets that are used to meet human spiritual, cultural, and moral needs, and social development [40] (Table 2).

Table 2. Indicators for assessing cultivated land resources value.

Account	First-Level Indicators	Second-Level Indicators
Physical Account	Extent	Area
	Biomass provision	Crop Production
Conditional Account	Site conditions	Elevation Slope
	Landscape index	Patch Density (PD) Edge Density (ED) Area-Weighted Mean Shape Index (AWMSI) Fragmentation Index of Patch Numbers (FN) Fragmentation Shape Index (FS) Aggregation Index (AI)
Monetary Account	Direct value	Crop Market Value
	Indirect value	Gas Regulation Climate Regulation Environmental Purification Hydrological Regulation Soil Conservation Maintenance of Nutrient Cycles Biodiversity Aesthetic Landscape

The direct value is the gross output value of various grains, tubers, oil crops, vegetables, and other crops in Guizhou Province. The indirect value is the sum of the value equivalent for each ecological function. The annual cropland resources value is the direct value plus the indirect value.

$$V_T = V_D + V_{ID} \tag{1}$$

$$V_T = \sum_{i=1}^{n} V_{Di} \tag{2}$$

$$V_{ID} = \sum_{i=1}^{n} V_{IDi} \tag{3}$$

In the formula, V_T is the total monetary value of cropland resources, V_D is the gross output value, and V_{ID} is the total value equivalent of cropland ecological function.

2.3.2. Landscape Index

Cropland fragmentation refers to the fragmentation, dispersion, and size of the cultivated land due to natural or human factors, and the area of each cultivated land is relatively small, showing a decentralized and disorderly pattern, which is a long-term dynamic process [41–43]. For the karst mountain areas, the high mountains and deep valleys lead to obvious cutting terrain, and cultivated land can only be distributed on gentle slopes or small flat land. Therefore, the degree of cultivated land fragmentation is a very typical quality evaluation index in karst areas and plays a decisive role in the realization of the value of cultivated land resources [44,45].

Research on the impact of cultivated land fragmentation on the landscape scale of cultivated land can directly reflect changes in cultivated land fragmentation. In this study, we used the open-source Python library to compute landscape metrics, and the following six indicators were selected to measure the cultivated land landscape [46]:

- Patch Density (PD)

This indicator refers to the number of cultivated land patches per unit area in the study area, and it has an important impact on biological protection, material, and energy distribution. This index reflects the situation in which the concentrated and contiguous cultivated land is divided into small patches, which directly reflects upon the connotation of cultivated land landscape fragmentation [47].

$$\mathrm{PD} = n/A \tag{4}$$

n is the number of the patches; A is the total area.

- Edge Density (ED)

This is an index that is used to analyze the shape of land patches, revealing the degree of cropland segmentation, as well as being a direct reflection of the degree of cultivated land fragmentation. The greater the edge density, the higher the degree of cultivated land division, and the more scattered the layout [48].

$$\mathrm{ED} = P/A \tag{5}$$

P is the total perimeter of all cropland patches; A is the total area.

- Fragmentation Index of Patch Numbers (FN)

The patch size is the most basic spatial feature and it directly affects the mechanization level of agricultural production. As such, this index is used to measure the degree of fragmentation of the landscapes.

$$\mathrm{FN} = (N-1)/MPS \tag{6}$$

MPS is the mean patch size; N is the number of cropland patches.

- Area-Weighted Mean Shape Index (AWMSI)

Since an irregular shape leads to a reduction in the actual planting area within the total area, the farming production cost per unit area will be increased. However, with an increase in the patch size, the impact caused by the irregular shape will gradually weaken. Considering this phenomenon, AWMSI is taken to be one of the indicators to measure the degree of the cultivated landscape.

$$\text{AMWSI} = \sum_{i=1}^{n} \left[\left(\frac{0.25P_i}{\sqrt{a_i}} \right) (a_i/A) \right] \tag{7}$$

n is the number of cropland patches; P_i is the perimeter of the patches; a_i is the area of the patches; A is the total area of cropland.

- Fragmentation Shape Index (FS)

This index is used to reflect the internal combination of cultivated land patches. The distribution of cultivated land patches becomes more scattered as the index increases. Additionally, the internal combination simultaneously becomes more complex.

$$\text{FS} = 1 - 1/MSI \tag{8}$$

$$MSI = \sum_{i=1}^{n} (0.25P_i/\sqrt{a_i})/N \tag{9}$$

MSI is the mean shape index; a_i is the patch area; P_i is the perimeter of the patch; N is the number of cropland patches.

- Aggregation Index (AI).

This index reflects the degree of patch agglomeration within the landscape type. When the value is larger, the landscape is composed of a few large patches, and when the value is smaller, the landscape is composed of many small patches [49].

$$\text{AI} = \frac{e_i}{\text{max\_}e_i} \times 100 \tag{10}$$

$$\text{max\_}e_i = \begin{cases} 2n(n-1), & m = 0 \\ 2n(n-1) + 2m - 1, & m \le n \\ 2n(n-1) + 2m - 2, & m > n \end{cases}, \quad \left(m = A_i - n^2 \right) \tag{11}$$

e_i is the number of edges that the patches have in common; max_e_i is the maximum number of edges that the patches have in common; P_i is the perimeter of the patch; n is the edge length of the largest integer square that does not exceed the total area of the cropland area.

2.3.3. Revisions of the Ecological Value Equivalent Factors

Costanza et al. proposed the principle and method of ecosystem service value estimation [50], but their methods were criticized because they resulted in the ecological value of the cultivated land being significantly low. Therefore, Chinese researchers such as Xie Gaodi revised Costanza's assessment framework based on China's economic situation, land use, and vegetation types, and developed an assessment method for China's ecosystem service value based on the unit area value equivalence factor [51–53] (Appendix A). As the ecological function value consequently varies with the internal structure and external form of ecosystem, constantly changing within different regions or different periods, we conducted two revisions to obtain the final ecosystem service value equivalent of Guizhou [54]:

1. Previous studies have shown that the ecosystem function is positively correlated with NPP and precipitation. As such, we used two temporal and spatial factors (NPP and precipitation) to modify the ecosystem service value equivalent table of China for each year.

$$F_i = \begin{cases} P_i \times F_{n1} \\ R_i \times F_{n2} \end{cases} \quad (12)$$

F_i refers to the unit area value equivalent of the ecological service function for each year; P_i refers to the NPP regulation factor; R_i refers to the precipitation regulation factor; F_{n1} represents the value equivalent per unit area of China for gas regulation, climate regulation, environmental purification, nutrient conservation, and biodiversity maintenance; and F_{n2} represents the value equivalent of China's unit area of hydrological regulation function.

2. According to Costanza's research, the economic value of ecological service value equivalent factors is 54 USD/hm^2 (1997). Combined with China's grain production income, Chinese scholars have calculated that the economic value of an ecological service value equivalent factor in China is 449 CNY/hm^2 (58.5 USD/hm^2 in 2007), using the shadow land rent method. However, the price index and grain yield vary interannually, and so to reflect the indirect value change of cultivated land resources more accurately, we revised the economic value by year to form the final economic value of the ecological function, to make it suitable for the study area [55].

$$E_{Vi} = \frac{1}{7}\sum_{i=1}^{n} \frac{m_i p_i q_i}{M} \quad (13)$$

E_{Vi} refers to the economic value of an ecological service value for equivalent factors of cropland resources in each year; m_i refers to the area of crops; p_i refers to the average price of crops; q_i represents the output of agricultural products; n represents the types of crop products.

3. Results

3.1. Physical Account Changes

3.1.1. Spatial Changes of Guizhou Province

Through the analysis of the land cover data of the study area from 2001 to 2020, it was found that the cropland resources in Guizhou Province experienced a small increase from 2001 to 2003, and they have then decreased year-by-year since 2004 (Figure 2). By 2020, the cropland resources had reduced to 3768.34 km^2, which means that the number had decreased by 55.52% compared to 2001. At the same time, it is easy to see that the cultivated land resources in Guizhou Province are very scarce. The proportion of cultivated land resources only accounted for 5.35% at the highest level (2003), while this figure reduced to 2.14% in 2020 (Table 3). Moreover, with the increase in the population, the percapita cultivated land resources in Guizhou Province show absolute scarcity, from 223.04 m^2 in 2001 to 97.68 m^2 in 2020.

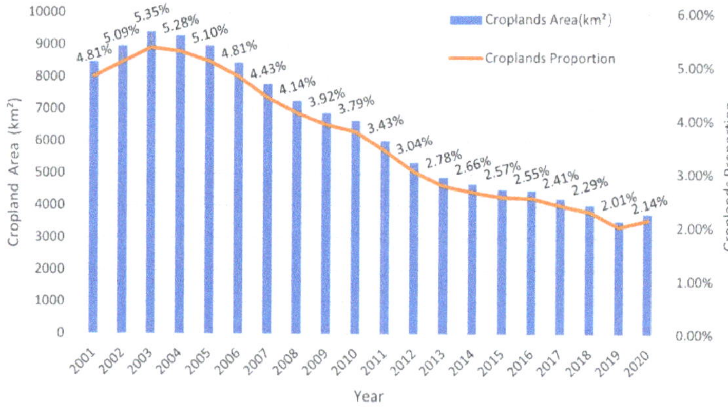

Figure 2. Cropland resources area of Guizhou Province.

Table 3. Area changes of cropland resources.

	2001	2002	2003	2004	2005	2006	2007	2008	2009	2010
Cropland Area (km^2)	8473.22	8968.61	9416.08	9294.78	8980.94	8469.23	7800.17	7286.11	6902.27	6667.85
Croplands Proportion	4.81%	5.09%	5.35%	5.28%	5.10%	4.81%	4.43%	4.14%	3.92%	3.79%
Croplands per capita (m^2)	223.04	233.74	243.31	238.08	240.78	229.52	214.76	202.62	195.14	191.66
	2011	**2012**	**2013**	**2014**	**2015**	**2016**	**2017**	**2018**	**2019**	**2020**
Croplands Area (km^2)	6047.527	5356.4	4894.82	4683.90	4526.89	4494.66	4236.63	4032.42	3543.72	3768.34
Croplands Proportion	3.43%	3.04%	2.78%	2.66%	2.57%	2.55%	2.41%	2.29%	2.01%	2.14%
Croplands per capita (m^2)	174.33	153.74	139.77	133.52	128.24	119.60	111.40	105.51	92.09	97.68

Secondly, each patch of land cover data was calculated, and the time series changes of each pixel was analyzed for different years, with the finding that the area of cropland resources experienced both transfer-in and transfer-out in the same year. During this period, the positive area changes in the cultivated land area are in a "U" shape, while the negative area changes are represented by a wave form. Moreover, with the transfer proportions of −9.43% (−798.82 km^2), −9.28% (−786.03 km^2), and −8.55% (−724.78 km^2), 2007, 2012 and 2019 became troughs. The overall distribution of the total change area was similar to that of the negative area (Figure 3).

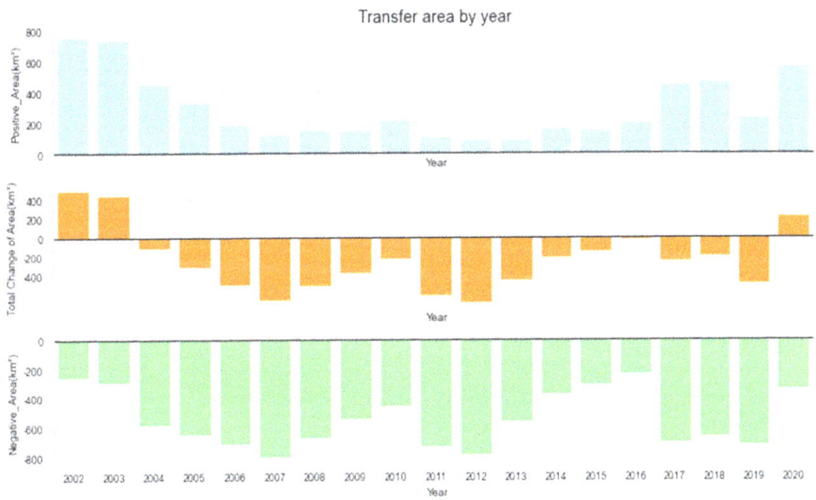

Figure 3. Cropland resources transfer area of Guizhou Province.

According to the International Geosphere-Biosphere Programme (IGBP) classification (Appendix A), there are 15 types of land cover in Guizhou Province. As can be seen from Figure 4, the cultivated land resources in Guizhou Province are mainly transferred with grasslands, savannas, and cropland/natural vegetation mosaics. Recent land cover data over the last 20 years show that the transfer of cropland/natural vegetation mosaics account for an average of 66% of the total transfer area. The type of cropland/natural vegetation mosaics are mosaics of small-scale cultivation, with 40–60% of natural tree, shrub, or herbaceous vegetation in a pixel. The increase in cropland/natural vegetation mosaics shows that the fragmentation of cultivated land resources in Guizhou Province increased from 2014 to 2019 (Table 4).

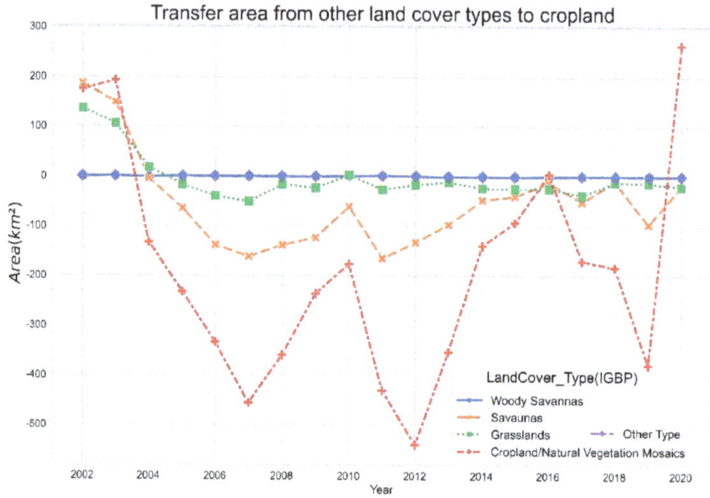

Figure 4. Transfer area of each landcover type in Guizhou Province.

Table 4. Transfer area and proportion of main landcover types.

		Woody Savannas	Savannas	Grasslands	Cropland/Natural Vegetation Mosaics
2002	Area (km^2)	−0.32	187.22	135.42	175.06
	Proportion	0.06%	37.79%	27.34%	35.34%
2003	Area (km^2)	0.45	148.45	105.53	193.05
	Proportion	0.10%	33.17%	23.58%	43.14%
2004	Area (km^2)	−1.23	−4.68	17.78	−133.17
	Proportion	0.79%	2.98%	11.34%	84.90%
2005	Area (km^2)	0.34	−64.31	−16.98	−232.89
	Proportion	0.11%	20.47%	5.40%	74.13%
2006	Area (km^2)	−0.22	−138.36	−39.78	−333.35
	Proportion	0.04%	27.04%	7.77%	65.14%
2007	Area (km^2)	−0.22	−161.69	−50.45	−456.26
	Proportion	0.03%	24.17%	7.54%	68.19%
2008	Area (km^2)	−0.22	−138.62	−16.06	−359.17
	Proportion	0.04%	26.97%	3.12%	69.87%
2009	Area (km^2)	−0.56	−123.74	−23.01	−236.31
	Proportion	0.15%	32.24%	6.00%	61.56%
2010	Area (km^2)	0.38	−60.77	2.67	−176.47
	Proportion	0.16%	25.27%	1.11%	73.37%
2011	Area (km^2)	1.63	−165.25	−26.13	−430.57
	Proportion	0.26%	26.50%	4.19%	69.05%
2012	Area (km^2)	−0.11	−133.43	−17.06	−540.31
	Proportion	0.02%	19.31%	2.47%	78.18%
2013	Area (km^2)	−0.76	−96.93	−11.00	−352.89
	Proportion	0.17%	21.00%	2.38%	76.45%
2015	Area (km^2)	−0.11	−12.31	−24.25	4.43
	Proportion	0.35%	25.13%	15.53%	58.99%
2016	Area (km^2)	−0.11	−12.31	−24.25	4.43
	Proportion	0.26%	29.95%	59.01%	10.78%
2017	Area (km^2)	0.16	−50.65	−37.40	−170.15
	Proportion	0.06%	19.60%	14.48%	65.86%
2018	Area (km^2)	0.28	−10.01	−11.73	−182.74
	Proportion	0.13%	4.89%	5.73%	89.25%
2019	Area (km^2)	0.00	−96.45	−12.99	−379.26
	Proportion	0.00%	19.74%	2.66%	77.61%
2020	Area (km^2)	0.12	−19.17	−20.16	263.83
	Proportion	0.04%	6.32%	6.65%	86.99%

3.1.2. Crop Production Changes in Guizhou Province

According to the research presented, there are two dimensions of changes in the output of agricultural products in Guizhou Province. The first is the change in quantity. The total output of agricultural products increased from 12.67 million tons up to 42.34 million tons, from 2001 to 2020, with an increase rate of 234.19%. The second is that the planting structure changed greatly, which is reflected in the changes in the crop types within the same crop type, and the quantitative changes between the different types.

The main agricultural products in Guizhou Province can be divided into grains, potatoes, oil crops, and others, of which the output of vegetables far exceeds other products, reaching 29.9087 million tons in 2020. Rice and corn are the main grain, showing little interannual change and fluctuating in the range of 605,940. Tubers increased slightly; Irish potatoes are the main crop and showed obviously changes. The median output from 2001 to 2020 was 1.535 million tons, and the third quarter was 2.335 million tons. Rapeseed is the main oil plant crop, accounting for more than 70%. As the economy has continued to develop, ramie has been completely replaced by other types of crops (Figure 5).

The proportion of grain compared to the total agriculture products of Guizhou Province increased from 72.55% in 2001 and plummeted to 16.34% in 2020; at the same time, the output of oil crops and tubers has also decreased by nearly half in 20 years, while other high-value-added crops that increased from 916.1 thousand tons (7.23%) in 2001 surged to 30.73 million tons (72.58%) in 2020 (Table 5).

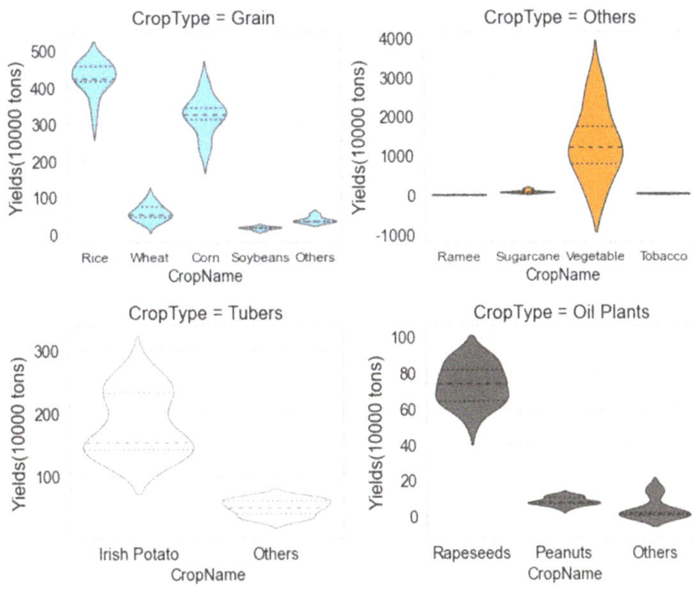

Figure 5. Crop production distribution in Guizhou Province (2001–2020).

Table 5. Crops production of Guizhou Province (10,000 tons).

	Grain		Oil Plants		Others		Tubers		Total Yields
2001	919.2	72.55%	71.32	5.63%	91.61	7.23%	184.9	14.59%	1267.03
2002	829.7	68.49%	72.48	5.98%	104.73	8.65%	204.5	16.88%	1211.41
2003	903.5	45.33%	72.31	3.63%	816.72	40.97%	200.8	10.07%	1993.33
2004	939.32	44.69%	82.71	3.94%	869.38	41.37%	210.26	10.00%	2101.67
2005	906.24	41.59%	84.89	3.90%	942.08	43.23%	245.82	11.28%	2179.03
2006	820.07	41.23%	68.24	3.43%	882.59	44.38%	217.93	10.96%	1988.83
2007	869.73	40.53%	69.66	3.25%	975.53	45.46%	231.13	10.77%	2146.05

Table 5. *Cont.*

	Grain		Oil Plants		Others		Tubers		Total Yields
2008	911.67	39.18%	68.39	2.94%	1100.62	47.30%	246.33	10.59%	2327.01
2009	918.76	37.87%	78.68	3.24%	1179	48.60%	249.51	10.29%	2425.95
2010	901.9	36.60%	60.34	2.45%	1291.3	52.41%	210.4	8.54%	2463.94
2011	605.13	26.52%	78.85	3.46%	1326.15	58.12%	271.77	11.91%	2281.9
2012	804.91	29.73%	87.38	3.23%	1540.9	56.91%	274.59	10.14%	2707.78
2013	718.88	25.46%	91.53	3.24%	1701.53	60.27%	311.11	11.02%	2823.05
2014	790.33	25.78%	98.05	3.20%	1829.23	59.67%	348.17	11.36%	3065.78
2015	815.89	25.48%	101.34	3.16%	1920.9	59.99%	364.11	11.37%	3202.24
2016	828.38	24.09%	113.66	3.31%	2132.24	62.01%	364	10.59%	3438.28
2017	808.94	22.11%	109.82	3.00%	2370.18	64.78%	369.6	10.10%	3658.54
2018	732.59	18.93%	112.62	2.91%	2698.65	69.72%	327.11	8.45%	3870.97
2019	707.57	17.81%	103.01	2.59%	2819.22	70.95%	343.67	8.65%	3973.47
2020	692.04	16.34%	103.4	2.44%	3073.27	72.58%	365.59	8.63%	4234.3

3.2. Conditional Account Changes

3.2.1. Changes in Site Conditions

We used the GEE to calculate the DEM data for the cultivated land resources in Guizhou Province, which showed that the cultivated land resources in Guizhou Province are mainly distributed near the elevations of 1320 m and 2220 m. The mean elevation increased by approximately 130 m from 2001 to 2020, but the standard deviation decreased significantly, implying that the elevation of the cultivated land resources in Guizhou Province is gradually concentrated to the average value (Table 6). Therefore, it can also be judged that the elevation of the cultivated land resources in Guizhou Province have shown an overall increase.

Table 6. Statistical results of the elevation of the cultivated land resources in Guizhou Province.

	Mean	Median	Std-Dev	Mix	Max
2001	1577.95	1461	526.76	299	2831
2002	1557.5	1440	524.91	229	2831
2003	1540.92	1426	527.19	229	2831
2004	1529.13	1415	534.46	229	2831
2005	1516.96	1403	538.91	229	2831
2006	1521.53	1402	536.46	229	2834
2007	1535.18	1407	531.34	229	2815
2008	1549.44	1418	527.26	229	2815
2009	1571.46	1433	524.21	229	2815
2010	1587.78	1448	517.44	229	2834
2011	1623.42	1483	516.99	229	2834
2012	1648.67	1516	520.09	229	2834
2013	1674.58	1564	516.2	229	2834
2014	1682.55	1576	510.25	261	2834
2015	1689.86	1585	506.38	261	2834
2016	1682.65	1559	500.84	260	2834
2017	1659.46	1508	493.31	260	2811
2018	1661.55	1520	494.06	260	2811
2019	1700.18	1633	492.13	241	2769
2020	1693.96	1587	475.21	262	2757

Meanwhile, through the statistics of the slope of each cultivated land pixel, it was found that 70–80% of the cultivated land resources in Guizhou Province are distributed in areas with a slope of less than 10°. With the evolution of the distribution pattern of the cultivated land resources, the changes in the slope of the cultivated land resources can be divided into two stages (Figure 6). First, from 2001 to 2004, the number of cultivated land pixels with a slope of less than 25° continued to increase. In 2004, 46812 pixels were distributed in areas below 5°, accounting for 50.68% of the total area of cultivated land resources. Additionally, there were 28556 pixels with a slope of between 5° and 10°, accounting for 29.75% of the total area of cultivated land resources in that year. Secondly, from 2005 to 2020, the area of cultivated land with a slope of more than 25° decreased significantly, with a maximum change rate of more than 80%, and the cultivated land area with a slope above 40° completely disappeared (Table 7).

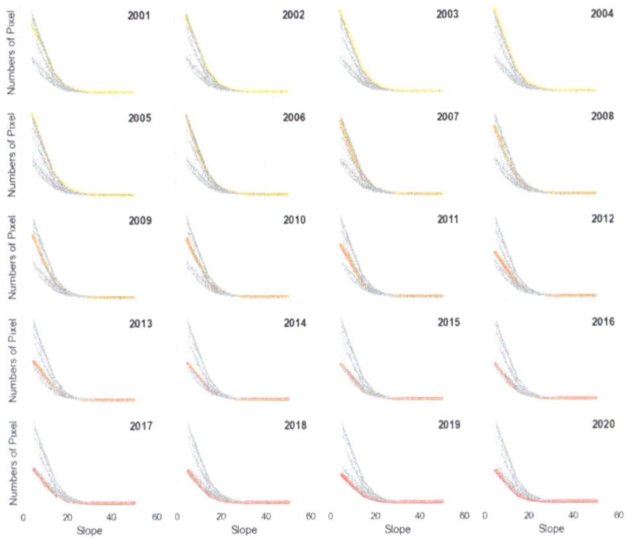

Figure 6. Slope changes of cropland resources in Guizhou Province.

Table 7. Pixels counts of the slopes of cropland resources in Guizhou Province.

Year	0–5°	5°–10°	10°–15°	15°–20°	20°–25°	25°–30°	30°–35°	35°–40°	40°–45°	45°–50°
2001	38,599	26,799	11,162	3935	1391	324	86	32	8	2
2002	42,765	27,879	11,562	4118	1422	338	91	34	8	2
2003	46,059	29,252	12,208	4396	1533	373	106	38	11	2
2004	46,812	28,556	11,993	4438	1553	363	94	31	10	2
2005	46,240	27,146	11,490	4369	1540	340	84	21	6	2
2006	44,636	25,236	10,535	3907	1335	286	74	19	6	2
2007	41,474	22,768	9360	3399	1127	256	64	17	6	2
2008	38,024	21,051	8620	3068	1024	233	59	16	6	2
2009	35,303	20,253	8232	2857	948	220	55	16	6	2
2010	33,253	19,613	7985	2790	936	218	57	16	6	2
2011	29,369	17,972	7294	2559	868	209	57	16	6	2
2012	25,028	15,550	6414	2260	781	187	48	15	8	2
2013	22,197	14,107	5833	2103	720	181	49	12	8	1
2014	20,752	13,127	5398	1925	636	147	38	10	5	1
2015	19,857	12,350	5021	1783	563	133	32	8	4	1
2016	19,717	11,945	4768	1597	496	117	32	7	4	1
2017	19,346	10,659	4098	1380	463	85	31	7	4	1
2018	17,932	9610	3763	1242	406	72	28	2	3	0
2019	15,385	8921	3498	1162	368	68	20	2	0	0
2020	17,405	9697	3497	1056	312	54	19	3	0	0
Max Change rate	62.82%	66.85%	71.35%	76.20%	79.89%	85.11%	82.08%	92.11%	100.00%	100.00%

3.2.2. Landscape Index Changes

By calculating the six dimensions of the landscape index for the cultivated land resources in Guizhou Province from 2001 to 2020 (Table 8), we found that the patch density (PD), edge density (ED), and aggregation index (AI) in Guizhou Province increased first, and then decreased. Meanwhile, the area-weighted mean shape index (AWMSI) showed a negative trend by year. Additionally, the change trend in the fragmentation index of the patch numbers (FN) was negatively correlated with the fragmentation shape index (FS). According to the calculation results, the PD decreased from 0.0117 in 2001 to 0.007 in

2020, indicating that the fragmentation of cultivated land resources in Guizhou Province improved, and the ED decreased from 0.9973 in 2001 to 0.5548 in 2020, indicating that the shape of the cultivated land gradually became more regular [47]. The AWMSI decreased from 7.954 in 2001 to 6.001 in 2020, indicating that the distribution of cultivated land plots tends to be centralized. It can be seen that the vulnerability of cultivated land in Guizhou Province has been reduced.

Table 8. Landscape index changes of cultivated land resources in Guizhou Province.

Year	PD	ED	FN	AWMSI	FS	AI
2001	0.0117	0.9973	14.802622	7.954	0.28310273	79.5801
2002	0.012	1.0353	14.666524	7.9734	0.28356498	79.9096
2003	0.0119	1.0653	13.813862	7.6421	0.29263634	80.2707
2004	0.0113	1.0362	12.507077	7.6585	0.29358576	80.6063
2005	0.0106	0.9917	11.415985	7.6782	0.29088073	80.8496
2006	0.0097	0.9384	10.262608	7.3696	0.29567545	80.878
2007	0.0093	0.8818	10.142569	6.8898	0.29473165	80.5931
2008	0.0089	0.846	9.9882939	6.5778	0.29903267	80.132
2009	0.0088	0.8114	10.413669	6.5818	0.29173454	79.9631
2010	0.0089	0.7977	10.807041	6.3748	0.2917847	79.6394
2011	0.0084	0.7385	10.757928	6.4825	0.28861066	79.3885
2012	0.0081	0.6836	11.313965	6.597	0.28310273	78.6491
2013	0.0078	0.6439	11.563548	6.4575	0.27917538	78.1715
2014	0.0081	0.638	12.935312	6.6323	0.27436325	77.4089
2015	0.0083	0.6324	13.99703	6.729	0.26691592	76.8763
2016	0.0085	0.6407	14.919812	6.7727	0.26524614	76.3702
2017	0.0081	0.6163	14.078087	6.3173	0.27028605	76.0021
2018	0.0083	0.6109	15.651382	6.0193	0.26975318	75.0247
2019	0.0071	0.54	13.104289	5.8462	0.27103076	75.385
2020	0.007	0.5548	11.886296	6.0001	0.274942	76.1123

PD: patch density. ED: edge density. FN: patch numbers. AI: aggregation index. AWMSI: area-weighted mean shape index. FS: fragmentation shape index.

3.3. Monetary Account Changes

From 2001 to 2020, with the development of the economy, the direct economic value of cultivated land resources in Guizhou Province increased rapidly. The production value increased from CNY 27,995 million per year to CNY 180,025 million per year, with an increase of 543%. It could be evidenced (Table 9) that in the past 20 years, with the adjustment of the industrial structure, the main labor force flowed to the secondary and tertiary industries with a high added value and high income, reducing the number of agricultural employees in Guizhou. In 2001, there were 1.36 million agricultural employments, while in 2020, only 0.634 million people were employed in agriculture. At the same time, the per capita output value increased from CNY 2046.42/y to CNY 28,395.11/y, an increase of 12.87-fold. However, according to the calculation of the price index of agricultural products in "The National Compilation of Cost-benefit data of Agricultural Products", the sales price of agricultural products in China only increased by 181.96% from 2001 to 2020. In other words, the direct economic value of cultivated land resources in Guizhou Province still improved significantly after removing the influence of the interannual differences in prices.

Table 9. Direct economic value of cropland resources in Guizhou Province.

Year	Cross Output Value (Million Yuan)	Agriculture Employment (10^4)	Cross Output Value per Capita (CNY)
2001	27,995	1368	2046.42
2002	27,888	1354	2059.68
2003	46,672	1322	3530.41

Table 9. *Cont.*

Year	Cross Output Value (Million Yuan)	Agriculture Employment (10^4)	Cross Output Value per Capita (CNY)
2004	52,464	1288	4073.29
2005	33,353	1268	2630.36
2006	34,797	1247	2790.46
2007	39,220	1388	2825.65
2008	30,848	1350	2285.04
2009	33,050	1299	2544.26
2010	38,561	1210	3186.86
2011	43,084	1194	3608.38
2012	56,132	1189	4720.94
2013	64,612	1180	5475.59
2014	85,189	1171	7274.89
2015	109,654	1162	9436.66
2016	119,650	883	13,550.35
2017	130,643	828	15,778.11
2018	143,929	765	18,814.19
2019	156,647	700	22,378.14
2020	180,025	634	28,395.11

As for the indirect value, according to the revised ecological value per unit area of farmland ecosystems and the equivalent of the ecological service value per unit area in Guizhou Province (Table 10), we calculated the indirect value of cultivated land resources (Table 11). Under the dual influence of cultivated land resource area falling and the grain price index increasing, the indirect economic value of cultivated land resources in Guizhou Province first increased, and then decreased. Among them, it reached a peak of CNY 7775.25 million in 2009, but the overall decrease was no more than 3%, indicating that the ecological function of the cultivated land resources in Guizhou Province is still well-protected while the economy is developing (Table 12).

Table 10. Ecosystem service equivalent value per unit area of cropland ecosystem in Guizhou.

	Regulating Services				Supporting Services			Cultural Services
	Gas Regulation	Climate Regulation	Environmental Purification	Hydrological Regulation	Soil Conservation	Maintenance of Nutrient Cycle	Biodiversity	Aesthetic Landscape
2001	3.25	1.70	0.49	5.27	1.90	0.57	0.62	0.27
2002	3.01	1.57	0.46	5.36	1.76	0.52	0.57	0.25
2003	2.77	1.45	0.42	4.64	1.62	0.48	0.53	0.23
2004	2.70	1.41	0.41	5.40	1.58	0.47	0.52	0.23
2005	2.91	1.52	0.44	4.44	1.70	0.51	0.56	0.25
2006	2.98	1.56	0.45	5.00	1.74	0.52	0.57	0.25
2007	3.09	1.61	0.47	5.47	1.80	0.54	0.59	0.26
2008	3.00	1.57	0.46	5.51	1.75	0.52	0.57	0.25
2009	3.03	1.58	0.46	4.49	1.77	0.53	0.58	0.26
2010	2.72	1.42	0.41	4.63	1.59	0.47	0.52	0.23
2011	2.84	1.48	0.43	4.17	1.66	0.49	0.54	0.24
2012	2.86	1.50	0.43	4.56	1.67	0.50	0.55	0.24
2013	3.15	1.64	0.48	4.16	1.84	0.55	0.60	0.27
2014	2.95	1.54	0.45	5.99	1.72	0.51	0.56	0.25
2015	3.03	1.58	0.46	5.48	1.77	0.53	0.58	0.26
2016	3.08	1.61	0.47	4.84	1.80	0.54	0.59	0.26
2017	3.07	1.60	0.47	4.97	1.79	0.53	0.59	0.26
2018	2.75	1.44	0.42	4.81	1.61	0.48	0.53	0.23
2019	3.19	1.66	0.48	5.39	1.86	0.55	0.61	0.27
2020	2.86	1.49	0.43	5.83	1.67	0.50	0.55	0.24

Table 11. Indirect value of cropland resources in Guizhou Province (million CNY).

	Regulating Services				Supporting Services			Cultural Services
	Gas Regulation	Climate Regulation	Environmental Purification	Hydrological Regulation	Soil Conservation	Maintenance of Nutrient Cycle	Biodiversity	Aesthetic Landscape
2001	1042.31	544.58	158.10	1687.38	608.99	181.53	199.09	87.83
2002	1037.58	542.11	157.39	1850.00	606.23	180.70	198.19	87.44
2003	1185.91	619.61	179.89	1989.32	692.89	206.54	226.52	99.94
2004	1353.88	707.36	205.36	2708.69	791.03	235.79	258.61	114.09
2005	1337.41	698.76	202.87	2036.25	781.41	232.92	255.46	112.70
2006	1443.82	754.36	219.01	2424.08	843.58	251.45	275.79	121.67
2007	1735.44	906.72	263.24	3075.90	1013.97	302.24	331.49	146.25
2008	1587.42	829.38	240.79	2913.23	927.48	276.46	303.22	133.77
2009	1699.16	887.76	257.74	2512.78	992.77	295.92	324.56	143.19
2010	1560.85	815.50	236.76	2655.25	911.96	271.83	298.14	131.53
2011	1012.34	528.92	153.56	1486.50	591.48	176.31	193.37	85.31
2012	1321.58	690.49	200.46	2102.56	772.16	230.16	252.44	111.37
2013	1121.78	586.10	170.16	1483.78	655.42	195.37	214.27	94.53
2014	1069.32	558.69	162.20	2173.93	624.77	186.23	204.25	90.11
2015	908.11	474.46	137.75	1645.79	530.58	158.15	173.46	76.53
2016	788.70	412.07	119.63	1239.54	460.81	137.36	150.65	66.46
2017	841.59	439.71	127.66	1364.21	491.72	146.57	160.75	70.92
2018	661.68	345.71	100.37	1157.39	386.60	115.24	126.39	55.76
2019	698.34	364.86	105.93	1181.51	408.02	121.62	133.39	58.85
2020	924.71	483.14	140.27	1885.77	540.28	161.05	176.63	77.93

Table 12. Changes of economic value of cropland resources in Guizhou Province (million CNY).

Year	Direct Value	Indirect Value	Total Value
2001	27,995.00	4509.80	32,504.80
2002	27,888.00	4659.64	32,547.64
2003	46,672.00	5200.60	51,872.60
2004	52,464.00	6374.81	58,838.81
2005	33,353.00	5657.76	39,010.76
2006	34,797.00	6333.76	41,130.76
2007	39,220.00	7775.25	46,995.25
2008	30,848.00	7211.75	38,059.75
2009	33,050.00	7113.88	40,163.88
2010	38,561.00	6881.84	45,442.84
2011	43,084.00	4227.79	47,311.79
2012	56,132.00	5681.21	61,813.21
2013	64,612.00	4521.42	69,133.42
2014	85,189.00	5069.51	90,258.51
2015	109,654.00	4104.83	113,758.83
2016	119,649.56	3375.23	123,024.78
2017	130,642.73	3643.13	134,285.85
2018	143,928.56	2949.13	146,877.69
2019	156,647.00	3072.52	159,719.52
2020	180,025.00	4389.76	184,414.76

4. Discussion

4.1. Analysis of Reasons for the Change in Physical and Conditional Account

Based on the results of this paper, the cultivated land resources in Guizhou Province declined continually after a short increase, and the reduced area was mainly transformed into natural vegetation and grassland, especially as the steep slope terraces disappeared from 2001 to 2020 [56]. These changes are closely related to the continuous implementation of the policy mandating the return of farmland to forest and grassland land types in Guizhou Province [57,58]. In particular, the intensity of returning farmland to forest and grassland in poverty-stricken areas of Guizhou has increased during a critical period of poverty alleviation, such as farmland with a slope of more than 25 degrees, severely sandy farmland, sloped farmland of 15–25 degrees in areas with important water sources, and steep sloped terraces, the conversion of all of which are examples of remarkable achievements [44,59]. At the same time, it is worth mentioning that on the premise of the obvious outflow of the physical account of the cultivated land resources, the output of agricultural products has still shown a huge increase [60,61]. It is not difficult to find that the output of grain crops decreased, but that the output of high value-added crops such as tobacco and vegetables increased. On the one hand, as people's quality of life improves, people's eating habits tend to become more diversified and healthier, leading to an increase in the demand for more value-added commodities in human society. Namely, the relationship between supply and demand in the market has guided farmers to the crop types that they choose to grow. On the other hand, in order to get rid of poverty in Guizhou Province, the government has implemented relevant policies with regard to the adjustment of the agricultural planting structure to improve the income of farmers in karst mountainous areas [62].

Meanwhile, six dimensions of cropland landscape indicators, such as edge density and the area-weighted means shape index were used as a measure of cultivated land fragmentation, to analyze the landscape change of cultivated land resources in Guizhou Province from 2001 to 2020 [63]. These indicators have decreased significantly, indicating that the fragmentation of cultivated land resources has been alleviated through land consolidation and ecological restoration projects [64].

4.2. Analysis of Reasons for the Change in Monetary Account

As for the results of the analysis on the monetary account, the monetary value of cultivated land resources in Guizhou Province has increased greatly over the past 20 years. It is interesting that, with the obvious outflow of the physical quantity account of cultivated land resources, the growth rate of the monetary value of agricultural products is still significant. More importantly, the settlement of the issue cannot be achieved by simply expanding the cultivated area or by increasing the employed population. It can be seen from the above data that agricultural employment and the cultivated land area in Guizhou Province have decreased by more than 50%, but the value of the agricultural products created per capita has increased by 12-fold. The improvement of cropland quality and the development of technology have led to a rise in cropland resource value in Guizhou Province. While the direct value has increased, the indirect value has not fallen sharply, indicating that the ecological environment has been protected during economic development [65].

In addition, we collected government expenditure data from the study area over the past 20 years (Appendix B). In 2001, the local government spent CNY 4.25 billion on farming, forestry, and water conservation, which has increased to CNY 10.431 billion in 2020, and this investment had increased 23-fold. This included the giving of subsidies to encourage farmers to adjust their planting structure, increasing the construction of water conservation facilities to ensure irrigation conditions, and conducting corresponding education on agricultural technology to improve farmers' planting skills. These policies ensure that the adjustment of planting structure can be quickly completed within the study area [66,67].

At the same time, with high mountains and steep slopes, the construction cost of roads and bridges is very high, which makes the transportation and sales of agricultural products inconvenient. The local government increased its investment in transportation and other infrastructure from CNY 4.31 billion in 2001 to CNY 34.15 billion in 2020, realizing the County-to-County Expressway and the "village to village" hardened road in Guizhou Province. From the results of this study, it seems that reasonable policy guidance and sustained high-level financial investment have led to a significant increase in the value of cultivated land resources in this area [68,69].

Affected by karst landforms, Guizhou Province has serious soil erosion, serious rocky desertification, and a lack of cultivated land resources [70]. In order to improve rocky desertification and soil erosion, Guizhou Province has conducted large-scale rocky desertification prevention and control projects. Meanwhile, the financial investment for environmental protection has increased from CNY 2.667 billion in 2006 to CNY 14.615 billion in 2020. By accounting for the cultivated land resource assets in Guizhou Province over the past 20 years, it has been found that even under relatively bad natural conditions, the asset value of cultivated land resources can apparently be improved and realized in a win-win situation of economic development and ecological protection, through the guidance of reasonable land use methods and scientific land policies [71].

4.3. Shortcomings/Uncertainties of This Research

However, the landcover data selected is of 250 m resolution in this study. For the karst mountainous areas, some sloped croplands of small areas may not have been identified, or they could have been identified as cropland/natural vegetation mosaics, which may lead to deviations in the evaluation results. In addition, only site conditions and landscape indexes are selected for conditional accounting. For cropland resources, soil quality, soil physical and chemical properties, and obstacle factors are also important measures. In future research, multiple measures should be added to the conditional account, so as to more comprehensively develop knowledge regarding the quality changes in cultivated land resources.

5. Conclusions

In this paper, multi-remote sensing data were used to calculate the physical and conditional account changes of the cultivated land resources in Guizhou Province at the pixel level, which may make up for the deficiency of traditional accounting of natural capital by presentation. At the same time, according to the characteristics of karst landforms in the study area, landscape factors were added to the conditional account, which will assist us with precisely analyzing the reasons for the change of monetary account. Through this research, we drew the following conclusions.

1. In the physical account, the cultivated land resources in Guizhou Province showed an obvious downward trend, but the planting structure of agricultural products showed obvious changes, and the gross output increased significantly. This shows that the value of the cultivated land is not strongly related to the size of the land area.
2. In the condition account, the quality of the cultivated land resources in Guizhou Province improved. Specifically, the fragmentation of the cultivated land improved, and the area of cultivated land on steep slopes decreased. This shows that the local governance policy on cultivated land is effective.
3. In the monetary account, the monetary value of the cropland resources in Guizhou Province increased greatly and rapidly. Additionally, an increase in economic value did not place negative impacts upon the ecological value of the cultivated land. This shows that reasonable policy and financial investment are of positive significance for the sustainable utilization of the cultivated land resources.

Based on the above conclusions, we believe that it is very necessary to introduce additional representative factors into the accounting of cultivated land resource value in the study area. Evaluation and research into the value of cultivated land resources in the karst mountainous areas in Southwest China can provide a good reference for scholars of related fields. Moreover, in this case, reasonable policies, such as returning farmland to forest and adjusting agricultural planting structure have very positive impacts on the value of cultivated land resources and the improvement of farmers' benefits in this area. This is not only an evaluation of the effect of land policy implementation through quantitative methods, but it is also is a useful demonstration for leaders in other areas with similar difficulties; an active exploration of the sustainable utilization of cultivated land resources.

There are still many deficiencies in this study, such as the low accuracy of land use classification, the factor of the condition account being imperfect, and so on. This is the direction in which we will continue to study in the future. It is hoped that a more perfect and universal accounting framework that is suitable for karst areas can be developed in the future, so that the evaluation results can better guide sustainable land use in the study area.

Author Contributions: Conceptualization, Z.Z. and L.Z.; methodology, Q.C.; software, Q.F. and L.Z.; formal analysis, L.Z. and Q.C.; investigation, L.W., D.L., and T.W.; data curation, Q.F.; writing—original draft preparation, L.Z.; writing—review and editing, Z.Z., L.Z., and Q.C.; visualization, L.Z.; supervision, Q.C.; project administration, L.Z.; funding acquisition, Z.Z. All authors have read and agreed to the published version of the manuscript.

Funding: This research was funded by the NSFC regional project, "Research on the coupling mechanism between ecological assets and regional poverty in karst rocky desertification areas (41661088)", by "Guizhou Province's high-level innovative talent training plan 'hundred' level talents (Qiankehe platform talents [2016] 5674)" and a special study of Guizhou Provincial Department of natural resources on the "Construction of evaluation system of real estate economic operation system in Guizhou Province" (520000215RSUFG5DLMENO).

Institutional Review Board Statement: Not applicable.

Informed Consent Statement: Not applicable.

Data Availability Statement: Not applicable.

Conflicts of Interest: The authors declare no conflict of interest.

Appendix A

Table A1. MCD12Q1 International Geosphere-Biosphere Programme (IGBP) legend and class descriptions.

Name	Value	Description
Evergreen Needleleaf Forests	1	Dominated by evergreen conifer trees (canopy > 2 m). Tree cover > 60%.
Evergreen Broadleaf Forests	2	Dominated by evergreen broadleaf and palmate trees (canopy > 2 m). Tree cover > 60%.
Deciduous Needleleaf Forests	3	Dominated by deciduous needleleaf (larch) trees (canopy > 2 m). Tree cover > 60%.
Deciduous Broadleaf Forests	4	Dominated by deciduous broadleaf trees (canopy > 2 m). Tree cover > 60%.
Mixed Forests	5	Dominated by neither deciduous nor evergreen (40–60% of each) tree type (canopy > 2 m). Tree cover > 60%.
Closed Shrublands	6	Dominated by woody perennials (1–2 m height), > 60% cover.
Open Shrublands	7	Dominated by woody perennials (1–2 m height), 10–60% cover.
Woody Savannas	8	Tree cover 30–60% (canopy > 2 m).
Savannas	9	Tree cover 10–30% (canopy > 2 m).
Grasslands	10	Dominated by herbaceous annuals (<2 m)
Permanent Wetlands	11	Permanently inundated lands with 30–60% water cover and >10% vegetated cover.
Croplands	12	At least 60% of area is cultivated cropland.
Urban and Built-up Lands	13	At least 30% impervious surface area, including building materials, asphalt, and vehicles.
Cropland/Natural Vegetation Mosaics	14	Mosaics of small-scale cultivation, 40–60% with natural trees, shrubs, or herbaceous vegetation.
Permanent Snow and Ice	15	At least 60% of area is covered by snow and ice for at least 10 months of the year.
Barren	16	At least 60% of area is non-vegetated barren (sand, rock, soil) areas with less than 10% vegetation.
Water Bodies	17	At least 60% of area is covered by permanent water bodies. Unclassified 255 Has not received a map label because of missing inputs.

Table A2. Ecosystem service equivalent value per unit area.

Ecosystem Classification		Provisioning Services			Regulating Services				Supporting Services		Cultural Services	
Primary Classification	Secondary Classification	Food Production	Raw Material Production	Water Supply	Gas Regulation	Climate Regulation	Environmental Purification	Hydrological Regulation	Soil Conservation	Maintenance of Nutrient Cycles	Biodiversity	Aesthetic Landscape
Crop land	Dryland	0.85	0.4	0.02	0.67	0.36	0.1	0.27	1.03	0.12	0.13	0.06
	Paddy field	1.36	0.09	−2.63	1.11	0.57	0.17	2.72	0.01	0.19	0.21	0.09
	Coniferous	0.22	0.52	0.27	1.7	5.07	1.49	3.34	2.06	0.16	1.88	0.82
Forest	Mixed coniferous	0.31	0.71	0.37	2.35	7.03	1.99	3.51	2.86	0.22	2.6	1.14
	Broad-leaved	0.29	0.66	0.34	2.17	6.5	1.93	4.74	2.65	0.2	2.41	1.06
	Shrub	0.19	0.43	0.22	1.41	4.23	1.28	3.35	1.72	0.13	1.57	0.69
	Grass	0.1	0.14	0.08	0.51	1.34	0.44	0.98	0.62	0.05	0.56	0.25
Grassland	Scrub	0.38	0.56	0.31	1.97	5.21	1.72	3.82	2.4	0.18	2.18	0.96
	Meadow	0.22	0.33	0.18	1.14	3.02	1	2.21	1.39	0.11	1.27	0.56
Wetland	Wetlands	0.51	0.5	2.59	1.9	3.6	3.6	24.23	2.31	0.18	7.87	4.73
Desert	Desert	0.01	0.03	0.02	0.11	0.1	0.31	0.21	0.13	0.01	0.12	0.05
	Bare ground	0	0	0	0.02	0	0.1	0.03	0.02	0	0.02	0.01
Waters	Water system	0.8	0.23	8.29	0.77	2.29	5.55	102.24	0.93	0.07	2.55	1.89
	Glacial snow	0	0	2.16	0.18	0.54	0.16	7.13	0	0	0.01	0

Appendix B

Table A3. Statistics of financial expenditure in Guizhou Province (section) (unit: CNY billion).

Unit: Billion CNY	General Public Budget Expenditure	Farming, Forestry and Water Conservancy	Transportation	Energy Saving and Environment Protection
2001	27.52	4.25	4.31	-
2002	31.67	4.86	3.64	-
2003	33.24	4.53	3.16	-
2004	41.84	7.23	3.83	-
2005	52.07	7.66	4.12	-
2006	61.041	6.155	4.193	-
2007	79.54	8.75	4.88	2.67
2008	105.54	12.17	4.94	4.04
2009	137.23	20.41	12.08	5.53
2010	163.15	24.68	10.96	5.43
2011	224.94	27.85	30.52	5.55
2012	275.57	36.19	28.86	6.57
2013	308.266	40.031	29.979	6.644
2014	354.28	44.719	43.201	8.534
2015	393.95	53.426	39.225	9.649
2016	426.236	62.938	28.997	12.709
2017	461.252	61.205	33.691	12.539
2018	502.968	66.484	38.149	13.438
2019	594.874	99.89	34.779	18.853
2020	573.95	102.431	34.15	14.615

References

1. Blignaut, J.; Aronson, J.; de Groot, R. Restoration of natural capital: A key strategy on the path to sustainability based on lucc. *Ecol. Eng.* **2014**, *65*, 54–61. [CrossRef]
2. Development, United Nations; European Union; Food and Agriculture Organization of the United Nations; Organisation for Economic Co-Operation; World Bank Group. System of Environmental Economic Accounting 2012—Experimental Cosystem Accounting. 2014. Available online: https://seea.un.org/home/Natural-Capital-Accounting-Project (accessed on 1 December 2021).
3. Liu, L. Comparison and implications of natural resource asset accounting at home and abroad. *Stat. Decis. Mak.* **2019**, *35*, 9–12.
4. Liu, G.; Hu, L.; Gao, X.; Du, L.; Li, J.; Xiao, N. Research on the connotation of natural capital and its accounting. *Ecoeconomics* **2018**, *34*, 63, 153–157.
5. Geng, J.; Hu, T.; Liu, Z. A preliminary study on the preparation and application of china's national balance sheet and natural resource balance sheet: An analysis of SNA 2008 and SEA 2012. *J. Account. Res.* **2015**, *1*, 15–24, 96.
6. Hu, W.; Shi, D. A study on the balance sheet framework system of natural resources in China: An idea based on SEEA 2012, SNA 2008 and national balance sheets. *Chin. Resour. Environ.* **2015**, *25*, 1–9.
7. Liu, S.; Yang, S. Comparison of accounting for natural resource assets at home and abroad and its implications. *Land Resour. Inf.* **2020**, *12*, 24–31.
8. Xiang, S.; Zheng, R. A study of the category of assets in the balance sheet of natural resources. *Stat. Res.* **2015**, *32*, 3–11.
9. Li, N. Discussion on the construction of statistical indicators of all natural resource assets owned by the whole people. *Land Resour. Inf.* **2020**, *11*, 16–21, 51.
10. Missemer, A. Natural capital as an economic concept, history and contemporary issues. *Ecol. Econ.* **2018**, *143*, 90–96. [CrossRef]
11. Virto, L.R.; Weber, J.L.; Jeantil, M. Natural capital accounts and public policy decisions: Findings from a survey. *Ecol. Econ.* **2018**, *144*, 244–259. [CrossRef]
12. Department of National Accounts, NBS. Outlines of Natural Capital Accounting and Valuation of Ecosystem Services Project in China. 2021. Available online: https://seea.un.org/content/china-0 (accessed on 1 March 2022).
13. He, L.; Shen, L.; Zhang, W.; Tao, J.; Fan, Z. Practical progress and theoretical system construction of natural resource accounting in China. *J. Nat. Resour.* **2020**, *35*, 2968–2979. [CrossRef]
14. Cheng, S.; Shen, L.; Feng, Z.; Zhong, S. The development history and prospect of natural resources research in China. *J. Nat. Resour.* **2020**, *35*, 1757–1772.
15. Tao, J.; Lv, Y.; He, L.; Shen, L. Research on land resource asset accounting and report preparation based on double-entry bookkeeping. *Chin. Resour. Environ.* **2020**, *30*, 22–29.
16. Li, Y.; Wu, D.; Liu, Y. Research hotspots and trends of arable land ecological risk assessment: Based on citespace visual analysis. *J. Agric. Resour. Environ.* **2019**, *36*, 502–512.
17. Li, X. Changes in China's arable land area in the past 20 years and its policy implications. *J. Nat. Resour.* **1999**, *4*, 329–333.

18. Wang, Y.J.; Xie, B.G.; Li, X.Q.; Zhao, W.Q.; Wang, J.Y.; Luo, H.B. Ecosystem carrying capacity balance of the Karst region in China, based on the supply and demand of cultivated land. *Acta Ecol. Sin.* **2017**, *37*, 7030–7038.
19. Wu, M.; Liu, S.J.; Ye, Y.Y.; Zhang, W.; Wang, K.L.; Chen, H.S. Spatial variability of surface soil organic carbon and its influencing factors in cultivated slopes and abandoned lands in a Karst peak-cluster depression area. *Acta Ecol. Sin.* **2016**, *36*, 1619–1627.
20. Li, D.; Yang, Y.; Du, G.; Huang, S. Understanding the contradiction between rural poverty and rich cultivated land resources: A case study of heilongjiang province in northeast China. *Land Use Policy* **2021**, *108*, 105673. [CrossRef]
21. Yang, Q.; Bi, G.; Chen, Z.; Zeng, L.; Yang, R. Spatial allocation of fallow land in karst ecologically fragile areas: A case study of Qinglong county, Guizhou province. *Acta Geogr. Sin.* **2018**, *73*, 2250–2266.
22. Chen, Q. Study on the mechanism of remote sensing assessment and temporal evolution of ecological assets in karst rocky desertification area. *Guizhou Norm. Univ.* **2021**, 20–25.
23. Liu, X.; Hu, G.; Chen, Y.; Li, X.; Xu, X.; Li, S.; Pei, F.; Wang, S. High-resolution multi-temporal mapping of global urban land using landsat images based on the google earth engine platform. *Remote Sens. Environ.* **2018**, *209*, 227–239. [CrossRef]
24. Su, W.; Zhu, W.; Xiong, K. Rocky desertification and its ecological and economic governance model in karst mountainous area of Guizhou province. *Karst China Karst* **2002**, *1*, 21–26.
25. Chen, H.; Zhu, D.; Chen, H.; Wen, Y.; Lin, D.; Wang, C. A comparative study on the characteristics of climate change in different water systems in the Yangtze River basin in Guizhou province. *Bull. Soil Water Conserv.* **2018**, *38*, 189–197, 203.
26. Xiong, K. *Remote sensing of karst rock desertification—A typical study of Gis—Aa case study in Guizhou Province*; Geological Publishing House: Beijing, China, 2002.
27. Zhang, Y.; An, Y.; Ma, L.; Li, X. Land use change on steep slopes in karst mountains of Guizhou Province from 1960 to 2010. *Prog. Geogr.* **2012**, *31*, 878–884.
28. Zhu, D.; Yang, Q.; Chen, H.; Chen, J.; Li, S. Temporal and spatial variation characteristics of rainfall erosion forces in different water systems in Guizhou Province from 1960 to 2017. *Bull. Soil Water Conserv.* **2021**, *41*, 6–14.
29. Zhang, Y.; Zhou, Z.; Huang, D.; Zhu, M.; Wu, Y.; Sun, J. Spatio-temporal evolution and impact factor analysis of cultivated land in karst mountainous areas. *Trans. Chin. Soc. Agric. Eng.* **2020**, *36*, 266–275.
30. Wang, L.; Diao, C.; Xian, G.; Yin, D.; Lu, Y.; Zou, S.; Erickson, T.A. A summary of the special issue on remote sensing of land change science with google earth engine. *Remote Sens. Environ.* **2020**, *248*, 112002. [CrossRef]
31. Liu, L.; Xiao, X.; Qin, Y.; Wang, J.; Xu, X.; Hu, Y.; Qiao, Z. Mapping cropping intensity in china using time series landsat and sentinel-2 images and google earth engine. *Remote Sens. Environ.* **2020**, *239*, 111624. [CrossRef]
32. Huang, H.; Chen, Y.; Clinton, N.; Wang, J.; Wang, X.; Liu, C.; Gong, P.; Yang, J.; Bai, Y.; Zheng, Y.; et al. Mapping major land cover dynamics in Beijing using all landsat images in google earth engine. *Remote Sens. Environ.* **2017**, *202*, 166–176. [CrossRef]
33. Robinson, N.P.; Allred, B.W.; Smith, W.K.; Jones, M.O.; Moreno, A.; Erickson, T.A.; Naugle, D.E.; Running, S.W. Terrestrial Primary Production for the Conterminous United States Derived from Landsat 30 M and Modis 250 M. *Remote Sens. Ecol. Conserv.* **2018**, *4*, 264–280. [CrossRef]
34. United Nations. Study on the Valuation Accounting Method of the Natural Resources Assets and Liabilities. 2020. Available online: https://seea.un.org/home/Natural-Capital-Accounting-Project (accessed on 1 December 2021).
35. Jiao, Z.; Wang, H.; Xu, X.; Yang, B. Natural resource balance sheet preparation design and application I.: Design. *J. Nat. Resour.* **2018**, *33*, 1706–1714.
36. Jiao, Z.; Wang, H.; Xu, X.; Yang, B. Natural resource balance sheet preparation design and application II: Application. *J. Nat. Resour.* **2018**, *33*, 1715–1724.
37. Shi, J.; Wang, P.; Li, N.; Li, Y. A Framework of ideas for the preparation of balance sheets of all natural resources for all people. *J. Nat. Resour.* **2020**, *35*, 2270–2282.
38. Hu, Y.; Yu, L.; Wu, J. Construction of a balance sheet accounting system for natural resources: A case study of forest resources. *For. Resour. Manag.* **2020**, 3, 22–25, 31.
39. Nations, United. Final Report on Ncaves Pilot Project in Guizhou Province. 2021. Available online: https://seea.un.org/content/china-0 (accessed on 1 March 2022).
40. Long, X.; Lin, H.; An, X.; Chen, S.; Qi, S.; Zhang, M. Evaluation and analysis of ecosystem service value based on land use/cover change in dongting lake wetland. *Ecol. Indic.* **2022**, *136*, 108619. [CrossRef]
41. Wan, W. Spatial differentiation of cultivated land fragmentation in Zhejiang province based on county scale. *Environ. Ecol.* **2021**, 3, 15–21, 48.
42. Hou, B.; Ma, R.; Jin, Y.; Zhang, Y.; Zhao, Y.; Dou, S. Spatial structure and audit zoning of land resource assets in Zhejiang province from the perspective of landscape ecology. *Bull. Soil Water Conserv.* **2018**, *38*, 268–276.
43. Li, W. Study on the Quality Change of Cultivated Land in the Suburban Area of Black Soil Based on the Standing Conditions and Localization Characteristics. Ph.D. Thesis, Jilin University, Jilin, China, 2018.
44. Liu, W.; Liao, H.; Wu, X.; Guo, Q.; Mao, X.; Li, C. A study on the spatial coupling relationship between arable land fragmentation and poverty in southwest karst region. *J. Southwest Univ.* **2019**, *41*, 10–20.
45. Zhang, X.; Cai, Z.; Li, G.; Sun, J.; Li, G.; Chen, Z. Analysis of the Causes of fragmentation of cultivated land landscape and its impact on rural income. *Sci. Surv. Mapp.* **2020**, *45*, 134–141.
46. Chen, W.; Feng, X.; Ma, R.; Hong, Q. Evaluation methods and empirical research on fragmentation of cultivated land: A case study of Ningbo city, Zhejiang province. *China Land Sci.* **2016**, *30*, 80–87.

47. Fu, B.; Ding, D. *Principles and Applications of Landscape Ecology*; Science Press: Beijing, China, 2002.
48. Forman, R.T.; Godron, M. *Landscape Ecology*; John & Wiley: New York, NY, USA, 1986.
49. Bosch, M. Pylandstats: An open-source pythonic library to compute landscape metrics. *PLoS ONE* **2019**, *14*, e0225734. [CrossRef] [PubMed]
50. Costanza, R.; d'Arge, R.; de Groot, R.; Farber, S.; Grasso, M.; Hannon, B.; Limburg, K.; Naeem, S.; ONeill, R.V.; Paruelo, J.; et al. The value of the world's ecosystem services and natural capital. *Nature* **1997**, *387*, 253–260. [CrossRef]
51. Xie, G.; Zhang, C.; Zhang, C.; Xiao, Y.; Lu, C. The value of ecosystem services in China. *Resour. Sci.* **2015**, *37*, 1740–1746.
52. Xie, G.; Zhen, L.; Lu, C.; Xiao, Y.; Chen, C. A value-based approach to ecosystem services based on expert knowledge. *J. Nat. Resour.* **2008**, *23*, 911–919.
53. Xie, G.; Lu, C.; Leng, Y.; Zheng, D.; Li, S. Valuation of ecological assets on the tibetan plateau. *J. Nat. Resour.* **2003**, *5*, 189–196.
54. Xie, G.; Zhang, C.; Zhang, L.; Chen, W.; Li, S. Improvement of valueization methods for ecosystem services based on value equivalence factor per unit area. *J. Nat. Resour.* **2015**, *30*, 1243–1254.
55. Gu, X.; Long, A.; Liu, G.; Yu, J.; Wang, H.; Yang, Y.; Zhang, P. Changes in ecosystem service value in the 1 km lakeshore zone of Poyang lake from 1980 to 2020. *Land* **2021**, *10*, 951. [CrossRef]
56. Wang, C.; Maclaren, V. Evaluation of economic and social impacts of the sloping land conversion program: A case study in Dunhua county, China. *For. Policy Econ.* **2012**, *14*, 50–57. [CrossRef]
57. Fan, M.; Xiao, Y.T. Impacts of the grain for green program on the spatial pattern of land uses and ecosystem services in mountainous settlements in Southwest China. *Glob. Ecol. Conserv.* **2020**, *21*, e00806. [CrossRef]
58. Cai, Z.; Jiang, Z.; Li, Y.; Zhang, L.; Jia, M.; Xie, Y. The Effectiveness of farmland reforestation policies and the existence of effective policies. *Resour. Environ.* **2015**, *25*, 60–69.
59. Wu, Y.; Zhou, Z.; Zhu, C.; Ma, G.; Huang, D. A study on the measurement and spatial differentiation of rural poverty in Karst mountainous areas: A case study of panzhou city. *Resour. Environ. Yangtze River Basin* **2020**, *29*, 1247–1256.
60. Hao, L.; Cai, Y.; Chen, R.; Chen, Q.; Xiang, Y. Evaluation of the effect of reforestation of farmland in southwest Karst based on vegetation remote sensing: A case study of Bijie area in Guizhou Province. *Acta Ecol. Sin.* **2011**, *31*, 3255–3264.
61. Shu, T.; Xiong, K.N.; Chen, L. Land use and landscape pattern changes under the control of rocky desertification. *Southwest China J. Agric. Sci.* **2022**, *35*, 446–452.
62. Chen, Q.; Xiong, K.; Dan, W.; Niu, L. Analysis of coupled characteristics of ecology and poverty in typical karst areas: A case study of 9000 provincial-level poor villages in Guizhou Province. *Acta Ecol. Sin.* **2021**, *41*, 2968–2982.
63. Zhang, T.; Shi, H.; Xu, Y.; Xue, J.; Zhu, J.; Geng, Q. Quantitative evaluation of the impact of returning farmland to forest on the landscape pattern of land use in karst areas. *J. Beijing For. Univ.* **2015**, *37*, 34–43.
64. Wang, X.; Yu, S.; Shi, Z.; Wang, L. Responses of ecosystem service trade-offs and synergies to the return of farmland to forests in the red soil area of southern China. *Chin. J. Ecol.* **2021**, *41*, 7002–7014.
65. Ding, Z.; Yao, S. Ecological effectiveness of payment for ecosystem services to identify incentive priority areas: Sloping land conversion program in China. *Land Use Policy* **2021**, *104*, 105350. [CrossRef]
66. Liu, Q.; Qiong, Q.; Xiao, H. The impact of farmland operation scale and financial support policies on agricultural carbon emissions. *Resour. Sci.* **2020**, *42*, 1063–1073.
67. Jin, F.; Jin, R. Spatial Effect Analysis of the Impact of Fiscal Support on changes in agricultural industrial structure. *J. Financ. Econ.* **2020**, *5*, 82–91.
68. Wu, H. A Study on the Impact of Local Fiscal Expenditure on Agricultural Production. Ph.D. Thesis, Nanjing Agricultural University, Nanjing, China, 2002; pp. 22–31.
69. Yang, L. The impact of transfer payments on local fiscal agricultural expenditure: A study based on prefecture-level panel data from 2003–2009. *Econ. Rev.* **2016**, *5*, 148–160.
70. Hu, Z.; Wang, S.; Bai, X.; Luo, G.; Li, Q.; Wu, L.; Yang, Y.; Tian, S.; Li, C.; Deng, Y. Changes in ecosystem service values in Karst areas of China. *Agric. Ecosyst. Environ.* **2020**, *301*, 107026. [CrossRef]
71. Gong, B.L.; Wang, S. The multi-channel impact of fiscal spending on China's agricultural growth. *Issues Agric. Econ.* **2021**, *1*, 54–68.

land

Article

Modeling Analysis on Coupling Mechanisms of Mountain–Basin Human–Land Systems: Take Yuxi City as an Example

Li Wu [1,2], Yanjun Yang [1] and Binggeng Xie [2,*]

[1] Department of Geography and Land Engineering, Yuxi Normal University, Yuxi 653100, China; wuli2009@yxnu.edu.cn (L.W.); yangyanjun@yxnu.edu.cn (Y.Y.)
[2] School of Geographic Sciences, Hunan Normal University, Changsha 410081, China
* Correspondence: xbgyb1961@163.com; Tel.: +86-138-0847-3925

Abstract: The result of a human–land relationship in geographical environment systems is a human–land coupling system, which is a comprehensive process of interaction and infiltration between human economic and social systems and the natural ecosystem. Based on the recognition that the human–land system is a nonlinear system coupled by multiple factors, a time delay fractional order dynamics model with a Holling-II-type transformation rate was constructed, the stability analysis of the system was carried out, the transformation times of different land classes were clarified, and the coupled dynamics model parameters of mountainous areas and basin areas were obtained by using the land-use change survey data and socio-economic statistical data in Yuxi City, respectively: the transformation parameter of the production and living land to the unused land in mountainous areas and basin areas (a_M, 0.0486 and a_B, 0.0126); the transformation parameter of unused land to production and living land in mountainous areas and basin areas (b_M 0.0062 and b_B, 0.0139); the transformation parameter of unused land to the forest and grass land in mountainous areas and basin areas (s_M, 0.0051 and s_B, 0.0028); the land area required to maintain the individual unit in mountainous areas and basin areas (h_M, 0.0335 and h_B, 0.0165); the average reclamation capacity in mountainous areas and basin areas (d_M, 0.03 and d_B, 0.05); the inherent growth rate of populations in mountainous areas and basin areas (r_M, 0.0563 and r_B, 0.151). Through analyzing the coupling mechanisms of human–land systems, the countermeasures for the difference between mountainous areas and basin areas in the future development are put forward. The mountainous area should reduce the conversion of forest and grass land to production and living land by reducing the average reclamation or development capacity, reducing the excessive interference of human beings on unused land, and speeding up its natural recovery and succession to forest and grass land. In addition to reducing the average reclamation or development capacity in basin areas, the reclamation or development rate of the idle land and degraded land should be increased, and the conversion of idle land and degraded land into productive and living land should be encouraged by certain scientific and technological means.

Keywords: mountain–basin human–land system; land-use change; land dynamical model with Holling-II type; coupling mechanism; Yuxi City

Citation: Wu, L.; Yang, Y.; Xie, B. Modeling Analysis on Coupling Mechanisms of Mountain–Basin Human–Land Systems: Take Yuxi City as an Example. *Land* **2022**, *11*, 1068. https://doi.org/10.3390/land11071068

Academic Editor: Xiaoyong Bai

Received: 9 June 2022
Accepted: 11 July 2022
Published: 13 July 2022

1. Introduction

Human–land relationship research is of great significance in geography, contributing to the duality of geography and the development of human geography [1]. For a long time, the study of "humans" and "land" has been carried out separately. However, as a series of environmental problems and food security problems brought about by human activities on the earth continue to affect the human system [2–4], the academic circle is paying more and more attention to the comprehensive research of "people" and "land" [5]. A variety of new comprehensive methods, including statistical methods, GIS and spatial analysis methods, simulation methods and hybrid methods have been applied [6–9]. As the questions raised by researchers increasingly involved the intersection of human activities and the earth's environmental system [10,11], the academic circle further recognized that the modeling of feedback between humans and the natural environment has become an urgent requirement [12,13]. In the 1980s, the modeling concept of coupled natural systems and human socio-economic systems was proposed, and the two-way coupling of positive and negative feedback and the coupling with human activities in the earth system model became the research object of academic community [14]. Synthetic integrated models that carried out bidirectional coupling and exchanges of information in certain forms have increasingly become a research hotspot [15]. Since the 1990s, 11 different Integrated Assessment Models (IAMs) have been developed worldwide [16]. The Integrated Model to Assess the Global Environment Framework (IMAGE) model developed by the Netherlands Environmental Assessment Agency is one of the representative models of comprehensive integration, in which the impacts of agricultural land expansion and changes in land-use types on the environment were evaluated by considering population density, resources, topography, etc. [17]. In addition, there are some models based on multi-agents [18] that analyze and explain the complex human–land coupling relationship and its coupling degree. Meanwhile, with the continuous development of computer technology, multi-source data-model fusion has made new progress, and the uncertainty of the human–land system coupling relationship has been further quantified [19,20].

With the deepening of studies on the human–land relationship, regional spatiality has attracted more and more attention [21,22], but most relevant studies on this complex issue focus on a single factor [23]. Mountainous areas and basin areas, as special geomorphic spaces in Yunnan Province, have not been strictly subdivided in existing studies, and the relationship between humans and land is rarely involved. The concept of "coupling" in geography originated from physics, which refers to the synergy of two or more systems through various interactions, or the dynamic relationship between the elements of the system [24]. Mountainous areas and basin areas mainly include flat land between mountains and surrounding mountains [25]. The two have a close genetic relationship in topography and geomorphology. Relying on their geographical proximity, they form a complex coupling system of mutual cooperation and constraints through continuous material circulation, energy flow and information transmission, including the two coupling relationships of near-range coupling and remote coupling [11,26,27]. In order to deeply reveal the interactions and feedback mechanisms between human activities and the natural environment in the mountain–basin human–land coupled system, it is necessary to conduct coupling simulations and predictions around the human–land system and build a comprehensive integrated human–land system dynamics model. By analyzing the interaction of element coupling and process coupling between two different geographic spaces, the complexity and dynamics of human–land systems coupling are revealed, and the mechanism and feedback paths of human activities such as social and economic development on land-use changes are explored. The human–land coupling system for mountain–basin has the nonlinear dynamics and chaotic characteristics of complex systems. To obtain a quantitative expression in the structure and function process, it is necessary to refer to a mature paradigm of the existing research and actively explore more integrated multivariate coupling models to dynamically resolve the interaction coupling relationship and dynamics mechanisms within the complex system and among subsystems based on an interdisciplinary perspective. An outstanding feature of human–land system dynamics models is that it can deal with nonlinear, complex, long-term and dynamic system coupling problems, and it is one of the main models to simulate human–land systems and other complex giant systems [28].

In 1997, Dobson published "Hopes for the Future: Restoration Ecology and Conservation Biology" in the journal Science [29], and proposed a dynamic land model to describe the transformation and restoration of natural habitats, which can explain the driving mechanism of increasing populations' agricultural demand on natural habitat transformation.

However, when a mathematical model needs to be established to solve many specific problems in reality, the time delay cannot be ignored, and it is also one of the essential characteristics of the evolution and interaction results of the human–land systems' elements. From the point of the dynamic system, the existence of a time delay can induce the stability of the system to switch, resulting in complex dynamic behaviors such as periodic oscillation and chaos. Therefore, it is quite necessary to consider the dynamic properties of the land dynamics model with a time delay [30]. In addition, fractional order calculus is an arbitrary generalization of integer order calculus in order, and calculus is widely used in the study of complex dynamic systems, such as the regulation of various ecosystems [31,32], secure communication [33,34], system controls [35,36] and stability issues [37]. Compared with the classical integer order model, fractional order calculus is more suitable for describing systems or processes with memory and hereditary characteristics, and can more accurately describe the physical and ecological phenomena in nature [38,39], which has attracted great attention from scholars at home and abroad [40–43].

Based on this, according to the relatively closed mountain–basin human–land system in Yuxi City, this study took advantage of the limitations on population density and introduced an appropriate land-use conversion rate to focus on analyzing the differences in land-use conversion and population changes over time in two different geographical spaces. On this basis of the land dynamics model and fractional calculus theory constructed by Dobson, a fractional human–land coupling dynamics model with a time delay was established to analyze the evolution mechanism of regional land-use systems and other issues, which is helpful and has important theoretical significance and a practical application value for the in-depth interpretation of the land-use system change mechanism with population development. It also provides reference for the differential human–land countermeasures of mountainous areas and basin areas in different development stages.

2. Materials and Methods

2.1. Study Area

Yuxi City is located in the central part of Yunnan Province, on the Yunnan Plateau at low latitudes. It belongs to the subtropical plateau monsoon climate, ranging from 23°19′ to 24°53′ north latitude and 101°16′ to 103°09′ east longitude (Figure 1). Yuxi is located in the core position of Yunnan Province, connecting the east to the west and connecting the north to the south. It is adjacent to the provincial capital, Kunming, which is to the northeast; Chuxiong Autonomous Prefecture in the north; Pu'er city in the southwest and Honghe Autonomous Prefecture in the southeast. The city covers an area of 15,285 km^2 and has jurisdiction over 75 townships (towns and streets) in 7 counties and 2 districts [44]. The terrain of Yuxi City is high in the northwest and low in the southeast. The western part is mainly deep-cut alpine and valley landforms, the central and eastern parts belong to the mountainous areas of central Yunnan and are dominated by mid-mountain landforms, and the eastern part is mainly plateau lake basin landforms. The Chengjiang, Jiangchuan and Tonghai lacustrine basins are formed around three plateau rifted lakes, the Fuxian Lake, Xingyun Lake and Qilu Lake, with flat and open terrain [45]. According to its special topography, combined with administrative regions, it can be divided into two types of geographical spaces: mountainous areas and basin areas [44]. Due to the complex terrain and large height difference, the mountainous area generally has more rainfall than the basin area. The cultivated land in the mountainous area is shallow and the soil fertility is low, but the basin area has fertile soil and more farmland with high and stable yields. From 1995 to 2018, the urban population growth and economic development in the basin area were significantly higher than those in the mountainous area, and the land-use change and social and economic development status differed significantly between the mountainous area and the basin area [27].

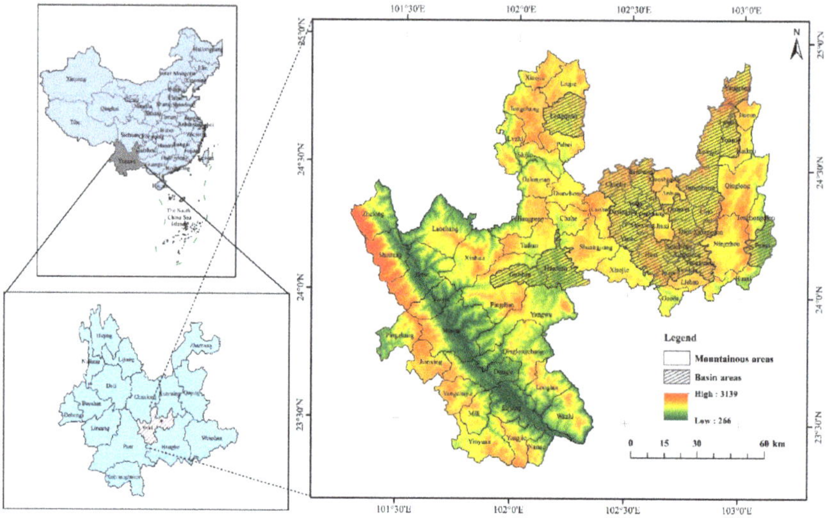

Figure 1. Location and elevation of the study area.

2.2. Data Sources

The land-use survey data in this study are mainly from the annual change survey data based on the second national land-use survey data (Table 1). The social and economic data involved are mainly from the statistical yearbook of Yunnan Province (1996–2019), the statistical yearbook of Yuxi City (1995–2018), the statistical yearbook of all counties and districts of Yuxi City and the statistical bulletin of national economic and social development from 1995–2018 (Table 1). The role of these data in the research is mainly to train and fit the parameters of human–land coupling dynamics models based on long time series data.

Table 1. Land-type area and population changes in mountainous and basin areas of Yuxi City from 1995 to 2018 (unit: hm^2, person).

Year	Mountainous Areas				Basin Areas			
	Forest and Grass Land	Production and Living Land	Unused Land	Population	Forest and Grass Land	Production and Living Land	Unused Land	Population
1995	875,820.94	207,305.42	85,826.37	931,088	179,807.93	93,783.09	51,990.05	974,706
1996	874,606.31	208,733.90	85,612.52	937,439	179,726.11	93,889.87	51,965.09	988,002
1997	873,145.48	210,799.69	85,007.56	944,984	179,675.47	93,964.25	51,941.36	1,001,365
1998	871,670.98	212,451.72	84,830.03	953,568	179,675.23	93,969.09	51,936.76	1,017,639
1999	871,436.37	213,042.87	84,473.49	961,434	179,625.42	94,070.12	51,885.53	1,029,912
2000	871,005.39	213,891.79	84,055.56	972,572	179,933.94	93,832.88	51,814.25	1,044,208
2001	872,014.87	212,901.29	84,036.57	979,242	180,761.39	93,006.43	51,813.26	1,054,823
2002	873,424.16	211,633.56	83,895.01	987,818	180,841.31	93,172.41	51,567.35	1,066,139
2003	875,598.05	210,385.32	82,969.37	990,615	181,235.14	92,977.79	51,368.14	1,077,858
2004	876,178.05	210,009.47	82,765.21	994,502	181,505.02	92,855.50	51,220.55	1,091,030
2005	875,941.02	210,234.29	82,777.43	993,708	181,290.67	93,097.38	51,193.02	1,097,941
2006	875,921.89	210,243.63	82,787.22	997,151	181,477.04	93,131.64	50,972.39	1,109,013
2007	876,001.57	209,665.68	83,285.49	1,002,592	182,061.55	92,746.59	50,772.93	1,119,930
2008	876,136.67	209,528.51	83,287.55	1,001,682	181,973.36	92,988.61	50,619.10	1,128,072
2009	876,243.64	209,477.35	83,231.74	1,006,236	180,517.24	95,251.30	49,812.53	1,137,356

Table 1. *Cont.*

Year	Mountainous Areas				Basin Areas			
	Forest and Grass Land	Production and Living Land	Unused Land	Population	Forest and Grass Land	Production and Living Land	Unused Land	Population
2010	876,416.81	209,309.41	83,226.51	1,006,097	179,974.29	95,977.57	49,629.21	1,139,411
2011	876,575.39	209,073.85	83,303.49	1,010,814	179,666.23	96,396.98	49,517.86	1,148,713
2012	877,034.45	208,619.57	83,298.71	1,010,315	179,505.91	96,597.58	49,477.58	1,155,451
2013	877,516.46	208,037.60	83,398.67	1,010,294	179,357.88	96,787.15	49,436.04	1,162,075
2014	877,662.39	207,725.11	83,565.23	1,016,317	179,146.49	97,031.90	49,402.68	1,168,698
2015	877,862.18	207,073.35	84,017.20	1,014,409	178,926.28	97,270.57	49,384.22	1,170,915
2016	878,148.01	206,650.20	84,154.52	1,018,955	178,707.29	97,537.97	49,335.81	1,181,352
2017	878,526.13	206,159.85	84,266.75	1,022,613	178,519.97	97,851.60	49,209.50	1,192,978
2018	879,014.29	205,543.20	84,395.24	1,025,623	178,183.32	98,277.02	49,120.73	1,202,647

2.3. Human–Land Coupling Model Construction

When discussing land-type transformation, Dobson only considered the direct transformation from the natural habitat to the agricultural land, but did not consider the direct transformation from the natural habitat to the construction land. According to the status of land-use changes in Yuxi City, this study has different definitions of land types based on the original model. Through the combination of land-use types in Yuxi City, it can be divided into the following three types: (1) Forest and grass land: they mainly represent the natural habitat and are set as the original state of land. The forest and grass land in this study are mainly the combination of forest and grass land. (2) Land for production and living: both the agricultural land and construction land transformed from the natural habitat under the current situation are taken into account. Therefore, the farmland, construction land and other necessary land for production and living are combined and collectively referred to as the production and living land. (3) Unused land: the land that cannot be used temporarily due to bad conditions, or after artificial reclamation or productive and living utilization, or long-term unmanaged and barren land. Let the area of forest and grass land in Yuxi be F, the production and living land be R, and the unused land be U, and $N = F + R + U = 1$. At the same time, the following assumptions are made:

a. The evolution of land use begins with the forest and grass land. The area of the forest and grass land at time t is $F(t)$. In order to maintain the survival of human beings, the forest and grass land needs to be reclaimed or developed and converted into production and living land (farmland or construction land). The area of the production and living land at time t is $R(t)$, and it is assumed that the land reclamation or development rate is related to both the individual reclamation or development capacity d and population density $P(t)$.
b. The unused land area at time t is $U(t)$. Assuming that the transformation rate of unused land b is only related to humans' ability to transform the land through science and technology, then b is controllable, that is, adjustable. Let h be the land area required to sustain a single individual.
c. Since there will be a time delay in the evolution of land use, the transfer mechanism of three land types is shown in Figure 2. The area of production and living land at time t is $R(t)$. Due to the abandoned farmland or construction land, it can be converted into unused land $U(t)$ within a period of $1/a$, and become forest and grass land through natural succession or ecological restoration after a period of $1/s$. The unused land can also be converted into production and living land after a time interval of $1/b$.

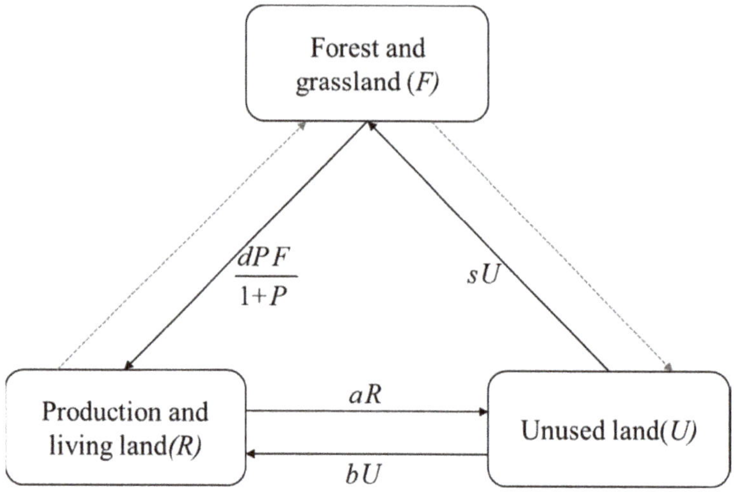

Figure 2. Evolution between different land-use types.

Since the unit of land-use-type area is not consistent with that of the population, the data are transformed into dimensionless data after normalization in the process of model construction and analysis. Through analyzing the data in Yuxi City over the years, it can be seen that the transformation function of population affecting land use is a nonlinear function, and the following function can be obtained through fitting:

$$f(t, P, F) = \frac{dPF}{1 + P} \tag{1}$$

where P and F are the population density and forest and grass land area at time t, respectively, and d is the average reclamation or development capacity of the land. This function is generally interpreted as a Holling-II functional response function in mathematical definitions. Assuming that population growth conforms to the Logistic Retarded Growth model:

$$\frac{dP}{dt} = rP\left(1 + \frac{P(t)}{P_{\max}}\right) \tag{2}$$

where r is the inherent growth rate of the population, and P_{max} (greater than 0) is the maximum population that the environment can carry.

Based on the transformation mechanism mentioned above, in the interval $(t, t + \Delta t)$, the forest and grass land, production and living land and unused land change with time t, and with the help of the population retardation growth model, the population also changes with time t under the constraints of the production and living land. Due to the need for both survival and population growth, more food is needed during population growth than during saturation, that is, the inherent growth rate of the population is a function of a time delay t-τ. Considering the coupling relationship between the population, forest and grass land, the production and living land, and unused land comprehensively, and only discussing the impact of population development on the time delay, the following fractional time delay human–land coupling dynamics model with a Holling-II functional response function can be obtained [30]:

$$\begin{cases} D^{\phi}F(t) = \frac{-dF(t)P(t-\tau)}{1+P(t-\tau)} + sU(t) \\ D^{\phi}R(t) = \frac{dF(t)P(t-\tau)}{1+P(t-\tau)} - aR(t) + bU(t) \\ D^{\phi}U(t) = aR(t) - sU(t) - bU(t) \\ D^{\phi}P(t) = rP(t)\left[1 - \frac{h}{R(t)}P(t-\tau)\right] \end{cases} \tag{3}$$

where $\phi \in [0, 1]$ is the fractional order, and r is the inherent growth rate of the population.

3. Results

3.1. Identification of Human–Land Coupling System Equilibrium Point

In order to make the application of the human–land coupling dynamics model meaningful in Yuxi City, it is necessary to find the sufficient conditions for the model's stability and discuss the stability of the model's equilibrium point and the sustainability of the model. Therefore, the model's equilibrium point is solved first, and the model's stability is further analyzed through the model's equilibrium point. When solving the fractional model, the Adama–Bashforth–Moulton predictive correction algorithm is applied [46]. To facilitate calculation, let the step size $\Delta t = 0.01$, $\phi = 0.9$, substitute $r = 0.0048$ into the model, and then through data analysis and simulation, parameters $a = 0.034$, $b = 0.012$, $s = 0.004$, $h = 0.05$, $d = 0.08$ can be obtained. After a normalized processing of the data for various types of areas and the population of Yuxi City in 1995, the initial values are $F(0)$, $R(0)$, $U(0)$, $P(0) = (0.07, 0.02, 0.4, 0.75)$; substitute them into Equation (3). Then, the equation

$$\begin{cases} 0 = \frac{-dF(t)P(t-\tau)}{1+P(t-\tau)} + sU(t) \\ 0 = \frac{dF(t)P(t-\tau)}{1+P(t-\tau)} - aR(t) + bU(t) \\ 0 = aR(t) - sU(t) - bU(t) \\ 0 = rP(t)\left(1 - \frac{h}{R(t)}P(t-\tau)\right) \end{cases}$$

is obtained, and the solution to this equation is

$$\begin{cases} F^* = 0.075617 \\ R^* = 0.021169 \\ U^* = 0.44984 \\ P^* = 0.423379 \end{cases}.$$

That is, the equilibrium point of the model is (0.075617, 0.021169, 0.44984, 0.423379), and the results are all positive. Therefore, it is consistent with the non-negative situation of land and population, which is systematic and meaningful and reflects the rationality of the model.

3.2. Visual Output and Expression of Human–Land Coupling Relationship

The basic reproduction number R_0 of the human–land coupling dynamics model is a very important parameter; it is said that in the state of balance, the amount of the increase in population brought by land-use changes is a sign that decides whether the land-use type changes or not, namely, only when $R_0 > 1$ does land-use-type transformation occur. If $R_0 < 1$, the transformation will tend to zero. Therefore, according to the calculation method of the basic regeneration number [47], the basic regeneration number of the model can be obtained after solving for the equilibrium point, $R_0 = \frac{bd+ds}{ash} = 18.8235 > 1$, indicating that the land-use types' transformation is significant in the model. Through calculation, $\omega_0 = 0.002358$, $\tau_0 = 728.403$ [30]. An arbitrary τ value which is less than τ_0 was chosen arbitrarily, and MATLAB software was used for numerical simulation. Let $\tau = 700$, and it can be seen through numerical simulation that the human–land change trend over time and the three-dimensional evolution of the human–land relationship can be obtained after a period of damped oscillation (Figures 3 and 4). As $\tau = 700 < \tau_0$, both the theoretical and numerical simulation show that the equilibrium point (0.075617, 0.021169, 0.44984, 0.423379) is locally asymptotically stable, that is, under the influence of the population, the development and change value of all kinds of land use fluctuates, but eventually tends to be stable around the equilibrium point of the model.

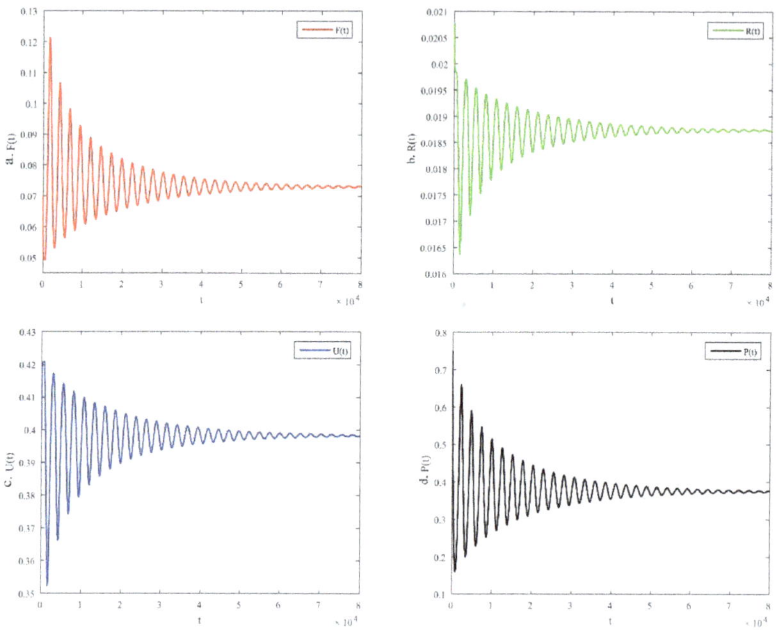

Figure 3. Evolution trend of people and land over time in Yuxi City when $\tau = 700$ ((**a**) change trends of forest and grass land over time; (**b**) change trends of production and living land over time; (**c**) change trends of unused land over time; (**d**) change trends of population over time).

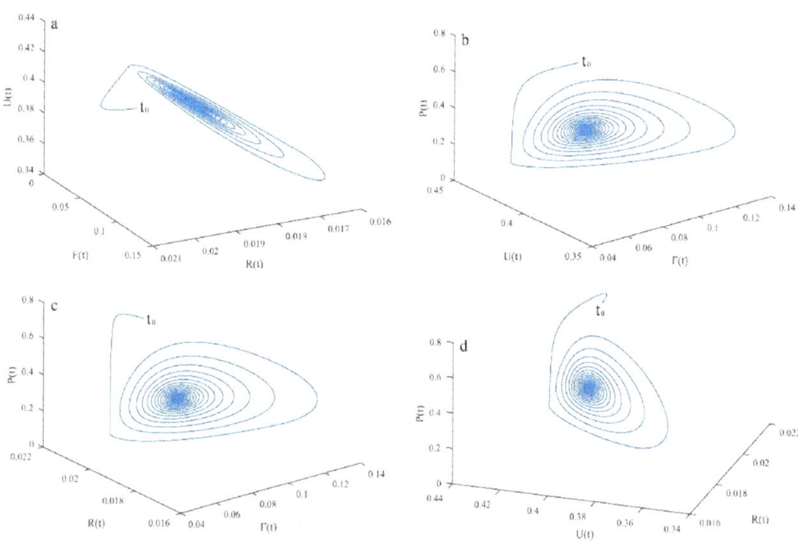

Figure 4. Three-dimensional map of human–land evolution in Yuxi City when $\tau = 700$ ((**a**) three-dimensional evolution of the unused land, forest and grass land and production and living land over time; (**b**) three-dimensional evolution of the population, unused land, and forest and grass land over time; (**c**) three-dimensional evolution of the population, production and living land and forest and grass land over time; (**d**) three-dimensional evolution of the population, unused land and production and living land over time).

As long as the time delay τ does not exceed τ_0 = 728.403, after a $t = 8 \times 10^4$ simulation running time, the land-type area and population in Yuxi City will reach a balance point with time, and the system tends to be in a dynamic stable state. As can be seen from Figure 3a, the forest and grass land in the study area keeps changing with time, increasing and decreasing, but eventually tends to a stable value (0.075617). Similarly, under the system's equilibrium state, the production and living land tends to 0.021169 (Figure 3b), and the unused land tends to 0.44984 (Figure 3c). Under the three land-type transformation and constraint conditions, the population also changes with time, and the final population tends to 0.423379 (Figure 3d). Figure 4 shows the human–land three-dimensional evolutionary relationship at τ = 700, which explains the dynamic characteristics of the human–land coupling system in a stable state. It can be seen that the system is in a stable state when the time delay is less than τ_0.

Similarly, a value of τ that is larger than τ_0 was arbitrarily selected for numerical simulation with MATLAB software. Let τ = 740, and after a period of oscillation, the trend of human–land changes over time and the three-dimensional evolution of the human–land relationship is obtained (Figures 5 and 6). It can be seen that when $\tau = 740 > \tau_0 = 728.403$, the equilibrium point (0.075617, 0.021169, 0.44984, 0.423379) is no longer stable, that is, under the influence of population, the development of different land types fluctuates greatly at first; however, after the time delay is greater than a certain value, the values of all land-use types will tend to oscillate within a certain period, that is, Hopf bifurcation occurs. At this time, the area of the forest and grass land, the area of the production and living land and the area of the unused land will all show a periodic decline and increase.

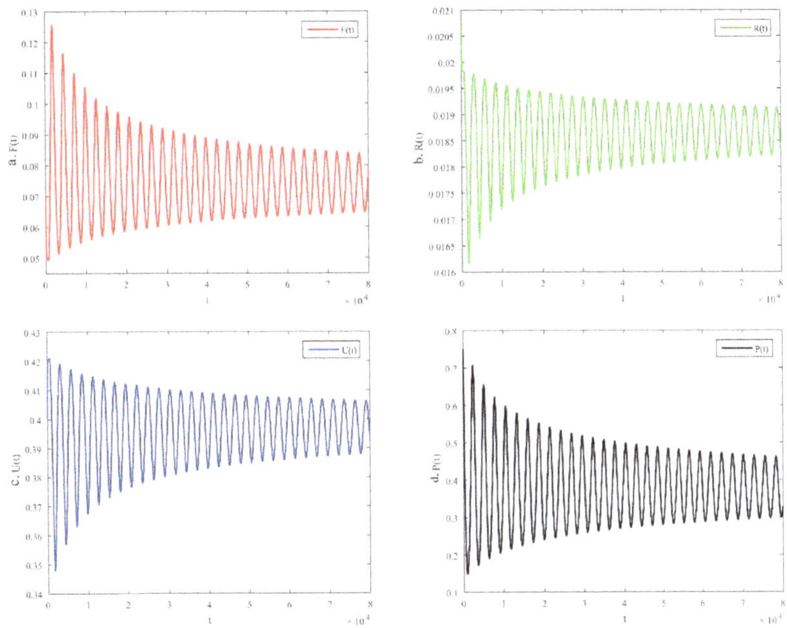

Figure 5. Evolution trend of people and land over time in Yuxi City when τ = 740 ((**a**) change trends of forest and grass land over time; (**b**) change trends of production and living land over time; (**c**) change trends of unused land over time; (**d**) change trends of population over time).

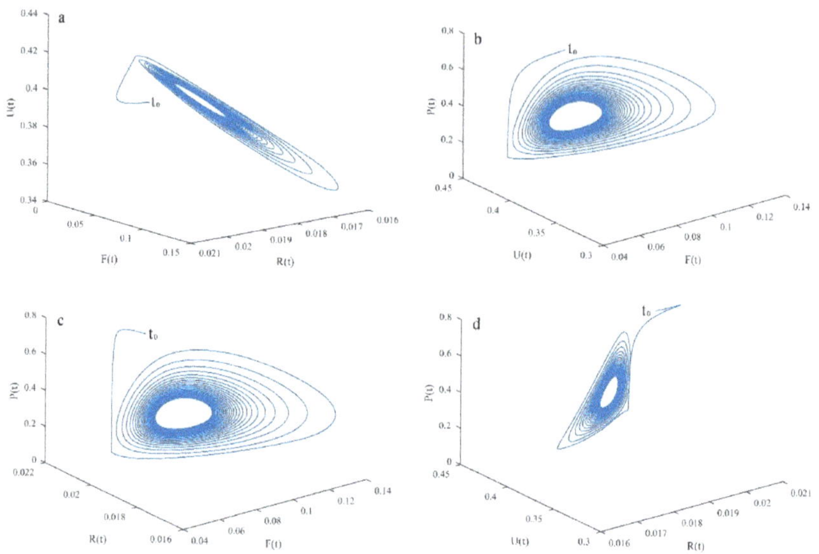

Figure 6. Three-dimensional map of human–land evolution in Yuxi City when τ = 740 ((**a**) three-dimensional evolution of the unused land, forest and grass land and production and living land over time; (**b**) three-dimensional evolution of the population, unused land, and forest and grass land over time; (**c**) three-dimensional evolution of the population, production and living land and forest and grass land over time; (**d**) three-dimensional evolution of the population, unused land and production and living land over time).

When the time delay τ exceeds τ_0 = 728.403, after the simulation running time of $t = 8 \times 10^4$, the population and land types change from the initial large fluctuations to periodic changes around the equilibrium points (0.075617, 0.021169, 0.44984, 0.423379) (Figure 5); that is, in an unstable state, it is significantly different from the situation when the time delay τ is less than τ_0 (Figure 3). When the time delay becomes larger, the area of forest and grass land $F(t)$, the area of production and living land $R(t)$ and the area of unused land $U(t)$ will change periodically. The forest and grass land change periodically around 0.075617 (Figure 5a), the production and living land change periodically around 0.021169 (Figure 5b), and the unused land changes periodically around 0.44984 (Figure 5c), and will not tend to a stable value. Figure 6 shows the three-dimensional evolution of the human–land relationship when τ = 740, which explains the dynamic characteristics of the human–land coupled system in the unstable state. It can be seen that when the time delay is greater than τ_0, the system exhibits an unstable periodic oscillation phenomenon.

Comparing Figure 4 with Figure 6, we could see that when $\tau > \tau_0$, the human–land evolution trend is more consistent with the actual situation, that is, with the periodic change in the population, the area of land types changes periodically. However, in order to tend to a stable state at $\tau < \tau_0$, that is, the equilibrium state of the human–land system, relevant policies can be formulated and implemented by the government, that is, parameters can be controlled and adjusted.

3.3. Coupling Spatiotemporal Parameters of Mountain–Basin Human–Land Relationship

With the migration of the population from mountainous areas to basin areas, the agglomeration of the population leads to the occupation of agricultural land and the expansion of urban land [45,48]. Early on, due to the limited population, the urbanization level is not high, and coupled with the influence of the policy constrains, the migration of the population is also rare. The main way of life is farming, and the ecological and environmental effects caused by population migration are not obvious. However, with the rapid improvement in the urbanization level and the acceleration of the migration of the population from rural to urban areas, land-use changes are accelerating. In mountainous areas, the arable land and construction land have been abandoned [49] and turned into a wasteland, that is, unused land. However, the wasteland will naturally recover into forest and grass land after a certain period of time. In basin areas, with the migration of the population, the demand for land is increasing, and the unused land will be gradually transformed into forest and grass land and farmland and construction land.

With the rapid increase in the population of the basin area, the land's resource-carrying capacity, industrial-supporting capacity and infrastructure are facing more challenges, driving the changes in the land-use pattern to meet the needs of population agglomeration. At the same time, the population in mountainous areas is decreasing, farmland and construction land will be abandoned, and ecological land has been restored. According to Model (3), two subsystems of mountainous areas and basin areas are distinguished, and human–land coupling evolution and development models of mountainous areas and basin areas are constructed, respectively:

$$\begin{cases} D^{\phi}F_M(t) = \frac{-d_M F_M(t) P_M(t-\tau_M)}{1+P_1(t-\tau_1)} + s_M U_M(t) \\ D^{\phi}R_M(t) = \frac{d_M F_M(t) P_M(t-\tau_M)}{1+P_M(t-\tau_M)} - a_M R_M(t) + b_M U_M(t) \\ D^{\phi}U_M(t) = a_M R_M(t) - s_M U_M(t) - b_M U_M(t) \\ D^{\phi}P_M(t) = r_M P_M(t)\left[1 - \frac{h_M}{R_M(t)} P_M(t-\tau_M)\right] \end{cases} \tag{4}$$

$$\begin{cases} D^{\phi}F_B(t) = \frac{-d_B F_B(t) P_B(t-\tau_B)}{1+P_B(t-\tau_B)} + s_B U_B(t) \\ D^{\phi}R_B(t) = \frac{d_B F_B(t) P_B(t-\tau_B)}{1+P_B(t-\tau_B)} - a_B R_B(t) + b_B U_B(t) \\ D^{\phi}U_B(t) = a_B R_B(t) - s_B U_B(t) - b_B U_B(t) \\ D^{\phi}P_B(t) = r_B P_B(t)\left[1 - \frac{h_B}{R_B(t)} P_B(t-\tau_B)\right] \end{cases} \tag{5}$$

In the formula, $\phi \in [0, 1]$ is the fractional order, and the relevant parameters are as follows: $F_M(t)$, $R_M(t)$, $U_M(t)$ and $F_B(t)$, $R_B(t)$, $U_B(t)$ respectively represent the forest and grass land, production and living land, unused land in mountainous areas and basin areas, and $F_M(t) + F_B(t) = F(t)$, $R_M(t) + R_B(t) = R(t)$, $U_M(t) + U_B(t) = U(t)$, $N = F(t) + R(t) + U(t) = 1$. $F_M(t)$, $R_M(t)$, $U_M(t)$ and $F_B(t)$, $R_B(t)$, $U_B(t)$ are the population density of mountainous areas and basin areas at time t, respectively. The production and living land [$R_M(t)$ and $R_B(t)$] will change into unused land [$U_M(t)$ and $U_B(t)$] during the period of $1/a_M$ and $1/a_B$, and then become forest and grass land [$F_M(t)$ and $F_B(t)$] through natural succession or ecological restoration after the interval of $1/s_M$ or $1/s_B$. The unused land can also be reclaimed or developed into production and living land after an interval $1/b_M$ and $1/b_B$. The average reclamation capacity of mountainous areas and basin areas is described by the constants d_M and d_B respectively. r_M and r_B are the natural growth rate of the population in mountainous areas and basin areas, respectively.

MATLAB software was used to further fit the land-type area and population for different periods of mountainous areas and basin areas, and the parameters of the human–land coupling dynamics model in mountainous areas and the human–land coupling dynamics model in basin areas were obtained, respectively (Table 2).

According to the existing data analysis, the conversion time parameters for production and living land to unused land of mountainous areas and basin areas in Yuxi City are $a_M = 0.0486$ and $a_B = 0.0126$, respectively, indicating that it takes about 20 years for production and living land in mountainous areas to be abandoned and then converted to unused land, while it takes about 80 years for basin areas. a_M is greater than a_B, indicating that the conversion time of productive and living land to unused land in the basin area is slower than that in the mountainous area, mainly because the population growth in the basin area is faster than that in the mountainous area, and the demand for productive and living land for economic construction is larger. This trend will continue for a long time in the future, making the conversion time of productive and living land to unused land much longer than that in the mountainous area.

Table 2. Fitting parameters of human–land coupling dynamics model in mountainous areas and basin areas of Yuxi City.

Description	Parameters in Mountainous Areas	Parameters in Basin Areas
Conversion of production and living land into unused land	$a_M = 0.0486, 1/a_M = 20.5761$	$a_B = 0.0126, 1/a_B = 79.3651$
Conversion of unused land into production and living land	$b_M = 0.0062, 1/b_M = 161.2903$	$b_B = 0.0139, 1/b_B = 71.9425$
Conversion of unused land into forest and grass land	$s_M = 0.0051, 1/s_B = 196.0784$	$s_B = 0.0028, 1/s_B = 357.1429$
Land area required to maintain the unit individual	$h_M = 0.0335$	$h_B = 0.0165$
Average reclamation capacity	$d_M = 0.03$	$d_B = 0.05$
Natural growth rate of population	$r_M = 0.0563$	$r_B = 0.151$

The time parameters of conversion from unused land to productive and living land in mountainous areas and basin areas are $b_M = 0.0062$ and $b_B = 0.0139$, respectively, indicating that unused land in mountainous areas will be reclaimed or developed into productive and living land again after about 160 years, while the time of conversion in basin areas is shorter, about 70 years. b_M is less than b_B, indicating that the conversion time of unused land to productive and living land in basin areas is faster than that in mountainous areas. With the rapid development of the society and economy and the growth of the population in the basin area, the demand for construction land is always on the rise. Under the background of the policy of vigorously protecting cultivated land, the demand for construction land is mainly solved by the transformation of forest and grass land and unused land, which allows the unused land in the basin area to be transformed quickly and in a shorter time than that in the mountainous area.

The conversion time parameters of unused land to forest and grass land for mountainous areas and basin areas are $s_M = 0.0051$ and $s_B = 0.0028$, respectively, indicating that it takes about 200 years for unused land to convert to forest and grass land by natural succession or ecological restoration in the mountainous area, while it takes about 350 years for the basin area. s_M is greater than s_B, showing that the conversion time from unused land to forest and grass land in the basin area is longer than that in the mountainous area. The main reason is that the social and economic development of the basin area has a radiation effect on the population and economy in the mountainous area, which makes the mountainous area gradually decline. The production and living land, such as farmland and construction land, is transformed into unused land and further transformed into forest and grass land due to their abandonment and extensive management, and the transformation speed is fast.

The land area parameters required for individual maintenance are $h_M = 0.0335$ and $h_B = 0.0165$, respectively. h_M is greater than h_B, indicating that the land area required for each unit in basin areas is smaller than that in mountainous areas. The main reason is that the land-use intensity in mountainous areas is significantly lower than that in basin areas, and the land yield rate is far lower than that in the basin area, so that the land area required by each unit is larger than that in the basin area. The average reclamation or development capacity in mountainous areas and basin areas are $d_M = 0.03$ and $d_B = 0.05$, respectively. d_M is less than d_B. This shows that the average reclamation and development capacity of the basin area is higher than that of the mountainous area, mainly because the economic development level, investment capacity and natural conditions of the basin area are better than those of the mountainous area.

4. Discussion

Under mathematical and geostatistical semantics, the order in a fractional differential equation can not only affect the dynamic characteristics of the fractional differential model, but also advance or delay the occurrence of the stability of the fractional differential model [50,51]. Therefore, the stability of the model can be improved by adjusting the model parameters. The influence of the time delay on the fractional order model is also dual. On the one hand, the time delay can make the fractional order model lose stability and lead to bifurcation. On the other hand, under certain conditions, the stability of the fractional order model can be improved with an appropriate time delay, and the occurrence of bifurcation can be further delayed [43]. Under the existing conditions of Yuxi City, $\tau_0 = 728.403$ is the critical point between the stability and instability of a human–land coupling system. The state of the human–land coupling system includes a stable equilibrium state and periodic oscillation state. The time delay can be used to determine the two states and how to adjust from periodic change to a stable state, or how to adjust to a stable state when periodic change is presented. Then, the time delay can be changed by changing the parameters a, b, s, h, d and r. The time delay is like an invisible hand, which can not only optimize the allocation of land resources, but also give a warning signal when the production and living land area tends to be unstable, reminding people to make the land-use evolution stay stable by regulating the average reclamation or development capacity of individuals (d) and the inherent growth rate of population changes (r). The rate by which the unused land is recultivated or developed (b) and an individual's average reclamation or development capacity (d) can be artificially controlled. When the human–land evolutionary system tends to be stable, $b \in [0.012, 1]$, $d \in [0.08, 1]$ and d is greater than b, indicating that the average reclamation or development capacity has a greater impact on the evolutionary stability of land use than the transformation rate of unused land. The optimal state of balancing a human–land coupling system is the optimal state of sustainable development, which is an ideal state. By adjusting these parameters, the system can be as close to the ideal state as possible.

Chen [47] constructed an integer order land dynamics model with a time delay in 2017, simulated the data of population and land-use change over the years in China, and found that the service life of subsistence land in China is about 25–70 years ($a = 0.04$), the time for reclaiming or developing the wasteland into subsistence land is about 100 years ($b = 0.011$), and the time for restoring the wasteland to original land is about 1000 years ($s = 0.005$). By comparison, there are similarities and differences with the parameters in this study, which are highlighted by the conversion time parameter s of unused land to forest and grass land. Theoretically, it will take a long time for the degraded land to be restored to the original forest and grass land through natural succession, but with the progress of science and technology, this time will be greatly shortened, especially in the mountainous area of Yuxi City with its better ecological environment. Of course, these time parameters are obtained from existing data analysis. The reclamation or development rate of unused land (b) and the average reclamation or development capacity (d) are both controllable factors, and the average reclamation or development capacity has a greater impact on the stability of human–land systems.

This study based on systems integration thought that by analyzing the relationship between population and land based on coupled differential equations of forest and grass land, production and living land, unused land and population, it could build a fractional order human–land coupling dynamics mode with a Holling-II-type land conversion rate and time delay. The human–land coupling mechanism of the mountain–basin system is quantitatively described, and a new simulation direction is provided for coordination and optimization. However, the application of the method still needs to be improved in the future. First, since the human–land coupling relationship is a long-term process, the accuracy of parameter estimations depends on long-term sequence data. However, the existing collected data are only 24 years old, which is too small compared to the land conversion cycle. The model built is a high-dimensional and nonlinear differential system, and the estimation of its parameters is inherently difficult. In addition, the lack of data makes it more difficult to achieve accurate model parameters such as for population growth rate and land conversion rate, etc. Secondly, in the classification process, there are problems such as the inaccuracy of various land types. For example, forest and grass land are assumed as primitive land in the model, but in fact, the existing forest and grass land have been transformed by human beings for a long time and lost their original nature. Finally, the coupling relationship between humans and land is complex. Although population is the main influencing factor for land-use changes, there are still more disturbance variables. The differential effects of special regional policies (such as urbanization in nearby areas, relocation of poor people from inhospitable areas, rural revitalization, development of plateau agriculture, ecological protection, etc.), technical means and micro-farmer behaviors on the coupled evolution of mountain–basin human–land systems should also be considered. Therefore, how to further build a multi-factor land-use dynamics model for specific regions and explore the coupled evolution mechanism of mountain–basin human–land systems should be the future direction of efforts.

5. Conclusions

A human–land coupling system is a nonlinear system, which is a differential equation composed of several coupling factors. In this study, the coupling factors of forest and grass land, production and living land, unused land and population were considered to construct a fractional order dynamics model with a time delay based on a Holling-II transformation rate. The stability of the system and its regulation mechanisms are discussed based on the solution of the equilibrium point of the system. It is known that the coupling state of human–land systems includes a stable equilibrium state and periodic oscillation state, and the two states of the system can be determined according to the time delay. How to adjust from periodic change to a stable state, or how to adjust to a stable state when periodic change is presented, the time delay can be changed by changing the parameters of the model.

On the basis of constructing the human–land coupling evolution model of Yuxi City, the human–land coupling evolution and development models of mountainous areas and basin areas were respectively constructed based on the internal relationship between the mountainous area and the basin area. The parameters of the human–land coupling evolution and development models were obtained by simulation analysis, using the existing data. According to the analysis of the parameters, the mountainous area and the basin area playing different roles in the process of human–land evolution systems tend to a stable state, i.e., an equilibrium state, and there are significant differences in the transformation time among different land types. Therefore, different land-use regulation strategies should be selected in different regions. In mountainous areas, the average reclamation or development capacity should be reduced to lower the conversion of forest and grass land to production and living land, and at the same time, excessive interference of human beings should be reduced to speed up the natural recovery and succession of unused land to forest and grass land. In basin areas, in addition to reducing the average reclamation or the development capacity, the reclamation or development rate of the idle land and degraded land should be increased, and the conversion of idle land into productive and living land should be encouraged with the help of certain scientific and technological means.

Author Contributions: Conceptualization, L.W.; methodology, L.W.; software, L.W.; validation, L.W. and B.X.; formal analysis, L.W.; investigation, L.W.; resources, L.W.; data curation, L.W.; writing—original draft preparation, L.W.; writing—review and editing, L.W., Y.Y. and B.X.; supervision, B.X.; funding acquisition, L.W. All authors have read and agreed to the published version of the manuscript.

Funding: This research was funded by the National Natural Science Foundation of China, grant number 42161041.

Institutional Review Board Statement: Not applicable.

Informed Consent Statement: Not applicable.

Data Availability Statement: Not applicable.

Conflicts of Interest: The authors declare no conflict of interest.

References

1. Kőszegi, M.; Bottlik, Z. Human-environment relationships in modern and postmodern geography. *Hung. Geogr. Bull.* **2015**, *64*, 87–99. [CrossRef]
2. Bohle, M. Handling of human-geosphere intersections. *Geosciences* **2016**, *6*, 3. [CrossRef]
3. Gong, S.H.; Wang, S.J.; Bai, X.Y.; Luo, G.J.; Wu, L.H.; Chen, F.; Qian, Q.H.; Xiao, J.Y.; Zeng, C. Response of the weathering carbon sink in terrestrial rocks to climate variables and ecological restoration in China. *Sci. Total Environ.* **2021**, *750*, 141525. [CrossRef] [PubMed]
4. Song, F.J.; Wang, S.J.; Bai, X.Y.; Wu, L.H.; Wang, J.F.; Li, C.J.; Chen, H.; Luo, X.L.; Xi, H.P.; Zhang, S.R.; et al. A new indicator for global food security assessment: Harvested area rather than cropland area. *Chin. Geogr. Sci.* **2022**, *32*, 204–217. [CrossRef]
5. Zvoleff, A.; Li, A. Analyzing human-landscape interactions: Tools that integrate. *Environ. Manag.* **2014**, *53*, 94–111. [CrossRef]
6. Li, A. Modeling human decisions in coupled human and natural systems: Review of agent-based models. *Ecol. Model.* **2012**, *229*, 25–36. [CrossRef]
7. French, C. People, societies, and landscapes. *Science* **2010**, *328*, 443–444. [CrossRef]
8. Liu, S.C.; Ma, L.B.; Yao, Y.; Cui, X.J. Man-land relationship based on the spatial coupling of population and residential land-A case study of Yuzhong County in Longzhong Loess Hilly Region, China. *Land Use Policy* **2022**, *116*, 106059. [CrossRef]
9. Magliocca, N.R. Agent-based modeling for integrating human behavior into the food-energy-water nexus. *Land* **2020**, *9*, 519. [CrossRef]
10. Zhang, S.R.; Bai, X.Y.; Zhao, C.W.; Tan, Q.; Luo, G.J.; Wu, L.H.; Xi, H.P.; Li, C.J.; Chen, F.; Ran, C.; et al. China's carbon budget inventory from 1997 to 2017 and its challenges to achieving carbon neutral strategies. *J. Clean. Prod.* **2022**, *347*, 130966. [CrossRef]
11. Ge, Q.S.; Fang, C.L.; Jiang, D. Geographical missions and coupling ways between human and nature for the Beautiful China Initiative. *Acta Geogr. Sin.* **2020**, *75*, 1109–1119. [CrossRef]
12. Palmer, P.I.; Smith, M.J. Earth systems: Model human adaptation to climate change. *Nature* **2014**, *512*, 365–370. [CrossRef] [PubMed]
13. Hu, Z.N.; Li, X.; Lou, S.Y.; Kang, J.R. Multi-scenario simulation of land use structure of Yangzhou City based on systems dynamics mode. *Bulletion Soil Water Conserv.* **2017**, *37*, 211–218. [CrossRef]
14. Motesharrei, S.; Rivas, J.; Kalnay, E.; Asrar, G.R.; Busalacchi, A.J.; Cahalan, R.F.; Cane, M.A.; Colwell, R.R.; Feng, K.; Franklin, R.S.; et al. Modeling sustainability: Population, inequality, consumption, and bidirectional coupling of the earth and human systems. *Natl. Sci. Rev.* **2016**, *3*, 470–494. [CrossRef]
15. Voinov, A.; Shugart, H.H. 'Integronsters', integral and integrated modeling. *Environ. Model. Softw.* **2013**, *39*, 149–158. [CrossRef]
16. Calvin, K.; Bond-Lamberty, B. Integrated human-earth system modeling-state of the science and future directions. *Environ. Res. Lett.* **2018**, *13*, 063006. [CrossRef]
17. Stehfest, E.; van Vuuren, D.; Bouwman, L.; Kram, T. *Integrated Assessment of Global Environmental Change with IMAGE 3.0: Model Description and Policy Applications*; Netherlands Environmental Assessment Agency (PBL): The Hague, The Netherlands, 2014.
18. Manson, S.M.; Peng, S.S.; Bonsal, D. *Agent-Based Modeling and Complexity*; Springer: Dordrecht, The Netherlands, 2012; pp. 125–139.
19. Zhu, G.F.; Li, X.; Su, Y.H.; Zhang, K.; Bai, Y.; Ma, J.Z.; Li, C.B.; Hu, X.L.; He, J.H. Simultaneously assimilating multivariate data sets into the two-source evapotranspiration model by Bayesian approach: Application to spring maize in an arid region of northwestern China. *Geosci. Model Dev.* **2014**, *7*, 1467–1482. [CrossRef]
20. Clark, M.P.; Kavetski, D.; Fenicia, F. Pursuing the method of multiple working hypotheses for hydrological modeling. *Water Resour. Res.* **2011**, *47*, W09301. [CrossRef]
21. Song, C.Q.; Cheng, C.X.; Shi, P.J. Geography complexity: New connotations of geography in the new era. *Acta Geogr. Sin.* **2018**, *73*, 1204–1213. [CrossRef]
22. Zhang, S.R.; Bai, X.Y.; Zhao, C.W.; Tan, Q.; Luo, G.J.; Cao, Y.; Deng, Y.H.; Li, Q.; Li, C.J.; Wu, L.H.; et al. Limitations of soil moisture and formation rate on vegetation growth in karst areas. *Sci. Total Environ.* **2022**, *810*, 151209. [CrossRef]
23. Dong, G.L.; Zhang, W.X.; Xu, X.L.; Jia, K. Multi-dimensional feature recognition and policy implications of rural human–land relationships in China. *Land* **2021**, *10*, 1086. [CrossRef]
24. Wu, Z.P.; Li, T.; Heavens, N.G.; Newman, C.E.; Richardson, M.I.; Yang, C.Y.; Li, J.; Cui, J. Earth-like thermal and dynamical coupling processes in the Martian climate system. *Earth-Sci. Rev.* **2022**, *229*, 104023. [CrossRef]
25. Yang, Z.M.; Zhao, Y.L.; Xue, C.L. Land use evolution and spatial differentiation characteristics of mountain-basin system in karst area. *Chin. J. Agric. Resour. Reg. Plan.* **2020**, *41*, 153–162. [CrossRef]
26. Fang, C.L.; Wang, J. A theoretical analysis of interactive coercing effects between urbanization and eco-environment. *Chin. Geogr. Sci.* **2013**, *23*, 147–162. [CrossRef]
27. Wu, L. Study on Land Use Change and Human-Land Coupling Mechanism for Mountain-Basin System at Township Scale. Ph.D. Thesis, Hunan Normal University, Changsha, China, 2022.

28. Lu, D.D. Theoretical studies of man-land system as the core of geographical science. *Geogr. Res.* **2002**, *21*, 135–145.
29. Dobson, A.P.; Bradshaw, A.D.; Baker, A.J.M. Hopes for the future: Restoration ecology and conservation biology. *Science* **1997**, *277*, 515–522. [CrossRef]
30. Wu, L.; Li, Z.H.; Zhang, Y.; Xie, B.G. Complex behavior analysis of a fractional-order land dynamical model with Holling-II type land reclamation rate on time delay. *Discret. Dyn. Nat. Soc.* **2020**, *2020*, 1053283. [CrossRef]
31. Ahmed, E.; El-Sayed, A.M.A.; El-Saka, H.A.A. Equilibrium points, stability and numerical solutions of fractional-order predator-Cprey and rabies models. *J. Math. Anal. Appl.* **2017**, *325*, 542–553. [CrossRef]
32. El-Saka, H.A.; Ahmed, E.; Shehata, M.I.; El-Sayed, A.M.A. On stability, persistence, and Hopf bifurcation in fractional order dynamical systems. *Nonlinear Dyn.* **2009**, *56*, 121–126. [CrossRef]
33. Luo, C.; Wang, X.Y. Chaos generated from the fractional-order complex chen system and its application to digital secure communication. *Int. J. Mod. Phys. C* **2013**, *24*, 1350025. [CrossRef]
34. Li, Y.; Wang, H.P.; Tian, Y. Fractional-order adaptive controller for chaotic synchronization and application to a dual-channel secure communication system. *Mod. Phys. Lett. B* **2019**, *33*, 1097–1102. [CrossRef]
35. Wang, Z.; Xie, Y.K.; Lu, J.W.; Li, Y.X. Stability and bifurcation of a delayed generalized fractional-order prey-predator model with interspecific competition. *Appl. Math. Comput.* **2019**, *347*, 360–369. [CrossRef]
36. Matignon, D. Stability results for fractional differential equations with applications to control processing. In Proceedings of the Computational Engineering in Systems and Application Multiconference, Beijing, China, 4–6 October 2016; pp. 963–968.
37. Laskin, N. Fractional quantum mechanics. *Phys. Rev. E* **2000**, *62*, 3135–3145. [CrossRef] [PubMed]
38. Podlubny, I. *Fractional Differential Equations*; Academic Press: Washington, DC, USA, 1999.
39. Kilbas, A.A.; Srivastava, H.M.; Trujillo, J.J. *Theory and Application of Fractional Differential Equations*; Elsevier: New York, NY, USA, 2006.
40. Zhao, T.; Zhang, Z.Z.; Upadhyay, R.K. Delay-induced Hopf bifurcation of an SVEIR computer virus model with nonlinear incidence rate. *Adv. Differ. Equ.* **2018**, *2018*, 256. [CrossRef]
41. Vargas-De-Leon, C. Volterra-type Lyapunov functions for fractional-order epidemic systems. *Commun. Nonlinear Sci. Numer. Simul.* **2015**, *24*, 75–85. [CrossRef]
42. Deshpande, A.S.; Daftardar-Gejji, V.; Sukale, Y.V. On Hopf bifurcation in fractional dynamical systems. *Chaos Solitons Fractals* **2017**, *98*, 189–198. [CrossRef]
43. Wang, H.; Yu, Y.G.; Wen, G.G.; Zhang, S. Stability analysis of fractional-order neural networks with time delay. *Neural Process. Lett.* **2015**, *42*, 479–500. [CrossRef]
44. Wu, L.; Xie, B.G.; Xiao, X.; Xue, B.; Li, J.Z. Classification method and determination of mountainous area types at township scales: A case study of Yuxi City, Yunnan Province. *Complexity* **2020**, *2020*, 3484568. [CrossRef]
45. Wu, L.; Xie, B.G. The variation differences of cultivated land ecological security between flatland and mountainous areas based on LUCC. *PLoS ONE* **2019**, *14*, e0220747. [CrossRef]
46. Bhalekar, S.; Varsha, D. A predictor-corrector scheme for solving nonlinear delay differential equations of fractional order. *J. Fract. Calc. Its Appl.* **2011**, *1*, 1–9. [CrossRef]
47. Chen, S.S.; Hua, J.; Zhang, F.X.; Li, Y.M. Dynamical analysis of land model with Holling-II land reclamation rate. *J. Syst. Sci. Complex.* **2017**, *37*, 819–827.
48. Yi, D.; Guo, X.; Han, Y.; Guo, J.; Ou, M.H.; Zhao, X.M. Coupling ecological security pattern establishment and construction land expansion simulation for urban growth boundary delineation: Framework and application. *Land* **2022**, *11*, 359. [CrossRef]
49. Shao, J.A.; Zhang, S.C.; Li, X.B. Farmland marginalization in the mountainous areas: Characteristics, influencing factors and policy implications. *J. Geogr. Sci.* **2015**, *25*, 701–722. [CrossRef]
50. Yang, L.X.; He, W.S.; Liu, X.J. Synchronization between a fractional-order system and an integer order system. *Comput. Math. Appl.* **2011**, *62*, 4708–4716. [CrossRef]
51. Odai, M.; Hori, Y. Controller design robust to nonlinear elements based on fractional order control system. *IEEJ Trans. Ind. Appl.* **2010**, *120*, 11–18. [CrossRef]

land

MDPI

Article

Dynamic Evolution and Quantitative Attribution of Soil Erosion Based on Slope Units: A Case Study of a Karst Plateau-Gorge Area in SW China

Chuhong Shen [1,2], Kangning Xiong [1,2,*] and Tian Shu [1,2,3]

1 School of Karst Science, Guizhou Normal University, Guiyang 550001, China; 20010170525@gznu.edu.cn (C.S.); 17030170025@gznu.edu.cn (T.S.)
2 State Engineering Technology Institute for Karst Desertification Control, Guiyang 550001, China
3 Institute of Science and Technology Information, Guizhou Academy of Agriculture Sciences, Guiyang 550006, China
* Correspondence: xiongkn@gznu.edu.cn

Abstract: Exploring the dynamics of soil erosion and identifying its driving mechanisms is key to understanding soil erosion processes, particularly in karst areas. In this study, the RUSLE model, optimized on the basis of rocky desertification factors, was used to estimate soil erosion in a karst plateau gorge area in SW China. The spatial and temporal dynamics of soil erosion in the region over the past 20 years were analyzed on the basis of slope units, while the relationship between soil erosion and elevation, slope, fractional vegetation cover (FVC), karst rocky desertification (KRD), rainfall, and land use cover/change (LUCC) was identified quantitatively by the geographical detector on the basis of spatial heterogeneity. The results showed that: (1) The no erosion area decreased from 2000 to 2020, with the highest proportion of light to medium erosion and an increasing trend of soil erosion. (2) Soil erosion conversion mainly occurred between no erosion, slight erosion, and light erosion. (3) The hotspots of erosion occurred in high slope–low elevation and high slope–high elevation units, while the coldspots of erosion occurred in low slope–low elevation units. (4) Soil erosion was positively correlated with FVC and slope, and negatively correlated with KRD. (5) The dominant factor of soil erosion changed from KRD-slope to LUCC-slope and finally to elevation-slope, while the q value of rainfall-elevation had the most significant increase throughout the study period. This study will help to advance the goal of sustainable development of soil and water conservation in karst areas.

Keywords: karst soil erosion; topographical units; RUSLE; geographical detector; influencing factors

Citation: Shen, C.; Xiong, K.; Shu, T. Dynamic Evolution and Quantitative Attribution of Soil Erosion Based on Slope Units: A Case Study of a Karst Plateau-Gorge Area in SW China. *Land* **2022**, *11*, 1134. https://doi.org/10.3390/land11081134

Academic Editor: Xiaoyong Bai

Received: 10 June 2022
Accepted: 20 July 2022
Published: 24 July 2022

1. Introduction

Soil erosion is considered to be the greatest threat to land degradation, seriously affecting terrestrial ecosystem security [1,2]. Soil erosion disrupts the soil agglomeration structure [3], resulting in the redistribution of soil nutrients [4], which changes soil carbon transport [5] and affects soil ecological service functions [6]. Soil erosion also reduces soil resources, exacerbates the scarcity of land resources [3], affects vegetation growth [7] and food security, and hinders sustainable socio-economic development [8]. To effectively control soil erosion and its negative socio-environmental impacts, the formation process, dynamic evolution, and hazards of soil erosion must be robustly assessed [9,10]. China is one of the countries most severely affected by soil erosion in the world, especially in the karst region of southwestern China [11]. Soil erosion leading to rock desertification has become a major environmental disaster limiting people's production and development [12]. In recent years, numerous studies have focused on soil erosion in karst areas, including erosion processes, spatial and temporal evolution, driving mechanisms [13], dynamic modeling [14], sensitivity evaluation [15], and control measures [16]. However, constrained by the highly heterogeneous geographical environment and complex erosion patterns [17], many research methods are difficult to perform and apply in karst areas, and little is known about soil erosion processes and drivers in karst areas. These analyses are fundamental requirements for combating soil erosion, and in particular, understanding the long-term dynamic evolution of soil erosion and its interactive drivers is essential for land managers to assess soil erosion changes and formulate soil and water conservation policies.

With the increasing abundance of environmental data, scientists have developed several mathematical models to emulate soil erosion processes at different spatial and temporal scales on the basis of topographic, climatic, soil, land use, and vegetation cover data [18,19]. Among the existing erosion models, the RUSLE model [16], the SWAT model [20], and the WEPP model [21] have been proven to be applicable at different spatial scales. They are widely used in complex topographic units due to their simple structure and GIS compatibility [22], such as in the karst areas of southwest China [23] or Cuba [24]. However, many studies have not considered the control of soil erosion by karst conditions and the direct use of the RUSLE model may overestimate soil erosion in karst areas [25]. Karst landscapes have a double-layer structure of surface and subsurface. Large exposures of carbonate rocks on the surface alter surface runoff velocity and flow patterns and intercept sediment runoff; secondary pore spaces are developed underground and contribute to the rapid transport of runoff sediment. Dai found a correlation coefficient of -0.076 ($p < 0.01$) between soil erosion and bedrock exposure on the basis of an artificially designed soil trough device with a double-layer spatial structure and simulated rock desertification [26]. Gao and Wang optimized the RUSLE model by introducing the rock desertification factor on this basis [23], and the results showed that the simulation accuracy of the RUSLE model was significantly improved after optimization. Therefore, this study used the RUSLE optimized to estimate soil erosion in karst areas.

Appropriate study units are an important prerequisite for scientific spatial analysis. In current soil erosion studies, the common study units are administrative units, grid units, and geographical feature units. However, the assessment results based on these study units are difficult to meet the requirements for fine-grained soil erosion assessment or control, which may make it difficult to carry out accurate soil and water conservation work. The slope cell, proposed by Carrara, is a topographic unit cut by a combination of ridgelines, valley lines, terrace boundaries, and valley bottom boundaries [27]. Slope units are constructed according to hydrological processes, ensuring maximum homogeneity within the unit and maximum heterogeneity between different units [28], and are currently widely used in the spatial distribution of landslides [29], sensitivity analysis [30], and prediction studies [31], among others, with the slope unit being a more sophisticated unit than traditional units. Compared with traditional units, the slope unit has higher classification performance and more stable estimation coefficients, which better reflect the actual geographical environment and reduce the uncertainty of control factors [32]. Using slope units as a basis for analyzing soil erosion can help in the analysis of dominant factors of soil erosion.

In this study, we aimed to investigate the long-term soil erosion evolution patterns in karst areas and the interaction of their driving forces. To achieve the objectives, the following analyses were made: (1) simulation of soil erosion by the RUSLE model optimized by the rocky desertification factor; (2) analysis of the long-term spatial and temporal dynamic evolution pattern of soil erosion; (3) identification of the dominant and interacting factors of soil erosion evolution. The results of the study can provide a scientific reference for determining suitable soil erosion control schemes in karst areas, and this contribution will also help to advance the sustainable development goals of soil and water conservation in karst areas.

2. Materials and Methods

2.1. Study Area

The karst zone of southwest China, centered on Guizhou, is the largest continuous zone of karst and has the most typical and complex karst landscape in the world. The study area (105°34′59″~105°43′06″ E, 25°37′18″~25°42′37″ N) was selected in southwest Guizhou, south of Guanling County and north of Zhenfeng County, on both sides of the Beipanjiang River Gorge, with a total area of 51.6197 km^2 (Figure 1). The altitude range of the area is between 443 and 1366 m a.s.l., which is a typical plateau canyon landform. The region is under a dry and hot southern subtropical river valley climate, with warm and dry winters and springs, and high temperatures and rain in summer and autumn [33]. Meteorological data show no significant increase or decrease in temperature or precipitation, both varied within regular ranges from 2000 to 2020. The average annual temperature is 18.4 °C, and the average annual precipitation is 1100 mm, with May to October accounting for more than 80% of the total annual precipitation. The lithology is mainly Middle Tertiary limestone and dolomite, and the soil is calcareous [34].

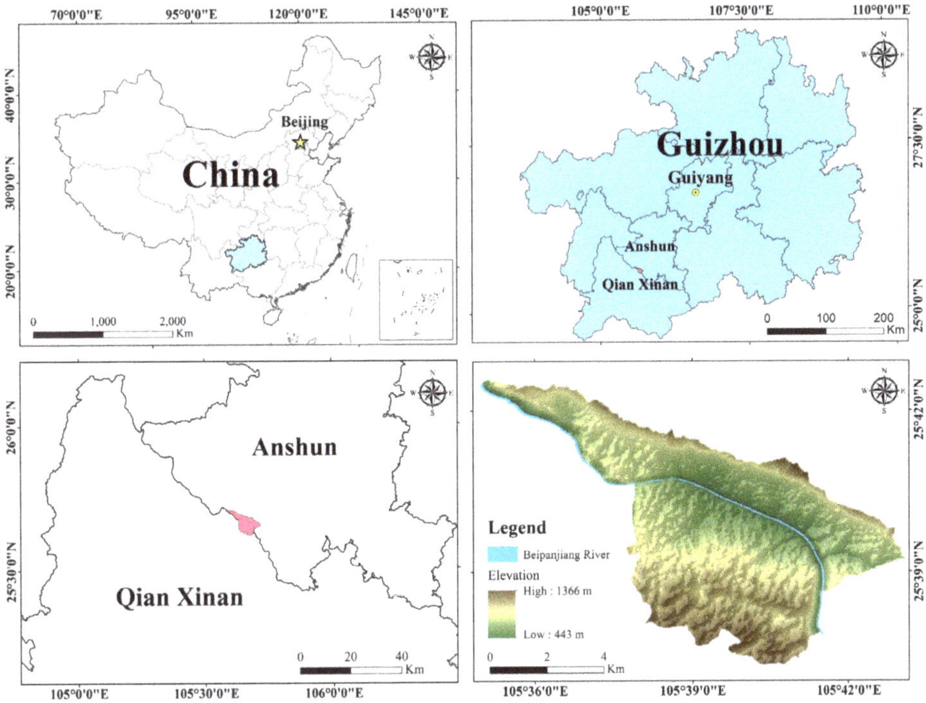

Figure 1. Location and elevation of the study area.

The research region is characterized by a rocky desert landscape, with fragmented and shallow soils, being prone to soil erosion in the presence of water, and having an extremely fragile ecological environment. Coupled with extensive deforestation and agricultural activities, the region has been caught in a vicious cycle of "environmental fragility–resource shortage-poverty, resource plunder–environmental degradation–further poverty" [35]. Since the beginning of the 21st century, the study area has been designated as a model area for the integrated management of karstic desertification ecology and environment, implementing natural restoration measures such as returning farmland to forest and grass. Therefore, the selected study area is typical, representative, and exemplary in the management of karst soil erosion.

2.2. Data

The data required for this study included remote sensing images, rainfall data, land use data, topographic and geomorphological data, and soil type data. The consistency and reliability of all data were strictly checked and controlled by the data production department.

(1) Remote sensing images and topographic data were obtained from the Geospatial Data Cloud (http://www.gscloud.cn (accessed on 7 October 2020)), with a resolution of 30 m. Remote sensing images without clouds in the study area were selected as the data source. (2) Rainfall data were obtained from the China Meteorological Data Network (http://data.cma.cn (accessed on 15 October 2020)), using ArcGIS10.2 to spatially interpolate and rasterize the rainfall dataset of the meteorological stations. (3) Land use data were obtained in two parts: The data from 2000 and 2005 were obtained from remote sensing images as the source data due to the long period and accuracy problems, and the initial land use data were obtained through supervised classification and manual interpretation at a later stage. The data from 2010 to 2020 were obtained through remote sensing image interpretation and correction by the research team through long-term field investigation. (4) Soil type data were obtained from the Resource and Environment Science and Data Centre of the Chinese Academy of Sciences (http://www.resdc.cn (accessed on 15 October 2020)) and calibrated concerning the 1:50,000 soil type map of Guizhou Province and the results of the team's field soil sample collection.

2.3. Methods

2.3.1. RUSLE Model

The RUSLE model is a modified version of the Universal Loss Equation (USLE). Due to its simple structure and ease of operation, scholars worldwide widely use it to estimate soil erosion. Its equation is as follows:

$$A = R \times K \times L \times S \times C \times P \quad (1)$$

where A is the average annual soil erosion ($t{\cdot}ha^{-1}{\cdot}a^{-1}$), which means the average annual soil loss from fine gully or inter-gully erosion on slopes caused by rainfall and runoff. R is the rainfall erosivity factor ($MJ{\cdot}mm{\cdot}ha^{-1}{\cdot}h^{-1}{\cdot}a^{-1}$), K is the soil erodibility factor ($t{\cdot}ha{\cdot}h{\cdot}MJ^{-1}{\cdot}mm^{-1}{\cdot}ha^{-1}$), L is the slope length factor (dimensionless), S is the slope gradient factor (dimensionless), C is the vegetation cover and management factor (dimensionless), and P is the soil and water conservation measures factor (dimensionless).

Gao modified the RUSLE model based on the correlation coefficients between exposed bedrock and surface sediments [23], with the following expressions:

$$A = (1 - 0.076^2 \times \alpha)\, R \times K \times L \times S \times C \times P \quad (2)$$

where α is a correction factor whose value is determined by the average of the bedrock exposure rates for the different rocky desertification classes (Table 1). Different equations were chosen to simulate soil erosion depending on the lithology of the study area. Equation (1) was selected for non-karst regions, and Equation (2) was selected to simulate soil erosion in karst regions.

Table 1. α values for different karst rocky desertification classes.

Rocky Desertification	None	Potential	Light	Moderate	High	Severe
Bedrock exposure rate (%)	20<	20–30	30–50	50–70	70–90	>90
α	10	25	40	60	80	90

The K factor is a measure of soil erosion resistance and reflects the sensitivity of the soil to erosion. The *K* factor is calculated using the erosion-productivity impact calculator (EPIC) model (Equation (3)) developed by Williams [36].

$$K = \left\{0.2 + 0.3\exp\left[0.0256SAN\left(1 - \frac{SIL}{100}\right)\right]\right\}\left(\frac{SIL}{CLA+SIL}\right)^{0.3} \left[1.0 - \frac{0.25C}{C+\exp(3.72-2.95)}\right]\left[1.0 - \frac{0.7SN_1}{SN_1+\exp(-5.51+22.9SN_1)}\right] \quad (3)$$

where K is the soil erodibility factor in t·hm^2·h·MJ^{-1}·mm^{-1}·hm^{-2}. SAN is the sand content; SIL is the silt content; CLA is the clay content; C is the organic carbon content, $SN_1 = 1 - SAN/100$.

The *R* factor is the driving force behind soil erosion and reflects the potential for soil loss through precipitation. As there are no weather stations in the study area, the nearest weather station to the study area was selected and the *R* factor for the study area was obtained by interpolation using the rainfall erosion equation (Equation (4)) created by Arnoldus [37]. The equation is as follows:

$$R = \sum_{i=1}^{12} 1.735 * 10^{\left[1.5*\log\left(\frac{P_i^2}{P}\right)-0.8188\right]} \quad (4)$$

where R is the rainfall erosion factor, P_i is the monthly rainfall (mm), and P is the annual rainfall (mm).

The *L* factor and *S* factor are closely related to topography and accumulated flow, and *L* factor and *S* factor together reflect the influence of topographic features on soil erosion. LS represents the ratio of soil loss on a given slope length and gradient to soil loss on a typical slope in a standard runoff plot, all other things being equal, and the LS value is proportional to soil loss, playing an accelerating role in soil erosion [38]. We chose to calculate the *L* factor [39] and *S* factor [40] by applying the formulae to the soil and rocky mountainous areas of southwest China as follows:

$$L = \left(\frac{\lambda}{22.13}\right)^m \quad (5)$$

$$m = \frac{\beta}{\beta + 1} \quad (6)$$

$$\beta = (sin\theta/0.0896)/[3 * (sin\theta)^{0.8} + 0.56] \quad (7)$$

$$S = \begin{cases} 10.80 * sin\theta + 0.03\ (\theta < 5^\circ) \\ 16.80 * sin\theta - 0.50\ (5^\circ \leq \theta \leq 10^\circ) \\ 20.204 * sin\theta - 1.2404\ (10^\circ \leq \theta > 25^\circ) \\ 29.585 * sin\theta - 5.6079\ (\theta > 25^\circ) \end{cases} \quad (8)$$

where λ is the sum of the slope lengths in the horizontal direction, m is the slope length factor, β is a factor related to the slope value, and θ is the slope angle extracted on the basis of the digital elevation.

C factor refers to the ratio of soil loss under a particular crop or vegetation cover to that of continuous recreational land after cultivation, all other factors being equal, and this factor measures the inhibitory effect of vegetation cover and management on soil erosion [41,42]. Currently, there are two forms for obtaining *C* factor values, equation calculation and assigning values on the basis of land use type. In this paper, the formula created by Durigon [43] was chosen to calculate the C value as follows:

$$C = \frac{-NDVI + 1}{2} \quad (9)$$

where $NDVI$ is the normalized vegetation index, calculated from the near-infrared band and visible red band in remote sensing imagery.

P factor refers to the ratio of soil erosion with soil conservation measures to soil erosion with down-slope planting, and its value is distributed between 0 and 1. When $p = 1$, it means that no water conservation measures are taken in the unit area; when $p = 0$, it means that no soil erosion will theoretically occur in the unit area, such as water bodies, construction sites, etc. The p-values were obtained by summarizing the results of previous authors in the karst region of southern China [23,44,45]. The results are presented in the Table 2.

Table 2. Land use types and p-value.

Land Use	**Arable Land**	**Garden Land**	**Wood Land**	**Shrub Land**	**Grass Land**	**Bare Land**	**Construction Land**	**Water**	**Bare Rock**
p-value	0.4	0.7	1	1	1	1	0	0	0

2.3.2. Geographical Detector

The geographical detector is a statistical method that allows for the exploration of spatial dissimilarity and its drivers [46]. The principle of the geographical detector is based on the idea that if the independent variable has a significant effect on the dependent variable, then there should be some similarity in the spatial distribution of the two [47]. It consists of four models, namely, factor detector, interaction detector, risk detector, and ecological detector. In this paper, we analyzed the spatial heterogeneity with the factor detector, interaction detector, and risk detector.

The factor detector detects the extent to which the independent variable can explain the spatial divergence of the dependent variable, and the q value can measure the extent of explanation with the following expression:

$$q = 1 - \frac{SSW}{SST} \tag{10}$$

$$SSW = \sum_{h=1}^{L} N_h \sigma_h^2,\ SST = N\sigma^2 \tag{11}$$

where h = 1,2,..., L is the partition of the independent and dependent variables; N_h and N are the number of cells in stratum h and all partitions; σ_h^2 and σ^2 are the variances of the dependent variables in stratum h and all partitions. SSW and SST are the sum of the within-stratum variances and the total variance of all partitions, respectively. q has a value of [0,1], with larger values indicating that the independent variable explains more of the dependent variable.

The interaction detector can be used to detect the interaction between factors X_S, i.e., the change in the degree of explanation of the dependent variable Y when X_1 and X_2 act together. The principle is to calculate the q values of X_1 and X_2 separately, then to calculate the new layer $q(X_1 \cap X_2)$ values by superimposing the X_1 and X_2 layers and comparing them among the three. The comparison between the two factors can be divided into the following results: non-linearly diminished, one-factor non-linearly diminished, two-factor enhanced, independent, and non-linearly enhanced.

The risk detector can be used to count the mean values of attributes between sub-regions of a single factor and to determine whether the heterogeneity between sub-regions is significant. The mean attribute results are expressed as numerical data, while whether the means are significantly different is expressed as 'N/Y' binary data.

2.3.3. Slope Units

The slope cell delineation method based on the principle of curvature-based watershed segmentation [48] proceeds as follows (Figure 2).

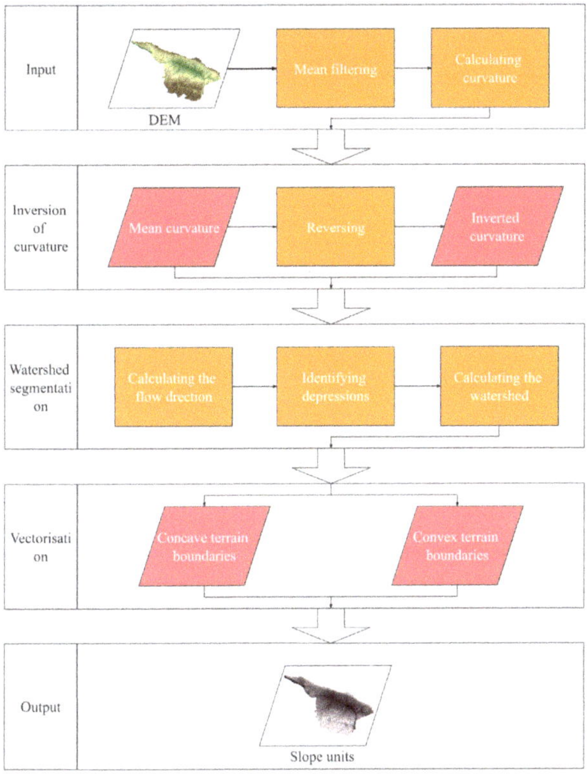

Figure 2. Flow chart of slope unit division.

3. Results

3.1. Spatial and Temporal Dynamics of Soil Erosion

The research region contains both karstic and non-karstic areas. Therefore, the rationalized RUSLE model was used to estimate soil erosion in karst areas, and the conventional RUSLE model was used for soil erosion in non-karst areas. Soil erosions in the study area were 6.11, 9.35, 7.49, and 8.88 t·ha^{-1}·a^{-1} in 2000, 2005, 2010, 2015, and 2020, respectively. This result was close to the average soil erosion in the Beipanjiang basin for the last 20 years published in the Guizhou Water Resources Bulletin (http://mwr.guizhou.gov.cn/ (accessed on 28 October 2020)) of 5.29 t·ha^{-1}·a^{-1}.

On the basis of the Standards for Classification and Gradation of Soil Erosion (SL190-2007) issued by the Ministry of Water Resources of the People's Republic of China, soil erosion in the study area was classified into seven classes: no erosion (construction land, water, and exposed bedrock), slight, light, medium, strong, very strong, and severe (Figure 3). The no erosion zone was the most widespread, accounting for approximately 50% of the total area, and was mainly located in the central, western, and north-western parts of the study area. Starting in 2010, there was a significant decrease in no erosion in the west and north-west, with no erosion mainly in the central region. This was followed by slight erosion and light erosion, accounting for about 40% of the total area, which was more evenly distributed in the study area. Medium erosion and strong erosion were less frequent and were mainly distributed along the southeastern edge, northwestern edge, and northern edge of the study area. Very strong and severe erosion accounted for less than 1% of the study area, with almost no extreme or severe erosion occurring in the research region.

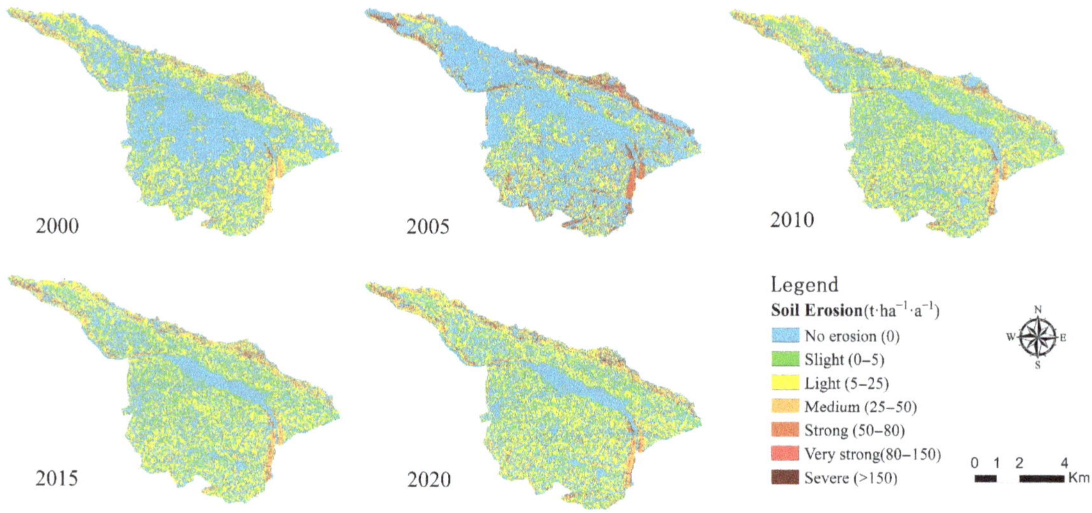

Figure 3. Spatial and temporal characteristics of soil erosion in 2000–2020.

As shown in Table 3, regional soil erosion fluctuated and increased during the study period, with the highest percentage of light and medium soil erosion overall, and the most extensive area with no erosion and light erosion. The ratio of different intensities of erosion showed an increase in medium, strong, and very strong erosion, with the greatest increase in intensity and extreme intensity in 2005, with 23.39% and 8.77% increases, respectively, compared to 2000. The amount of light erosion decreased, especially by 31.2% in 2005, compared to 2000. The area of no erosion showed the greatest variation in the different intensities of erosion, with a general trend of increasing and then decreasing; the area of no erosion increased by 7.26% from 2000 to 2005, while it decreased by 21.28% from 2005 to 2020. Unlike the erosion-free area, the erosion area ratio generally tended to increase at all other intensities over the study period. In addition, we found that in 2005, the area without erosion was the most extensive, while the total soil erosion was the highest, with medium and strong erosion accounting for more than 60% of the total regional erosion, and the area with strong erosion increased by 3.72% compared to 2000.

Table 3. The proportion of soil erosion characteristics.

	2000		2005		2010		2015		2020	
	Erosion Ratio	**Area Ratio**	**Erosion Ratio**	**Area Ratio**	**Erosion Ratio**	**Area Ratio**	**Erosion Ratio**	**Area Ratio**	**Erosion Ratio**	**Area Ratio**
No erosion	0%	59.12%	0%	66.38%	0%	46.29%	0%	45.37%	0%	45.10%
Slight	3.39%	8.04%	0.94%	3.58%	4.81%	13.84%	5.06%	14.24%	3.34%	11.29%
Light	54.99%	25.58%	23.79%	15.15%	48.20%	30.18%	50.81%	31.43%	45.61%	31.72%
Medium	36.64%	6.80%	37.36%	9.77%	38.75%	8.71%	36.72%	8.08%	40.16%	10.38%
Strong	3.80%	0.40%	27.19%	4.12%	6.85%	0.90%	6.27%	0.80%	8.93%	1.37%
Very strong	0.86%	0.05%	9.63%	0.96%	0.80%	0.06%	0.92%	0.06%	1.42%	0.13%
Severe	0.32%	0.01%	1.09%	0.04%	0.59%	0.02%	0.22%	0.01%	0.54%	0.02%
Total soil erosion ($t{\cdot}a^{-1}$)	327,735.00		501,600.55		410,090.06		401,732.04		475,961.02	

3.2. Regional Differentiation Based on Different Slope Units

The study area was divided into 2491 slope units on the basis of hydrological processes (Figure 4). The minimum cell size was 1.17×10^2 m^2, while the maximum cell size was 9.72×10^4 m^2 with an average cell size of 2.07×10^4 m^2.

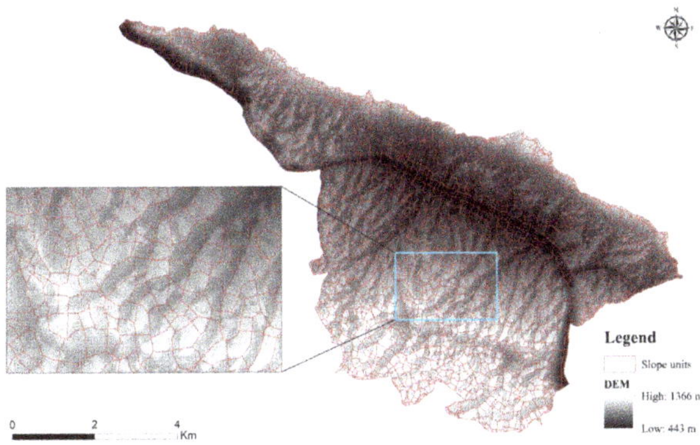

Figure 4. Division of slope units.

3.2.1. Soil Erosion Class Transfer Based on Slope Units

The mean soil erosion values in terms of slope units are available in six classes: no erosion, slight, mild, medium, strong, and very strong. To further understand the quantitative changes in soil erosion in the region, we produced soil erosion grade transfer maps for four time periods: 2000–2005, 2005–2010, 2010–2015, and 2015–2020 (Figure 5). The result shows that the transfer in soil erosion levels during the study period occurred mainly between no erosion, slight erosion, and light erosion, with medium, strong, and very strong erosion remaining relatively stable. We found that 17.5% and 9.8% of no erosion changed to slight and light erosion, respectively, during 2000–2005. Slight erosion converted mainly to no erosion and light erosion by 18.66% and 31.9%, respectively. From 2005 to 2010, 37.87% and 24.56% of no erosion transformed to minor and minor erosion, respectively. Slight and medium erosion moved mainly to light erosion, with 27.81% and 71.65% transfers, respectively. Soil erosion transferred in a similar direction for both the 2010–2015 and 2015–2020 periods. The soil erosion classes were relatively stable, except for some of the slight erosion transferring to light erosion, with soil erosion transition occurring mainly between the same soil erosion grade.

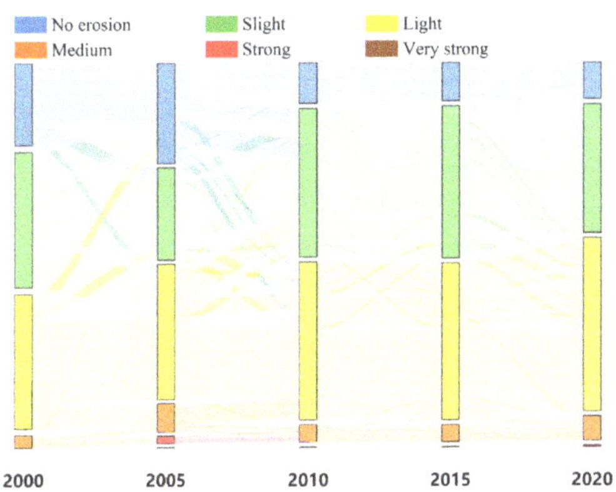

Figure 5. Soil erosion class transfer.

3.2.2. Hotspot Analysis Based on Different Types of Slope Units

We used a slope of 25° and an elevation of 896 m as the distinction. If the slope was less than or equal to 25°, then it was defined as low slope, and otherwise as high slope. An elevation less than or equal to 896 m was defined as low elevation, otherwise as high elevation. The slope units were divided into four unit types: low slope–low elevation, low slope–high elevation, high slope–low elevation, and high slope–high elevation. The four unit types were overlaid with the results of the soil erosion hotspot analysis as shown in Figure 6. The result shows that the hotspots of erosion remained relatively stable and the erosion coldspots decreased significantly during the study period. In general, erosion hotspots were mainly located in the high slope–low elevation and high slope–high elevation units along the northern, north-western, and south-eastern edges of the study area, with a few erosion hotspots also located in the high slope–high elevation units in the south. Coldspot areas of erosion were mainly found in the central and western low slope–low elevation units. From 2000 to 2020, the erosion coldspots decreased from the central and western low slope–low elevation units to the central low slope–low elevation units, and the erosion cold point confidence level decreased from 95% to 90%.

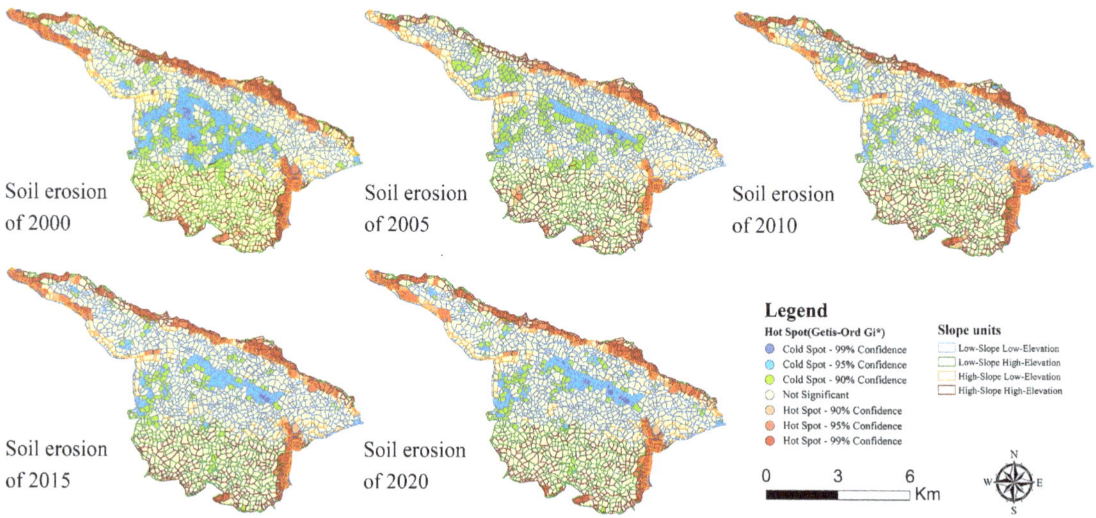

Figure 6. Hotspot analysis under different slope units.

3.3. *Quantitative Attribution of Soil Erosion Variability*

The mean values of soil erosion and the dominant values of the environmental factors within each cell were assigned to the corresponding cell. The contribution of each environmental factor to soil erosion (q) was quantified with the help of the geographical detector, and the results showed significant ($p < 0.05$) confidence in the q values for all factors.

3.3.1. Soil Erosion Risk Analysis

The risk detector of the geographical detector can detect potential relationships between factor variation and soil erosion risk by analyzing the mean soil erosion values for each interval within the factor. As shown in Figure 7, the differences in soil erosion risk under each factor sub-interval were significant. There was no significant pattern in the mean soil erosion values within the rainfall intervals, and the differences in soil erosion risk were not significant. Soil erosion risk increased with increasing vegetation cover, but the trend of increasing soil erosion risk decreased after the vegetation cover exceeded 80%, and the maximum soil erosion value did not exceed 15 t·ha^{-1}·a^{-1}. The risk of soil erosion increased and then decreased with increasing altitude. When the elevation was below 1170 m a.s.l., the risk of soil erosion increased as elevation increased. When the elevation was higher than 1170 m a.s.l., the risk of soil erosion decreased as the elevation increased. Soil erosion risk was graded between different land use zones, with soil erosion risk ranked as grassland > wood land > utilized land > water > arable land > construction land. The risk of soil erosion gradually changed from high to low from no rocky desertification to heavy rocky desertification. The average soil erosion was 11.76 t·ha^{-1}·a^{-1} in areas without rocky desertification and 6.46 t·ha^{-1}·a^{-1} in areas with heavy rocky desertification. The soil erosion risk increased with slope, and the growth of increase in soil erosion risk increased when the slope increased, with a maximum value of 38.57 $t{\cdot}ha^{-1}{\cdot}a^{-1}$.

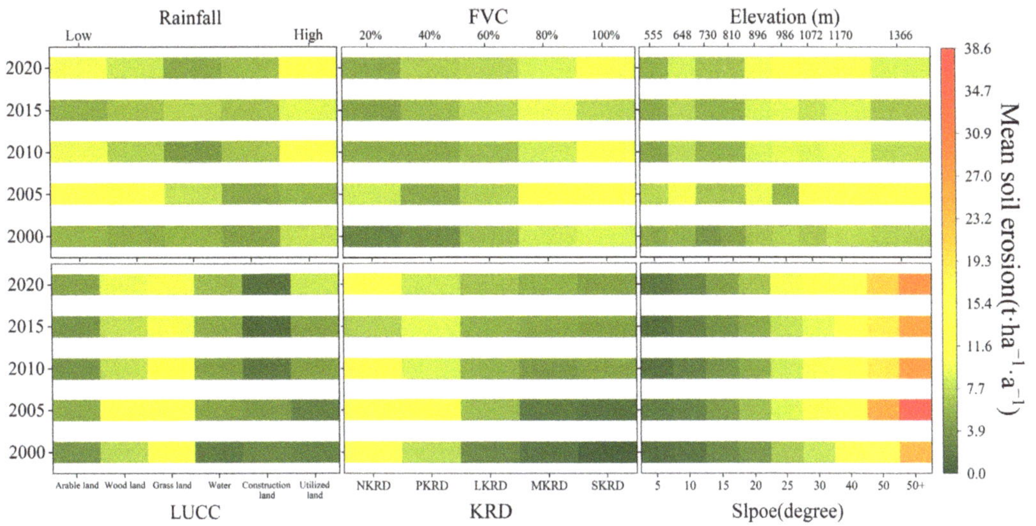

Figure 7. Factor classification and mean soil erosion values.

3.3.2. Single-Factor Analysis

Elevation, rainfall, slope, fractional vegetation cover (FVC), karst rocky desertification (KRD), and land use cover/change (LUCC) were analyzed by the geographic detector. As shown in Figure 8, the overall contribution of the six individual factors to soil erosion results during the study period was slope > LUCC > KRD > FVC > rainfall > elevation, but there were minor differences between years. The q value of KRD gradually decreased but also contributed to the spatial diversity of soil erosion only less than slope and LUCC. In 2000, the q value for KRD was 0.15, which was only lower than the slope. From 2005 onwards, the contribution of KRD to soil erosion results was lower than that of LUCC. LUCC was the environmental factor that had the greatest influence on soil erosion results apart from the slope, with a relatively stable q value. From 2000 to 2005, the q value for FVC decreased from 0.1 to 0.03. Moreover, from 2005 to 2020, the q value for FVC increased from 0.03 to 0.05. The contribution of rainfall to soil erosion results in the study area was low, but in individual years, it had an important influence. For example, the q value for rainfall in 2020 was as high as 0.08, contributing more to the soil erosion results than LUCC, KRD, and FVC. The effect of elevation alone on soil erosion results was not significant.

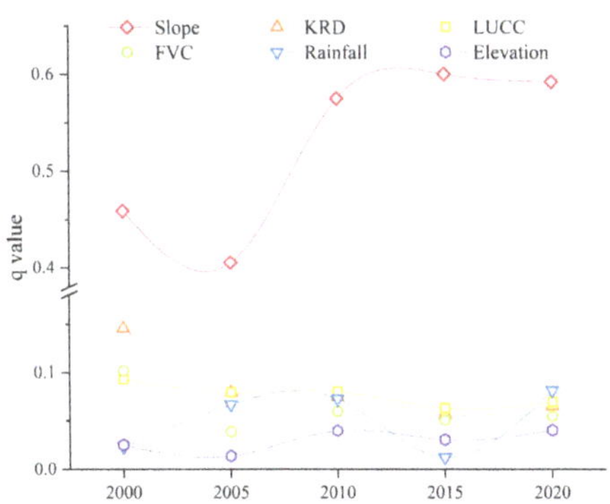

Figure 8. Single-factor trends.

3.3.3. Factor Interaction Analysis

The interaction results showed a decreasing trend in the q values of the interaction of KRD with the other four factors, except the slope factor. Their contribution to the spatial divergence of soil erosion gradually decreased. The interaction of rainfall with slope, elevation, and LUCC, which had a very low q value, increased the q value significantly and had a weaker and stronger effect on the spatial variation of soil erosion. The interactions of both LUCC-slope and vegetation-slope also gradually increased. In 2000 and 2005, the interactions of KRD-slope, elevation-slope, and rainfall-slope were the main influencing factors on the spatial variation of soil erosion. However, from 2010 to 2020, the interaction of slope-KRD was not significant and was replaced by slope-LUCC. The factor combinations with the highest q values in each period and the most significant increase in q values compared to the sum of the q values of the single factors were selected and are shown in Figure 9. X1 was the sum of the q values of the two factors and X2 was the q value after the factor interaction. The dominant factor influencing soil erosion varied between years. The years 2000 and 2005 were KRD-slope, explaining 60% and 49% of the spatial distribution of soil erosion, respectively. The year 2010 was for LUCC-slope, with a q value of 0.64. Meanwhile, the years 2015 and 2020 were for elevation-slope, and the q values after the interaction were higher than the sum of the single factors. In comparison with the sum of the q values of the individual factors, the q value of the interaction between elevation and rainfall increased most significantly throughout the whole period.

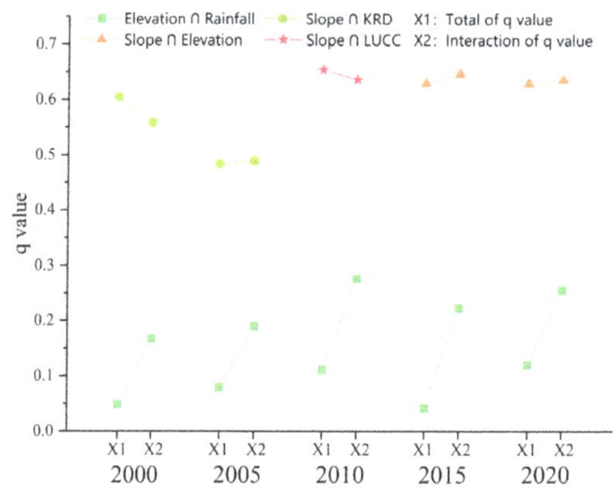

Figure 9. Factor interaction variation. X1 is the sum of the two-factor q values. X2 is the q value of the interaction of the two-factor.

4. Discussion

Soil erosion processes and driving mechanisms in karst areas are not yet understood due to the special karst structure and complex erosion patterns. Most of the existing modeling studies do not take into account the grade of rocky desertification, leading to large errors in results. In addition, the current soil erosion results based on administrative divisions and raster networks can hardly meet the requirements of refined soil erosion control. In this study, soil erosion in karst areas was estimated using an optimized RUSLE model with the karst rocky desertification factor. On this basis, the spatial and temporal dynamics of soil erosion in the study area in the last two decades were studied on the basis of slope units, while the soil erosion driving factors were quantitatively identified with the geographic detector.

4.1. Spatial and Temporal Dynamics of Karst Soil Erosion

Understanding the dynamic evolution of soil erosion is not only the basis and prerequisite for the prevention and control of soil erosion but also has great significance in the conservation of soil resources and ecological restoration. In this study, the erosion area of the study area was mainly no erosion or light erosion, while medium erosion and strong erosion were less. It differs from Guizhou as a whole, which is predominantly light to medium. This difference may be influenced by the level of rocky desertification development at different study scales. The limited erodible soil sources in the medium-intensity rocky desertification area constrain the development of soil erosion. There is a need for small-scale studies in karst areas with highly heterogeneous geography, and the results of these studies are important in tailoring soil and water conservation efforts to local conditions. The no erosion area decreased during the study period, with the highest proportion of light to medium erosion. The reduction of no erosion area was the ecological restoration effect of the national implementation of the comprehensive rocky desertification management project [49]. The highest percentage of light to medium erosion shows that controlling light to medium erosion is the key to effectively managing soil erosion in the region. The high proportion of light erosion is mainly due to the large proportion of light erosion areas, and it is recommended to reduce unreasonable human actions to allow the natural ecosystem to repair itself. The small area of moderate erosion or high erosion required soil and water conservation measures to maintain the stability of the soil in the area.

Soil erosion fluctuated and increased in the last two decades, unlike the findings of some karst areas where soil erosion had been decreasing. This difference might reflect the unreasonable human activities in the early part of the research area and the good management effect of the later management project. Soil erosion tended to increase, but the natural ecosystem developed benignly. The area was ecologically fragile and had low agricultural productivity. In the early years, people plundered land resources, causing massive soil loss, and eventually there was not even soil left to erode. In the later period, with the rocky desertification control project, soil resources gradually recovered. The highest soil erosion in 2005 was related to the high number of heavy rainfall events in that year. Extreme precipitation is often considered to be an important factor influencing erosion processes. Soil erosion under extreme precipitation conditions may account for the majority of annual soil erosion [50]. This is also confirmed on the basis of the proportion of sediment produced during the 3 and 10 largest erosion events [51].

The conversion in soil erosion classes over the whole period occurred mainly between no erosion, slight erosion, and light erosion. The conversion from no erosion to slight erosion and light erosion was related to the decrease in the area of rock desertification and the increase in eroded soil sources as described above. The reciprocal transfer between slight and light erosion may have been influenced by long-term rocky desertification management projects. The high slope unit is a susceptible area for soil erosion, and the local government should coordinate with multiple departments to implement the following to restore agricultural land to forest, alleviate the ecological carrying capacity, and reduce soil erosion. The erosion cold spot areas occurred in the low slope-low elevation unit, mainly because the soil resources were lost due to unreasonable human activities in the early years, and it was difficult to restore the soil resources in the short term. The progression of the erosion cold spot confidence level from 95% to 90% is evidence of the effectiveness of ecological restoration over time. In the case of meeting human needs, it is appropriate to convert the low-slope units into terraces; strengthen agroforestry cultivation [52], which can increase the harvested area of arable land [53]; and meet living needs while enhancing water and soil conservation measures.

4.2. Soil Erosion Impact Driver Analysis

Determining the driving factors of soil erosion helps to elucidate the driving mechanisms behind changes in soil erosion and is a key link for researchers to formulate soil protection policies scientifically and rationally. In the new form of synergistic economic and ecological development, the driving mechanism of soil erosion in karst areas has also changed.

The risk detector indicates that there is no trend correlation between rainfall intervals and soil erosion. Rainfall heterogeneity is not significant at small scales, and spatial heterogeneity of soil erosion is mainly influenced by other factors. Soil erosion in karst tends to increase with increasing vegetation cover but does not exceed a maximum of 15 $t{\cdot}ha^{-1}{\cdot}a^{-1}$. Erodible soil sources are extremely limited in exposed karst areas and increase with increasing vegetation cover. With 80% vegetation cover, the erosive power of rainfall and runoff on the soil is significantly reduced. The risk of soil erosion becomes less after an altitude of more than 1170 m a.s.l. and is related to the extent of human activity. Middle- and low-altitude areas are rich in water and heat resources and have a significant impact on soil erosion due to the concentration of human activities [54]. Higher altitudes have limited hydrothermal conditions and constrained human activities, which do not have a significant impact on soil erosion. Other studies have found that soil erosion in cropland > grassland > woodland. However, in this research area, it is grassland > woodland > cropland [54]. The plant roots formed a root–soil compound with the surrounding soil, which reduced soil erosion [55]. The low risk of soil erosion on arable land is due to the significant loss of soil resources caused by early over-farming. Despite recent efforts to combat soil erosion, soil resources in karst areas are difficult to recover in the short term [56]. The risk of soil erosion on arable land may appear to be low. However, once soil erosion has occurred on arable land, it would become a serious threat to the security of agricultural production. The gradual restoration of arable soil resources and the reduction of soil erosion from arable land should be the long-term goal of soil and water conservation work. The intensity of rocky desertification is negatively correlated with the risk of soil erosion. Similarly, a low risk of soil erosion does not mean a low hazard. On the contrary, the risk of erosion is much higher than that of light rocky desertification. Long-term natural restoration measures such as returning farmland to forests and grasses also aim to control soil erosion in areas of mild rocky desertification and to restore soil resources in areas of intense rocky desertification. Soil erosion risk is positively correlated with slope, and the higher the slope, the greater the increase in soil erosion risk. It is recommended to increase the vegetation cover on steep slopes to improve the stability of the soil.

The factor detector shows that slope explains 60% of the spatial variation in soil erosion. Effective measures to control soil erosion for different slopes can solve most of the soil problems. Some studies have found that sediment is mainly from sloping land [57], suggesting that sloping land above 25° should be converted to woodland. We recommend that conversion of land use be accompanied by the establishment of soil erosion protection zones because soils in disturbed soils are highly susceptible to erosion and soils in karst areas are difficult to recover effectively in the short term. The q value between LUCC and KRD gradually decreases but still plays a major role in the spatial differentiation of soil erosion. It can be seen that the impact of KRD and LUCC on soil erosion is far-reaching and long-lasting. Both karst rock desertification and soil erosion are closely related to human activities. Effectively improving people's production and living standards is also an important means of combating soil erosion. The role of vegetation cover in soil erosion shows a decreasing and then increasing trend, indicating that the rocky desertification control project in the last 20 years has achieved remarkable results. With continued ecological management, we infer that the role of vegetation cover will outweigh that of rocky desertification and land use in the coming decades. The insignificant effect of rainfall and elevation alone on regional soil erosion is related to the small scale of the study.

The results of the interaction of the factors show that the q value increases significantly after the interaction such as slope-rainfall and LUCC-rainfall. Rainfall is a direct driver of soil erosion occurrence and an important influencing factor for soil erosion [58]. The weak influence of rainfall alone in this study is mainly restricted by the scale of the study. The interaction of slope and rainfall increases the flow rate of the slope surface created by rainfall, intensifying scour and increasing erosion. Land use destroys the natural soil structure and weakens the soil's resistance to erosion. Land use also disturbs stable slope flow, which in combination with rainfall increases soil erosion. During the study period, the dominant factor in soil erosion changed from KRD-slope to LUCC-slope, and finally to elevation-slope. From 2000 to 2005, karst rocky desertification was very serious and was second only to slope in its impact on soil erosion, so the interaction of KRD and slope was the dominant factor in soil erosion. As the area of rocky desertification decreased, the influence of KRD diminished. Therefore, the dominant factor became LUCC-slope. In the course of long-term natural ecological restoration, the land use structure is optimized and the dominant factor changes again to elevation-slope. In comparison with the sum of the q values of the individual factors, the q values of the interaction between elevation-rainfall increased most significantly throughout the study period. The study area is a typical plateau valley landscape with a relative elevation difference of 923 m. Differences in altitude provide a potential energetic base for rainfall runoff and therefore increase the ability to erode the soil [59].

4.3. Challenges and Perspectives

The RUSLE model, based on the elements of topography, soil, and vegetation, can simulate regional soil erosion effectively. However, the model should be modified according to the actual environment. Considering the unique rocky desertification phenomenon in the karst area, the RUSLE model with the introduction of rocky desertification factors was chosen for an attempt. However, there are still shortcomings and further research needs to be strengthened. Factors such as data accuracy, the algorithm of each factor within the model, and the karst environment all add to the uncertainty of the model simulation and subsequent analysis. In this paper, the results of RUSLE calculations for the rocky desertification factors introduced due to data limitations could not be compared with the results of field surveys. To reduce the uncertainty of RUSLE models in karst areas, the following aspects should receive attention in future studies. Firstly, the study of soil erosion should be analyzed dynamically with local economies, national strategies [60], and climate change [61]. Secondly, there are biases in the results of each factor of the RUSLE model due to different algorithms, and different sets of equations have been developed by scholars in different regions [62]. Analysis of equation factor algorithms should be strengthened in future research to select the most appropriate algorithm for the study area. Finally, there are both surface loss and subsurface leakage in karst areas, and future research could focus on subsurface leakage.

5. Conclusions

This paper used a modified RUSLE model to estimate soil erosion in the karst plateau-gorge area over the past 20 years. The spatial and temporal evolution of soil erosion was quantified on the basis of slope units, and the influence of single and interactive factors on soil erosion was investigated using the geographical detector.

Soil erosion determined on the basis of slope units can more accurately reflect soil erosion in the actual environment and provide better decision support for regional erosion control and management. In 2000–2020, regional soil erosion showed an increasing trend. The results of this study further confirm the significant effect of the regional perennial rocky desertification control project. Soil erosion is serious in the high slope–low elevation and high slope–high elevation units, and relevant authorities should pay more attention to these areas to improve soil erosion control.

The contribution of the six factors to soil erosion fluctuates, but in the last 20 years, the pattern was slope > LUCC > KRD > FVC > rainfall > elevation. Slope played a dominant role in soil erosion differentiation on the karst plateau, while rock desertification and land use, which are closely related to human activities, also had a stronger influence on soil erosion. The influence of multiple factors on soil erosion is significantly stronger than that of single factors, and the dominant interaction factor varies with changes in rock desertification and land use. The dominant combination of soil erosion changed from KRD-slope (2000,2005) to LUCC-slope (2010) and finally to elevation-slope (2015, 2020). On the basis of this study, we recommend adjusting unreasonable human activities, insisting on natural restoration measures such as returning farmland to forest, and at the same time establishing soil consolidation projects for areas where the soil is gradually recovering.

Author Contributions: Conceptualization, C.S. and K.X.; methodology, T.S.; software, C.S.; formal analysis, C.S.; resources, T.S.; data curation, C.S. and T.S.; writing—original draft preparation, C.S.; writing—review and editing, C.S., K.X., and T.S.; visualization, C.S.; supervision, K.X.; project administration, K.X.; funding acquisition, K.X. All authors have read and agreed to the published version of the manuscript.

Funding: This research was supported by the Key Science and Technology Program of Guizhou Provence: Poverty Alleviation Model and Technology demonstration for Ecoindustries Derivated from the karst desertification control (no. 5411 2017 QianKehe Pingtai Rencai), the World Top Discipline Program of Guizhou Provence: "Karst Eco-environment Science (no. 125 2019 Qianjiao Keyan Fa), and the China Overseas Expertise Introduction Program for Discipline Innovation: Overseas Expertise Introduction Center for South China Karst Eco-environment Discipline Innovation (D17016).

Institutional Review Board Statement: Not applicable.

Informed Consent Statement: Not applicable.

Data Availability Statement: Not applicable.

Conflicts of Interest: The authors declare no conflict of interest.

References

1. Chen, M.Y.; Ma, L.; Shao, M.G.; Wei, X.R.; Jia, Y.H.; Sun, S.C.; Zhang, Q.Y.; Li, T.C.; Yang, X.; Gan, M. Chinese zokor (Myospalax fontanierii) excavating activities lessen runoff but facilitate soil erosion—A simulation experiment. *Catena* **2021**, *202*, 105248. [CrossRef]
2. Teng, M.J.; Huang, C.B.; Wang, P.C.; Zeng, L.X.; Zhou, Z.X.; Xiao, W.F.; Huang, Z.L.; Liu, C.F. Impacts of forest restoration on soil erosion in the Three Gorges Reservoir area, China. *Sci. Total Environ.* **2019**, *697*, 134164. [CrossRef] [PubMed]
3. Lin, J.K.; Guan, Q.Y.; Tian, J.; Wang, Q.Z.; Li, Z.J.; Wang, N. Assessing temporal trends of soil erosion and sediment redistribution in the Hexi Corridor region using the integrated RUSLE-TLSD model. *Catena* **2020**, *195*, 104756. [CrossRef]
4. Munodawafa, A. Assessing nutrient losses with soil erosion under different tillage systems and their implications on water quality. *Phys. Chem. Earth Parts A/B/C* **2007**, *32*, 1135–1140. [CrossRef]
5. Luo, X.L.; Bai, X.Y.; Tan, Q.; Ran, C.; Chen, H.; Xi, H.P.; Chen, F.; Wu, L.H.; Li, C.J.; Zhang, S.R.; et al. Particulate organic carbon exports from the terrestrial biosphere controlled by erosion. *Catena* **2022**, *209*, 105815. [CrossRef]
6. Li, Y.F.; Wang, Z.G.; Zhao, J.L.; Lin, Y.M.; Chen, A. Characterizing soil losses in China using data of 137Cs inventories and erosion plots. *Catena* **2021**, *203*, 105296. [CrossRef]
7. Zhang, S.R.; Bai, X.Y.; Zhao, C.W.; Tan, Q.; Luo, G.J.; Cao, Y.; Deng, Y.H.; Li, Q.; Li, C.J.; Wu, L.H.; et al. Limitations of soil moisture and formation rate on vegetation growth in karst areas. *Sci. Total Environ.* **2022**, *810*, 151209. [CrossRef]
8. Sirjani, E.; Sameni, A.; Moosavi, A.A.; Mahmoodabadi, M.; Laurent, B. Portable wind tunnel experiments to study soil erosion by wind and its link to soil properties in the Fars province, Iran. *Geoderma* **2018**, *333*, 69–80. [CrossRef]
9. Rahman, M.R.; Shi, Z.H.; Cai, C.F. Soil erosion hazard evaluation—An integrated use of remote sensing, GIS and statistical approaches with biophysical parameters towards management strategies. *Ecol. Modell.* **2009**, *220*, 1724–1734. [CrossRef]
10. Selmy, S.A.H.; Al-Aziz, S.H.A.; Jiménez-Ballesta, R.; Garcí a-Navarro, F.J.; Fadl, M.E. Modeling and Assessing Potential Soil Erosion Hazards Using USLE and Wind erosion in Integration with GIS Technique: Dakhla Oasis, Egypt. *Agriculture* **2021**, *11*, 1124. [CrossRef]
11. Karalis, S.; Karymbalis, E.; Mamassis, N. Effects of lithology and geomorphology on sediment yield in karst mountainous catchments. *Geomorphology* **2019**, *343*, 119–128.
12. Cao, Z.H.; Zhang, K.L.; He, J.H.; Yang, Z.C.; Zhou, Z.L. Linking rocky desertification to soil erosion by investigating changes in soil magnetic susceptibility profiles on karst slopes. *Geoderma* **2021**, *389*, 114949. [CrossRef]
13. Wang, M.; Wang, H.S.; Jiang, C.; Sun, J.X. Spatial Soil Erosion Patterns and Quantitative Attribution Analysis in Southwestern China Based on RUSLE and Geo-Detector Model. *J. Basic Sci. Eng.* **2021**, *29*, 1386.
14. Zhang, X.Q.; Hu, M.C.; Guo, X.Y.; Yang, H.; Zhang, Z.K.; Zhang, K.L. Effects of topographic factors on runoff and soil loss in Southwest China. *Catena* **2018**, *160*, 394–402. [CrossRef]
15. Feng, T.; Chen, H.S.; Polyakov, V.O.; Wang, K.L.; Zhang, X.B.; Zhang, W. Soil erosion rates in two karst peak-cluster depression basins of northwest Guangxi, China: Comparison of the rusle model with 137cs measurements. *Geomorphology* **2016**, *253*, 217–224. [CrossRef]
16. Li, Y.C.; Liu, C.X.; Zhao, C.Y.; Wang, C.J.; Zhang, H.; Ming, J.; Wang, Y. Assessment and spatial differentiation of sensitivity of soil erosion in Three Gorges Reservoir area of Chongqing. *Acta Ecol. Sin.* **2009**, *29*, 788–796.

17. Zhu, Z.; Wang, J.X.; Hu, M.G.; Jia, L. Geographical detection of groundwater pollution vulnerability and hazard in karst areas of Guangxi Province, China. *Environ. Pollut.* **2018**, *245*, 627–633. [CrossRef]
18. Morgan, R.P.C.; Nearing, M.A. *Handbook of Erosion Modelling*; Blackwell Publishing Ltd.: Hoboken, NJ, USA, 2011.
19. Nearing, M.A. Soil erosion and conservation. In *Environmental Modelling: Finding Simplicity in Complexity*, 2nd ed.; Wiley: Hoboken, NJ, USA, 2013; pp. 365–378.
20. Tian, Y.C.; Wang, S.J.; Bai, X.Y.; Luo, G.J.; Yan, X. Trade-offs among ecosystem services in a typical Karst watershed, SW China. *Sci. Total Environ.* **2016**, *566*, 1297–1308. [CrossRef]
21. Long, M.Z.; Wu, K.H.; Xiong, K.N. Assessment of the applicability of the WEPP model(Hill slope Version) for soil erosion in karst rock desertification area, Guizhou Province. *Carsol. Sin.* **2014**, 33, 201–207.
22. Qin, W.; Guo, Q.K.; Cao, W.H.; Zhe, Y.; Yan, Q.H.; Shan, Z.J.; Zheng, F.L. A new RUSLE slope length factor and its application to soil erosion assessment in a Loess Plateau watershed. *Soil Tillage Res.* **2018**, *182*, 10–24. [CrossRef]
23. Gao, J.B.; Wang, H. Temporal analysis on quantitative attribution of karst soil erosion: A case study of a peak-cluster depression basin in Southwest China. *Catena* **2019**, *172*, 369–377. [CrossRef]
24. Febles-Gonzalez, J.M.; Vega-Carreno, M.B.; Tolon-Becerra, A.; Lastra-Bravo, X. Assessment of soil erosion in karst regions of Havana, Cuba. *Land Degrad. Dev.* **2012**, *23*, 465–474. [CrossRef]
25. Xu, Y.Q.; Shao, X.M.; Kong, X.B.; Peng, J.; Cai, Y.L. Adapting the RUSLE and GIS to model soil erosion risk in a mountains karst watershed, Guizhou Province, China. *Environ. Monit. Assess.* **2008**, *141*, 275–286. [CrossRef]
26. Dai, Q.H.; Peng, X.D.; Yang, Z.; Zhao, L.S. Runoff and erosion processes on bareslopes in the karst rocky desertification area. *Catena* **2017**, *152*, 218–226. [CrossRef]
27. Carrara, A.; Cardinali, M.; Detti, R.; Guzetti, F.; Pasqui, V.; Reichenbach, P. GIS techniques and statistical models in evaluating landslide hazard. *Earth Surf. Process. Landforms* **1991**, *16*, 427–445. [CrossRef]
28. Jia, N.; Mitani, Y.; Xie, M.W.; Tong, J.X.; Yang, Z.H. GIS deterministic model-based 3d large-scale artificial slope stability analysis along a highway using a new slope unit division method. *Nat. Hazards* **2015**, *76*, 873–890. [CrossRef]
29. Wang, K.; Xu, H.; Zhang, S.J.; Weri, F.Q.; Xie, W.L. Identification and Extraction of Geomorphological Features of Landslides Using Slope Units for Landslide Analysis. *ISPRS Int. J. Geo-Inf.* **2020**, *9*, 274. [CrossRef]
30. Ba, Q.Q.; Chen, Y.M.; Deng, S.S.; Yang, J.X.; Li, H.F. A comparison of slope units and grid cells as mapping units for landslide susceptibility assessment. *Earth Sci. Inform.* **2018**, *11*, 373–388. [CrossRef]
31. Tanyas, H.; Rossi, M.; Alvioli, M.; van Westen, C.J.; Marchesini, I. A global slope unit-based method for the near real-time prediction of earthquake-induced landslides. *Geomorphology* **2019**, *327*, 126–146. [CrossRef]
32. Jacobs, L.; Kervyn, M.; Reichenbach, P.; Rossi, M.; Marchesini, I.; Alvioli, M.; Dewitte, O. Regional susceptibility assessments with heterogeneous landslide information: Slope unit vs. pixel-based approach. *Geomorphology* **2020**, *356*, 107084. [CrossRef]
33. Li, Y.; Liu, Z.Q.; Liu, G.H.; Xiong, K.N.; Cai, L.L. Dynamic variations in soil moisture in an epikarst fissure in the karst rocky desertification area. *J. Hydrol.* **2020**, *591*, 125587. [CrossRef]
34. Legrain, X.; Berding, F.; Dondeyne, S.; Schad, P.; Chapelle, J. *World Reference Base for Soil Resources 2014: International Soil Classification System for Naming Soils and Creating Legends for Soil Maps*; FAO: Rome, Italy, 2014.
35. Hu, F.; An, Y.L.; Xu, J. Preliminary discussion on concept and landscape features of "Semi-Karst": A case study of Guizhou Province. *Geogr. Res.* **2015**, *34*, 1569–1580.
36. Williams, J.R.; Jones, C.A.; Kiniry, J.R.; Spanel, D.A. The EPIC Crop Growth Model. *Trans. ASAE* **1989**, *32*, 0497–0511. [CrossRef]
37. Arnoldus, H.M.J. *An Approximation of the Rainfall Factor in the Universal Soil Loss Equation*; John Wiley and Sons Ltd.: Hoboken, NJ, USA, 1980; pp. 127–132.
38. Renard, K.G.; Foster, G.R.; Weesies, G.A.; McCool, D.; Yoder, D. Predicting soil erosion by water: A guide to conservation planning with the revised universal soil loss equation (RUSLE). In *Agriculture Handbook*; United States Government Printing: Washington, DC, USA, 1997; pp. 110–135.
39. Liu, B.Y.; Nearing, M.A.; Shi, P.J.; Jia, Z.W. Slope Length Effects on Soil Loss for Steep Slopes. *Soil Sci. Soc. Am. J.* **2000**, *64*, 1759–1763. [CrossRef]
40. Liu, B.T.; Song, C.F.; Shi, Z.; Tao, H.P. Correction Algorithm of Slope Factor in Universal Soil Loss Equation in Earth-Rocky Mountain Area of Southwest China. *Soil Water Consevation China* **2015**, *8*, 49–51.
41. Jong, S.M.D. Derivation of vegetative variables from a landsat tm image for modelling soil erosion. *Earth Surf. Proc. Land* **1994**, *19*, 165–178. [CrossRef]
42. Patil, R.J.; Sharma, S.K. Remote Sensing and GIS based modeling of crop/cover management factor (C) of USLE in Shakker river watershed. In Proceedings of the International Conference on Chemical, Agricultural and Medical Sciences (CAMS-2013), Kuala Lumpur, Malaysia, 29–30 December 2013; pp. 29–30.
43. Durigon, V.L.; de Carvalho, D.F.; Antunes, M.A.H.; Oliveira, P.T.; Fernandes, M.M. NDVI time series for monitoring RUSLE cover management factor in a tropical watershed. *Int. J. Remote Sens.* **2014**, *35*, 441–453. [CrossRef]
44. Li, Y.; Bai, X.Y.; Zhou, Y.C.; Qin, L.Y.; Tian, X.; Tian, Y.C.; Li, P.L. Spatial–temporal evolution of soil erosion in a Typical mountainous karst basin in SW China, based on GIS and RUSLE. *Arab. J. Sci. Eng.* **2016**, *41*, 209–221. [CrossRef]
45. Zeng, C.; Wang, S.J.; Bai, X.Y.; Li, Y.B.; Tian, Y.C.; Yue, L.; Wu, L.H.; Luo, G.J. Soil erosion evolution and spatial correlation analysis in a typical karst geomorphology using RUSLE with GIS. *Solid Earth* **2017**, *8*, 721–736. [CrossRef]

46. Wang, J.F.; Li, X.H.; Christakos, G.; Liao, Y.L.; Zhang, T.; Gu, X.; Zheng, X.Y. Geographical detectors-based health risk assessment and its application in the neural tube defects study of the Heshun region, China. *Int. J. Geogr. Inf. Sci.* **2010**, *24*, 107–127. [CrossRef]
47. Wang, J.F.; Xu, C.D. Geodetector: Principle and prospective. *Acta Geogr. Sin.* **2017**, *72*, 116–134.
48. Yan, Y.J.; Dai, Q.H.; Yuan, Y.F.; Peng, X.D.; Zhao, L.S.; Yang, J. Effects of rainfall intensity on runoff and sediment yields on bare slopes in a karst area, SW China. *Geoderma* **2018**, *330*, 30–40. [CrossRef]
49. Xiong, K.N.; Zhu, D.Y.; Peng, T.; Yu, L.F.; Xue, J.H.; Li, P. Study on Ecological industry technology and demonstration for Karst rocky desertification control of the Karst Plateau-Gorge. *Acta Ecol. Sin.* **2016**, *36*, 7109–7113.
50. Estrany, J.; Garcia, C.; Batalla, R.J. Hydrological response of a small mediterranean agricultural catchment. *J. Hydrol.* **2010**, *380*, 180–190. [CrossRef]
51. Gonzalez-Hidalgo, J.; Batalla, R.; Cerdà, A.; De Luis, M. Contribution of the largest events to suspended sediment transport across the USA. *Land Degrad. Dev.* **2010**, *21*, 83–91. [CrossRef]
52. Karamage, F.; Zhang, C.; Kayiranga, A.; Shao, H.; Fang, X.; Ndayisaba, F.; Nahayo, L.; Mupenzi, C.; Tian, G. USLE-based assessment of soilerosion by water in the Nyabarongo River catchment, Rwanda. *Int. J. Environ. Res. Public Health* **2016**, *13*, 835. [CrossRef]
53. Song, F.J.; Wang, S.J.; Bai, X.Y.; Wu, L.H.; Wang, J.F.; Li, C.J.; Chen, H.; Luo, X.L.; Xi, H.P.; Zhang, S.R.; et al. A new indicator for global food security assessment: Harvested area rather than cropland area. *Chin. Geogr. Sci.* **2022**, *32*, 204–217. [CrossRef]
54. Gao, J.B.; Wang, H.; Zuo, L. Spatial gradient and quantitative attribution of karst soil erosion in Southwest China. *Environ. Monit. Assess.* **2018**, *190*, 730. [CrossRef]
55. Fu, J.T.; Li, G.Y.; Hu, X.T.; Li, Q.Z.; Yu, D.M.; Zhu, H.L.; Hu, X.G. Research status and development tendency of vegetation effects to soil reinforcement and slope stabilization. *J. Eng. Geol.* **2014**, *22*, 1135–1146.
56. Xiong, K.N.; Li, J.; Long, M.Z. Features of soil and water loss and key issues in demonstration areas for combating karst rocky desertification. *Acta Geogr. Sin.* **2012**, *67*, 878–888.
57. Gan, Y.X.; Dai, Q.H.; Fu, W.B.; Yan, Y.J.; Peng, X.D. Characteristics of soil erosion on Karst slopes under the artificial rainfall experiment conditions. *Chin. J. Appl. Ecol.* **2016**, *27*, 2754–2760.
58. Xiong, K.N.; Chi, Y.K. The problems in southern China karst ecosystem in southern of China and its countermeasures. *Ecol. Econ.* **2015**, *31*, 23–30.
59. Jiang, Y.; Gao, J.; Yang, L.; Wu, S.; Dai, E. The interactive effects of elevation, precipitation and lithology on karst rainfall and runoff erosivity. *CATENA* **2021**, *207*, 105588. [CrossRef]
60. Zhang, S.R.; Bai, X.Y.; Zhao, C.W.; Tan, Q.; Luo, G.J.; Wu, L.H.; Xi, H.P.; Li, C.J.; Chen, F.; Ran, C.; et al. China' s carbon budget inventory from 1997 to 2017 and its challenges to achieving carbon neutral strategies. *J. Clean. Prod.* **2022**, *347*, 130966. [CrossRef]
61. Zhang, S.R.; Bai, X.Y.; Zhao, C.W.; Tan, Q.; Luo, G.J.; Wang, J.F.; Li, Q.; Wu, L.H.; Chen, F.; Li, C.J.; et al. Global CO_2 consumption by silicate rock chemical weathering: Its past and future. *Earth's Futur.* **2021**, *9*, e2020EF001938. [CrossRef]
62. Ghosal, K.; Bhattacharya, S.D. A review of RUSLE model. *J. Indian Soc. Remote Sens.* **2020**, *48*, 689–707. [CrossRef]

land

MDPI

Article

The Spatial and Temporal Evolution of Ecological Environment Quality in Karst Ecologically Fragile Areas Driven by Poverty Alleviation Resettlement

Zhongfa Zhou [1,2,*], Qing Feng [1,2,3], Changli Zhu [1,2], Wanlin Luo [4], Lingyu Wang [1,2], Xin Zhao [1,2] and Lu Zhang [1,2]

1 School of Karst Science, Guizhou Normal University, Guiyang 550001, China; 20030170040@gznu.edu.cn (Q.F.); 18030090025@gznu.edu.cn (C.Z.); 18030170034@gznu.edu.cn (L.W.); 21030090039@gznu.edu.cn (X.Z.); sophiazl@gznu.edu.cn (L.Z.)
2 National Engineering Research Center for Karst Rocky Desertification Control, Guiyang 550001, China
3 College of Tourism and Aviation Culture, Guizhou City Vocational College, Guiyang 550046, China
4 The Engineering Branch of the Third Institute of Surveying and Mapping of Guizhou Province, Guiyang 550001, China; 19010170430@gznu.edu.cn
* Correspondence: fa6897@gznu.edu.cn

Abstract: Many scholars are skeptical about the poverty reduction effect and the ecological effect of poverty alleviation resettlement (PAR). This study evaluates the spatial and temporal evolution of the ecological environment quality (EEQ) to analyze the effectiveness of ecological restoration from PAR. Based on cloud computing using the Google Earth Engine platform, remote-sensing data were obtained and reconstructed from 2000 to 2020. The remote-sensing ecological index (RSEI) was used to analyze the spatial and temporal evolution of EEQ. The results show that the RSEI of the study area increased by 13.07% after the implementation of PAR, and the rate of increase was higher than that in the period before PAR; the Pu'an and Qinglong areas improved most obviously, in terms of the fragile ecological environment and the prominent contradiction between peasants and land. The residual trends method indicated that the contribution rate of improvement in RSEI due to PAR was 70.56%, 88.38%, and 82.96% in 2017, 2018, and 2020, respectively. An increase in RSEI was more obvious in the area with a greater relocated population and a higher corresponding coupling coordination level. PAR has a promoting effect on EEQ improvement but does not have ecological restoration benefits in every region. It is not satisfactory in terms of the degeneration of the LST indicator and the ecological impact of human wells.

Keywords: poverty alleviation resettlement; ecological environment quality; remote sensing ecological index; karst ecologically fragile areas

Citation: Zhou, Z.; Feng, Q.; Zhu, C.; Luo, W.; Wang, L.; Zhao, X.; Zhang, L. The Spatial and Temporal Evolution of Ecological Environment Quality in Karst Ecologically Fragile Areas Driven by Poverty Alleviation Resettlement. *Land* **2022**, *11*, 1150. https://doi.org/10.3390/land11081150

Academic Editor: Xiaoyong Bai

Received: 1 June 2022
Accepted: 22 July 2022
Published: 26 July 2022

1. Introduction

Climate change and human activities have an overall impact on global ecology [1,2]: on the one hand, human activities such as urban expansion and deforestation affect ecosystem degradation in most parts of the world [3,4]; on the other hand, human activities such as ecological restoration can improve ecosystems to a certain degree [5,6]. Ecological environment quality (EEQ) is the degree of suitability of the ecological environment for human survival and sustainable social–economic development within a certain space–time range [7]. Scientific monitoring and the evaluation of the impact of human activities on EEQ and temporal and spatial changes have shown important theoretical and practical significance to coordinate the relationship between human activities and the ecological environment, and to promote the sustainable development of society.

The United Nations identified poverty eradication as the primary goal of sustainable development, having invested 600 billion CNY in PAR from 2016 to 2020, involving 10 million extremely poor farmers; this is one of the flagship projects to eliminate poverty in

China [8]. The area of PAR is typical of the vicious circle of poverty and ecological environment deterioration, with a high overlap of ecological fragility and extreme poverty [9,10]. At the same time, the PAR areas also belong to the spatial poverty trap [11], and the task of eradicating poverty in situ is extremely difficult. The PAR is a better development opportunity for farmers who live below the poverty line in ecologically fragile areas, by moving them to cities and towns.

The Chinese government believes that China's PAR has positive significance for eradicating poverty and improving the ecological environment [12]. Should we consider PAR as an effective human activity to improve ecology? However, the existing research on PAR has basically focused on the social effectiveness of its mechanisms of participation and poverty reduction [13,14], with few studies being related to ecological restoration. Similar to the ecological resettlement policy, most scholars are skeptical of the ecological protection effect of the large-scale implementation of ecological resettlement in ecologically fragile areas, despite the effectiveness of ecological restoration or the welfare of relocated farmers [15,16]. However, different from ecologically fragile areas such as Tibet and Inner Mongolia, the climate conditions in the southwest karst mountainous area where PAR is mainly implemented are more suitable for plant growth (the annual average rainfall is approximately 1100–1300 mm, and the average annual temperature is approximately 16 °C), so the dominant biophysical limiting factor is not the climate, but the soil resources [17], and human interference is the main factor influencing ecological restoration [18,19]. According to peasants–land coordination theory, in regions with limited resources, human constraints and natural interference are the best choices to coordinate the relationship between peasants and land [20]. PAR reduces human disturbance to natural resources (soil resources), resulting in ecological improvement and the rapid shrinkage of inefficient agricultural production space in the relocated area, which is in line with the Environmental Kuznets Curve theory of ecological economics [21,22]. This paper aims to explore whether the PAR really contributes to the improvement of EEQ as a human activity and actually improves the living environment for human beings. If the EEQ improves the study area, can we consider it to be caused by PAR? We need to quantify the effectiveness of PAR-driven eco-environmental improvements, among which we must distinguish the influence of human activities from natural factors, which will be the focus of this paper.

Due to remote-sensing data being timely and effective, covering a wide area, and being objective and sustainable, the application of remote-sensing technology in ecological environment assessment has increasingly attracted attention from scholars [23]. Using the normalized difference vegetation index (NDVI) to assess ecological environment is the most common method [24], and most scholars use land surface temperature (LST) in evaluating the effect of the urban heat island [25]. Similarly, the wet components form tasseHed captransform (WET), and the normalized difference impervious surface index (NDISI) indicators are the most important indicators for the intuitive human perception of ecological conditions [26,27]. Compared with a single indicator, the ecological status reflected by the comprehensive indicator is more complex and diverse. RSEI, which is based on the Ecological Index (EI) from the Ministry of Ecology and Environment of China, reflects the Technical Criterion for Ecosystem Status Evaluation (HJ 192e2015). EI is authoritative and extensive in regional EEQ assessment in China [28]. Many scholars have verified that RSEI and EI are highly comparable in the ecological sense [29]. The RSEI (Remote Sensing Ecological Index) model integrates intuitive and key influence factors, including greenness, wetness, dryness, and heat. It has the advantages of real and effective evaluation data sources, objective and fair evaluation conclusions, and intuitive and visual evaluation results [30,31]. Additionally, many scholars have evaluated the EEQ improvement effect by using RSEI as a technical means in projects such as Northwest Beijing Ecological Containment Area [32], the Northern Sand-Prevention Belt [33], and the Three-North Shelter Forest Program [34]. Existing research results show that the model of the trend of normalized residuals enables distinguishing between climatic factors and ecological effects caused by human activities [35]. It is necessary to clarify the turning point where PAR causes obvious changes in RSEI; a regression model was constructed using weather factors and human-activity factors during the time before the turning point to predict the RSEI trends during the period of PAR implementation, which was not affected by PAR, and the residual between the observed RSEI and the predicted RSEI, which was thought to be caused by PAR. Because the conclusion of the regression model is predictive, no PAR data are involved, so a correlation analysis model needs to be constructed with PAR implementation data and period RSEI variables to further explain the EEQ effect of PAR. The overall objective of this study is to evaluate the effectiveness of PAR on long-term EEQ dynamics in the study area, by the following means: (1) the analysis of the spatial and temporal evolution trend of long-series RSEI; (2) the elimination of the impact of climate factors and ecological restoration projects such as the Karst Rocky Desertification Restoration Project, to analyze the ecological contribution of the PAR; and, (3) establishing a coupling model between RSEI changes and the village-level PAR population to determine the association.

Southwest Guizhou Autonomous Prefecture, located in the Yunnan, Guizhou, Guangxi, concentrated, contiguous special-hardship area, is one of the most ecologically fragile regions in China, with few resources, a low environmental carrying capacity, a fragile ecological environment and human–land conflict [36]. From 2016 to 2019, 74,600 households with 338,600 people were relocated for the purposes of poverty alleviation, accounting for 3.38% of the total relocated population in China. This paper takes Southwest Guizhou Autonomous Prefecture as the study area, and first applies the Google Earth Engine (GEE) processing platform and cloud computing to obtain and reconstruct remote sensing data from 2000 to 2020, then applies the RSEI model to quantitatively evaluate the spatial and temporal evolution of EEQ in the study area. The ecological contribution of the PAR to the study area is quantified using the model of the trend of normalized residuals, and the association between the PAR and RSEI changes is further determined with a coupled model. Furthermore, we quantitatively reveal the changes in EEQ spatial distribution and the trend of ecological environment improvement caused by PAR, and provide theoretical support for coordinating ecological environment protection and social and economic development in ecologically fragile regions to achieve harmonious development between man and nature.

2. Materials and Methods

2.1. Study Area

The Guangxi, Yunnan, and Guizhou areas are located in the karst mountains of southwest China, with a combined population of 220 million people. They span across 0.54 million km^2 of carbonate rock area, which is one of the most ecologically fragile and densely populated areas in the world [37]. The Southwest Guizhou Autonomous Prefecture is located in the southwestern part of Guizhou Province between 104°35′–106°32′ E and 24°38′–26°11′ N. It has eight counties under its jurisdiction and a land area of 16,800 km^2. The area belongs to the subtropical humid monsoon climate and has the most widely distributed carbonate rock layer containing magnesium in the Triassic marine. The karst area in the region is spread across 10,200 km^2, accounting for 60.28% of the total land area. Dominant ecological problems in Southwest Guizhou Autonomous Prefecture are stone desertification and soil erosion. The potential stone desertification is spread across 0.21 million km^2, where the known stone desertification area is 0.50 million km^2, accounting for 42.51% of the land (Figure 1), making it one of the most severely ecologically compromised areas of China [38]. With a rural-poor population of 432,300 in 2015 and a poverty incidence rate of 13.75%, the problem of poverty is relatively prominent. Southwest Guizhou Autonomous Prefecture is a region with high overlap between ecological fragility and extreme poverty.

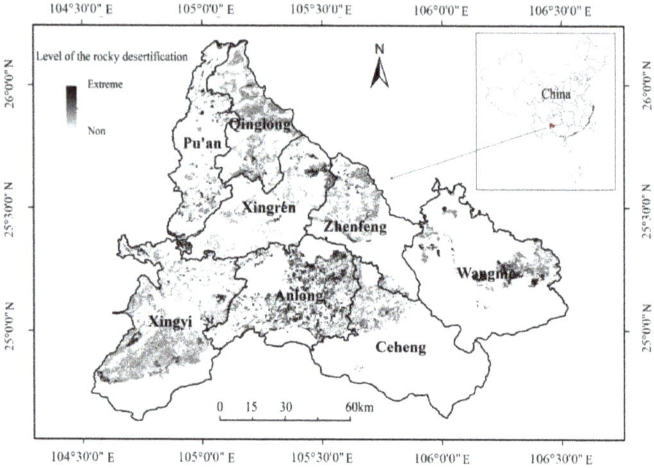

Figure 1. Distribution of degree of stone desertification in the study area.

PAR is the flagship project of the Chinese government's poverty alleviation project. For people living in areas where local resources cannot effectively carry them out of poverty, they relocate to urban areas with better education, medical care, transportation, communication, employment and other improved conditions. Here, they enjoy the favorable development resources of the city, which help them to become free of poverty. The Chinese government concluded that PAR has 9 key achievements, including improved living conditions, broader employment prospects, and the relief of ecological environment pressure [12]. PAR in Southwest Guizhou Autonomous Prefecture involved 1222 villages, 74,600 households and 338,600 people from 2016 to 2019, all of whom were farmers living below the poverty line. The areas from which people were relocated, from low-resource areas to high-resource areas, were Xingyi, Xingren, Anlong, Pu'an, Zhenfeng, Wangmo, Qinglong, and Ceheng, which are concentrated in the north and southeast (Figure 2). The relocated farmers were resettled in 65 resettlement sites in cities and towns, accounting for 99.74% of the total relocated population (the other 0.26% were resettled in centralized rural areas, being relatively scattered), and 26 resettlement sites, with more than 5000 people each, resettled a total of 242,567 people, accounting for 71.66% of the total resettlement. Southwest Guizhou Autonomous Prefecture is dominated by centralized resettlement in cities (Figure 2).

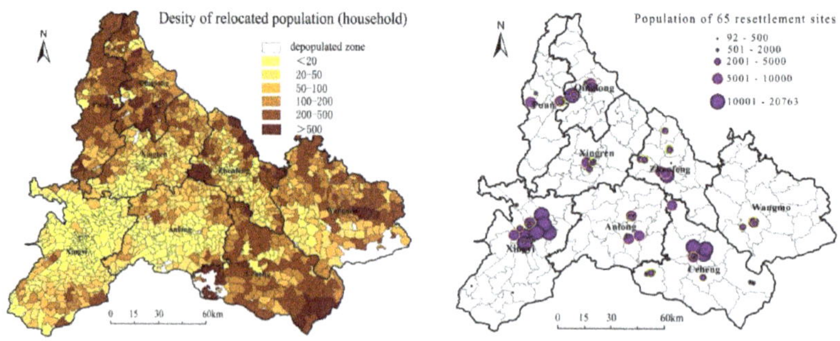

Figure 2. Density of relocated household and distribution of 65 resettlement sites of PAR.

2.2. Data Resources and Pre-Processing

The remote-sensing data were mainly obtained from Landsat in the GEE platform database, including Landsat 8 (OLI) data for 2013–2020 and Landsat 5 (TM) data for

2000–2012, with a spatial resolution of 30m and a temporal resolution of 16 d. In addition, considering the cloudy climatic attributes of Guizhou Province, effective remote-sensing images of low cloud cover could not be collected in the summer. Therefore, the study screened automatic synthetic Landsat images from April to October for the target years. The GEE official programming algorithm was then used to pre-process data, and to complete geometric correction, radiation correction and atmospheric correction. We achieved cloud-removal processing through the cloud-mask algorithm. In addition, the updated mask was implemented by code to avoid the effect of water area on the load distribution of the principal components. After using the GEE programming calculus to obtain standardized remote-sensing data on the study area from April to October, vector data on the administrative regions at all levels in Southwest Guizhou Autonomous Prefecture State were obtained from the Resource and Environment Science Data Center of the Chinese Academy of Sciences (http://www.resdc.cn, 15 December 2021), the Grain to Green Project, the Karst Rocky Desertification Restoration Project from the Master Plan of National Forestry and Grassland Administration (http://www.forestry.gov.cn, 12 March 2022), and the National Development and Reform Commission (http://www.rdrc.gov.cn, 12 March 2022). The meteorological data were obtained from the China Meteorological Administration Network (http://www.cma.gov.cn, 12 March 2022), and population data for the PAR were obtained from the Ecological Migration Bureau of the Guizhou Province (10 September 2021).

2.3. Methodology

The RSEI model first proposed by Xu et al., was related to four indicators—greenness, wetness, heat, and dryness—which can be visually determined and are widely used to understand the quality of the ecological environment.

The entire data calculation process was based on the GEE online platform, ArcGIS software, and ENVI software. GEE is the most critical platform for original data acquisition, preprocessing, and RSEI calculation. The GEE operation process was as follows: determine the scope and timeliness; clarify the image type; function cloud mask; function remove cloud; calibrated radiance; function normalization; use unit scale to normalize the pixel values; calculate the NDWI; calculate the NDVI, WET, LST, and NDISI (formula 1–9); collection merge; visualization; map; function PCA model; eigenvalue, eigenvector; return result; normalize the RSEI; and export image to drive (see attachment for original data). ArcGIS and ENVI further process the original data in order to meet the research needs.

(1) Calculation of component indexes. Among the four indexes, the greenness index reflects the regional vegetation coverage, and the normalized difference vegetation index (NDVI) is closely related to the leaf area index and vegetation coverage. The wetness indicators (WET) reflect the moist conditions of the regional surface objects and are expressed as the wet components of a tasseHed captransform of the surface vegetation, soil, etc. Here, the formulae used to calculate EM and OLI were somewhat different. The dryness index reflected the surface drying condition; the soil index and the index-based built-up index IBI were expressed as the normalized difference impervious surface index (NDISI). Heat indicators reflected the surface temperature conditions and were expressed by the land surface temperature (LST). Based on the existing research results, the four indicators were calculated as follows:

$$NDVI = \frac{B_{NIR} - B_{red}}{B_{NIR} + B_{red}} \tag{1}$$

$$WET_{TM} = 0.0315B_{blue} + 0.2021B_{green} + 0.3102B_{red} + 0.1594B_{NIR} - 0.6806B_{SWIR1} - 0.6109B_{SWIR2} \tag{2}$$

$$WET_{OLI} = 0.1511B_{blue} + 0.1972B_{green} + 0.3283B_{red} + 0.3407B_{NIR} - 0.7117B_{SWIR1} - 0.4559B_{SWIR2} \tag{3}$$

$$NDBSI = \frac{SI + IBI}{2} \tag{4}$$

$$SI = \frac{(B_{SWIR1} + B_{red}) - (B_{NIR} + B_{blue})}{(B_{SWIR1} + B_{red}) + (B_{NIR} + B_{blue})} \tag{5}$$

$$IBI = \{2 \times B_{SWIR1}/(B_{SWIR1} + B_{NIR}) - [B_{NIR}/(B_{NIR} + B_{red}) + B_{green}/\{(B_{green} + B_{WIR1})]\}/2 \times B_{SWIR1}/(B_{SWIR1} + B_{NIR}) + [B_{NIR}/(B_{NIR} + B_{red}) + B_{green}/(B_{green} + B_{WIR1})]\} \quad (6)$$

$$LST = \frac{K_2}{\ln(K_1/B(T_S) + 1)} \quad (7)$$

$$B(T_S) = \frac{L_{10} - L_{up} - \tau_{10}(1 - \varepsilon_{10})L_{down}}{\tau_{10}\varepsilon_{10}} \quad (8)$$

$$L_{10} = \tau_{10}[\varepsilon_{10}B(T_S) + (1 - \varepsilon_{10})L_{down}] + L_{up} \quad (9)$$

Here, *NDVI* indicates greenness, *WET* indicates humidity, *NDISI* indicates dryness, and *LST* indicates heat. The specific meaning of each variable in the equation is referred to in reference [29].

(2) Calculation of RSEI. Based on the results of the four indicators, NDVI, WET, NDSI, and LST, they were first normalized by using positive normalization to standardize their values. The initial RSEI values were calculated using NEVI software by principal component analysis of the standardized data from the four indicators for PC1. Indicator weights that were not dictated by humans were considered in the calculation of the initial RSEI value. The final RSEI value was obtained using the forward normalization process, where the value was between 0 and 1. The higher the numerical value, the better the EEQ.

$$NI = \frac{I - I_{min}}{I_{max} - I_{min}} \quad (10)$$

$$RSEI_0 = PC1[f(NDVI, WET, NDBSI, LST)] \quad (11)$$

$$RSEI = \frac{RSEI_0 - RSEI_{0-min}}{RSEI_{0-max} - RSEI_{0-min}} \quad (12)$$

In the above formula, *NI* denotes the standard index value after processing; *I* is the index value; and I_{max} and I_{min} are the maximum and minimum values of the index, respectively. *RSEI* denotes the final remote sensing ecological index; $RSEI_0$ denotes the primary remote sensing ecological index. $RSEI_{0\text{-}max}$ and $RSEI_{0\text{-}min}$ are the maximum and minimum values of the primary remote-sensing ecological index in the current period, respectively. *PC1* denotes the first principal component.

(3) Residual trends method. RSEI changes are influenced by climatic conditions and human activities. We used the residual trends method to calculate the extent to which PAR contributes to RSEI. Based on the mean value of RSEI from 2000 to 2020, the turning point of RSEI was determined according to linear trend analysis. A multiple linear regression analysis model was established with RSEI (dependent variable) and four factors (independent variables). These were the natural factors—mean annual temperature (MAT) and annual total precipitation (ATP)—and the human activity factors, outlined in the Grain to Green Project and the Karst Rocky Desertification Restoration Project, that may affect EEQ changes during the reference period (before the turning point).

$$V_{D,T}(RSEI) = a \times V_{I,T}(MAT) + b \times V_{I,T}(ATP) + c \times V_{I,T}(\text{funds of the Grain to Green Project}) + d \times V_{I,T}(\text{funds of the Karst Rocky Desertification Restoration Project}) \quad (13)$$

where $V_{D,T}$ denotes the dependent variables at a specific time, $V_{I,T}$ denotes the independent variables at a specific time, *a* and *b* are multivariate regression standardization coefficients of climate conditions, while *c* and *d* are multivariate regression standardization coefficients of human activities. Formula (13) was used to calculate the predicted RSEI value during project implementation (only affected by four factors) with a 95% confidence interval; the residual between the observed RSEI mean value and the predicted RSEI mean value was calculated. If the residual value was positive, PAR was assumed to have a positive impact on EEQ over time. If the residual was negative, PAR was assumed to harm the local ecology.

(4) Analysis of coupling coordination degree. To test the degree of spatial and temporal correlation between PAR and EEQ, this study used the coupling coordination degree to analyze and describe the spatial characteristics between PAR and ecological environment variables. The range of values was [0, 1], where a larger coupling coordination degree indicated a stronger correlation between the two variables [39].

$$C = \sqrt{\frac{U_1 \times U_2}{[(U_1 + U_2)/2]^2}} \tag{14}$$

$$D = \sqrt{C \times T};\ T = \alpha \times U_1 + \beta \times U_2 \tag{15}$$

Using formula C in particular, the system coupling degree was calculated. Here, U_1 and U_2 are the RSEI variables, and the data were normalized for the size (number of people) of the relocated population. T is the comprehensive evaluation index of U_1 and U_2; α and β are the undetermined coefficients, where $\alpha = \beta = 0.5$; D is the coupling coordination degree. The coupling coordination degree is based on SPSS software, which normalizes and standardizes the analysis data. The conclusion of calculations is between 0 and 1, which the software automatically divides into 10 levels. According to the trend in the level distribution and the existing method of dividing the research results, the coupling coordination level is divided into 4 levels: extreme detuning (0–0.3), general detuning (0.3–0.5), general coordination (0.5–0.7), and extreme coordination (0.7–1) [40].

3. Results

3.1. Ecological Environment Quality, Spatial and Temporal Evolution, and Driving Forces

3.1.1. Evolution Trend Analysis

According to the statistical analysis of the average EEQ value from 2000 to 2020 (Figure 3), the overall EEQ of the study area showed an increasing trend. The mean value of RSEI increased from 0.5329 in 2000 to 0.6363 in 2020, i.e., an increase of 0.1034 or 19.40% in 20 years. The overall trend of EEQ in the research area is still improving steadily. The overall trend of RSEI from 2000 to 2011 was relatively stable, although there were some fluctuations. From 2011 to 2016, the RSEI began to improve, although, again, there were still some fluctuations. The overall trend showed a steady improvement. From 2016 to 2020, RSEI in the study area began to improve significantly and remained stable. According to the evolution trend, the turning point of RSEI in the study area was determined to be 2016. In combination with the human activity in the area, we focused on analyzing the spatial and temporal distribution of RSEI to achieve the research objectives of this paper. The Grain to Green Project began in 2000; the Karst Rocky Desertification Restoration Project was implemented in 2008; and the PAR was implemented in 2016. Considering the time interval of the study period and that the effect of ecological restoration projects is somewhat delayed, we focused on data of the spatial and temporal distribution of RSEI for the years 2000, 2005, 2010, 2015, and 2020.

3.1.2. Spatial–Temporal Evolution Analysis of EEQ

For convenient comparison, the average RSEI value was divided into five grades according to the average value: poor (0–0.2), relatively poor (0.2–0.4), moderate (0.4–0.6), good (0.6–0.8), and excellent (0.8–1.0) [41]. The spatial distribution of RSEI values in the study area was characterized by higher EEQ grades in the southeastern part of Ceheng and Wangmo, and all regions were optimized after 2010 to some extent.

During 2000–2010, the EEQ values of the study areas were generally similar, i.e., excellent and good grades accounted for ~44% of the subject areas, particularly in the southeastern areas of Ceheng and Wangmo (Figure 4).

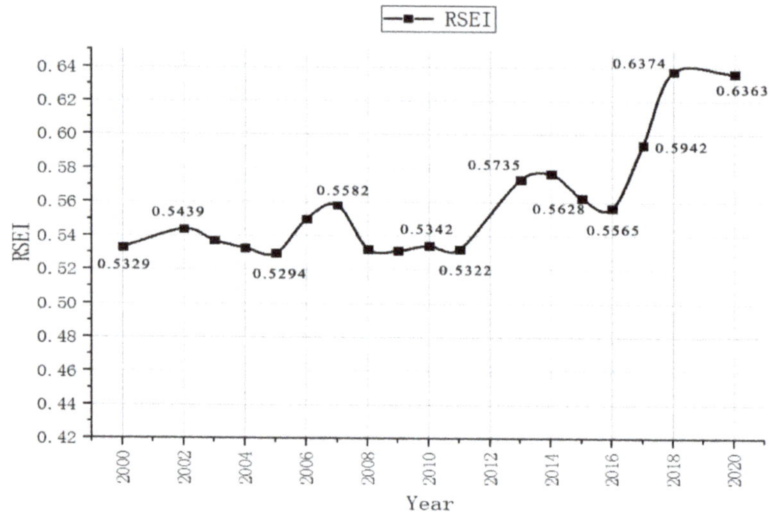

Figure 3. Mean RSEI value during 2000–2020.

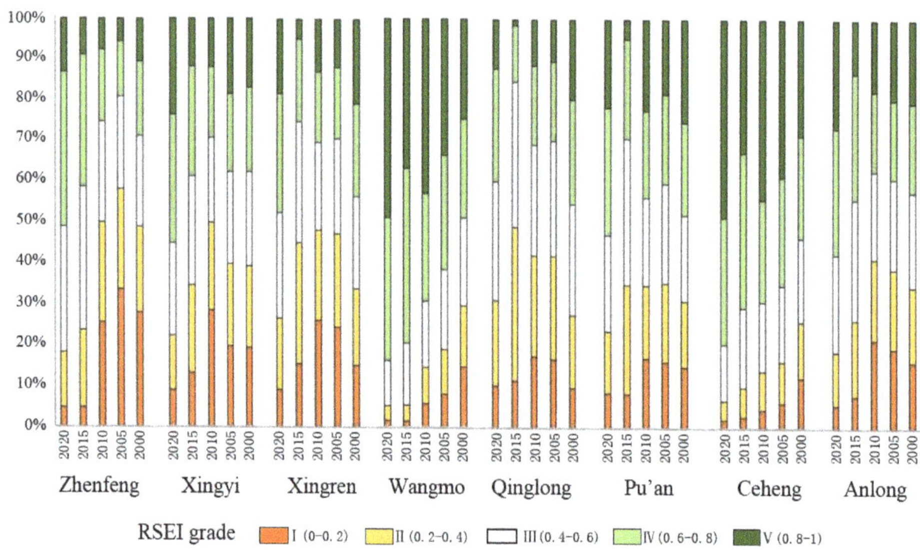

Figure 4. Eco-environment quality dynamics by classified RSEI of the counties from 2000 to 2020.

During the period 2010–2015, the proportion of EEQ attributable to poor and relatively poor grades decreased to 26.26%, which originated from areas mainly concentrated in north-central Xingren, south-central Pu'an, north-central Xinyi, and other regions. The proportion of excellent and good grades rose to 48.46%, which was attributable to the areas that were concentrated in the southeastern area of Ceheng and Wangmo. During this period, the EEQ grade remained the same in an area accounting for 42.48% of the total study area. The area that saw an EEQ reduction accounted for 24.80% of the study area. This area with reduced EEQ was mainly distributed in Wangmu, Ceheng, Pu'an, and Qinglong. In this period, the dominant EEQ grades were excellent and good (Figures 4 and A1, Table 1).

Table 1. The transition matrix of EEQ levels during 2010–2015.

EEQ Level	I	II	III	IV	V
I	830.13	1143.01	644.35	209.61	40.58
II	261.50	1038.92	1088.78	408.93	80.93
III	93.70	600.43	1471.60	1013.50	208.03
IV	32.88	185.00	892.54	1777.49	576.91
V	11.96	51.58	294.53	1679.01	1910.32

During the period from 2015 to 2020, the proportion of areas with an EEQ grade of poor or relatively poor decreased to 17.64%, and the distribution was mainly concentrated in the key areas of urban development. The proportion of areas with excellent and good grades further increased to 63.35%, with the most concentrated contiguous areas in Ceheng, Wangmo, and southeast of Zhenfeng. The EEQ improvement regions were most concentrated in Qinglong and Pu'an, which were the areas with the most fragile ecological environment areas and the most prominent peasants–land conflicts. The area of reduced EEQ was mainly concentrated in the urban development areas, such as Xingyi, where the population was further concentrated, and socio-economic development was more important (Figures 4 and A1, Table 2).

Table 2. The transition matrix of EEQ levels during 2015–2020.

EEQ Level	I	II	III	IV	V
I	395.94	407.33	303.77	98.76	22.56
II	337.98	830.39	1175.63	566.99	102.63
III	137.48	475.53	1383.03	1892.35	489.99
IV	41.54	127.90	497.46	2215.79	2177.92
V	10.00	29.22	99.47	541.48	2098.71

3.1.3. RSEI Result Test

The Pearson correlation coefficient was used to test the RSEI results. The results are presented in Table 3, where the average correlation of RSEI and the four indicators reached a maximum of 0.953, 0.964, −0.739 and −0.945, respectively. The correlation between RSEI and NDVI, WET, NDISI was significant at the 0.01 level, indicating strong significance, and the correlation with LST was significant at the 0.1 level, indicating general significance.

Table 3. Correlation matrix of indexes during 2000–2020.

Indicator	NDVI	WET	LST	NDISI	RSEI
NDVI	1	0.969	−0.869	−0.814	0.953
WET	0.969	1	−0.804	−0.831	0.964
LST	−0.869	−0.804	1	0.56	−0.739
NDISI	−0.814	−0.831	0.56	1	−0.945
RSEI	0.953	0.964	−0.739	−0.945	1
Sig.	0.012	0.008	0.154	0.015	

3.2. *Drivers of Change in EEQ*

Unlike other karst areas in the world, where the population density is low, the karst mountains in southwest China are populous and ecologically fragile. There is sparse coordination between peasant–land conflicts and high ecological pressure [42]. Various ecological restoration projects have been promoted since 2000, such as the Grain to Green Project (from 2000) and the Karst Rocky Desertification Restoration Project (from 2008) [43]. According to the mean values of the four indicators presented in Table 4, the overall indexes were relatively stable, and the average RSEI values were also relatively stable.

Table 4. Mean indicators of RSEI from 2000 to 2020.

	NDVI	WET	LST	NDISI	RSEI
2000	0.5555	0.5286	0.5627	0.5073	0.5329
2005	0.5430	0.5266	0.5008	0.5046	0.5294
2010	0.5398	0.5411	0.5518	0.5120	0.5342
2015	0.6689	0.5642	0.4279	0.5095	0.5628
2020	0.7511	0.5952	0.4307	0.4217	0.6363

During 2010–2015, the mean value of NDVI increased by 23.92%, and the WET mean value also increased significantly. The effectiveness of the ecological restoration project is also highlighted during this period, having contributed to a significant decrease in the mean value of LST. The ecological restoration project also contributed to an increase in the mean value of RSEI in the study area (Table 4).

From 2015 to 2020, PAR was promoted and completed. During this period, NDISI was reduced significantly, by 17.24%. Relocation to alleviate poverty caused many rural people to move to towns and cities, while their original home base was reclaimed and re-greened. In this case, the impact of human activities on rural areas was significantly reduced, which directly contributed to a significant reduction in the dryness index. The NDVI was increased, along with the WET, which means that the conflict between humans and land was fundamentally relaxed, and the effectiveness of various ecological restoration projects was maintained and further improved. PAR was a key factor in the improvement of RSEI in the study area (Table 4).

3.3. The Contribution of PAR to EEQ Changes

A multiple linear regression analysis model was used to calculate the residual trends in PAR implementation, whereby the turning point of RSEI change caused by PAR was determined to occur in 2016. The variation in RSEI was influenced by natural factors, including MAT and ATP, as well as human activities, including the Grain to Green Project and the Karst Rocky Desertification Restoration Project. We took these four factors as the independent variables, i.e., the influencing factors, and RSEI was adopted as the dependent variable, i.e., the resulting factor. Standardized coefficients of the regression models were analyzed by the SPSS with 95% confidence (Table A1). The results show that the observed cumulative probability and the predicted cumulative probability were normally distributed in the linear regression analysis model (Figure 5). Furthermore, the standardized residuals were randomly distributed without outliers (Figure 6), and the regression model significance was 0.038 at a significance level of 95%. Therefore, the multiple linear regression equation was verified to be stable. The analysis results show that the RSEI variable could mathematically be represented as Variable (RSEI) = 0.273 × Variable(MAT) + 0.285 × Variable(ATP) − 0.144 × Variable (funds of the Grain to Green Project) + 0.520 × Variable (funds of the Karst Rocky Desertification Restoration Project). This equation predicted the mean RSEI values in 2017, 2018, and 2020 to be 0.5676, 0.5660, and 0.5701, respectively, under the influence of the four factors. The observed RSEI values in 2017, 2018, and 2020 were 0.5942, 0.6374 and 0.6363, respectively. The residual difference between the observed and predicted mean values of RSEI was likely due to the PAR-driven RSEI improvement. The improvements in RSEI caused by PAR were 0.0266, 0.0715, and 0.0662 in 2017, 2018, and 2020, respectively, according to the actual increase of 0.0266, 0.0715, and 0.0662. Compared with the turning point of 2016, the PAR contributed to the RSEI growth contribution rate of 70.56%, 88.38%, and 82.96% in the three years, respectively. Collectively, after 2016, the RSEI values increased significantly, and from there on, the average RSEI reached good levels and remained relatively stable.

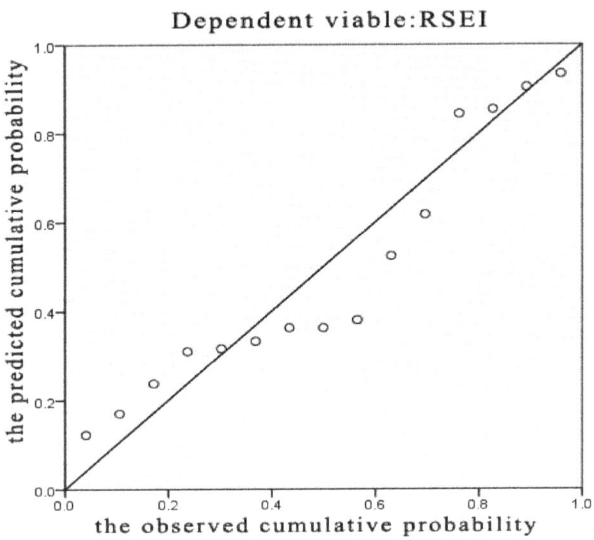

Figure 5. P-P plot of regression standardized residuals.

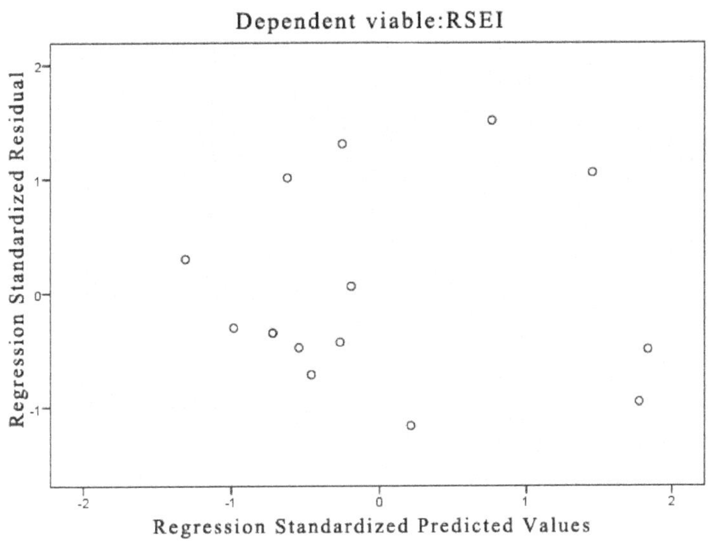

Figure 6. Scatter diagram.

3.4. Correlation between PAR and EEQ Changes

The correlation between relocation and EEQ changes was analyzed by taking the village area as the basic unit. A map showing the change in RSEI during 2015–2020 is shown in Figure 7.

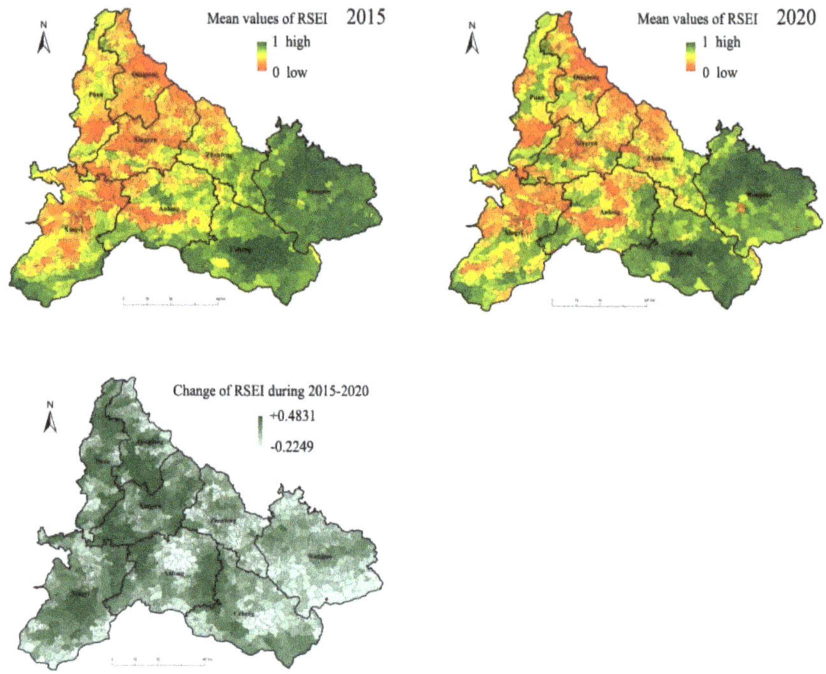

Figure 7. Change map of RSEI during 2015–2020.

There were 1330 administrative village units in the study area. The number of villages in the area with reduced RSEI was 205, accounting for 15.41% of the total. A total of 1125 villages exhibited improved RSEI. These villages were mostly located in the central and northern regions of Pu'an and Qinglong. The original EEQ of these regions was relatively low, and the effect of the upgrade in terms of RESI was obvious. The relocation of the five counties on the northeast side of the study area of Southwest Guizhou Autonomous Prefecture, including the counties of Ceheng, Qinglong, Wangmo, Zhenfeng, and Pu'an, was the most concentrated, whereby 74,600 households were relocated, accounting for 82.47% of the total (Figure 2).

Figure 8 shows the spatial distribution of the coupling coordination between the relocation population density and RSEI variables in 1222 administrative villages in eight counties. There were 247 villages with extreme coordination, accounting for 25.33% of the area, mainly concentrated in Ceheng, Wangmo, Qinglong, and Pu'an, which are the most densely relocated areas. There were 411 general coordination villages, accounting for 33.90% of the area, most of which were located in the northeast and southeast of the study area. There were 403 general detuning villages, accounting for 29.99% of the area, and 161 extreme detuning villages, accounting for 7.63% of the total villages. These villages were predominantly located in economically developed areas, such as Xinyi. After the spatial analysis of the coupling coordination between relocation population density and RSEI variables, a significant coupling coordination relationship was found between EEQ enhancement and PAR in the study area.

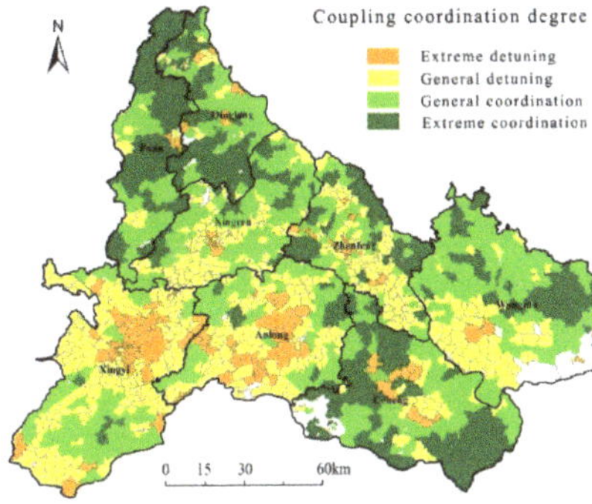

Figure 8. Coupling coordination degree of density of relocated population, and change in RSEI.

3.5. Forecast of Future RSEI

The RSEI of the study area was predicted after the implementation of the PAR to analyze its sustained impact on regional EEQ. The ARIMA model starts from the time-series itself and forecasts future data based on past behavioral data [44]. The expert modeler was used for prediction, where the autocorrelation and partial correlation coefficients of the model were all within the confidence zone of 95%. After prediction, the mean value of RSEI in the study area was predicted to improve to 0.7355 in 2025, and to 0.8603 in 2030. Continuous improvement and optimization were maintained from thereon. The spatial distribution effects are shown in Figure 9.

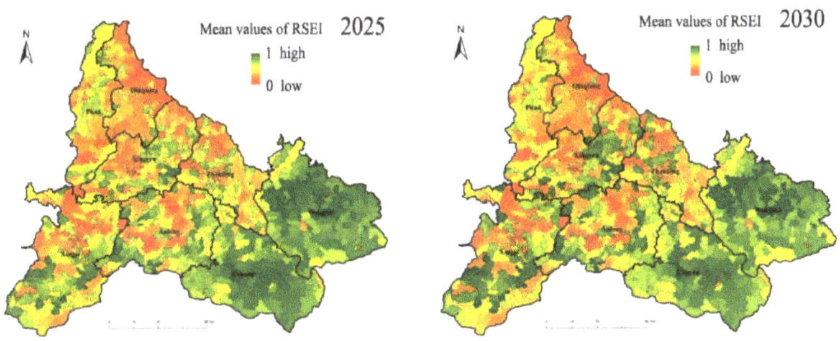

Figure 9. Forecast of the mean values of RSEI.

4. Discussion

The results show that not all areas with a large number of relocated people have significantly increased RSEI. In northern Pu'an, Qinglong, and other severely rocky desertification areas (Figure 1) where the population density is relatively large (Figure A2) and the contradiction between peasants and land is prominent, the RSEI increase is more obvious in areas with a large number of people involved in PAR (Figure 7). However, in non-rocky desertification areas such as Ceheng and Wangmo in the southeast, where the population density is relatively small, the RSEI does not have an obvious increase, although the number of people relocated is relatively large. This shows that the implementation of PAR is more effective in areas with relatively poor ecology and a prominent contradiction between peasants and land, while in areas where the ecology is better, the contradiction between peasants and land is not prominent, and the ecological effect is not obvious (Table 5).

Table 5. Summary table of positive and negative effects of PAR.

Indicators		Positive Effects	Negative Effects	Remark
Society		Improved infrastructure, education, employment and health care, etc.	Increased cost of living, changed way of life, blocked cultural inheritance, etc.	Will not be discussed
Sustainability		The remaining farmers will have more production resources, and will obtain more employment opportunities.	Traditional culture is destroyed, increasing living costs, and increased burden on the government.	
Ecology	Spatial features	The RSEI of the entire study area significantly improved.	On the southeast side where the contradiction between peasants and land is not prominent, the RSEI improvement is not obvious.	Unbalanced RSEI promotion
	Indicator structure	Increase in NDVI and decrease in NDISI.	The increase in WET is small, and the negative indicator Lst increases.	
	Biodiversity	Human activity is reduced in abandoned areas, which is beneficial to biodiversity.	After PAR, the planting structure will tend to be single-species, thus affecting the biodiversity.	
	Human well-being	The overall EEQ of the study area improved, and the entire area became greener.	There is a certain distance of EEQ between the ecologically improved area and the gathering area of the relocated farmers.	
	Applicability	PAR can promote ecological improvement in karst ecologically fragile areas where the contradiction between peasants and land is prominent.	PAR may not be effective in ecological restoration in areas where climate is the main limiting factor (such as water resources in semi-arid regions).	
	Other aspects	The relocation area can make more convenient and intensive use of land, and more effectively promote the implementation of various ecological restoration projects.	May possibly cause single-species forest and the waste of water resources (in the context of global warming, extreme dry weather has already occurred in 2011).	

In terms of the effectiveness of EEQ, the improvement in RSEI in the study area contributed by PAR is mainly due to the increase in NDVI and the decrease in NDISI. After relocation, the large-scale management of forestland will become a trend, with farmers moving away from contracted land, and farmland being transformed to forest, which contributes to the substantial increase in NDVI. Whether the substantial increase in NDVI will cause single-species forest and excessive water consumption needs further consideration [45]. The increase in WET is significantly smaller than the increase in NDVI, which is further verified. The RSEI of the study area is indeed significantly improved, while the RSEI of the urban areas to which farmers are moved is reduced. Does such an increase in RSEI really improve the living environment for local humans? During the implementation of PAR, the overall increase in RSEI in the study area was partially reduced (with a decrease in the more concentrated areas of population) and the comprehensive indicators showed improvement. Whether such improvement is conducive to sustainable social development is something our follow-up study needs to consider deeply (Table 5).

Biodiversity has an important impact on ecosystem services, and the impact of human activities on biodiversity is the focus of various scholars and organizations such as the United Nations Environment Programme (UNEP) [46,47]. After PAR, the number of land managers decreased and land-use patterns changed, which shifted the trend towards large-scale land management. For the purpose of easy management and economic efficiency, the planting structure of the land tends to be homogeneous after large-scale management, which leads to an impact on biodiversity and the weakening of symbiosis among plants, thus further affecting the ecosystem service function [48]. In contrast, some areas become similar to ecological reserves, where the disturbance of human activities is reduced, and biological succession proceeds in an orderly manner, which is beneficial to biodiversity. There are many influencing factors and mechanisms of internal change of biodiversity, such as human disturbance, environmental factors, management methods, etc. The impact of PAR on biodiversity needs to be further studied in future work [49,50] (Table 5).

The global trend in terms of helping rural areas is to promote traditional and sustainable farming with nature-friendly measures, rather than relocating the rural population to cities [51]. PAR is the transfer of farmers from rural to urban areas, which destroys traditional rural culture, increases their cost of living and changes their way of life. The fact that farmers move to cities to take up jobs they are not good at, and that employment training is not sufficient to help low-ability farmers relocate successfully, also greatly increases the burden on the government, in addition to the fact that the minimum cost of living for relocated households is approximately 70% higher in cities than in rural areas. PAR poses a great challenge to the sustainability of farmers [17]. The owners of the relocated rural areas are not those executing PAR, but some of the poorest farmers. The most crucial role of PAR is to alleviate the contradiction between peasants and land. The remaining farmers will have more production resources (such as renting the land of the relocated farmers), and may also obtain more employment opportunities (there will be some vacancies for forest-protection work and road-cleaning work after the relocation) and development opportunities. Relocating farmers to live together in a concentrated area will also facilitate the establishment of infrastructure such as medical care, education, training, factories, etc., to a certain extent, and it is easier to accept new knowledge, which will help farmers to achieve sustainability (Table 5).

The GEE platform was used to address the problem of the difficult acquisition of effective Landsat images under cloudy and foggy weather in Guizhou. With its powerful processing capabilities, GEE provided a foundation for the accurate analysis of the temporal and spatial patterns and evolution of EEQ. The RSEI model allows the objective analysis of EEQ. The effects of PAR and climatic factors and other ecological restoration projects on the EEQ changes could be distinguished scientifically using the residual trends. The results could then be used to establish a high and low series of coupling coordination levels between the number of relocated populations and EEQ variables, and to clarify the effective degree of correlation between PAR and EEQ enhancement. Furthermore, based on the ARIMA model, a prediction of future ecological trends concluded that the relocation of impoverished residents has a significant and sustainable driving effect on the promotion of regional EEQ. However, the RSEI model needs to use a water body mask to ensure the normalization accuracy of RSEI [30]. The Southwest Guizhou Autonomous Prefecture belongs to the ecological protection barrier in the upper reaches of the Pearl River, and 1.86% of the water area has important ecological services that unfortunately cannot be effectively reflected [41]. Due to the influence of cloudy weather in Guizhou, it is difficult to collect remote-sensing data in the same period for different years. The base data used in this paper were Landsat data from April to October of the target years. The time span was large, and there may be bias in the measurement of RSEI each year. There are other shortcomings, which provide a basis for the key research directions of the future.

PAR in karst mountains can improve local EEQ to a certain extent, but there have been some controversies. In later studies, researchers further broadened their research scope to investigate the effect of PAR on the regional carbon neutral effect [52] from the direction of carbon neutrality [53].

5. Conclusions

The residual trends method eliminated the confounding effects of other influencing factors and clarified the contribution of PAR to RSEI growth. The PAR made a contribution to RSEI improvement with a 13.07% increase in the study area during 2015–2020; this RESI was significantly higher than the increase noted during 2000 to 2015. The EEQ of the most ecologically vulnerable areas, such as Qinglong and Pu'an, was most significantly improved after the PAR. However, the EEQ of economically developed areas, such as Xingyi, showed a decreasing trend.

After the implementation of PAR, a large number of rural houses were dismantled and the land reclaimed as green regions. Additionally, many rural construction projects did not advance, contributing to an obvious decrease in the mean value of NDISI (to 17.24%). Furthermore, the core population of farmers moved from cultivated land to urban employment, leaving much farmland abandoned or available for planting trees, contributing to a sharp increase in the NDVI (to 12.28%). The residual trends model predicted that the mean values of RSEI in 2017, 2018, and 2020 were 0.5676, 0.5660, and 0.5701, respectively, under the influence of a multitude of factors, except for the PAR. The observed mean values of RSEI were 0.5942, 0.6374, and 0.6363, respectively. The improvement in RSEI caused by PAR was 0.0266, 0.0715, and 0.0662 in 2017, 2018, and 2020, respectively.

The spatial distribution of the coupling coordination between the relocation population density and RSEI variables showed that there was a significant positive correlation between the increase in RSEI and the relocation population density. The larger the relocated population, the greater and more significant the increase in RSEI in the region. This also led to a higher corresponding level of coupling coordination. As measured by the ARIMA prediction model, the EEQ of the study area, with many people undergoing PAR and in a cluster shape, will continue to evolve for the better. Furthermore, the EEQ of the developed urban areas represented by Xingyi will continue to decrease.

Author Contributions: Conceptualization, Z.Z.; data curation of PAR, W.L.; data curation of Landsat, L.W. and X.Z.; writing—original draft preparation, Q.F.; writing—review and editing, Z.Z. and Q.F.; visualization, Q.F.; supervision, C.Z.; project administration, L.Z. All authors have read and agreed to the published version of the manuscript.

Funding: This research was funded by the National Natural Science Foundation of China (41661088), jointly funded by the Program in Guizhou Planning of Philosophy and Social Science (21GZZD39) and the High-level Innovative Talents Training Program in Guizhou Province (2016-5674).

Institutional Review Board Statement: Not applicable.

Informed Consent Statement: Not applicable.

Data Availability Statement: The data presented in this study are available on request from the corresponding author. Some of the data are not publicly available due to privacy constraints.

Conflicts of Interest: The authors declare no conflict of interest.

Appendix A

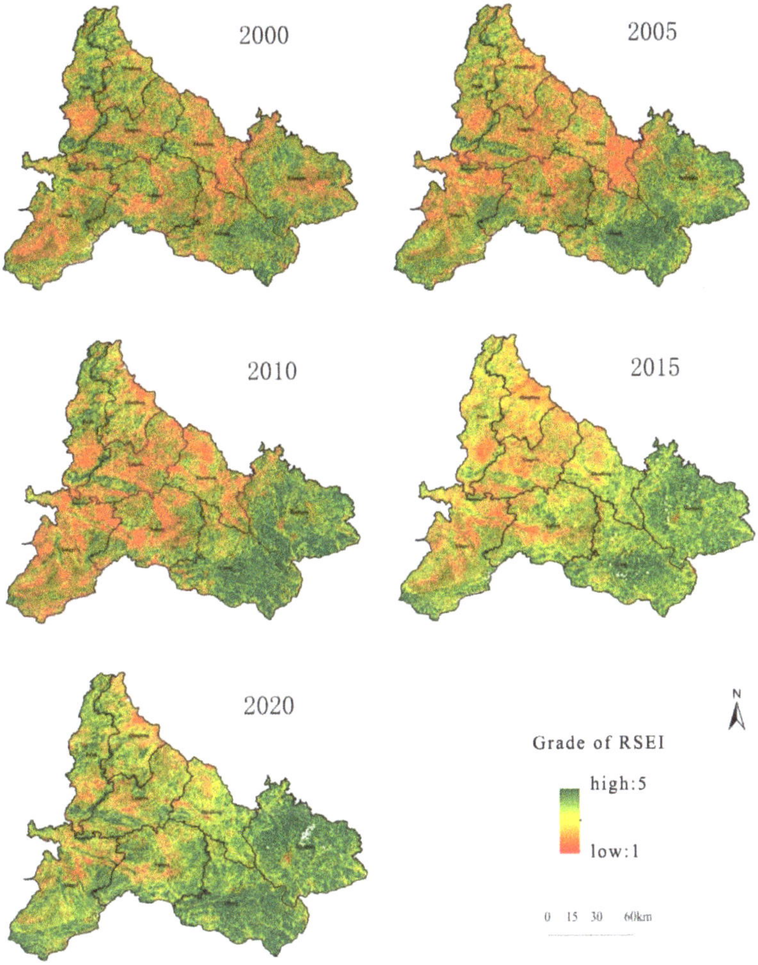

Figure A1. Distribution of RSEI from 2000 to 2020.

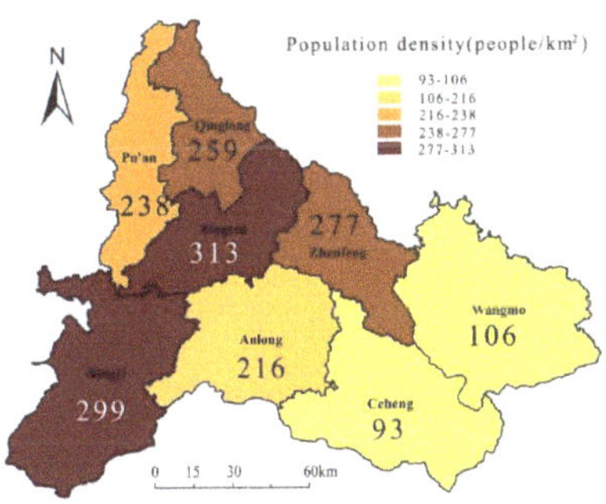

Figure A2. Population density of 8 counties (2016).

Table A1. Statistical table of changing EEQ image factors.

Year	RSEI	MAT (°C)	ATP (mm)	Grain to Green (10^{10} CNY)	Karst Rocky Desertification Restoration (10^{10} CNY)
2000	0.5329	15.67	1395.22	0.09	
2001		16.35	1459.53	0.10	
2002	0.5439	16.66	1218.52	1.02	
2003	0.537	16.81	1233.95	1.73	
2004	0.5326	15.92	1121.14	1.60	
2005	0.5294	16.03	1259.33	1.68	
2006	0.5499	16.38	1226.83	1.48	
2007	0.5582	16.36	1428.46	1.37	
2008	0.5322	15.61	1489.99	1.25	0.44
2009	0.5311	16.61	1011.30	1.15	0.63
2010	0.5342	16.66	1291.39	0.78	0.63
2011	0.5322	15.36	811.22	0.39	4.25
2012		15.83	1206.55	0.28	4.25
2013	0.5735	16.53	941.18	0.15	4.25
2014	0.5767	16.43	1540.23	0.88	4.25
2015	0.5628	16.93	1484.35	0.95	4.25
2016	0.5565	16.65	1270.80	0.95	6.25
2017	0.5942	16.46	1359.94	0.95	6.25
2018	0.6374	16.42	1309.58	0.95	6.25
2019		16.80	1438.02	0.95	6.25
2020	0.6363	16.43	1489.82	0.95	6.25

References

1. Schwalm, C.R.; Anderegg, W.R.L.; Michalak, A.M.; Fisher, J.B.; Biondi, F.; Koch, G.; Litvak, M.; Ogle, K.; Shaw, J.D.; Wolf, A. Global patterns of drought recovery. *Nature* **2017**, *548*, 202–205. [CrossRef] [PubMed]
2. Foley, J.A.; Ramankutty, N.; Brauman, K.A.; Cassidy, E.S.; Gerber, J.S.; Johnston, M.; Mueller, N.D.; O'Connell, C.; Ray, D.K.; West, P.C. Solutions for a cultivated planet. *Nature* **2011**, *478*, 337–342. [CrossRef] [PubMed]
3. Mahmoud, S.H.; Gan, T.Y. Impact of anthropogenic climate change and human activities on environment and ecosystem services in arid regions. *Sci. Total Environ.* **2018**, *633*, 1329–1344. [CrossRef] [PubMed]
4. Joppa, L.N.; Boyd, J.W.; Duke, C.S.; Hampton, S.; Jackson, S.T.; Jacobs, K.L.; Kassam, K.A.S.; Mooney, H.A.; Ogden, L.A.; Ruckelshaus, M. Government: Plan for ecosystem services. *Science* **2016**, *351*, 1037. [CrossRef]
5. Bryan, B.A.; Gao, L.; Ye, Y.; Sun, X.; Connor, J.D.; Crossman, N.D.; Smith, M.S.; Wu, J.; He, C.; Yu, D.; et al. China's response to a national land-system sustainability emergency. *Nature* **2018**, *559*, 193–204. [CrossRef]
6. Chen, C.; Park, T.; Wang, X.; Piao, S.; Xu, B.; Chaturvedi, R.K.; Fuchs, R.; Brovkin, V.; Ciais, P.; Fensholt, R.; et al. China and India lead in greening of the world through land-use management. *Nat. Sustain.* **2019**, *2*, 122–129. [CrossRef]
7. Steffen, W.; Broadgate, W.; Deutsch, L.; Gaffney, O.; Ludwig, C. The trajectory of the Anthropocene: The Great Acceleration. *Anthr. Rev.* **2015**, *2*, 81–98. [CrossRef]
8. Yang, Y.; Sherbinin, D.A.; Liu, Y. China's poverty alleviation resettlement: Progress, problems and solutions. *Habitat Int.* **2020**, *98*, 102135. [CrossRef]
9. Dasgupta, S.; Deichmann, U.; Meisner, C.; Wheeler, D. Where is the poverty–environment nexus? Evidence from Cambodia, Lao PDR, and Vietnam. *World Dev.* **2005**, *33*, 617–638. [CrossRef]
10. Cavendish, W. Empirical regularities in the poverty-environment relationship of rural households: Evidence from Zimbabwe. *World Dev.* **2000**, *28*, 1979–2003. [CrossRef]
11. Bird, K.; Shepherd, A. Livelihoods and chronic poverty in semi-arid Zimbabwe. *World Dev.* **2003**, *31*, 591–610. [CrossRef]
12. National Development and Reform Commision. The 13th Five-Year Plan for China's poverty alleviation resettlement: Great achievements and practical experience. *Macroecon. Manag.* **2021**, *9*, 4–9.
13. Rogers, S.; Xue, T. Resettlement and climate change vulnerability: Evidence from rural China. *Glob. Environ. Chang.* **2015**, *35*, 62–69. [CrossRef]
14. Liu, Y.; Liu, J.; Zhou, Y. Spatio-temporal patterns of rural poverty in China and targeted poverty alleviation strategies. *J. Rural. Stud.* **2017**, *52*, 66–75. [CrossRef]
15. Yeh, E.T. Tibetan Range Wars: Spatial Politics and Authority on the Grasslands of Amdo. *Dev. Chang.* **2003**, *34*, 499–523. [CrossRef]
16. Bossio, D.; Geheb, K.; Critchley, W. Managing water by managing land: Addressing land degradation to improve water productivity and rural livelihoods. *Agric. Water Manag.* **2010**, *97*, 536–542. [CrossRef]

17. Fan, M.; Li, Y.; Li, W. Solving one problem by creating a bigger one: The consequences of ecological resettlement for grassland restoration and poverty alleviation in Northwestern China. *Land Use Policy* **2015**, *42*, 124–130. [CrossRef]
18. Brandt, M.; Yue, Y.; Wigneron, J.P.; Tong, X.; Tian, F.; Jepsen, M.R.; Xiao, X.; Verger, A.; Mialon, A.; Al-Yaari, A.; et al. Satellite-Observed Major Greening and Biomass Increase in South China Karst During Recent Decade. *Earth's Future* **2018**, *6*, 1017–1028. [CrossRef]
19. Zhang, S.; Bai, X.; Zhao, C.; Tan, Q.; Luo, G.; Wang, J.; Li, Q.; Wu, L.; Chen, F.; Li, C.; et al. Global CO_2 Consumption by Silicate Rock Chemical Weathering: Its Past and Future. *Earth's Future* **2021**, *9*, e2020EF001938. [CrossRef]
20. World Commission on Environment and Development. *Our Common Future*; Oxford University Press: Oxford, UK, 1987.
21. Vina, A.; McConnell, W.J.; Yang, H.B.; Xu, Z.C.; Liu, J.G. Effects of conservation policy on China's forest recovery. *Sci. Adv.* **2016**, *2*, 1500965. [CrossRef]
22. Kuznets, S. Economic growth and income inequality. *Am. Econ. Rev.* **1955**, *45*, 1–28.
23. Mohammadi, A.; Costelloe, J.F.; Ryu, D. Application of time series of remotely sensed normalized difference water, vegetation and moisture indices in characterizing flood dynamics of large-scale arid zone floodplains. *Remote Sens. Environ.* **2017**, *190*, 70–82. [CrossRef]
24. Vickers, H.; Hogda, K.A.; Solbo, S.; Karlsen, S.R.; Tommervik, H.; Aanes, R.; Hansen, B.B. Changes in greening in the high Arctic: Insights from a 30 year AVHRR max NDVI dataset for Svalbard. *Environ. Res. Lett.* **2016**, *11*, 105044. [CrossRef]
25. Coutts, A.M.; Harris, R.J.; Phan, T.; Livesley, S.J.; Williams, N.S.G.; Tapper, N.J. Thermal infrared remote sensing of urban heat: Hotspots, vegetation, and an assessment of techniques for use in urban planning. *Remote Sens. Environ.* **2016**, *186*, 637–651. [CrossRef]
26. Crist, E.P. A TM Tasseled Cap equivalent transformation for reflectance factor data. *Remote Sens. Environ.* **1985**, *17*, 301–306. [CrossRef]
27. Xu, H. Analysis of Impervious Surface and its Impact on Urban Heat Environment using the Normalized Difference Impervious Surface Index (NDISI). *Photogramm. Eng. Remote Sens.* **2010**, *76*, 557–565. [CrossRef]
28. Man, W.; Liu, M.; Li, X.; Wang, Z.; Jia, M.; Li, X.; Mao, D.; Ren, C.; Ouyang, L. Assessment on the dynamics of eco- environmental statuses from 1990 to 2015 in the ecological function zone of Sanjiang Plain. *J. Arid. Land Resour. Environ.* **2018**, *32*, 136–141.
29. Shan, W.; Jin, X.; Ren, J.; Wang, Y.; Zhou, Y. Ecological environment quality assessment based on remote sensing data for land consolidation. *J. Clean. Prod.* **2019**, *239*, 118–126. [CrossRef]
30. Xu, H.; Wang, M.; Shi, T.; Guan, H.; Fang, C.; Lin, Z. Prediction of ecological effects of potential population and impervious surface increases using a remote sensing based ecological index (RSEI). *Ecol. Indic.* **2018**, *93*, 730–740. [CrossRef]
31. Xu, H.Q.; Wang, Y.F.; Guan, H.D.; Shi, T.T.; Hu, X.S. Detecting Ecological Changes with a Remote Sensing Based Ecological Index (RSEI) Produced Time Series and Change Vector Analysis. *Remote Sens.* **2019**, *11*, 2345. [CrossRef]
32. Xu, J.Q.; Wang, J.; Xiong, N.N.; Chen, Y.H.; Sun, L.; Wang, Y.T.; An, L.K. Analysis of Ecological Blockage Pattern in Beijing Important Ecological Function Area, China. *Remote Sens.* **2022**, *14*, 1151. [CrossRef]
33. Chen, A.; Yang, X.C.; Guo, J.; Xing, X.Y.; Yang, D.; Xu, B. Synthesized remote sensing-based desertification index reveals ecological restoration and its driving forces in the northern sand-prevention belt of China. *Ecol. Indic.* **2021**, *131*, 108230. [CrossRef]
34. Mu, H.W.; Li, X.C.; Ma, H.J.; Du, X.P.; Huang, J.X.; Su, W.; Yu, Z.; Xu, C.; Liu, H.L.; Yin, D.Q.; et al. Evaluation of the policy-driven ecological network in the Three-North Shelterbelt region of China. *Landsc. Urban Plan.* **2022**, *218*, 104305. [CrossRef]
35. Evans, J.; Geerken, R. Discrimination between climate and human-induced dryland degradation. *J. Arid. Environ.* **2004**, *57*, 535–554. [CrossRef]
36. Zhang, S.; Bai, X.; Zhao, C.; Tan, Q.; Luo, G.; Cao, Y.; Deng, Y.; Li, Q.; Li, C.; Wu, L.; et al. Limitations of soil moisture and formation rate on vegetation growth in karst areas. *Sci. Total Environ.* **2021**, *810*, 151209. [CrossRef]
37. Gong, S.; Wang, S.; Bai, X.; Luo, G.; Wu, L.; Chen, F.; Qian, Q.; Xiao, J.; Zeng, C. Response of the weathering carbon sink in terrestrial rocks to climate variables and ecological restoration in China. *Sci. Total Environ.* **2021**, *750*, 141525. [CrossRef]
38. Tong, X.W.; Wang, K.L.; Yue, Y.M.; Brandt, M.; Liu, B.; Zhang, C.H.; Liao, C.J.; Fensholt, R. Quantifying the effectiveness of ecological restoration projects on long-term vegetation dynamics in the karst regions of Southwest China. *Int. J. Appl. Earth Obs. Geoinf.* **2017**, *54*, 105–113. [CrossRef]
39. Ariken, M.; Zhang, F.; Chan, N.W.; Kung, H.T. Coupling coordination analysis and spatio-temporal heterogeneity between urbanization and eco-environment along the Silk Road Economic Belt in China. *Ecol. Indic.* **2021**, *121*, 107191. [CrossRef]
40. Li, J.; Sun, W.; Li, M.; Meng, L. Coupling coordination degree of production, living and ecological spaces and its influencing factors in the Yellow River Basin. *J. Clean. Prod.* **2021**, *298*, 126803. [CrossRef]
41. Yuan, B.; Fu, L.; Zou, Y.; Zhang, S.; Xie, Y. Spatiotemporal change detection of ecological quality and the associated affecting factors in Dongting Lake Basin, Based on RSEI. *J. Clean. Prod.* **2021**, *302*, 126995. [CrossRef]
42. Yue, Y.M.; Liao, C.J.; Tong, X.W.; Wu, Z.B.; Fensholt, R.; Prishchepov, A.; Jepsen, M.R.; Wang, K.L.; Brandt, M. Large scale reforestation of farmlands on sloping hills in South China karst. *Landsc. Ecol.* **2020**, *35*, 1445–1458. [CrossRef]
43. Hoek, J.V.D.; Ozdogan, M.; Burnickic, A.; Zhu, A.-X. Evaluating forest policy implementation effectiveness with a cross-scale remote sensing analysis in a priority conservation area of Southwest China. *Appl. Geogr.* **2014**, *47*, 177–189. [CrossRef]
44. Aasim; Singh, S.N.; Mohapatra, A. Repeated wavelet transform based ARIMA model for very short-term wind speed forecasting. *Renew. Energy* **2019**, *136*, 758–768. [CrossRef]
45. Zastrow, M. China's tree-planting drive could falter in a warming world. *Nature* **2019**, *573*, 474–475. [CrossRef]

46. UNEP-WCMC. Biodiversity and Poverty Reduction: The Importance of Biodiversity for Ecosystem Services. Available online: http://www.unep-weme.org/latenews (accessed on 12 May 2022).
47. Tilman, C.D. Conserquences and ethics of biodiversity. *Nature* **2000**, *405*, 208–211. [CrossRef]
48. Maxwell, S.L.; Cazalis, V.; Dudley, N.; Hoffmann, M.; Rodrigues, A.S.L.; Stolton, S.; Visconti, P.; Woodley, S.; Kingston, N.; Lewis, E.; et al. Area-based conservation in the twenty-first century. *Nature* **2020**, *586*, 217–227. [CrossRef]
49. Tilman, D.; Downing, J.A. Biodiversity and stability in grasslands. *Nature* **1994**, *367*, 363–367. [CrossRef]
50. Rajaniemi, T.K. Explaining productivity-diversity relationships in plants. *Oikos* **2003**, *101*, 449–457. [CrossRef]
51. Baird, I.G.; Shoemaker, B. Unsettling Experiences: Internal Resettlement and International Aid Agencies in Laos. *Dev. Chang.* **2007**, *38*, 865–888. [CrossRef]
52. Zhang, S.R.; Bai, X.Y.; Zhao, C.W.; Tan, Q.; Luo, G.J.; Wu, L.H.; Xi, H.P.; Li, C.J.; Chen, F.; Ran, C.; et al. China's carbon budget inventory from 1997 to 2017 and its challenges to achieving carbon neutral strategies. *J. Clean. Prod.* **2022**, *347*, 130966. [CrossRef]
53. Luo, X.L.; Bai, X.Y.; Tan, Q.; Ran, C.; Chen, H.; Xi, H.P.; Chen, F.; Wu, L.H.; Li, C.J.; Zhang, S.R.; et al. Particulate organic carbon exports from the terrestrial biosphere controlled by erosion. *Catena* **2022**, *209*, 105815. [CrossRef]

land

MDPI

Article

Identifying the Multi-Scale Influences of Climate Factors on Runoff Changes in a Typical Karst Watershed Using Wavelet Analysis

Luhua Wu [1,2,3], Shijie Wang [2,3], Xiaoyong Bai [2,3,*], Fei Chen [2,3], Chaojun Li [2,3], Chen Ran [2,3] and Sirui Zhang [2,3]

1 School of Economics and Management, Tongren University, Tongren 554300, China
2 State Key Laboratory of Environmental Geochemistry of Geochemistry, Institute of Geochemistry, Chinese Academy of Sciences, Guiyang 550081, China
3 Puding Karst Ecosystem Observation and Research Station, Chinese Academy of Sciences, Puding 562100, China
* Correspondence: baixiaoyong@vip.skleg.cn

Citation: Wu, L.; Wang, S.; Bai, X.; Chen, F.; Li, C.; Ran, C.; Zhang, S. Identifying the Multi-Scale Influences of Climate Factors on Runoff Changes in a Typical Karst Watershed Using Wavelet Analysis. *Land* **2022**, *11*, 1284. https://doi.org/10.3390/land11081284

Academic Editor: Ilan Stavi

Received: 6 July 2022
Accepted: 8 August 2022
Published: 10 August 2022

Abstract: Identifying the impacts of climatic factors on runoff change has become a central topic in climate and hydrology research. This issue, however, has received minimal attention in karst watersheds worldwide. Multi-resolution analysis (MRA), continuous wavelet transform (CWT), cross wavelet transform (XWT) and wavelet transform coherence (WTC) are used to study the teleconnection in time and frequency between climate change and hydrological processes in a typical karst watershed at different time scales. The main results are: (1) All climatic factors exhibit a main cycle at 12-month time scales with runoff changes, but the main periodic bandwidth of rainfall on runoff changes is much wider than that of temperature and evaporation, indicating that rainfall is the main factor affecting runoff changes. (2) In other cycles, the impact of rainfall on runoff changes is the interlacing phenomena with positive and negative, but the impact of temperature and evaporation on runoff change is mainly negative. (3) The response of runoff to rainfall is in time in the high-energy region and the low-energy significant-correlation region and has shown a positive correlation with a smaller phase angle, but it is slightly lagged at 16-month time scales. Moreover, the runoff change lags behind temperature and evaporation for 1–2 months in those regions. (4) It has been found that there is a strong effect of rainfall over runoff, but a lesser effect of temperature and evaporation over runoff. The study sheds light on the main teleconnections between rainfall, evapotranspiration and surface runoff, which in turn might help to attain the better management of water resources in typical karst watersheds.

Keywords: karst; watershed; runoff change; climate factors; wavelet analysis

1. Introduction

The changes in hydrology and water resources caused by climate change have stimulated hydrologists to pay attention to and study the impacts in this field, which have become one of the popular issues at home and abroad [1–3]. Runoff change is mainly affected by climate change and underlying surface conditions [4–6]. The most important manifestations of climate change on runoff are the changes in rainfall amount and temporal and spatial distributions [7]. Climatic and hydrological processes have highly non-linear and unstable characteristics due to the complex exchange process of the Earth's atmosphere system [8–10] and the difference in watersheds' geographical and human environment characteristics; these characteristics cause great difficulties in the simulation and prediction of hydrological changes. Therefore, studying the multi-scale evolution law in the interaction process between meteorology and hydrology has scientific significance for the management and optimal regulation of water resources in river watersheds.

Most of the previous studies have focused on the correlation between meteorological factors and runoff, but limited analysis has been conducted on the multi-temporal characteristics of runoff and major meteorological elements [6,11–17]. In addition, the traditional hydrological research methods, such as linear correlation, the Pearson correlation coefficient, the linear trend method and the multiple regression method, can only reveal the variation characteristics on a single time scale. The evolution relationship and interaction characteristics of the two hydrometeorological variables on multi-time-scales are impossible to demonstrate. Spectral analysis and spatial coherency can reveal the scale-dependent relationships between the variables but are only applicable to stationary systems. Hydrometeorological series behavior belongs to non-stationary systems [18]. Wavelet analysis can deal with non-stationary data series and thereby provide an opportunity to analyze the temporal patterns of hydrometeorological series at multiple scales.

The climatic factor is the driving factor of runoff change; therefore, different time scale correlations must exist between runoff and the climatic factor in its oscillation frequency. Wavelet analysis, especially cross wavelet analysis, had been gradually applied to the analysis and study of runoff changes and meteorological factors in river watersheds on multi-time-scales in recent years [19–21] and had also been used to determine the overall and scale-dependent similarities of the temporal patterns of soil moisture [22]. Multiresolution analysis (MRA) can study signals represented at different resolutions [23]. This method can be used to decompose a signal into a progression of successive approximations and details in increasing order of resolution [24]. Continuous wavelet transform (CWT) is a common tool for analyzing localized intermittent oscillations in time series, and it is often desirable to examine two time series together that may be expected to be linked in some way. We can show the strength of sequence signals at different time scales by analyzing the wavelet power spectrum. cross wavelet transform (XWT) will expose their common power and relative phase in time–frequency space and can reflect that two sequences have the same energy spectrum region after wavelet transform, thus revealing the significance of the interaction between the two sequences in different time–frequency domains. Wavelet transform coherence (WTC) can find significant coherence although the common power is low [19]. In this field, CWT has been recently used to determine the effect of climatic phenomena on stream flow regimes [25–28], runoff processes [29–31], surface–groundwater interactions and the hydrogeological behavior of karst systems [32,33]. Furthermore, XWT, which has strong signal coupling and resolution ability, can show the common high-energy region and phase correlation of two time-series data. However, XWT has a great unsolved shortcoming that cannot find significant coherence when analyzing the low-energy regions of two time series' data in the time–frequency domain, and its functional defects in the low-energy areas must be compensated for by WTC [34]. In view of this, the coupling of MRA, CWT, XWT and WTC will be generally applied in the field of hydrometeorology.

At present, the implications of climate and anthropic pressures on the short- to long-term changes in the water resources of a Mediterranean karst were assessed by using wavelet analysis [23], and the non-stationary relationships of ocean and atmosphere mean conditions and freshwater discharge, which were integrated at the continental scale, were studied by using XWT [27]. In addition, the impacts of rainfall, air temperature and evapotranspiration on the annual runoff in the source region of the Yangtze River were investigated in the time domain by using wavelet analysis and multiple regression [17]. WTC was used to determine the overall and scale-dependent similarities of the temporal patterns of soil moisture in the karst catchments of Southwestern China [32]. CWT analysis was used to detect the trends and periodicity in sediment discharge, whilst WTC was used to detect the temporal covariance between sediment discharge and water discharge, rainfall, potential evapotranspiration and vegetation index in two typical karst watersheds in southwest China [33] and to assess the relative importance of catchment properties in modulating streamflow and modes of variability in West Africa and Central Africa [35].

Karst landforms are developed in highly heterogeneous carbonate rocks that are easily eroded by flowing water, widely distributed in Southwest China and generally have different hydrogeological characteristics from non-karst areas [36]. Thus, the karst watersheds are characterized by broken surfaces, low runoff coefficients, serious underground leakage, thin surface soil and poor regulation and water conservation capacities [37–40]. The unique two-dimensional/three-dimensional hydrogeological structure can accelerate the hydrological process [41,42]. In particular, rainfall drains rapidly to underground systems through numerous cracks and fissures [40,43–46]. The soil–epikarst system plays important roles in runoff generation due to the large storage capacity and high infiltration rate of karst carbonate fissures and fractures [2]; consequently, runoff changes in karst areas are sensitive to climatic factors, and small climatic fluctuations will cause large fluctuations in runoff. The appearance, storage and circulation of water in karst aquifers are apparently different from those of water in non-karst areas. The special hydrological process in karst areas will lead to the influence of climate change on runoff with time lag or advance at different time scales. In addition, the hydrometeorological evolution in karst areas has obvious seasonal and multi-scale characteristics. Significant differences exist in the evolution and influence relationships at different scales, especially the intrinsic relationship between monthly and seasonal rainfall, evaporation, temperature and runoff; their vibration energy distribution characteristics and correlations in time and frequency domains are extremely complex. Numerous studies have focused solely on non-karst watersheds. On the contrary, the impacts of climatic factors on runoff (surface runoff) changes have rarely been identified for karst watersheds. Specifically, the research on the time-varying characteristics of climate and runoff and their coupling relationship in karst trough valley watersheds is scarce. Therefore, the objectives of this study are to (1) analyze the multiscale temporal variability effects of runoff with climatic factors, (2) characterize the coupling relationship between runoff and climatic factors in common high- and low-energy regions and high-correlation regions at different time scales and (3) provide a theoretical basis and technological support for water resource safety management in karst watersheds.

2. Study Site

The Yinjiang River watershed (108°21′21″–108°47′27″ E, 27°53′17″–28°13′57″ N), which is located in northeast of Guizhou Province (Figure 1a), is a typical karst watershed of a trough valley, SW China. It covers an area of 691.56 km^2, with the karst area of 376.77 km^2 and non-carbonate rocks area of 314.79 km^2, accounting for 54.68% and 45.32% of the total watershed area, respectively. Elevation in the study area decreases from southeast to northwest, ranging in a large scope with an elevation range of 439–2466 m above sea level and a mean elevation of 1033 m above sea level (Figure 1b).

The southeast part of the watershed is dominated by non-karst areas, and the karst is widely distributed in the middle and northwest parts of the watershed. A small number of banded non-karst regions are concentrated in the western, central and northern parts of the watershed. Six types of lithology are present, namely, homogenous limestone, interbedded limestone and clastic rock, clastic rock of limestone interlayer, non-carbonatite, homogenous dolomite, mixture of homogeneous limestone and dolomite (Figure 1c). A karst valley with a geographical background of a syncline structure in the center of the valley with steep bedding slopes exists on both sides. The land surface is steep and broken with numerous underground cracks, causing a severe underground loss of rainfall and runoff. The middle part of the watershed is a typical deep-cut karst trough valley, and the middle part of the trough is a karst valley with a synclinal structure as its geological background. Both sides of the trough are steep beddings or inversion slopes, and the top of the trough is over 1000 m above sea level; thus, it has a good ecological three-dimensional climate characteristic.

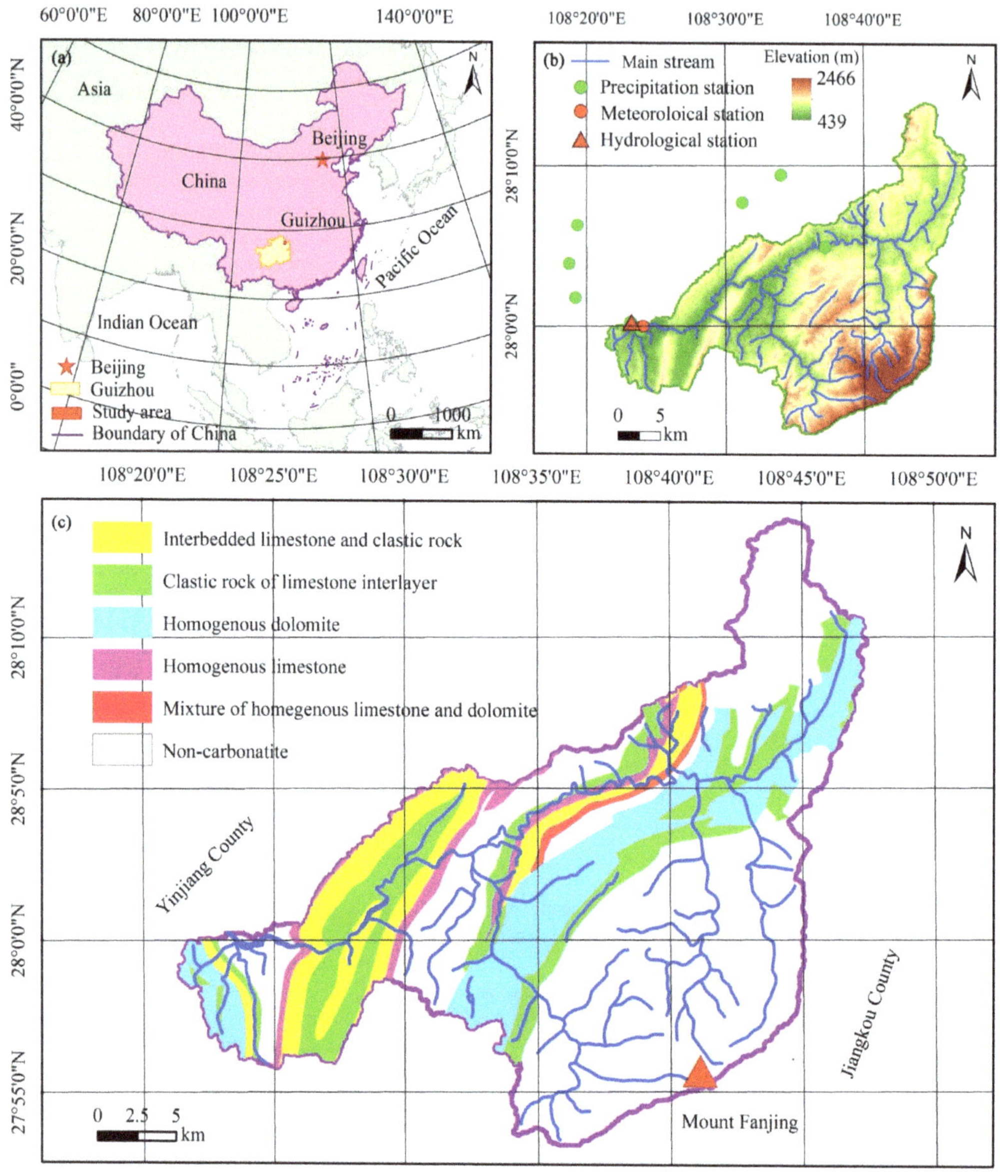

Figure 1. Location and overview of the study area. Study area location in China (**a**), topography (**b**) and lithology (**c**) in Yinjiang River watershed.

3. Materials

The monthly rainfall (P) data of 8 rainfall observation stations and the monthly runoff (Q) data (long data from January 1984 to December 2015) at the hydrological station were collected from the Guizhou Provincial Hydrology and Water Resources Bureau (http://www.gzswj.gov.cn/hydrology_gz_new/index.phtml) (accessed on 15 September 2016). The monthly evaporation (E) data (January 1984 to December 2015), which were also derived from the Guizhou Provincial Hydrology and Water Resources Bureau, were measured by the evaporation dish of the hydrological observation station and represented the water evaporation of the water surface or soil. The monthly temperature (T) data of the weather station corresponding to the runoff time series were obtained from the China Meteorological Data Sharing Service System (http://cdc.cma.gov.cn/) (accessed on 5 August 2021). The average annual rainfall data for the watershed were interpolated by the Kriging method with the annual rainfall data of all 8 rainfall stations (both inside and outside the watershed). The DEM data with a spatial resolution of 30 m were obtained from the International Scientific and Technical Data Mirror Site, Computer Network Information Center, Chinese Academy of Sciences, which could be downloaded from the Geospatial Data Cloud (http://www.gscloud.cn) (accessed on 12 September 2021). The lithology data were derived from the Karst Scientific Data Center (http://www.karstdata.cn/) (accessed on 12 September 2021), Institute of Geochemistry, Chinese Academy of Sciences.

4. Methodology

A series of wavelet analysis methods were used to identify the multi-scale impact of climate factors on runoff change in the Yinjiang River watershed. These wavelet analysis methods included MRA [47], CWT [19–21,48,49], XWT [19–21], cross wavelet phase angle (CWPA) [50] and WTC [19]. MRA was carried out using a free MATLAB software package provided by the WaveLab Development Team and available at http://statweb.stanford.edu/~wavelab/, (accessed on 12 September 2021). Other methods, including CWT, XWT, WTC, and CWPA, were carried out using a free MATLAB software package (Mathworks, Natick, MA, USA) kindly provided by Grinsted et al. [19] at http://noc.ac.uk/using-science/crosswavelet-wavelet-coherence, (accessed on 12 September 2021). The package includes code originally written by Torrence and Compo [20] of the University of Alaska, available at http://paos.colorado.edu/research/wavelets/, (accessed on 12 September 2021). The flowchart for identifying the multi-scale influences of climate factors on runoff changes is shown in Figure 2.

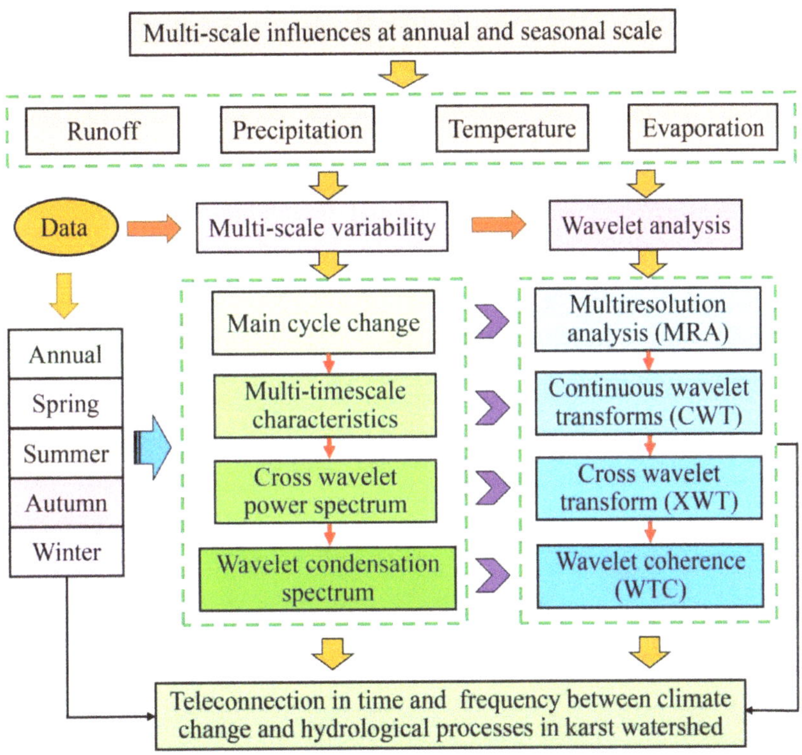

Figure 2. Flowchart for identifying the multi-scale influences of climate factors on runoff changes.

4.1. MRA

Choosing the particular values of $a_0 = 2$ and $s_0 = 1$ corresponds to the dyadic case used in MRA. The aim is to reduce/increase the resolution by a factor of 2 between two scales. Therefore, the approximation of a signal *x*(*t*) at a resolution *j*, denoted by A_x^j, and the detail of the same function at a resolution *j*, denoted by D_x^j, are defined by:

$$A_x^j(t) = \sum_{k=-\infty}^{+\infty} C_{j,k}\psi_{j,k}(t) \tag{1}$$

$$D_x^j(t) = \sum_{k=-\infty}^{+\infty} D_{j,k}\Phi_{j,k}(t) \tag{2}$$

where $\Phi_{j,k}(t)$ is a scaled and translated basis function called the scaling function [47], which is determined with $\psi_{j,k}(t)$ when a wavelet is selected. $C_{j,k}$ is the scaling coefficient given the discrete sampled values of *x*(*t*) at resolution *j* and location *k*. It is calculated from $\Phi_{j,k}(t)$ in a similar way for the wavelet coefficient $D_{j,k}$ from $\psi_{j,k}(t)$ for detailed mathematical expressions [49].

The signal *x*(*t*) can be reconstructed from the approximation and detail components as:

$$x(t) = A_x^j(t) + \sum_{j=1}^{J} D_x^j \tag{3}$$

where *J* is the highest resolution level considered. Since the MRA ensures variance is well captured in a limited number of resolution levels, the analysis of energy distribution in the sampling time series across scales gives a good idea of the energy distribution across frequencies.

4.2. CWT

The wavelet transform can be seen as a bandpass filter of uniform shape and varying location and width [20]. The continuous wavelet transform (CWT) $W_x(a,\tau)$ of a time series *x*(*t*) is given as follows:

$$W_x(a,\tau) = \int_{-\infty}^{+\infty} x(t)\psi^*(t;a,\tau)d_t \tag{4}$$

$$\psi(t;a,\tau) = \frac{1}{\sqrt{a}}\psi(\frac{t-\tau}{a}) \tag{5}$$

$$\psi(t) = \sqrt[4]{\frac{1}{\pi}}\cos(kt)e^{-\frac{t^2}{2}} \tag{6}$$

where $W_x(a,\tau)$ represents a group of wavelet functions, $W_{(a,b)}$, based on a mother wavelet ψ which can be scaled and translated, modifying the scale parameter a and the translation parameter τ, respectively. $\psi^*(t;a,\tau)$ corresponds to the complex conjugate of $\psi(t;a,\tau)$. $\psi(t)$ is the Morlet wavelet function, *k* is the non-dimensional frequency, here taken to be 6 to satisfy the admissibility condition, and *t* is time.

The statistical significance of wavelet power can be assessed relative to the null hypothesis that the signal is generated by a stationary process with a given background power spectrum (P_η). Many geophysical time series have distinctive red noise characteristics that can be modeled very well by a first-order autoregressive (AR1) process. The Fourier power spectrum of an AR1 process with lag-1 autocorrelation α [48] is given by

$$P_\eta = \frac{1-\alpha^2}{|1-\alpha e^{-2i\pi\eta}|^2} \tag{7}$$

The probability that the wavelet power of a process with a given power spectrum (P_η) [19] is greater than p is

$$D\left(\frac{|W_n^X(s)|^2}{\sigma_X^2} < p\right) = \frac{1}{2}P_\eta \chi_v^2(p) \tag{8}$$

where η is the Fourier frequency index. v is equal to 1 for real and 2 for complex wavelets.

4.3. XWT

The XWT of two time series X and Y is defined as

$$W^{XY} = W^X W^{Y*} \tag{9}$$

where * denotes complex conjugation. We further define the cross wavelet power as $|W^{XY}|$. The complex argument arg(W^{XY}) can be interpreted as the local relative phase between X and Y in time–frequency space. The theoretical distribution of the cross wavelet power of two time series with background power spectra P_k^X and P_k^Y is given as

$$D\left(\frac{W_n^X(s)W_n^Y(s)}{\sigma_X \sigma_Y} < p\right) = \frac{Z_v(p)}{v}\sqrt{P_k^X P_k^Y} \tag{10}$$

where $Z_v(p)$ is the confidence level associated with the probability p defined by the square root of the product of two X^2 distributions [19].

We use the circular mean of the phase over regions with higher than 5% statistical significance that are outside the cone of influence (COI) to quantify the phase relationship. This is a useful and general method for calculating the mean phase. The circular mean of a set of angles (a_i, $i = 1 \ldots n$) is defined as

$$a_m = \arg(X, Y) \ with \ X = \sum_{i=1}^{n} \cos(a_i) \text{ and } Y = \sum_{i=1}^{n} \sin(a_i) \tag{11}$$

It is difficult to calculate the confidence interval of the mean angle reliably since the phase angles are not independent. The number of angles used in the calculation can be set arbitrarily high simply by increasing the scale resolution. However, it is interesting to know the scatter of angles around the mean. For this, we define the circular standard deviation as

$$S = \sqrt{-2\ln(R/n)} \tag{12}$$

where $R = \sqrt{X^2 + Y^2}$. The circular standard deviation is analogous to the linear standard deviation in that it varies from zero to infinity.

4.4. WTC

Cross wavelet power reveals areas with high common power. Another useful measure is how coherent the cross wavelet transform is in time–frequency space. We define the wavelet coherence of two time series as

$$R_n^2(s) = \frac{\left|S\left(s^{-1}W_n^{XY}(s)\right)\right|^2}{S\left(s^{-1}\left|W_n^X(s)\right|^2\right) \cdot S\left(s^{-1}\left|W_n^Y(s)\right|^2\right)} \tag{13}$$

where S is a smoothing operator. Notice that this definition closely resembles that of a traditional correlation coefficient, and it is useful to think of the wavelet coherence as a localized correlation coefficient in time–frequency space. We write the smoothing operator S as

$$S(W) = S_{scale}(S_{time}(W_n(s))) \tag{14}$$

where S scale denotes smoothing along the wavelet scale axis and S time denotes smoothing in time. It is natural to design the smoothing operator so that it has a similar footprint as the wavelet used. For the Morlet wavelet, a suitable smoothing operator is given by Torrence and Compo [20]:

$$S_{time}(W)\big|_s = \left(W_n(s) \cdot c_1^{\frac{-t^2}{2s^2}} \right)\Bigg|_s \tag{15}$$

$$S_{time}(W)\big|_s = \left(W_n(s) \cdot c_2\, \Pi(0.6s) \right)\big|_n \tag{16}$$

where c_1 and c_2 are normalization constants and Π is the rectangle function. The factor of 0.6 is the empirically determined scale decorrelation length for the Morlet wavelet [20]. In practice, both convolutions are conducted discretely and therefore the normalization coefficients are determined numerically.

In this study, the cross wavelet energy, wavelet correlation agglomeration and phase spectra were calculated for monthly temperature, evaporation and rainfall and monthly runoff series to analyze the multi-temporal correlation amongst temperature, evaporation, rainfall and runoff. The hydrometeorological variables were used as input and output signals to characterize the responses of runoff changes to climatic factors in the Yinjiang River watershed. Climatic factors (rainfall, temperature and evaporation) were taken as input signals, and runoff was taken as an output signal. The correlation between climatic factors and runoff signals in the frequency and time domains at different energies was analyzed by using XWT and WTC. We had focused our analysis on the P–Q relationship to assess the impact of rainfall on runoff changes. Besides that, the E–Q and T–Q relationship were used to assess the impacts of evaporation and temperature on runoff changes, respectively.

5. Result Analysis

5.1. Annual, Seasonal and Monthly Evolution of Runoff and Climatic Factors on Main Scales Analyzed by MRA

5.1.1. Monthly Variation

The main aim of this section is to visualize the distribution of energy across scales (or resolution levels) of the hydrogeological time series. The MRA was performed on monthly data, and the results for the first 10 MRA levels are shown in Figure 3. Overall, the energy is distributed variably across levels in the hydrogeological time series and has significant periodic characteristics in different time scales, especially at a large scale. Runoff and rainfall show high energy oscillation at all levels, but they fluctuate with time, which indicates that the high-energy differences in monthly runoff and rainfall explain most of the differences. The energy vibrations at all time scales have a high consistency, and the vibration consistency is significant at a large scale, which demonstrates that runoff is significantly affected by rainfall. Evaporation shows high-energy vibrations at all time scales. Evaporation oscillation that gradually diminishes at 16-, 32- and 64-month time scales, however, is consistent with runoff and rainfall at 128- and 256-month time scales. Several obvious abrupt changes are detected from temperature in the vibration characteristics of 1 month to 4 months with the mutation years of 1990 (70-month) and 2000 (200-month) according to energy distribution. The oscillation characteristics of temperature are consistent with those of runoff and rainfall at time scales of 16, 64, 128 and 256 months. After 128-month time scales, the oscillation characteristics of runoff, rainfall, evaporation and temperature all appear to have the same vibration characteristics at 10–20-year time scales.

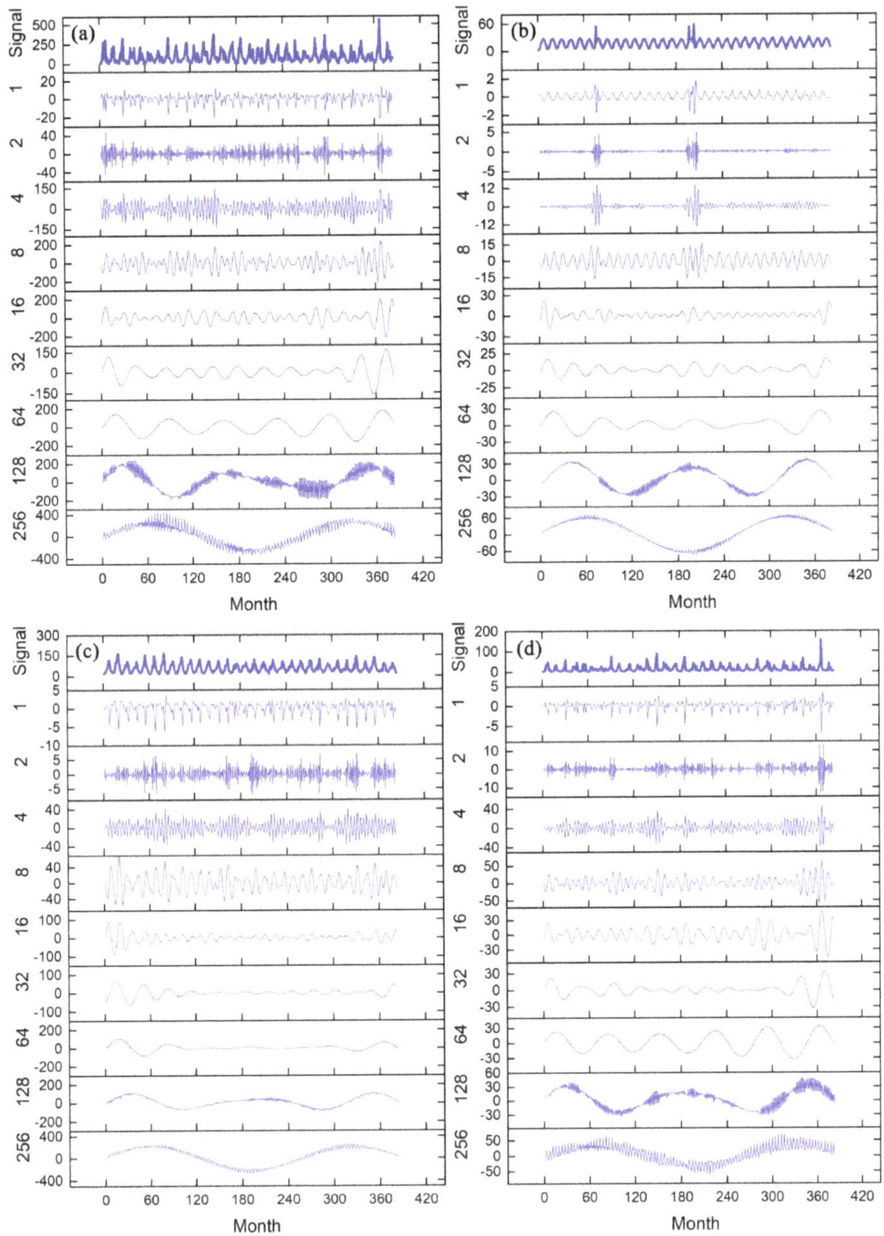

Figure 3. Multi-time-scale variations of monthly rainfall (**a**), temperature (**b**), evaporation (**c**) and runoff (**d**).

5.1.2. Seasonal Variation

As shown in Figures 4–7, there are some synchronization characteristics on different time scales for the evolution characteristics of surface runoff and climate factors on the annual scale and four seasonal scales after scale segmentation by MRA. Except for the small difference in runoff in autumn and winter, the evolution characteristics of runoff at other scales are basically synchronous, especially in summer. The amplitude of rainfall oscillation at each time scale is larger than that of runoff, but a minimal difference exists between them in spring, autumn and winter. Evaporation is asynchronous with temperature for the variation characteristics at 1-year scale in spring and winter and at less than 4-year scale in summer but relatively synchronous with insignificantly different amplitudes at other scales. Over 16-year time scales, the time series of each factor has shown the consistent evolution characteristics in four seasons and the amplitude of each factor increases with the increase in time scale.

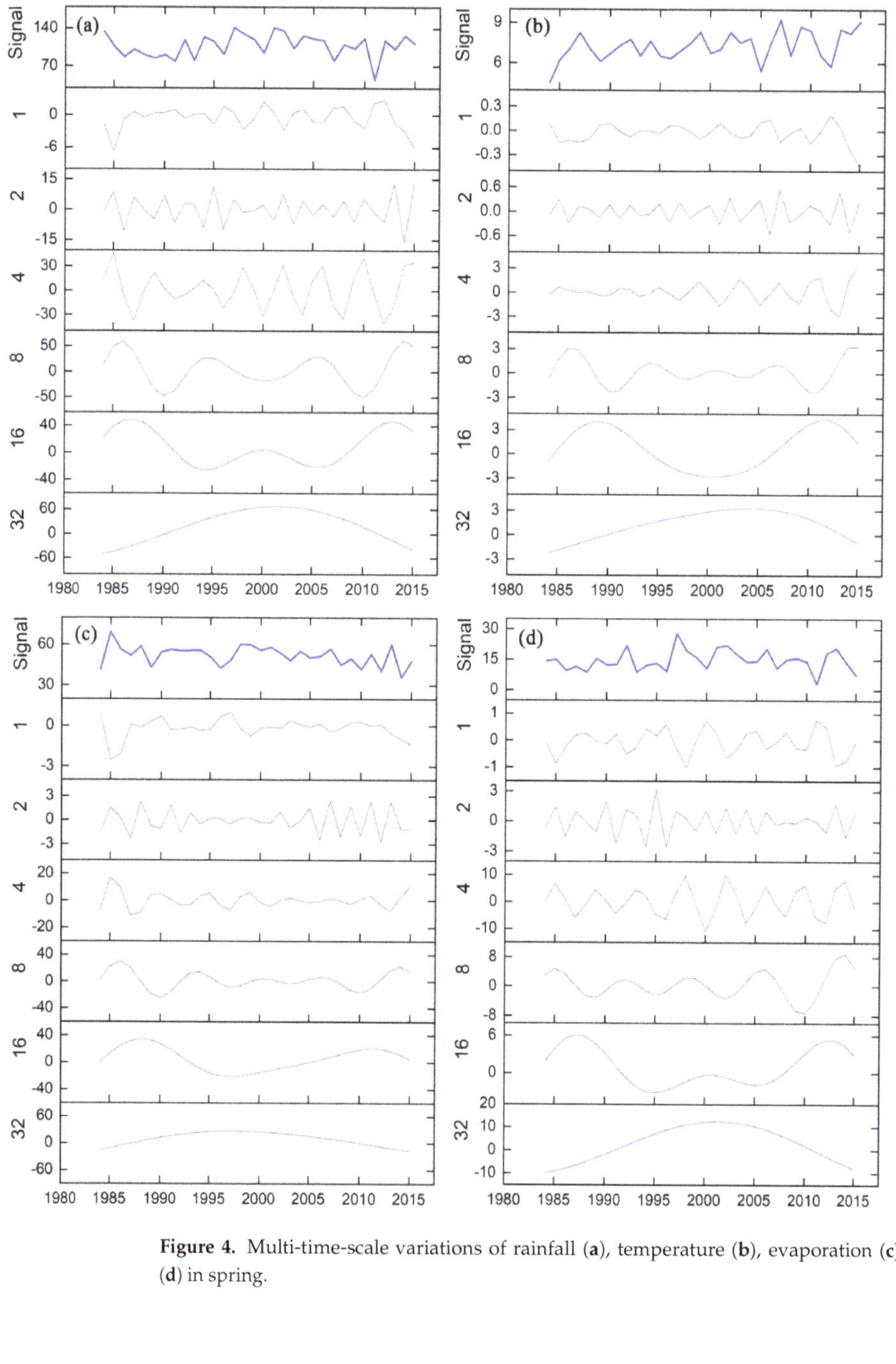

Figure 4. Multi-time-scale variations of rainfall (**a**), temperature (**b**), evaporation (**c**) and runoff (**d**) in spring.

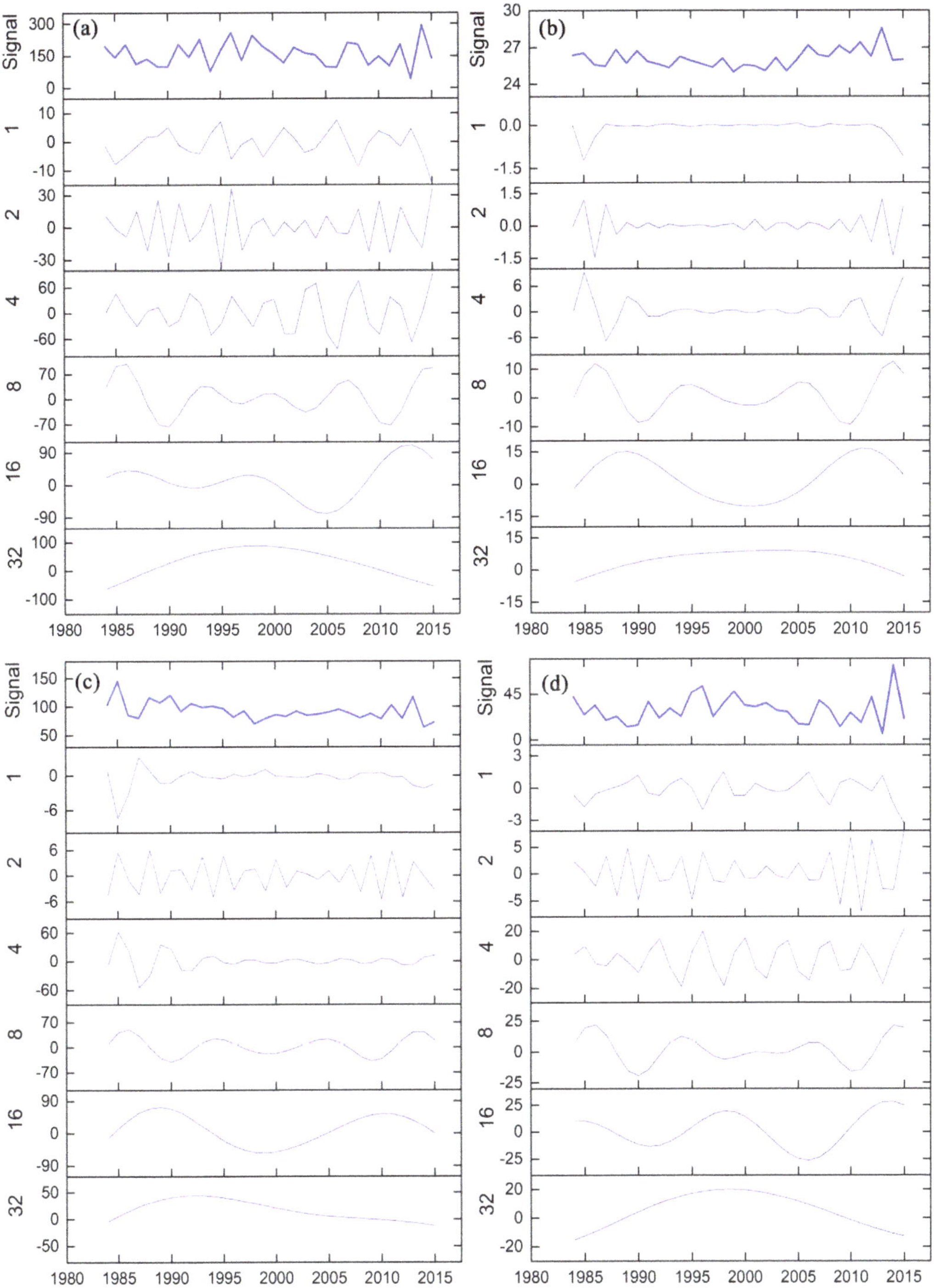

Figure 5. Multi-time-scale variations of rainfall (**a**), temperature (**b**), evaporation (**c**) and runoff (**d**) in summer.

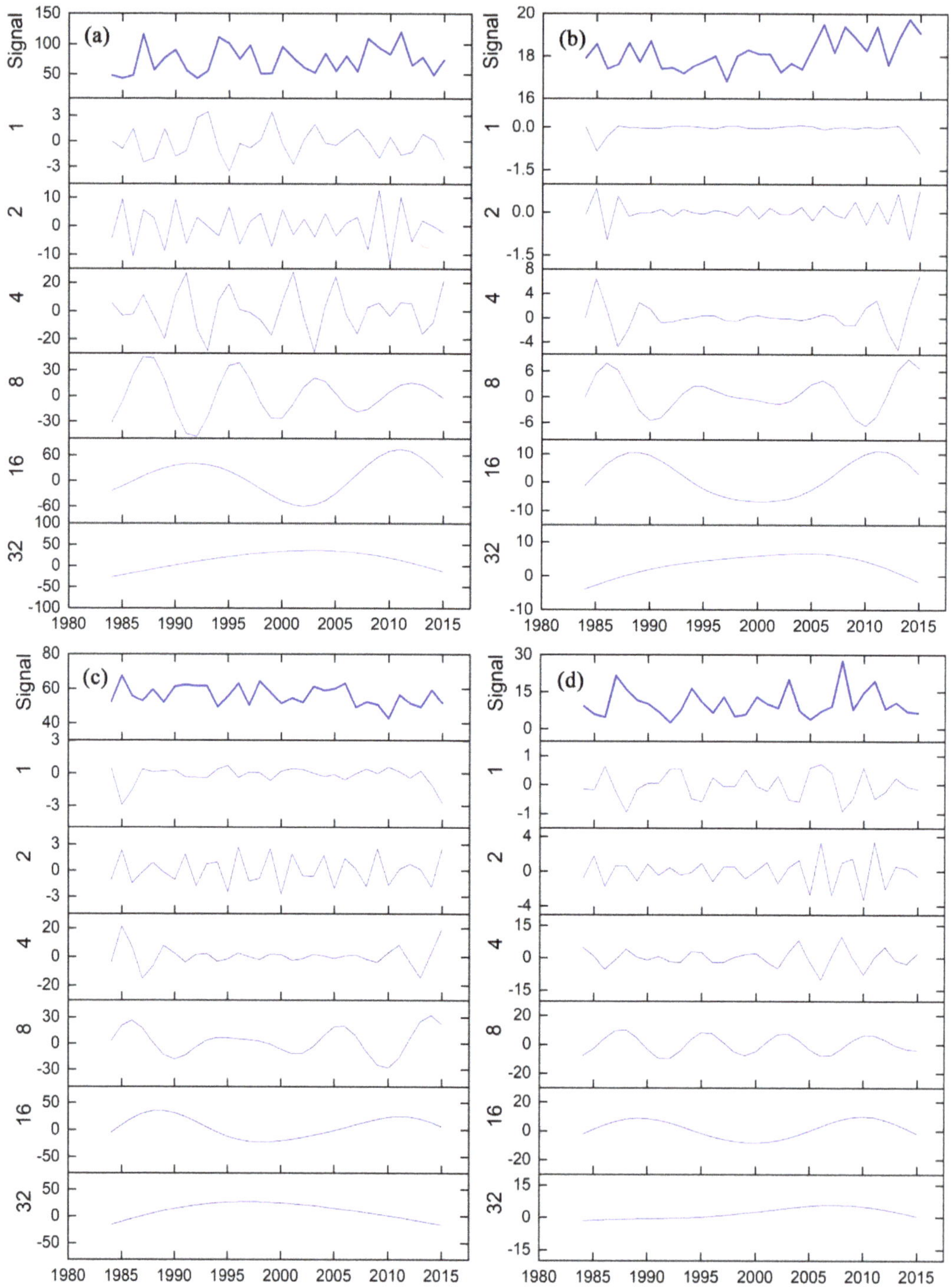

Figure 6. Multi-time-scale variations of rainfall (**a**), temperature (**b**), evaporation (**c**) and runoff (**d**) in autumn.

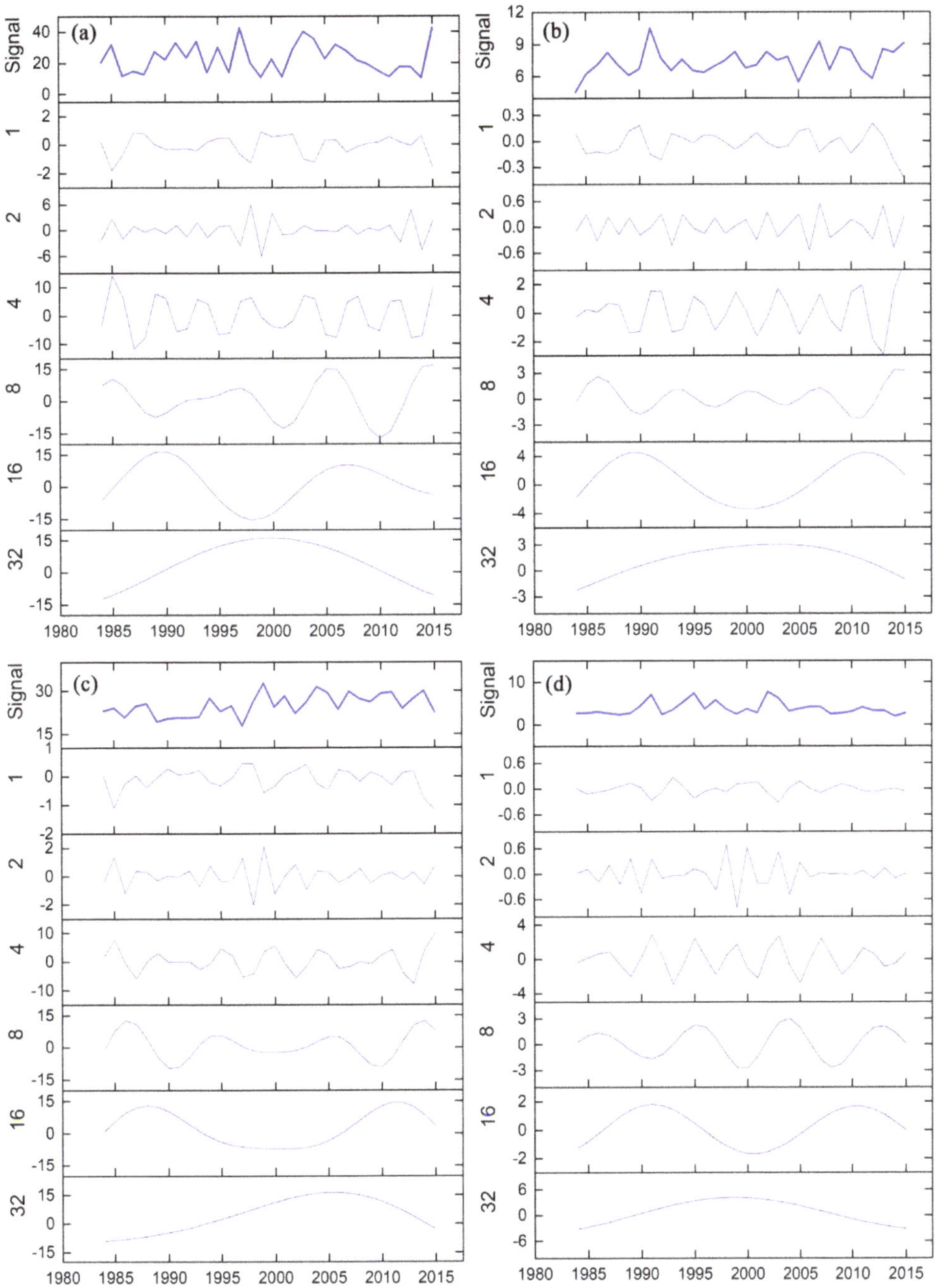

Figure 7. Multi-time-scale variations of rainfall (**a**), temperature (**b**), evaporation (**c**) and runoff (**d**) in winter.

5.1.3. Annual Variation

It has been found that rainfall and runoff show the same oscillation characteristics in different annual scales and the oscillation is obvious at 1–2- and 16–32-year time scales (Figure 8). Temperature and evaporation oscillate at 4–32-year sales. Runoff, rainfall, temperature and evaporation have the same oscillation characteristics at 4-, 8- and 32-year time scales. Runoff is mainly affected by rainfall, whereas evaporation is mainly affected by temperature. As a whole, it has been found that there is a strong effect of rainfall over runoff but a lesser effect of temperature and evaporation over runoff.

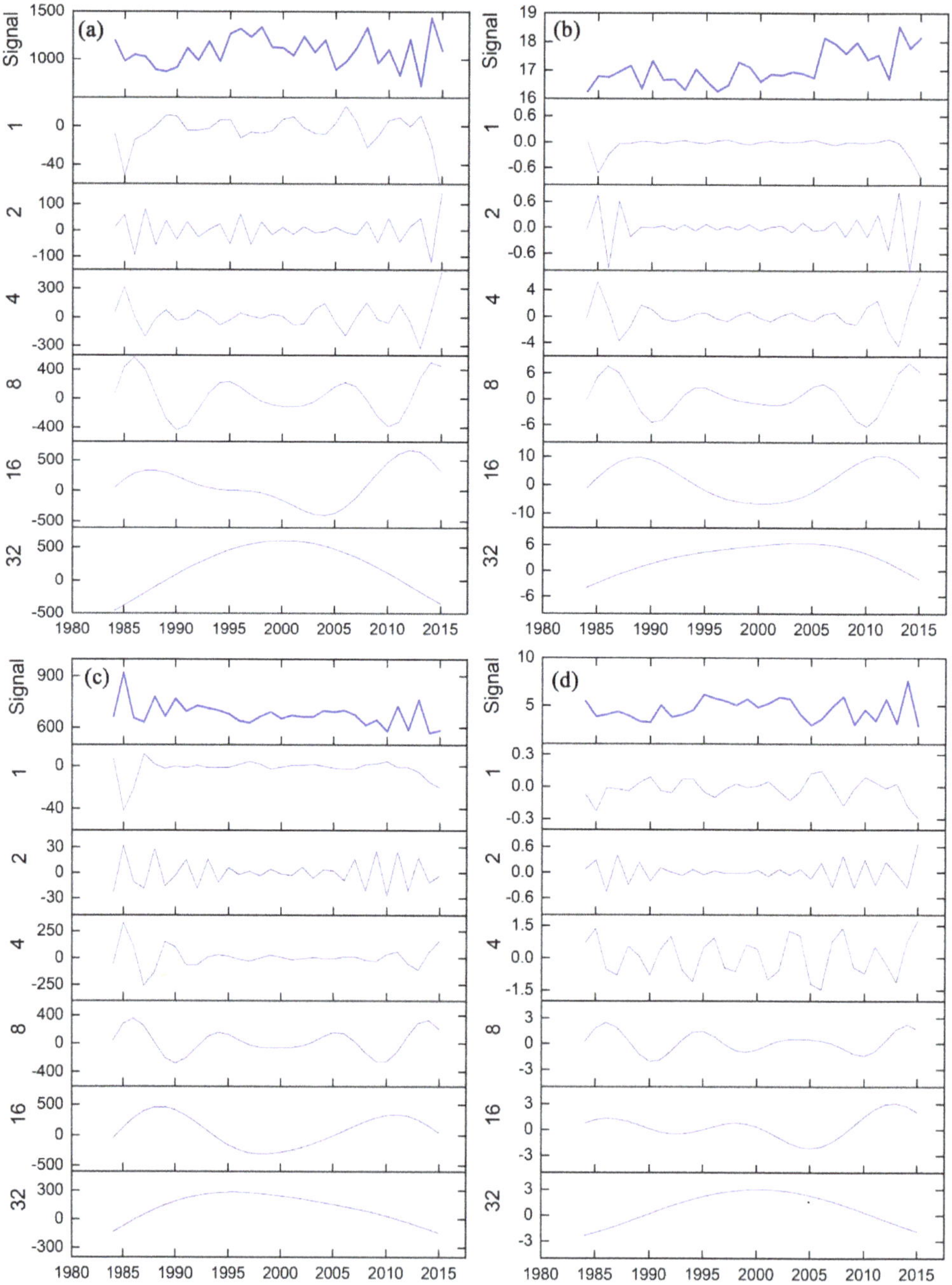

Figure 8. Multi-time-scale variations of annual rainfall (**a**), temperature (**b**), evaporation (**c**) and runoff (**d**).

5.2. Multi-Scale Evolution of Runoff and Climatic Factors Analyzed by CWT

As shown in Figure 9, the higher the spectral value (that is, the stronger the oscillation energy) is, the more significant the periodic oscillation passes the 0.05 confidence level. Except for the breakpoints at 8–16-month time scales in 1990 in Figure 9—Month Q, a main cycle always exists at 12-month time scales, which reflects the overall and significant periodic variation characteristics of monthly runoff. Several subcycles appear at approximately 36-month time scales (1990–2000) and 18–24-month time scales (1993–1997 and 2007–2012), which are related to the significant increase in rainfall in this period. The cycle at 4–6-month time scales (1984–2015) fluctuates in the time domain.

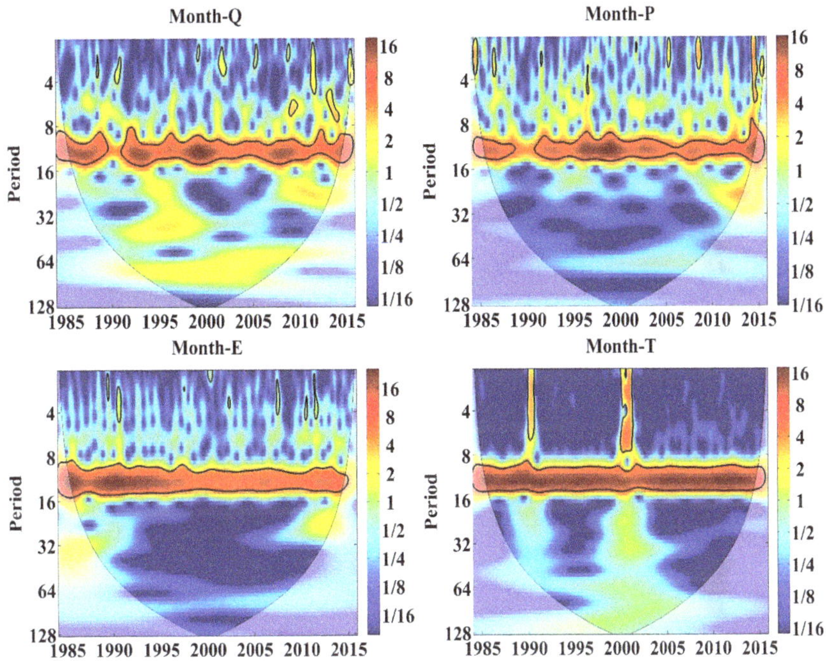

Figure 9. The continuous wavelet power spectra of monthly runoff (Month Q), rainfall (Month P), evaporation (Month E) and temperature (Month T) in the Yinjiang River watershed. The thick black contour designates the 5% significance level against red noise and the cone of influence (COI) where edge effects might distort the picture is shown as a lighter shade.

There are three subperiods with different significant levels of monthly rainfall in the time domain (Figure 9—Month P). The subperiod at 64-month time scales (1993–2008) indicates that the runoff has an important characteristic at 5-year time scales and a basically stable periodic variation. The subperiod at 24-month time scales (2007–2013) and at 18-month time scales (1993–1997) denotes that rainfall exerts a significant impact on runoff. Rainfall has similar fluctuation characteristics to runoff at 4–6 month time scales (1985–2015) and shows large fluctuations in the time domain.

There are three subcycles at 32-month time scales (1988–1992), 24-month time scales (2007–2013) and 6-month time scales (1984–2015) with different significant levels of monthly rainfall in the time domain (Figure 9—Month E). The fluctuation characteristics of evaporation are similar to those of runoff and rainfall at 6-month time scales. Obvious differences are also observed in the energy distribution characteristics in different periods, which indicates that rainfall and evaporation have obvious local characteristics of subperiodic variation consistent with those of runoff, but their energy is relatively weakened.

The monthly temperature is close to the monthly evaporation periodic bandwidth without interruption (Figure 9—Month T), which reflects the global and significant periodic variation characteristics of the monthly temperature and evaporation and indicates that the monthly temperature may affect the runoff mainly by changing the monthly evaporation. The monthly temperature has significant high-energy characteristics in the time domain around 1990 and 2000. The energy is strong at high frequencies below the scale of 1–8 months and weak at low frequencies after 8 months, but the periodic bandwidth increases with the scale.

It has also been found that a main period for runoff and climatic factors appears at 12-month time scales, which indicates that the periodic changes in hydrometeorology are mainly reflected in the annual scale. The discontinuous period and periodic bandwidth of climatic factors are basically consistent with the monthly runoff. The monthly runoff is consistent with the monthly rainfall, and the monthly temperature is consistent with the monthly evaporation. In addition, it has been found that the monthly runoff, rainfall and evaporation have the significant global fluctuation characteristics at high-frequency scales of below 8 months, whereas temperature exhibits only local fluctuation characteristics in 1990 and 2001 but significant impacts over 12-month time scales from 1995 to 2005. It can be concluded that rainfall is the main factor that affects runoff change in high-frequency regions and temperature and evaporation are the main factors in low-frequency regions.

As can be shown in Figures 10 and 11, runoff in spring has a main cycle at 4–6 years (1995–2000) and a subcycle at 1–2-year time scales (1990–1997). Runoff in summer is an insignificant period on the 4-year time scales (1990–2000). There are global insignificant characteristics for the main period of runoff in autumn at 1–2-year time scales (2006–2010) and the subperiod at 7-year time scales (1993–2005).

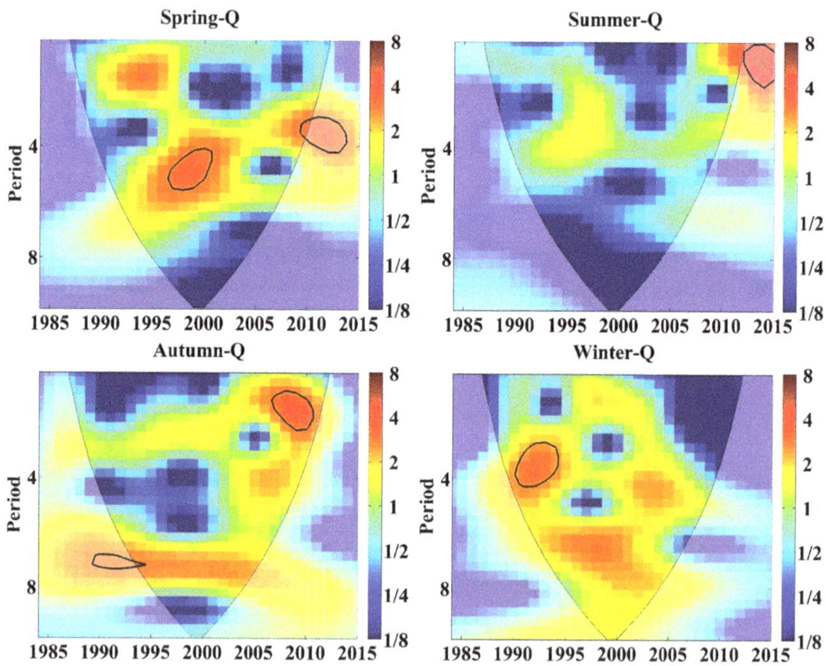

Figure 10. The continuous wavelet power spectra of seasonal runoff in the Yinjiang River watershed. The thick black contour designates the 5% significance level against red noise and the cone of influence (COI) where edge effects might distort the picture is shown as a lighter shade.

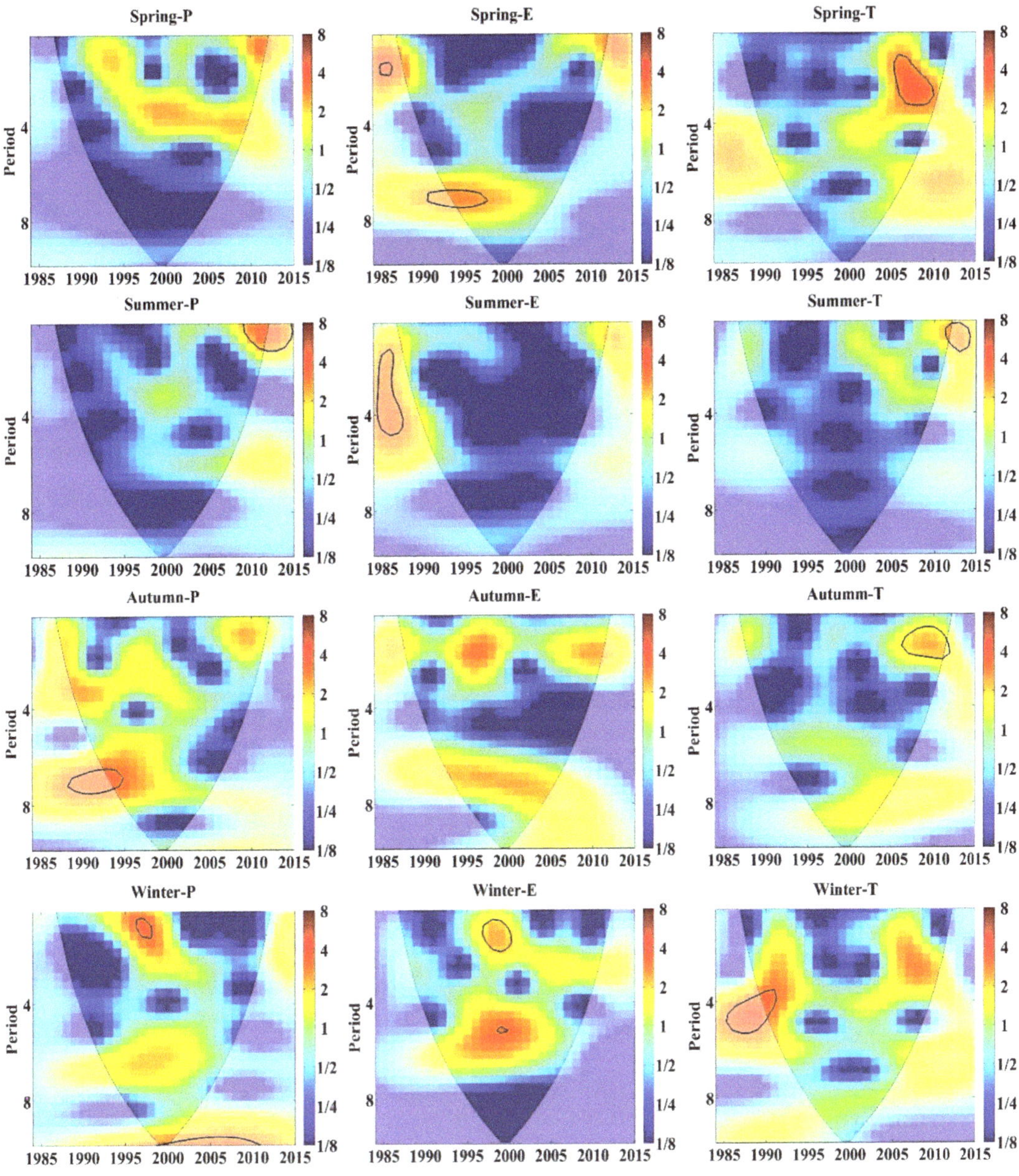

Figure 11. The continuous wavelet power spectra of seasonal rainfall, evaporation and temperature in Yinjiang River watershed. The thick black contour designates the 5% significance level against red noise and the cone of influence (COI) where edge effects might distort the picture is shown as a lighter shade.

The evolution characteristics of rainfall, temperature, evaporation and runoff vary greatly in spring, autumn and winter, but they are relatively stable in summer. They have great variations in high-frequency scales in spring and in low- and high-frequency scales in autumn, as well as periodicity in middle- and high-frequency scales in winter. Moreover, no obvious periodicity is detected in summer, whereas the significant periodic characteristics exist in other seasons.

From the annual scale (Figure 12), annual runoff has a main period at 6-year time scales (1995–2007), and the main period of rainfall is insignificant on this scale in the same period. However, the high-power spectrum of rainfall at this scale shows that it has an important impact on runoff change. No significant main period of evaporation is observed in the entire valid spectrum period, but significant periodic variations in temperature occur at 1–6-year time scales in 1997–2003. The power spectrum value of annual temperature at 6–8-year time scales remains high but insignificant. Over the 8-year time scales, the periodicity is significant, but the period bandwidth is narrowed.

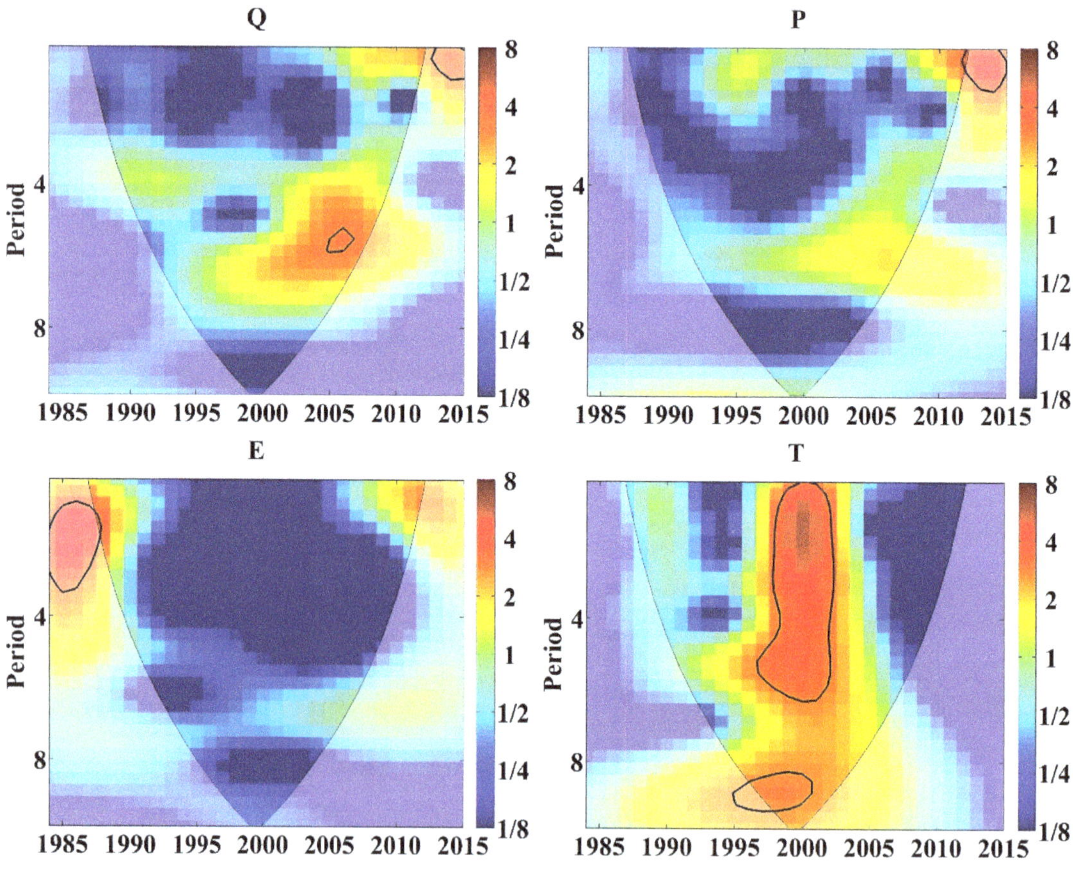

Figure 12. The continuous wavelet power spectra of annual rainfall, evaporation and temperature in Yinjiang River watershed. The thick black contour designates the 5% significance level against red noise and the cone of influence (COI) where edge effects might distort the picture is shown as a lighter shade.

5.3. Response Characteristics of Runoff Changes to Climatic Factors at Different Time Scales

5.3.1. Response of Runoff Changes to Climatic Factors on Monthly Scale

The cross wavelet power spectra of monthly rainfall and runoff, as shown in Figure 13, illustrate that the interaction between monthly rainfall and runoff is mainly concentrated in the main cycle at 12-month time scales from 1984 to 2015, which indicates a significant correlation between them at 1-year scale. The interaction between monthly rainfall and runoff is also shown in two subcycles at approximately 24-month time scales (1993–1996 and 2007–2013) and 72-month time scales (1993–2008). The energy difference in main cycles and subcycles in the time domain reflects that rainfall in different years has different regulating effects on runoff change.

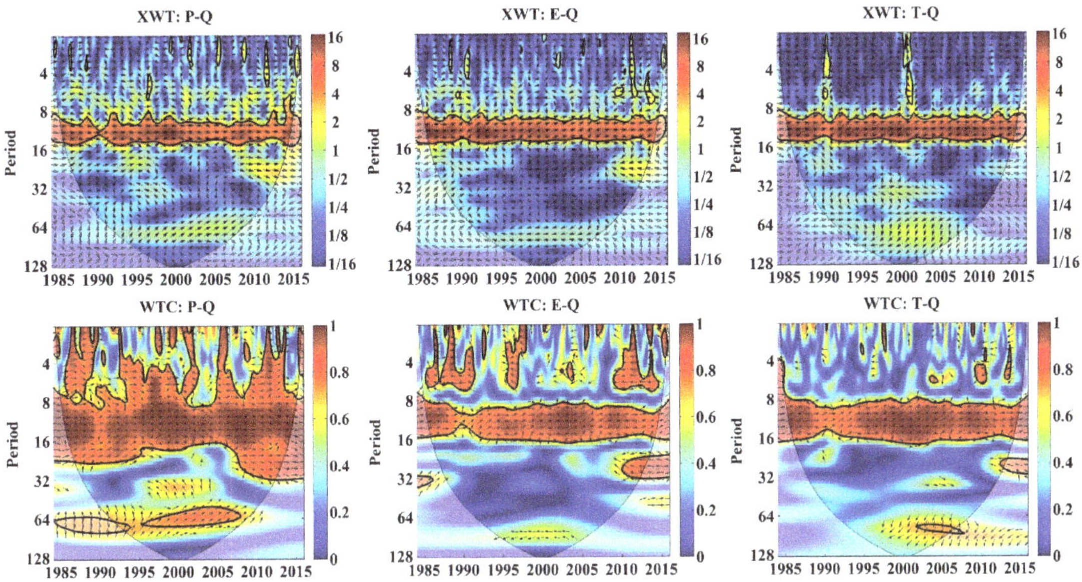

Figure 13. Cross wavelet spectra and wavelet coherence spectra of monthly rainfall, temperature and evaporation with runoff in the Yinjiang River watershed from 1984 to 2015, respectively. The thick black contour designates the 5% significance level against red noise and the cone of influence (COI) where edge effects might distort the picture is shown as a lighter shade. Arrows denote relative phase difference: The arrows from left to right indicate that the influencing factors and runoff are in the same phase, which implies a positive correlation; the arrows from right to left indicate an inverse phase, which implies a negative correlation; the downward arrows indicate that the influence factor is 90° ahead of the runoff change and the upward arrows indicate that the influence factor is 90° lagging the runoff.

The cross wavelet coherence spectra can compensate for the lack of correlation analysis of the cross wavelet power spectra in the low-energy region. The cross wavelet condensation spectra has a larger time–frequency domain space compared with the cross wavelet power spectra. In addition to the positive correlation at 12-month time scales between 1984 and 2015, a positive correlation is also observed in subperiods between 1991 and 2008 (72-month time scales) and 1984 and 2015 (4–8 months). By contrast, a negative secondary cycle occurs in 1996–2004 (36-month time scales). The contribution of rainfall to runoff in the Yinjiang River watershed can also be determined from the phase relationship between the cross wavelet power spectra and the cross wavelet condensation spectra. The main periods of 4–8-, 12- and 72-month time scales show a significantly positive correlation. This result is mainly because soil moisture is easy to be saturated due to the increase in rainfall on this scale, thus accelerating the formation of slope runoff within the river watershed and forming an effective replenishment for the river. On the contrary, the phase is negatively correlated in the subcycle at 36-month time scales, which indicates that the influence of rainfall fluctuation on runoff changes is negative. On the one hand, runoff will increase with the increase in rainfall on the river recharge. On the other hand, the increasing rainfall will increase vegetation coverage, enhance water conservation, increase evaporation and eventually reduce river recharge. The high-correlation region of the cross wavelet condensation spectra is basically consistent with the high-energy region of the cross wavelet power spectra. A strong correlation exists between the monthly rainfall and runoff at 64-month time scales from 1995 to 2008, and the phase angle is 30°, which indicates that the rainfall lags behind the runoff for 1 month.

The monthly evaporation, temperature and runoff have main periods of 12-month time scales in high-energy areas. In low-energy regions, not only a high correlation at 12-month time scales but also a local significant correlation between the effects of temperature and evaporation on runoff changes at 1–8-month time scales are determined. Thus, temperature is mainly regulated indirectly by controlling evaporation. The influence of evaporation on runoff will superimpose the influence of temperature due to the indirect regulation of temperature on runoff because their influence on runoff is generally consistent in different scales but slightly different in different years.

The effects of monthly rainfall, temperature and evaporation on runoff are positively correlated in the main cycle at 12-month time scales, which indicates that their effects on runoff are positive and mainly at the annual scale. The main periodic bandwidth of rainfall on runoff changes is wider than that of temperature and evaporation, which indicates that rainfall is the main factor that affects runoff variation. In other cycles, the phases of rainfall's impacts on runoff changes are the interlacing phenomena of positive and negative, whereas the phases of temperature and evaporation that affect runoff changes are mainly negative. Accordingly, the impact of rainfall fluctuation on runoff changes on this scale is both positive and negative, whereas that of evaporation is always negative. However, temperature and evaporation have negative effects on runoff in each subcycle, which may be because evaporation increases with the increase in temperature, thus reducing runoff recharge.

The response of runoff to rainfall is timely in the high-energy region and the low-energy significant-correlation region. There has been shown a positive correlation with a smaller phase angle, but there is also a slight lag at 16-month time scales. The phase angles of evaporation, temperature and runoff range from 30° to 45°, which demonstrates that runoff changes have lagged behind temperature and evaporation for 1–2 months. The similarities of the effects of monthly temperature and evaporation on runoff changes have also proved that temperature indirectly affects runoff changes by changing evaporation.

5.3.2. Response of Runoff Changes to Climatic Factors on Seasonal Scale

The effect of rainfall on runoff changes in the four seasons has a significant high-energy region from the XWT power spectra (Figure 14). There are higher power spectrum values, stronger influences and wider time domains in spring and autumn. In spring, the related regions are mainly in the main cycle at 4-year time scales from 1998 to 2012 and the subcycle at 1-year scale from 1991 to 1995. In autumn, the relevant regions are mainly in the main cycle at 7-year time scales from 1992 to 2000 and the subcycle at 2-year time scales from 1995 to 2000 and then are invisible.

The effect of evaporation on runoff changes presents some significant high-energy zones in four seasons. Energy is strongest in autumn and weakest in summer. The evaporation effect is significant at 6–8-year time scales from 1993 to 2005. In spring, summer and autumn, the locations of significant high-energy zones on the same scale are basically the same. The effect of evaporation on runoff in spring and summer differs on different time scales. In autumn and winter, the energy distribution of the effect of evaporation on runoff changes is similar to that of rainfall on runoff changes, but the phase relation is opposite. Accordingly, the regulation of rainfall to runoff changes is positive, whereas that of evaporation is negative. This phenomenon may be caused by drought, less rain and strong evaporation, which can directly reduce runoff.

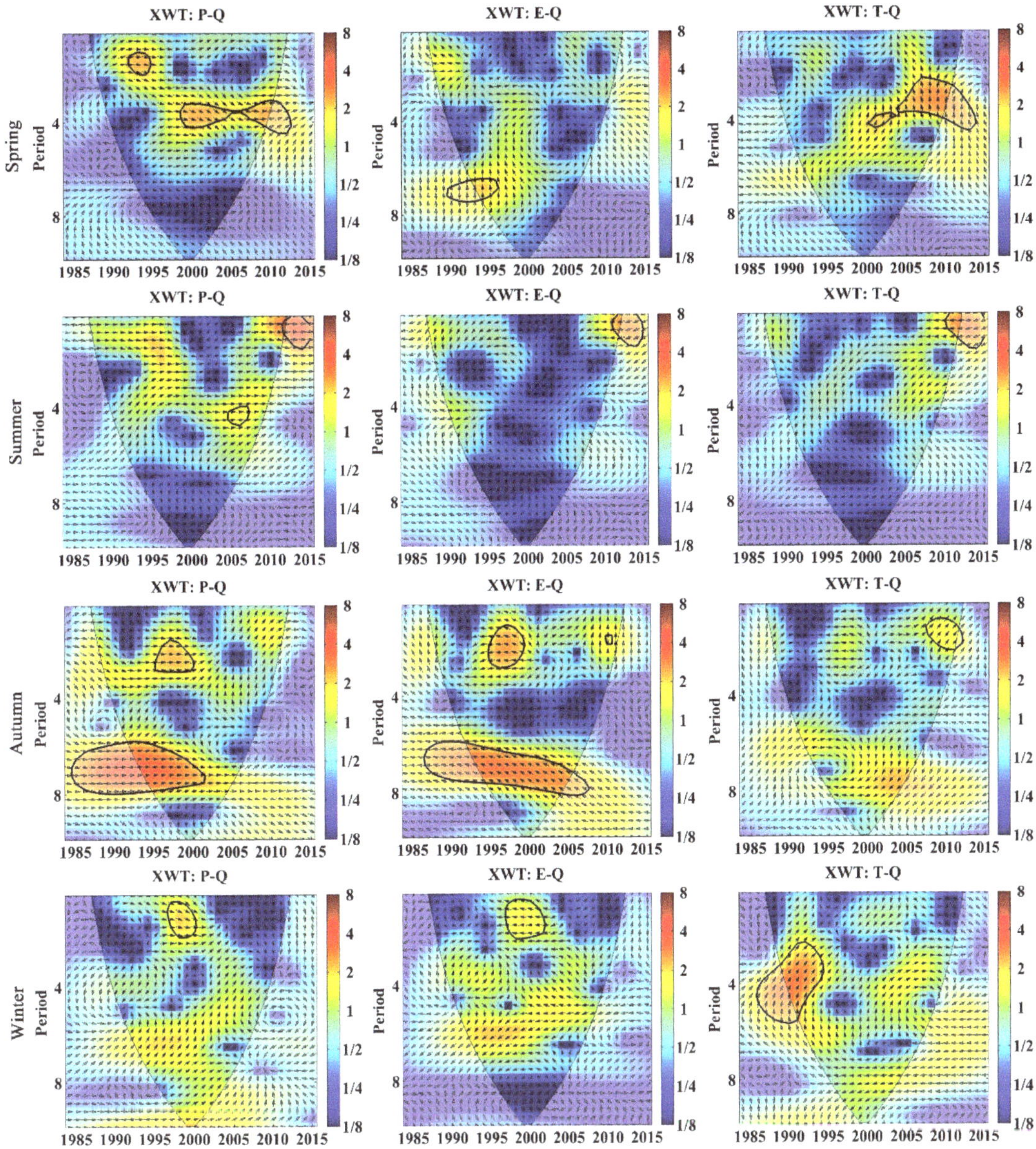

Figure 14. XWT between runoff and rainfall, evaporation and temperature on monthly scale. The thick black contour designates the 5% significance level against red noise and the cone of influence (COI) where edge effects might distort the picture is shown as a lighter shade. Arrows denote relative phase difference: The arrows from left to right indicate that the influencing factors and runoff are in the same phase, which implies a positive correlation; the arrows from right to left indicate an inverse phase, which implies a negative correlation; the downward arrows indicate that the influence factor is 90° ahead of the runoff change and the upward arrows indicate that the influence factor is 90° lagging the runoff.

The effect of temperature on runoff also has significant high-energy regions in the four seasons. The power spectrum value is high, and the influence is strong in spring and winter. The influence of temperature in spring mainly concentrates on the main cycle at 4-year time scales from 2000 to 2014, and it is consistent with the influences of rainfall and evaporation on runoff at more than 6-year time scales. The effect of temperature on runoff changes mainly concentrates on the main cycles at 1-year time scale in summer and at 2-year time scales in autumn from 2007 to 2010. However, the effect of temperature on runoff in autumn has a subcycle at approximately 8-year time scales with high energy from 1995 to 2005, and it is similar to that of rainfall and evaporation at the same scale in the same time domain. The changes in temperature and evaporation in summer are ahead of runoff change, but the lead time of temperature ahead of runoff is greater than that of evaporation in summer. In addition, the effect of temperature on runoff changes in winter has a significant high-energy region at 4-year scale from 1990 to 1995, and that of rainfall and evaporation on runoff is stronger in the same time–frequency domain. Seasonally, the influence of temperature on runoff is similar to that of rainfall on runoff in the energy distribution in spring, which indicates that the increase in temperature results in increased rainfall and thus increases rainfall supply to runoff. In autumn, the influences of temperature, evaporation and rainfall on runoff have a consistent feature in energy distribution, which also shows that temperature has an important impact on evaporation and rainfall and leads to the same effect on runoff.

From the WTC condensation spectra of rainfall, temperature and evaporation with runoff in the four seasons (Figure 15), the highly correlated area of rainfall impacts runoff changes with an increase in years and scales, and it changes from 1-year scale in 1990–2000 to 4-year scale in 1995–2005. The main period of the impact of rainfall on runoff in summer is concentrated on the high-frequency scales, and the bandwidth tends to widen, which indicates that the period tends to be stable. The effect of rainfall on runoff in autumn is mainly manifested in the main period at approximately 8-year time scales from 1992 to 2005, with a wide bandwidth and an extremely stable period. The significant-correlation area of rainfall on runoff in winter is concentrated at 1-year and 7-year time scales from 1993 to 2003, and the influence is relatively weak. The significant-correlation region of evaporation in each season is consistent with the high-energy region of the XWT power spectrum, but it is more significant in autumn over 4-year time scales. Although the bandwidth in autumn is narrowed from 1990 to 2006, it still has a wide periodic bandwidth, which is similar to the impact of rainfall on runoff in the same season, indicating that the runoff change is mainly affected by rainfall and evaporation in autumn. According to phase characteristics, the phase relation between rainfall and runoff is positive, which implies that the influence is always positive, whereas the potential correlation between evaporation and runoff is negative; hence, the influence is negative, and their impact on runoff has a common main period at 8-year time scales. The influence of temperature in spring, summer and winter on runoff is relatively significant, and it is concentrated at the 4-year scale from 2000 to 2010 in spring, at the 1–4-year time scales from 2004 to 2010 in summer and at the 4-year time scales from 1990 to 2000 in winter. This finding indicates that the influence of temperature on runoff has significant differences in scale and time domain in different seasons, that is, it has significant local characteristics rather than global characteristics. As a whole, the main periodic bandwidth of the significant-correlation region in each season tends to widen, and the time-domain range of the correlation region also increases. The frequency-domain structure of the significant-correlation region of the WTC is basically consistent with that of the XWT high-energy region.

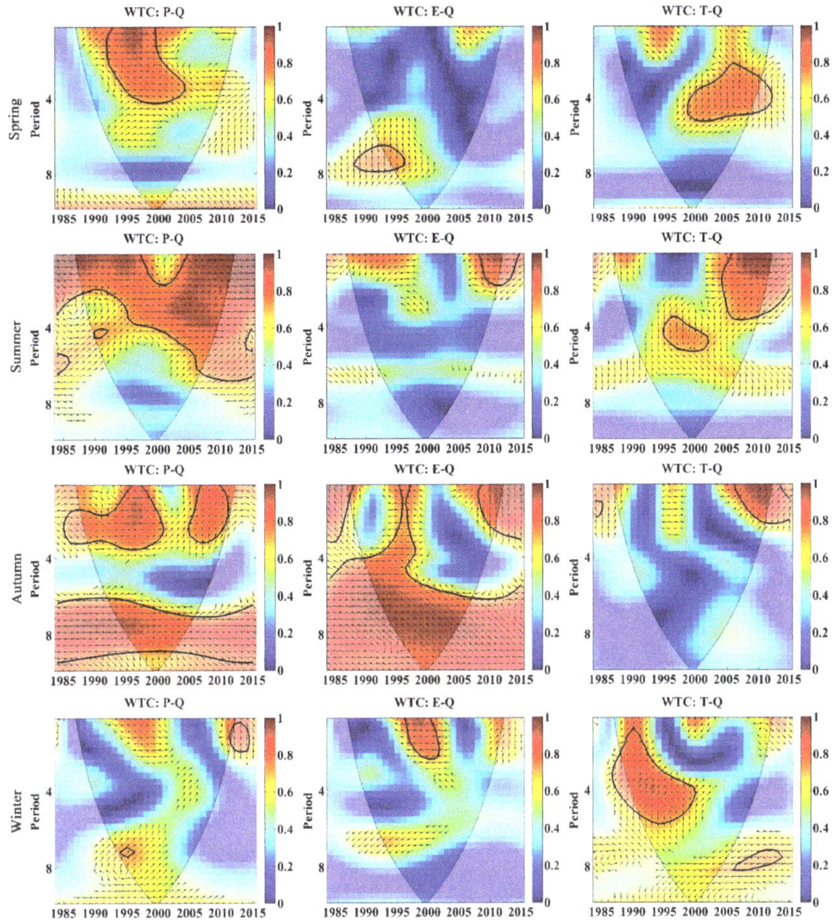

Figure 15. WTC between runoff and rainfall, evaporation and temperature on monthly scale. The thick black contour designates the 5% significance level against red noise and the cone of influence (COI) where edge effects might distort the picture is shown as a lighter shade. Arrows denote relative phase difference: The arrows from left to right indicate that the influencing factors and runoff are in the same phase, which implies a positive correlation; the arrows from right to left indicate an inverse phase, which implies a negative correlation; the downward arrows indicate that the influence factor is 90° ahead of the runoff change and the upward arrows indicate that the influence factor is 90° lagging the runoff.

In summary, runoff changes are mainly affected by rainfall and temperature in spring, mainly by direct rainfall recharge. That temperature increases rainfall and evaporation is the reason why its phase relation presents a positive and negative interlacing phenomenon in spring. In summer, runoff is mainly affected by direct rainfall recharge, the effect of evaporation on runoff changes is negative, and the positive effect is mainly reflected at more than 5-year time scales. In autumn, runoff change is affected by a small amount of rainfall supply and runoff loss is caused by evaporation. In winter, runoff is mainly affected by temperature because the rainfall in karst areas cannot form the effective recharge for runoff due to the drought and minimal rain; however, the temperature can indirectly adjust runoff changes by changing evaporation.

5.3.3. Response of Runoff Changes to Climatic Factors on Annual Scale

For XWT P–Q (Figure 16), a strong-influence period occurs in 2005, which indicates that the intensity of interaction between rainfall and runoff changes after the sudden variation in runoff in 2003. This phenomenon also implies that climatic factor is the main driving factor for the recent runoff increase. The high-intensity effects of P–Q present a significantly positive correlation and occur mainly at approximately 6-year time scales in the period from 2000 to 2010. The effect of P–Q passes the test of the red noise standard spectrum at the 0.05 significance level with a phase angle of 60°, which indicates that runoff is ahead of rainfall by 2 years on the 6-year time scales. The highly significant correlation after 2010 occurs on the 1–2-year time scales with consistent characteristics.

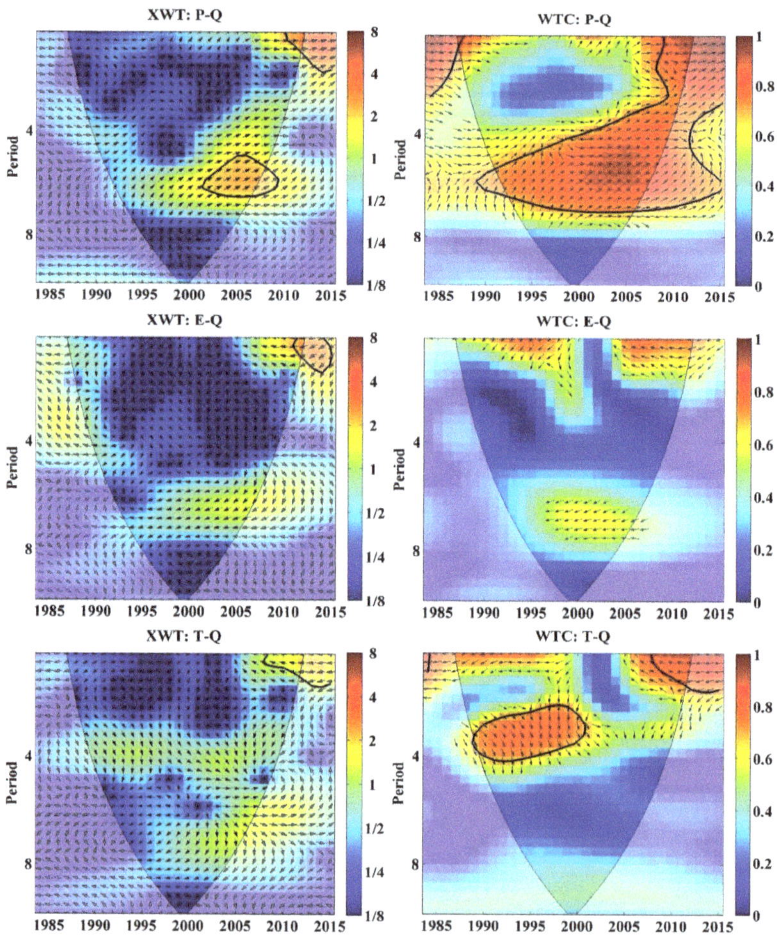

Figure 16. The XWT and WTC for annual rainfall (P) and runoff (Q), evaporation (E) and runoff (Q) and temperature (T) and runoff (Q) in the Yinjiang River watershed from 1984 to 2015. The thick black contour designates the 5% significance level against red noise and the cone of influence (COI) where edge effects might distort the picture is shown as a lighter shade. Arrows denote relative phase difference: The arrows from left to right indicate that the influencing factors and runoff are in the same phase, which implies a positive correlation; the arrows from right to left indicate an inverse phase, which implies a negative correlation; the downward arrows indicate that the influence factor is 90° ahead of the runoff change and the upward arrows indicate that the influence factor is 90° lagging the runoff.

For WTC P–Q, a significant high-correlation region on 4–7-year time scales exists during the entire period, which indicates that runoff is strongly affected by rainfall. From the phase diagram of P–Q, runoff has shown 2–3-year time scales ahead of rainfall in 1990–2000. Thus, the change in runoff is mainly affected by human activities. After the 2000s, the phase angle decreases gradually, indicating that runoff is gradually aggravated by rainfall. According to the results of the XWT and WTC of P–Q, the high-energy region and high-correlation region of P–Q are mainly concentrated around the middle of 2010, and the main cycle is mainly at 6-year time scales.

For XWT E–Q, since 2010, it has a highly significant correlation on the 1–2-year time scales, which demonstrates that evaporation has a significant impact on runoff at this time scale. For WTC E–Q, the E–Q cycle mainly concentrates at 1–2-year time scales during the period from 1990 to 2000 and from 2005 to 2015. The E–Q cycle mainly concentrates on 6–8-year time scales from 1995 to 2005, which indicates that the effect of evaporation on runoff is small. The E–Q phase correlation also shows that the E–Q phase correlation is an inverse phase with a phase angle of 45° from 1990 to the end of 2010 at 1–2-year time scales, which implies that evaporation is 1–2 years ahead of runoff. After 2005, evaporation remains in an inverse phase with runoff, with an initial phase angle of approximately 30 (at 1-year time scale), and then decreases and finally increases. The relationship between evaporation and runoff changes from lag to consistency to advance, which indicates that evaporation pays an important role on runoff changes. The results of XWT and WTC show that the phase angle on 7-year time scales is approximately 45° from 1995 to 2005, indicating that E is approximately 1.5 years ahead of Q.

For XWT T–Q, there is a high-energy region existing at 1–2-year time scales after 2005, which has passed the test of the standard spectrum of red noise at the 0.05 significant level. Therefore, the influence of temperature on runoff suddenly strengthens around 2005. However, the influence is relatively weak on the 4-year time scale and 6–8-year time scales, and it has not passed the test of red noise standard spectrum at the 0.05 significance level. The WTC results show two low-energy areas with the greatest impact. One is at the time of 3–4-year time scales in the period from 1990 to 2002, in phase with the phase angle between 60° and 70°, showing that temperature is ahead of runoff for more than 2 years. The other is at 1–2 year time scales after 2005, in which T–Q shows a negative correlation in the opposite phase with the initial phase angle of 45° and then gradually reduces to 0°. The above results show that the effect of temperature on runoff gradually changes from lag to consistency, indicating that the effect of temperature on runoff changes is increasingly obvious.

Overall, the interaction of rainfall with runoff changes at 6-year time scales across the entire period. However, the effects of temperature and evaporation on runoff changes are locally significant. The effect of evaporation on runoff changes is similar to that of temperature and has obvious local characteristics, mainly on small cycles.

6. Discussion

6.1. Multi-Scale Effects of Rainfall on Runoff Changes

Although the interaction between rainfall and runoff is positively correlated on the whole, the temporal effects are inconsistent in different time domains and scales. The effect of rainfall on runoff is ahead, lagging and consistent in time, ahead in high-frequency and low-frequency scales, lagging at medium-frequency scales (approximately 4 years) and consistent at 1-year scale and significant main periodic scales. For the leading effect, runoff may be mainly affected by early rainfall, which mainly occurs in rainy weather in spring and autumn. The surface and underground areas of karst are filled with soluble rocks with the main type of carbonate, which is vulnerable to erodible water that contains CO_2, thereby forming a large number of karst pipelines and fissures [51–53] over a long period of time and two sets of surface and underground hydrological systems [46]. Rainfall requires first to saturate soil water due to low soil moisture in karst areas [54–56]. Surface runoff is difficult to form with small rainfall due to the fragmented surface, steep and rugged slopes, low runoff coefficients in slope surfaces and small river network density [57–61]. Rainfall is the main factor for runoff formation, and its intensity, duration and areas have great influences on runoff change. When raining heavier, rainwater may hardly infiltrate and leak and thus then increase runoff. If the rainfall intensity is smaller, most of the rainwater infiltrates into the soil and leaks through enormous karst fissure pipelines, which can reduce the runoff. The longer the duration of rainfall and the larger the area of rainfall, the easier the soil moisture will be saturated, and the runoff generated on the slope will inevitably be larger. Runoff monitoring studies on karst slopes show that light rainfall intensities (15~30 mm/h) generate subsurface lateral flow and underground fissure flow, whereas great rainfall intensities generate surface runoff in addition to subsurface and underground flows [62]. However, only a single rainfall of more than 60 mm on a karst slope can produce stable runoff because once the atmosphere rains [63], it immediately runs off into the ground through a broken surface and underground fissure, with distribution ratios of 27.8–78.0%, dominating the total flow yield. Therefore, the loss of rainfall and the formation of runoff in the slope surfaces of karst areas are much more difficult than those in non-karst areas. Only the last rainfall may form slope surface runoff under repeated rainfall because of the recharge of soil moisture first and the loss through fragmented surface leakage. Pre-rainfall mainly supplements soil moisture or leaks down through the broken surface to the pipeline and fissure. For all that, it has been found that all climatic factors exhibit a main cycle at 12-month time scales with runoff changes, which may show that the hydrometeorological processes in karst watersheds represent the same characteristics as those in non-karst watersheds at 1-year time scale (12-month time scale) periodic variations. This may be mainly because the impact of karst characteristics on hydrometeorological processes is mainly manifested on the slope scale, and all flows in the watershed will eventually converge to the outlet of watershed [45,46], which leads to the same annual periodic characteristics as those of non-karst watersheds. If the interval of multiple rainfall is long and the cumulative rainfall is less than 60 mm, it may lead to rainfall changes ahead of runoff on a monthly scale. If the cumulative rainfall is large under the condition of multiple short-term rainfall, the last few small rainfall events will produce obvious runoff on the slope after the saturation of soil water, and then the rainfall before the saturation of soil water will produce a leading effect on runoff. The effect of rainfall after soil water saturation on runoff changes will be synchronous because surface runoff would only occur when both soil and carbonate fissures and fractures are fully saturated with water [36,39]. Most of the rainfall is transported to the groundwater system through carbonate fractures and fractures, while the rainfall that can form surface runoff is very small [53,64].

In addtion, some studies have shown that antecedent rainfall and rainfall intensity are the major factors that control rainfall–runoff and soil erosion processes [65]. Rainfall intensity, slope angle and groundwater porosity [57] are the influencing factors of runoff changes mainly because the runoff mechanism caused by rainfall is different in years with different soil water contents. Different soil moisture contents are present in the early stage and the runoff generated by rainfall is also different in the year of rainfall approaching due to the different soil moisture contents in the early stage. In this case, the annual runoff depth is related to the rainfall year. For some places where recharging soil moisture by rainfall is difficult, the annual runoff depth is even related to the last rainfall year or even the previous years.

The process from rainfall to runoff will undergo seepage storage, slope overflow and channel flow collection. In karst watersheds, each process will be accompanied by underground leakage and the broken surface will affect the time, which greatly lengthens the lag time of runoff change. The influence of the changes in underlying surface conditions on runoff is a gradual process, but the influence of human activities on runoff is a catastrophic process. Therefore, the main reason why rainfall lags behind runoff is that human activities lead to catastrophic changes in runoff, especially land use changes, which destroy the original runoff production and confluence conditions. Such human activities as pumping and storing or introducing water into farmland can also lead to catastrophic changes in runoff. Therefore, the effect of rainfall on runoff will be delayed. Due to the large amount of runoff that will be produced when a heavy rainfall falls, the runoff series will show a great jump. At this period, obvious runoff will be produced directly on karst slopes because soil moisture is absorbed and saturated in a short time due to the large amount of rainfall, and its response to rainfall is timely with the consistent variation relationship. On the contrary, the runoff series may jump negatively due to the lack of rainfall when the watershed suffers from years of rare drought, but the positional correlation between the two is positive. The effect of rainfall on runoff can be influenced by human activities, such as soil and water conservation, which may play an important role in reducing runoff. However, the role of soil and water conservation will become small or ineffective when encountering heavy rain or rainstorm. The effect of rainfall on runoff changes will change the relationship between rainfall and runoff because of different patterns, intensity or frequency when raining. Temporary water intake by human activities can also alter runoff, thereby resulting in different time effects of advance, synchronization and lag.

6.2. Multi-Scale Effects of Evaporation on Runoff Changes

In the process of rainfall, evaporation exerts a minimal effect on runoff but has a great impact on the water storage capacity of the basin before rainfall. The greater the evaporation intensity, the smaller the soil water content before the rain, which increases the infiltration loss of rainfall and reduces the small-diameter flow. This study has supported the previous conclusion in annual scale that the effect of evaporation on runoff change was significantly enhanced, showing a great contribution of 10–90% [32], but there were some new discoveries during the year. The effect of evaporation on runoff was only in the high-frequency scale in summer and the 6-year scale in winter. In other seasons or scales, most hysteresis effects with a few synchronous relationships in the time domain have shown at different time scales, which indicates that runoff changes are affected by evaporation. The evaporation is larger in summer; hence, short-term evaporation has a significant impact on runoff changes, which results in the changes in runoff lagging behind evaporation. The effect of runoff ahead of evaporation has been virtually masked by rainfall and human activities. The runoff changes are greatly influenced by abrupt rainfall and human activities, whilst evaporation shows a continuous stable process. Rainfall burst or human activities will contribute to the changes in the underlying surface of the watershed, which directly alters the evaporation conditions and volume that will cause the time dislocation in different time domains. The strong disturbance of human activities on runoff will directly lead to relatively stable and persistent evaporation lagging behind the change in runoff.

The essence of changing runoff by evaporation is to reduce the recharge of runoff and increase the evaporation of the river surface. In addition, evaporation shows a high impact on runoff change also because of the influence of the subtropical monsoon climate, abundant light and heat resources in Southwest China. The influence of evaporation on runoff varies obviously in different periods, which is mainly affected by the light, temperature, heat, climate and water content of underlying surface. However, runoff changes are affected not only by evaporation but also by rainfall and human activities, which makes it impossible for the evolution of runoff and evaporation to be completely consistent.

6.3. Multi-Scale Effects of Temperature on Runoff Changes

The influence of temperature on runoff is consistent with that of evaporation in both time and frequency domains and has the same multi-time-scale characteristics and time–frequency relationship. However, a negative correlation exists between temperature and runoff because an increase in temperature leads to the intensification of evaporation on slopes and rivers of the watershed and decreases air humidity, thus changing the runoff.

On monthly and annual scales, as well as in summer, the effect of temperature on runoff is mostly ahead of schedule, whereas it is mainly lagging behind in spring, autumn and winter. Thus, the regulation of temperature on runoff is mainly reflected on the season scale. In summer, the change in runoff is mainly caused by changing evaporation and increasing rainfall to recharge soil moisture, and hence its impact on runoff shows a longer lead time than that of rainfall and evaporation. In other seasons, human activities change runoff intensely because of the relatively minimal rainfall, which leads to the relative lag of temperature change. On annual scale, the temperature regulation effect before 2000 is relatively stable, which mainly changes the runoff by changing the roles of evaporation and rainfall, resulting in a leading effect. Overall, the inter-annual temperature regulation is gradually lost, and the temperature regulation during the year is relatively prominent, but this regulation remains affected by human activities.

7. Conclusions

In this study, the multi-scale influences of climate factors on runoff changes in the Yinjiang River watershed are identified by using wavelet analysis, and the evolution relationship of time and frequency between runoff changes and climatic factors is further revealed at different time scales. The main conclusions are as follows:

(1) All climatic factors exhibit a main cycle at 12-month time scales with runoff changes, but the main periodic bandwidth of rainfall on runoff changes is much wider than that of temperature and evaporation, indicating that rainfall is the main factor affecting runoff changes.

(2) In other cycles, the impact of rainfall on runoff changes is the interlacing phenomena with positive and negative, but the impact of temperature and evaporation on runoff change is mainly negative.

(3) The response of runoff to rainfall is timely in the high-energy region and the low-energy significant-correlation region and shows a positive correlation with a smaller phase angle, but it is slightly lagged at 16-month time scales, in which the runoff changes lag behind temperature and evaporation for 1–2 months.

(4) It has been found that there is a strong effect of rainfall over runoff but a lesser effect of temperature and evaporation over runoff.

(5) The interaction of rainfall with runoff changes at 6-year time scales across the entire period. The effect of evaporation on runoff changes is similar to that of temperature and exhibited obvious local characteristics, mainly at small cycles.

The study has revealed the evolution process of river runoff in typical karst basins and the interaction mechanism between river runoff and climatic factors on multiple time scales, providing theoretical inspiration for fully solving the regional water shortage and engineering water shortage problems in the karst areas of Guizhou Province.

Author Contributions: Conceptualization, methodology and validation, X.B. and S.W.; Software and data curation, L.W. and C.R.; Formal analysis, L.W. and C.L.; Investigation, L.W. and F.C.; Writing—original draft preparation, L.W., C.R. and S.Z. All authors have read and agreed to the published version of the manuscript.

Funding: This research work was supported jointly by the Western Light Cross-team Program of Chinese Academy of Sciences (No. xbzg-zdsys-202101), National Natural Science Foundation of China (No. 42077455 & No.42167032), Strategic Priority Research Program of the Chinese Academy of Sciences (No. XDB40000000 & No. XDA23060100), Guizhou Provincial Science and Technology Projects (No. 2022-198), High-level innovative talents in Guizhou Province (No. GCC[2022]015-1 & No. 2016-5648), Guizhou Provincial 2020 Science and Technology Subsidies (No. GZ2020SIG), Opening Fund of the State Key Laboratory of Environmental Geochemistry (No. SKLEG2021072001 & No. SKLEG2022206 & No. SKLEG2022208) and Doctoral Research Startup Fund Project of Tongren University (No. trxyDH2103).

Institutional Review Board Statement: Not applicable.

Informed Consent Statement: Not applicable.

Data Availability Statement: The data analyzed in this study are subject to the following licenses/restrictions: The dataset can only be accessed from China Meteorological Data Sharing Service System, Karst Scientific Data Center and Guizhou Provincial Hydrology and Water Resources Bureau. Requests to access these datasets should be directed to jgywlh@gztrc.edu.cn.

Acknowledgments: We would like to thank all the authors and reviewers for their great guidance and help in writing this manuscript.

Conflicts of Interest: The authors declare no conflict of interest.

References

1. Liu, M.; Xu, X.; Wang, D.; Sun, A.Y.; Wang, K. Karst catchments exhibited higher degradation stress from climate change than the non-karst catchments in southwest China: An ecohydrological perspective. *J. Hydrol.* **2016**, *535*, 173–180. [CrossRef]
2. Li, Z.W.; Xu, X.L.; Yu, B.F.; Xu, C.H.; Liu, M.X.; Wang, K.L. Quantifying the impacts of climate and human activities on water and sediment discharge in a karst region of southwest China. *J. Hydrol.* **2016**, *542*, 836–849. [CrossRef]
3. Tian, Y.C.; Wang, S.J.; Bai, X.Y.; Luo, G.J.; Xu, Y. Trade-offs among ecosystem services in a typical Karst watershed, SW China. *Sci. Total Environ.* **2016**, *566–567*, 1297–1308. [CrossRef] [PubMed]
4. Burns, D.A.; Klaus, J.; McHale, M.R. Recent climate trends and implications for water resources in the Catskill Mountain region, New York, USA. *J. Hydrol.* **2007**, *336*, 155–170. [CrossRef]
5. Sun, W.G.; Cheng, B.Y.; Li, R. Seasonal variations of runoff and its wavelet correlations with regional climate in source regional of the Yellow River. *J. Desert Res.* **2010**, *30*, 712–721.
6. Yonaba, R.; Biaou, A.C.; Koïta, M.; Tazen, F.; Mounirou, L.A.; Zouré, C.O.; Queloz, P.; Karambiri, H.; Yacouba, H. A dynamic land use/land cover input helps in picturing the Sahelian paradox: Assessing variability and attribution of changes in surface runoff in a Sahelian watershed. *Sci. Total Environ.* **2020**, *757*, 143792. [CrossRef]
7. Chen, Y.N.; Xu, C.C.; Yang, Y.H.; Hao, X.M. Hydrology and water resources variation and its response to regional climate change in Xinjiang. *Acta. Geogr. Sin.* **2009**, *64*, 1331–1341. [CrossRef]
8. Cannon, A.J.; McKendry, I.G. A graphical sensitivity analysis for statistical climate models: Application to Indian monsoon rainfall prediction by artificial neural networks and multiple linear regression models. *Int. J. Clim.* **2002**, *22*, 1687–1708. [CrossRef]
9. Liang, L.; Liu, Q. Streamflow sensitivity analysis to climate change for a large water-limited watershed. *Hydrol. Process.* **2014**, *28*, 1767–1774. [CrossRef]
10. Xu, X.; Du, Y.; Tang, J.; Wang, Y. Variations of temperature and rainfall extremes in recent two decades over China. *Atmos. Res.* **2011**, *101*, 143–154. [CrossRef]
11. Wu, S.F.; Liu, Z.H.; Han, P.; Zhu, Z.C. Impact of climate change on water resources of the Urumchi River watershed. *J. Glaciol. Geocryol.* **2006**, *28*, 703–706.
12. Li, Z.; Wang, W.; Zhang, M.; Wang, F.; Li, H. Observed changes in streamflow at the headwaters of the Urumqi River, eastern Tianshan, central Asia. *Hydrol. Process.* **2010**, *24*, 217–224. [CrossRef]
13. Lan, Y.; Shen, Y.; Zhong, Y.; Wu, S.F.; Wang, G. Sensitivity of the mountain runoff of Urumqi River to the climate changes. *J. Arid. Land Resour. Environ.* **2010**, *24*, 50–55.
14. Cui, Y.; Ye, B.; Wang, J.; Liu, Y.; Jing, Z. Analysis of the spatial-temporal variation of the positive degree-day factors on the glacier no.1 at the headwaters of the Urumqi River. *J. Glaciol. Geocryol.* **2010**, *32*, 265–274.
15. Duan, K.Q.; Yao, T.D.; Wang, N.L.; Liu, H.C. Numerical simulation of Urumqi Glacier No. 1 in the eastern Tianshan, central Asia from 2005 to 2070. *Sci. Bull.* **2012**, *57*, 4505–4509. [CrossRef]
16. Cui, Y.H. The runoff simulations for the Glacier No.1 hydrologic section at the headwaters of the Urumqi River on the different timescales. *J. Arid. Land Resour. Environ.* **2013**, *27*, 119–126.
17. Qian, K.; Wang, X.S.; Lv, J.; Wan, L. The wavelet correlative analysis of climatic impacts on runoff in the source region of Yangtze River, in China. *Int. J. Clim.* **2013**, *34*, 2019–2032. [CrossRef]
18. Si, B.C. Spatial Scaling Analyses of Soil Physical Properties: A Review of Spectral and Wavelet Methods. *Vadose Zone J.* **2008**, *7*, 547–562. [CrossRef]
19. Grinsted, A.; Moore, J.C.; Jevrejeva, S. Application of the cross wavelet transform and wavelet coherence to geophysical time series. *Nonlin. Process. Geophys.* **2004**, *11*, 561–566. [CrossRef]
20. Torrence, C.; Compo, G.P. A practical guide to wavelet analysis. *Bull. Am. Meteorol. Soc.* **1998**, *9*, 61–78. [CrossRef]
21. Hao, Y.; Liu, G.; Li, H.; Li, Z.; Zhao, J.; Yeh, T.C.J. Investigation of karstic hydrological processes of Niangziguan Springs (North China) using wavelet analysis. *Hydrol. Process.* **2011**, *26*, 3062–3069. [CrossRef]
22. Li, X.; Xu, X.; Liu, W.; He, L.; Zhang, R.; Xu, C.; Wang, K. Similarity of the temporal pattern of soil moisture across soil profile in karst catchments of southwestern China. *J. Hydrol.* **2017**, *555*, 659–669. [CrossRef]
23. Charlier, J.B.; Ladouche, B.; Maréchal, J.C. Identifying the impact of climate and anthropic pressures on karst aquifers using wavelet analysis. *J. Hydrol.* **2015**, *523*, 610–623. [CrossRef]
24. Labat, D.; Ababou, R.; Mangin, A. Rainfall–runoff relations for karstic springs. Part II: Continuous wavelet and discrete orthogonal multiresolution analyses. *J. Hydrol.* **2000**, *238*, 149–178. [CrossRef]

25. Labat, D.; Ronchail, J.; Guyot, J.L. Recent advances in wavelet analyses: Part 2—Amazon, Parana, Orinoco and Congo discharges time scale variability. *J. Hydrol.* **2005**, *314*, 289–311. [CrossRef]
26. Massei, N.; Durand, A.; Deloffre, J.; Dupont, J.P.; Valdes, D.; Laignel, B. Investigating possible links between the North Atlantic Oscillation and rainfall variability in northwestern France over the past 35 years. *J. Geophys. Res.* **2007**, *112*, D09121. [CrossRef]
27. Labat, D. Cross wavelet analyses of annual continental freshwater discharge and selected climate indices. *J. Hydrol.* **2010**, *385*, 269–278. [CrossRef]
28. Fu, G.B.; Charles, S.P.; Chiew, F.H.S. A two-parameter climate elasticity of streamflow index to assess climate change effects on annual streamflow. *Water Resour. Res.* **2007**, *43*, W11419. [CrossRef]
29. Lafrenière, M.; Sharp, M. Wavelet analysis of inter-annual variability in the runoff regimes of glacial and nival stream catchments, Bow Lake, Alberta. *Hydrol. Process.* **2003**, *17*, 1093–1118. [CrossRef]
30. Schaefli, B.; Maraun, D.; Holschneider, M. What drives high flow events in the Swiss Alps? Recent developments in wavelet spectral analysis and their application to hydrology. *Adv. Water Resour.* **2007**, *30*, 2511–2525. [CrossRef]
31. Wu, L.; Wang, S.; Bai, X.; Luo, W.; Tian, Y.; Zeng, C.; Luo, G.; He, S. Quantitative assessment of the impacts of climate change and human activities on runoff change in a typical karst watershed, SW China. *Sci. Total Environ.* **2017**, *601–602*, 1449–1465. [CrossRef] [PubMed]
32. Li, Z.W.; Xu, X.L.; Xu, H.; Liu, X.; Wang, K.L.; Yi, R. Monthly sediment discharge changes and estimates in a typical karst catchment of southwest China. *J. Hydrol.* **2017**, *555*, 95–107. [CrossRef]
33. Li, Z.W.; Xu, X.L.; Xu, C.H.; Liu, M.X.; Wang, K.L. Dam construction impacts on multiscale characterization of sediment discharge in two typical karst watersheds of southwest China. *J. Hydrol.* **2018**, *558*, 42–54. [CrossRef]
34. Liu, Y.C.; Liu, Z.F.; Hao, Y.H.; Han, T.D.; Shen, Y.P.; Jiao, K.Q.; Huo, X.L. Multi-time scale characteristics of the runoff in the upstream of Rümqi River, Tianshan Mountains, based on cross-wavelet transformation. *J. Glaciol. Geocryol.* **2013**, *35*, 1564–1572.
35. Sidibe, M.; Dieppois, B.; Eden, J.; Mahé, G.; Paturel, J.E.; Amoussou, E.; Anifowose, B.; Lawler, D. Interannual to Multi-decadal streamflow variability in West and Central Africa: Interactions with catchment properties and large-scale climate variability. *Glob. Planet. Chang.* **2019**, *177*, 141–156. [CrossRef]
36. Hartmann, A.; Goldscheider, N.; Wagener, T.; Lange, J.; Weiler, M. Karst water resources in a changing world: Review of hydrological modeling approaches. *Rev. Geophys.* **2014**, *52*, 218–242. [CrossRef]
37. Liu, C.Q. *Biogeochemical Processes and Cycling of Nutrients in the Earth's Surface: Cycling of Nutrients in Soil-Plant Systems of Karstic Environments*; Science Press: Beijing, China, 2009.
38. Liu, C.C.; Liu, Y.G.; Fan, D.Y.; Guo, K. Plant drought tolerance assessment for re-vegetation in heterogeneous karst landscapes of southwestern China. *Flora-Morphol. Distrib. Funct. Ecol. Plants* **2012**, *207*, 30–38. [CrossRef]
39. Wilcox, B.P.; Wilding, L.P.; Woodruff, C.M. Soil and topographic controls on runoff generation from stepped landforms in the Edwards Plateau of Central Texas. *Geophys. Res. Lett.* **2007**, *34*, L24S24. [CrossRef]
40. Williams, P.W. The role of the epikarst in karst and cave hydrogeology: A review. *Int. J. Speleol.* **2008**, *37*, 1–10. [CrossRef]
41. Nie, Y.P.; Chen, H.S.; Wang, K.L.; Yang, J. Water source utilization by woody plants growing on dolomite outcrops and nearby soils during dry seasons in karst region of Southwest China. *J. Hydrol.* **2012**, *420–421*, 264–274. [CrossRef]
42. Butscher, C.; Huggenberger, P. Modeling the Temporal Variability of Karst Groundwater Vulnerability, with Implications for Climate Change. *Environ. Sci. Technol.* **2009**, *43*, 1665–1669. [CrossRef] [PubMed]
43. Petrella, E.; Falasca, A.; Celico, F. Natural-gradient tracer experiments in epikarst: A test study in the Acqua dei Faggi experimental site, southern Italy. *Geofluids* **2008**, *8*, 159–166. [CrossRef]
44. Wilcox, B.P.; Taucer, P.I.; Munster, C.L.; Owens, M.K.; Mohanty, B.P.; Sorenson, J.R.; Bazan, R. Subsurface stormflow is important in semiarid karst shrublands. *Geophys. Res. Lett.* **2008**, *35*, L10403. [CrossRef]
45. Qin, L.Y.; Bai, X.Y.; Wang, S.J.; Zhou, D.Q.; Li, Y.; Peng, T.; Tian, Y.; Luo, G. Major problems and solutions on surface water resource utilisation in karst mountainous areas. *Agric. Water Manag.* **2015**, *159*, 55–65. [CrossRef]
46. Luo, G.J.; Wang, S.J.; Bai, X.Y.; Liu, X.M.; Cheng, A.Y. Delineating small karst watersheds based on digital elevation model and eco-hydrogeological principles. *Solid Earth* **2016**, *7*, 457–468. [CrossRef]
47. Burrus, C.S.; Gopinath, R.A.; Guo, H. *Introduction to Wavelets and Wavelet Transforms*; Prentice Hall: Hoboken, NJ, USA, 1998.
48. Allen, M.R.; Smith, L.A. Monte Carlo SSA: Detecting irregular oscillations in the presence of coloured noise. *J. Clim.* **1996**, *9*, 3373–3404. [CrossRef]
49. Kumar, P.; Foufoula-Georgiou, E. Wavelet analysis for geophysical applications. *Rev. Geophys.* **1997**, *35*, 385–412. [CrossRef]
50. Zar, J.H. *Biostatistical Analysis*; Prentice Hall: Hoboken, NJ, USA, 1999.
51. Wang, S.J.; Zhang, D.F. Karst rocky desertification in south western China: Geomorphology, landuse, impact and rehabilitation. *Land Degrad. Dev.* **2004**, *15*, 115–121. [CrossRef]
52. Williams, P.W. The epikarst: Evolution of understanding. In *Epikarst*; Jones, W.K., Culver, D.C., Herman, J.S., Eds.; Karst Waters Institute: Charles Town, WV, USA, 2004; pp. 8–15.
53. Peng, T.; Wang, S.J. Effects of land use, land cover and rainfall regimes on the surface runoff and soil loss on karst slopes in southwest China. *Catena* **2012**, *90*, 53–62. [CrossRef]
54. Zhang, S.R.; Bai, X.Y.; Zhao, C.W.; Tan, Q.; Luo, G.J.; Cao, Y.; Deng, Y.; Li, Q.; Li, C.; Wu, L.; et al. Limitations of soil moisture and formation rate on vegetation growth in karst areas. *Sci. Total Environ.* **2021**, *810*, 151209. [CrossRef]

55. Xiong, L.; Bai, X.Y.; Zhao, C.W.; Li, Y.B.; Tan, Q.; Luo, G.J.; Wu, L.; Chen, F.; Li, C.; Ran, C.; et al. High-Resolution Data Sets for Global Carbonate and Silicate Rock Weathering Carbon Sinks and Their Change Trends. *Earth's Feature* **2022**, *10*, e2022EF002746. [CrossRef]
56. Xi, H.P.; Wang, S.J.; Bai, X.Y.; Tang, H.; Luo, G.J.; Li, H.W.; Wu, L.; Li, C.; Chen, H.; Ran, C.; et al. The responses of weathering carbon sink to eco-hydrological processes in global rocks. *Sci. Total Environ.* **2021**, *788*, 147706. [CrossRef] [PubMed]
57. Yan, J.H.; Zhou, G.Y.; Shen, W.J. Grey correlation analysis of the effect of vegetation status on surface runoff coefficient of forest ecosystems. *Chin. J. Appl. Environ. Biol.* **2000**, *6*, 197–200.
58. Dong, H.B.; Zhang, J.J.; Zhang, B.; Zhang, R.; Zhou, X.X. Research on rain-runoff relationship in different land use types on the loess area in western Shanxi province. *J. Arid. Land Resour. Environ.* **2009**, *23*, 110–116.
59. Li, X.Y.; Sergio, C.; Albert, S.B.; Yolanda, C.; Francisco, D.; Roberto, L.; Henry, L.; Bas, V.W.; Juan, P. Controls of infiltration-runoff processes in Mediterranean karst rangelands in SE Spain. *Catena* **2011**, *86*, 98–109. [CrossRef]
60. Chen, H.S.; Yang, J.; Fu, W.; He, F.; Wang, K.L. Characteristics of slope runoff and sediment yield on karst hill-slope with different land-use types in northwest guangxi. Trans. *Chin. Soc. Agric. Eng.* **2012**, *28*, 121–126.
61. Chen, W.Y.; Wang, S.J.; Bai, X.Y.; Qin, L.Y.; Zhang, S.Y. Drainage network distribution and its master natural factors in Guizhou province. *Earth. Environ.* **2013**, *41*, 193–202.
62. Peng, X.D.; Dai, Q.H.; Li, C.L.; Zhao, L.S. Role of underground fissure flow in near-surface rainfall-runoff process on a rock mantled slope in the karst rocky desertification area. *Eng. Geol.* **2018**, *243*, 10–17. [CrossRef]
63. Peng, T.; Yang, T.; Wang, S.J.; Zhang, X.B.; Chen, B.; Wang, J.Y. Monitoring results of soil loss in karst slopes. *Earth Environ.* **2009**, *37*, 126–130. (In Chinese)
64. Fu, Z.; Chen, H.; Zhang, W.; Xu, Q.; Wang, S.; Wang, K. Subsurface flow in a soil-mantled subtropical dolomite karst slope: A field rainfall simulation study. *Geomorphology* **2015**, *250*, 1–14. [CrossRef]
65. Maass, M.; Ahedo-Hernández, R.; Salvador, A.; Verduzco, A.; Martínez-Yrízar, A.; Jaramillo, V.J.; Parker, G.; Pascual, F.; García-Méndez, G.; Sarukhán, J. Long-term (33 years) rainfall and runoff dynamics in a tropical dry forest ecosystem in western Mexico: Management implications under extreme hydrometeorological events. *For. Ecol. Manag.* **2018**, *426*, 7–17. [CrossRef]

land

MDPI

Article

Temporal Changes in Land Use, Vegetation, and Productivity in Southwest China

Xuan Li [1], Li Rong [1,*], Mengmeng Zhang [1], Wensong Yang [1], Zhen Zeng [1], Chengjun Yuan [2] and Qi Wang [1,2]

[1] School of Geography and Environmental Sciences, Guizhou Normal University, Guiyang 550000, China
[2] School of Karst Science, Guizhou Normal University, Guiyang 550001, China
* Correspondence: ronglit@gznu.edu.cn

Abstract: In recent decades, vegetation coverage and land use/land cover (LULC) have constantly changed, especially in southwest China. Therefore, it is necessary to conduct in-depth research into the temporal–spatial variation patterns of vegetation greening, LULC, and gross primary productivity (GPP). Here, we used remote sensing to analyze the spatial and temporal variation in the normalized difference vegetation index (NDVI) and GPP in the growing season under different LULCs in southwest China. Results showed: (1) From 2000–2019, the forest area in southwest China had increased by 2.1%, while the area of cropland and grassland had decreased by 3.2% and 5.5%, respectively. Furthermore, there are significant differences in spatial variation patterns. (2) NDVI and GPP in the growing season showed a general increasing trend ($p < 0.01$); vegetation coverage is dominated by high coverage to highest coverage and medium coverage to high coverage transfer. (3) Under different LULCs, the migration directions of NDVI and GPP were different. The center of gravity migration of highest and medium coverage shifted to the southeast by 1.69° and to the northwest by 1.81°, respectively. The results showed the ecosystem evolution and will help to guide the maintenance measure of ecosystem balance and sustainable development.

Keywords: southwest China; normalized difference vegetation index (NDVI); gross primary productivity (GPP); land use/land cover (LULC); center of gravity shift model

Citation: Li, X.; Rong, L.; Zhang, M.; Yang, W.; Zeng, Z.; Yuan, C.; Wang, Q. Temporal Changes in Land Use, Vegetation, and Productivity in Southwest China. *Land* **2022**, *11*, 1331. https://doi.org/10.3390/land11081331

Academic Editor: Xiaoyong Bai

Received: 19 July 2022
Accepted: 15 August 2022
Published: 17 August 2022

1. Introduction

The ecological environment of karst landforms in southwest China is fragile and has been significantly affected by climate and human activities in recent decades [1–3]. Changes in vegetation, LULC, and GPP affect biogeochemical cycles, and social effects in this region impact the area range of influence [4–7]. However, the spatial and temporal characteristics of LULC, vegetation, and carbon storage are not clear. This has a significant impact on ecological evolution and regional social development [8,9]. Therefore, there is a need to clarify the temporal change characteristics of LULC, vegetation, and carbon storage in southwest China.

Southwest China has a large number of karst ecosystems, which are hypersensitive and fragile. First, this area is one of the largest exposed areas of carbonate rock salts in the world [10], and in these environments, the soil formation rate is low, and the permeability is high due to the presence of interstitial fractures. Furthermore, it has unique and fragile geomorphological and hydrogeological features [11]. In recent decades, long-term and severe climate change and human activities have brought enormous pressure to the ecosystem in this area [12–14]. Rocky desertification has become one of the most serious environmental problems in karst areas [15,16]. Terrestrial vegetation types and compositions have changed due to climatic conditions, carbon dioxide fertilization effects, and LULC [17,18]. Second, under the background of population pressure and urbanization, the intensity of human activities has increased rapidly, and the land cover has undergone drastic changes [19]. Third, since the end of the 20th century, China has implemented a large number of ecological engineering constructions, which have achieved an increase in vegetation coverage and carbon storage through ecological restoration and improved ecosystem services [20–22]. In addition, the southwest region is an important ecological barrier and ecologically fragile, with extensive potential for various ecosystem services, such as soil and water conservation, climate regulation, and carbon balance [23–25], providing a huge contribution to social development, ecosystem balance and carbon sequestration [24,26,27].

At present, the spatial and temporal changes in LULC, vegetation cover, and gross primary productivity (GPP) in southwest China are not clear. Vegetation is an important factor affecting the ecological balance and is usually considered as a direct and obvious indicator to analyze the impact of natural seasonal changes and human activities on the ecological environment [28,29]. Gross primary productivity (GPP) is an important indicator reflecting vegetation status, ecosystem structure, and function [30] and plays a key role in carbon cycling in terrestrial ecosystems [31], and is an important factor in measuring the regional ecological value [32,33]. Therefore, clarifying temporal and spatial evolution processes is of great significance for understanding the value and sustainable development of ecosystems. Studies have found that China's vegetation has shown an overall greening trend in the past 30 years [24]. However, due to the vast heterogeneity of climate, topography, and human activities in the southwest, the spatial and temporal distributions of LULC, vegetation dynamics, and gross primary productivity (GPP) are significantly different [2,14,34]. Since 2000, the LULC change in southwest China has been mainly manifested in the expansion of forest land and the reduction in cropland [4,35]. The study found that NDVI increased significantly in low- to mid-altitude areas < 3400 m due to improved afforestation and agricultural productivity [36]. In the afforestation and grassland restoration areas, the direct contribution of forest land to the annual growth rate of GPP is 24.64% [37]. In addition, according to long-term remote sensing vegetation data, it is found that short-term extreme climate events respond differently to different land-use types, resulting in differences in regional ecological effects [38,39]. Therefore, it is of great significance to understand the temporal and spatial pattern characteristics and change processes of different LULC types, vegetation dynamics, and gross primary productivity in the region for correctly understanding the temporal dynamic changes and spatial changes in regional vegetation dynamics and gross primary productivity.

The changes in vegetation and productivity center can reflect the evolution of ecosystems influenced by human activities and climate change. Human activities affect vegetation and productivity changes, such as ecological engineering, which increases vegetation growth and carbon storage in southwest China [40,41], and positively contributes to vegetation productivity [42]. However, the expansion of arable land and the surge in population has also led to the degradation of vegetation [15]. Deforestation reduced the GPP and leaf area index in China between 1982 and 2011, and their centers of gravity shifted [43]. The spatial and temporal changes in vegetation cover and productivity in different regions have obvious uncertainties [6,44,45]. Natural evolution is also an important factor leading to the migration of its center of gravity; for example, the northward shift of the climatic zone makes the ecological center of gravity move northward [46,47]. In summary, combined with different LULC types, studying the temporal and spatial variation patterns of different levels of vegetation cover (NDVI) and its gross primary productivity (GPP) in southwest China can deepen the understanding of vegetation and productivity changes in southwest China. It has very important ecological value and practical significance for the balance and sustainable development of the ecosystem.

The purpose of this study is to clarify the temporal and spatial dynamic of LULC, vegetation cover, and GPP in southwest China and the migration pattern of the center of gravity. Combined with MODIS remote sensing, we analyzed the temporal and spatial changes of vegetation cover (NDVI) and gross primary productivity (GPP) under different LULC types in southwest China. Therefore, our aims in the study are: (1) to clarify the migratory direction of LULC in southwest China and the spatial and temporal change patterns of NDVI and GPP in the growing season; (2) to explore the change characteristics of

GPP under different land use types and different vegetation coverage levels; (3) to analyze the migration law of vegetation cover and GPP center of gravity.

2. Data and Methods

2.1. Study Area

Southwest China covers a region bounded by 20°54′–34°19′ N and 97°21′–112°04′ E (Figure 1). It includes five provinces (municipalities, autonomous regions): Chongqing, Sichuan, Guizhou, Yunnan, and Guangxi. The area has a total coverage of 1.38 × 106 km^2, which is about 14.34% of China. The study area is dominated by the subtropical monsoon climate, affected by the southeast and the southwest monsoon, with a mean annual temperature of 14.6 °C and an average annual precipitation of 1195 mm, of which 600 mm occurs in the growing season. The terrain is high in the west and low in the east, with a stepped distribution, diverse landform types, and high landscape heterogeneity [48]. Southwest China is the most concentrated area of karst landforms, accounting for 23.14% of China's karst distribution area. The forest has a large area and is widely distributed in the study area. Grassland is mainly distributed in the western Sichuan Plateau, and cropland is mainly distributed in the Sichuan Basin, the middle of the Guangxi hills, and the eastern part of the Yunnan–Guizhou Plateau. The three main LULC types are grassland, forest, and cropland, accounting for more than 98% of the total area of southwest China.

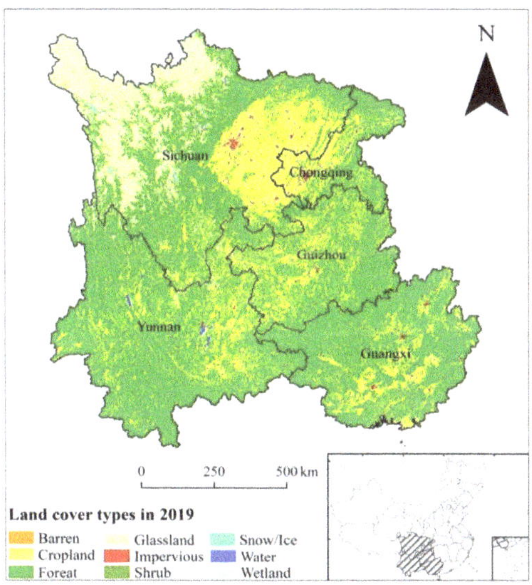

Figure 1. Study area location in China and land cover types in 2019.

2.2. Data

The normalized difference vegetation index (NDVI) data came from the study derived from the MOD13A2 data product, with a spatial resolution of 1 km and a temporal resolution of 16d (https://ladsweb.modaps.eosdis.nasa.gov/ (accessed on 6 August 2021)). These data were preprocessed on the Google Earth engine platform (https://earthengine.google.com (accessed on 6 August 2021)), and the NDVI data from May to September 2000–2019 were downloaded. The gross primary productivity (GPP) datasets were obtained from the MOD17A2HV6 data product with a spatial resolution of 500 m and a temporal resolution of 8d (https://lpdaac.usgs.gov/products/mod17a2hv006/ (accessed on 28 August 2021)), and we downloaded the contemporaneous datasets from the Google Earth engine platform (https://earthengine.google.com (accessed on 28 August 2021)). The LULC data came from the first Landsat-derived annual China Land Cover Dataset

(CLCD) from 1990 to 2019, with a spatial resolution of 30 m and an overall accuracy rate of 79.31%, better than MCD12Q1, ESACCI_LC, FROM_GLC, and GlobaLand30. The LULC types in these data were divided into nine categories: cropland, forest, shrub, grassland, water, snow/ice, barren, impervious, and wetland. The CLCD dataset introduced in this article is freely available at http://doi.org/10.5281/zenodo.4417810 (accessed on 10 August 2021) [49].

2.3. Methods

The normalized difference vegetation index (NDVI) can reflect the growth and coverage of surface vegetation [50,51]. Therefore, the NDVI values of different LULC types were extracted to represent the growth state of this vegetation type. Vegetation has obvious inter-annual and seasonal variation characteristics, and its NDVI value in the season of the most active growth stage can more accurately represent the vegetation growth state in this region [52]. In order to reduce the NDVI error caused by the seasonal changes in vegetation, this paper used NDVI and GPP values in the growing season (May–October) to analyze vegetation activities, which accurately reflected the status of vegetation cover and GPP in southwest China. To minimize the effect of cloud contamination and atmospheric variability, we calculated the annual growing season NDVI by using the maximum value composites (MVC) method [53]. The MODIS NDVI and GPP data in the study area were resampled to the monthly scale using ArcGIS10.4, with a spatial resolution of 1 km. Additionally, by averaging each pixel, the average values from May to October were obtained as the growing season NDVI and GPP for each year from 2000 to 2019, which meant in pixels of 500 × 500, there was, on average, one gC unit of GPP per 1 m^2.

Simple linear regression was used to analyze the annual trends of average NDVI and GPP values in the growing season of different LULC in southwest China from 2000–2019. All data were analyzed using Python software. Additionally, referring to related research [54], the vegetation coverage was divided into lowest coverage ($0 < NDVI \leq 0.35$), low coverage ($0.35 < NDVI \leq 0.55$), medium coverage ($0.55 < NDVI \leq 0.75$), high coverage ($0.75 < NDVI \leq 0.85$), and highest coverage ($0.85 < NDVI \leq 1$), five grades. The gravity center migration can show the change directions in the region's center of gravity. Therefore, the change directions in the region's NDVI and GPP gravity centers were calculated by gravity center migration. This, in turn, can reflect the import of human activities or climate change on the migration of vegetation zone. The main LULC types in southwest China were cropland, forest, and grassland, and other land types (shrub, water, snow/ice, barren, impervious, and wetland) accounted for only 2% of the total area (Figure 1). Therefore, this paper mainly analyzed the change characteristics of three LULC types of cropland, forest, and grassland, the temporal and spatial distribution of NDVI and GPP in the growing season of different LULC types, as well as the regular change in the center of gravity.

3. Results

3.1. Land Types Changes

During the study period, the composition of LULC types in the study area changed significantly, with a high spatial heterogeneity between the LULC types (Figure 1). The conversion of farmland into forests was the major driving force of LULC change. Overall, the forest area showed increasing trends after 2000, but it showed a decreasing trend first and then increased (1.04 $year^{-1}$, $p < 0.01$). Forest area increased by 16,180.0 km^2, accounting for 58.7% of the regional area from 2000 to 2019 (Figure 2, Table 1). The growing regions of forest were mainly distributed in northeastern Sichuan and southwestern Yunnan, and the decreasing regions were mainly distributed in southeastern Guizhou, northwestern Guangxi, and northwestern Yunnan. The cropland area first increased and then decreased, showing a significant downward trend overall (-0.707 $year^{-1}$, $p < 0.01$). The area of conversion of farmland into forests was 61,318.48 km^2, which was the LULC change type with the greatest land area transfer, mainly concentrated in the Sichuan Basin, northern Guizhou, and central and southern Guangxi (Figure 1). The grassland area showed a continuously decreasing trend (-0.64 year^{-1}, $p < 0.01$), of which 10,948.30 km^2 of the grassland area was converted into forest. The change regions were concentrated in the Hengduan Mountains in the northwest of Yunnan, Kunming–Zhaotong regions in the northwest, and the western Sichuan plateau region. A grassland area of 4231.28 km^2 was converted into cropland, the change regions were mainly concentrated in the northeast part of Yunnan, the western part of Guizhou, and there were also sporadic changes in southern Sichuan.

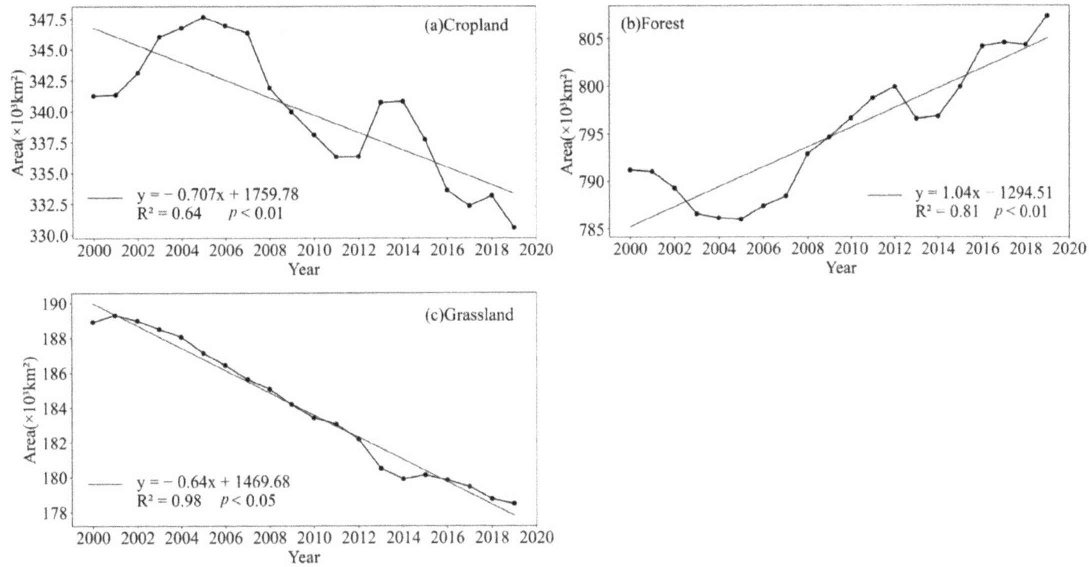

Figure 2. Changes in area of different LULC types: (**a**) cropland, (**b**) forest, (**c**) grassland. (Note: the black solid line represents the trend line, and the formula of cropland is $y = -0.707x + 1759.78$; the formula of forest is $y = 1.04x - 1294.51$; the formula of grassland is $y = -0.64x + 1469.68$).

Table 1. Transformation matrix of land cover types from 2000 to 2019 (km^2).

2000–2019	Cropland	Forest	Grassland
Cropland	264,799.33	61,318.48	3956.10
Forest	55,715.85	719,978.95	4073.26
Grassland	4231.28	10,948.30	166,551.06

3.2. The Characteristics of Inter-Annual Variation in NDVI

From 2000 to 2019, the average NDVI of vegetation in the growing season in the whole study area showed an overall upward trend with fluctuations (slope = 0.0023 year^{-1}, $p < 0.01$). The growing season average NDVI values of different land cover types all showed an increasing trend, but there were some differences. Among them, forest had the highest NDVI value, ranging from 0.79 to 0.84; cropland NDVI value ranged from 0.74 to 0.79; grassland vegetation NDVI varied from 0.70 to 0.73 (Figure 3b). The increase in the NDVI of forest was 0.0025 year^{-1}, and the trend of NDVI of cropland was 0.0023 year^{-1}, which was twice the increase in grassland (0.0011 year^{-1}).

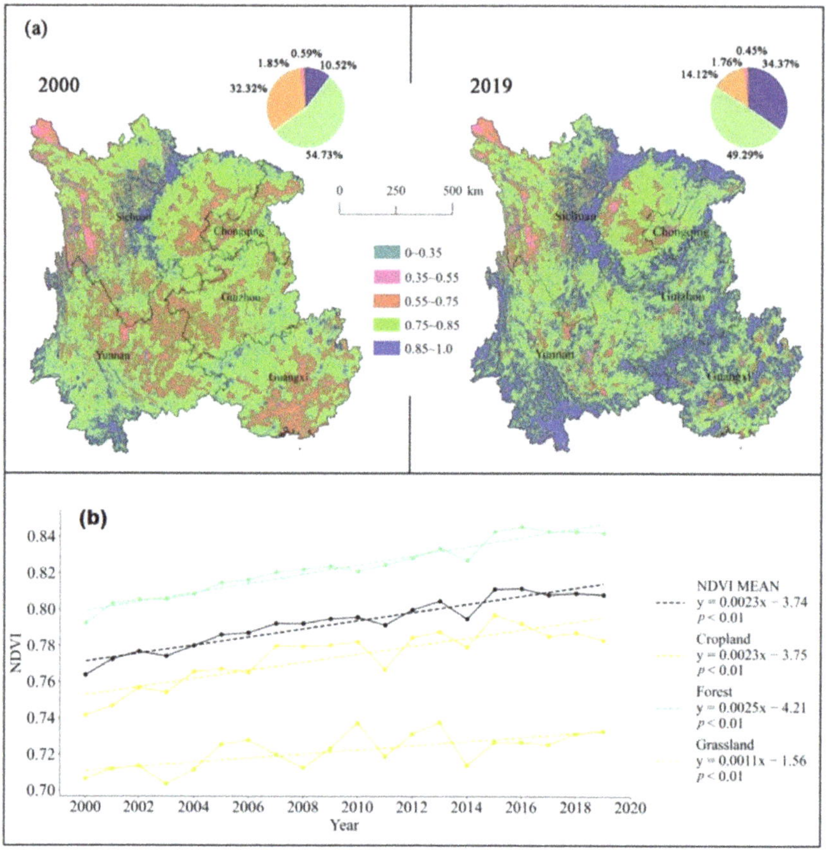

Figure 3. (**a**) Spatial distribution of NDVI in southwest China and (**b**) Changing trends in NDVI values from 2000 to 2019 for different land use types.

The vegetation coverage in southwest China in 2000 was dominated by medium and high coverage, accounting for 87.05% of the regional area. In 2019, the vegetation coverage was dominated by highest and high coverage, accounting for 83.66% of the regional area. In terms of spatial distribution, there was a large heterogeneity in the level of surface cover (Figure 3a). The coverage situation in the southwest region showed that the area of lowest coverage, low coverage, medium coverage, and high coverage decreased, but the area of highest coverage increased. Overall, the degree of the greening of vegetation was strengthened. From 2000 to 2019, the area of highest coverage (0.85~1.0) increased from 10.52% to 34.37%, and the main change region was concentrated on the edge of the Sichuan Basin and the three provinces of Yunnan, Guizhou, and Guangxi, with the land cover being forest. The area of high coverage (0.75~0.85) decreased from 54.73% to 49.30%, mainly concentrated in the Sichuan Basin, the eastern Yunnan-Guizhou Plateau, and the northern hills of Guangxi, with the land cover being mainly cropland and forest. The area of medium coverage (0.55~0.75) decreased the most from 32.32% to 14.12%. The main change region was distributed in the Yunnan–Guizhou Plateau, with the cover area being mainly grassland and cropland. There was little change in the low coverage (0.35~0.55) and lowest coverage (0~0.35), which decreased from 2.70% to 2.21%, with the main type of cover being grassland, distributed in the northwest of Sichuan. The vegetation cover types remained unchanged at 48.66% of the total area, while vegetation cover improved to 45.60%.

From 2000 to 2019, the vegetation cover in southwest China mainly shifted from "high coverage to highest coverage" and "medium coverage to high coverage and highest coverage" (Table A1). The specific transfer amount was as follows: the area from high coverage to highest coverage was 311,870.61 km^2, accounting for 44.17% of the total change, and the area from medium coverage to high coverage was 259,673.48 km^2, accounting for 36.77% of the total change. From the perspective of the final change, the transfer-out area of high coverage was the largest, which was 357,704.37 km^2, 1.26 times the transfer-in area, accounting for 50.66% of the total transfer-out area; the highest coverage transfer-out area was 23,192.33 km^2, accounting for 3.28% of the total transfer-out area; the medium coverage transfer-out area was 307,709.17 km^2, accounting for 43.58% of the total transfer-out area and 5.36 times the transfer-in area; the low coverage transfer-out area was 13,993.99 km^2, accounting for 1.98% of the total transfer-out area; and the transfer-out area of lowest coverage was 3544.33 km^2, accounting for 0.5% of the total transfer-out area. The transfer-in area of highest coverage was the largest, which was 351,279.44 km^2, accounting for 49.75% of the total transfer-in area, and 15.15 times the transfer-in area; the high coverage transfer-in area was 283,127.25 km^2, accounting for 40.10% of the total transfer-in area; the area of medium coverage transfer-in area was 57,361 km^2, accounting for 8.12% of the total transfer-in area; the transfer-in area of low coverage was 12,805.25 km^2, accounting for 1.81% of the total transfer-in area; and the transfer-in area of lowest coverage was 1570.49 km^2, accounting for the total transfer-in area of 0.22%. In this period, the vegetation coverage mainly showed a trend of gradual improvement, and the vegetation coverage also gradually increased.

3.3. GPP Changes

Since 2000, the overall trend of the annual mean value of GPP in the growing season in the study area increased significantly, with a change rate of 3.65 gC·m^{-2}year^{-1} ($p < 0.01$). From 2000 to 2019, the average GPP in the growing season increased from 553.77 gC·m^{-2} to 624.33 gC·m^{-2}, an increase of 12.74% (Figure 4a). Among them, the average value of GPP in the growing season was the lowest at 553.77 gC·m^{-2} in 2000, and the average value was the highest at 728.83 gC·m^{-2} in 2016. During the study period, the GPP of the three land cover types showed an increasing trend. Cropland and forest showed a significant increasing trend ($p < 0.01$), while grassland showed a slight upward trend ($p > 0.05$). The GPP of different land cover types showed a great difference in growing seasons, and the GPP of forest and cropland had a high coincidence with the overall GPP trend in southwest China. Forest was the land cover type with the highest gross primary productivity, and the highest average GPP in the growing season was 654.7 gC·m^{-2}; the second was cropland, which was 571.63 gC·m^{-2}; and the lowest value of grassland was 420 gC·m^{-2}. Since forest had the highest average GPP, the transfer of forest to other land cover types led to a decrease in GPP, and the conversion of cropland and grassland to forest led to an increase in GPP. In general, the carbon storage of land cover increased, indicating that the ecological environment in southwest China improved.

The average value of the growing season in southwest China from 2000 to 2019 was 594.61 gC·m^{-2}, which showed a decreasing trend from southeast to northwest, corresponding to the distribution of cropland, forest, and grassland, with significant spatial differences. The high-value regions of GPP were mainly distributed in northwestern Yunnan, southeastern Guangxi, Sichuan Basin, etc. In addition, the GPP was relatively large in central and southern Guizhou, which was covered by forest and cropland. However, the concentrated distribution region of grassland in northwestern Sichuan, the concentrated distribution region of cropland in the Sichuan Basin, and the GPP value were generally lower than the average value (Figure 4b).

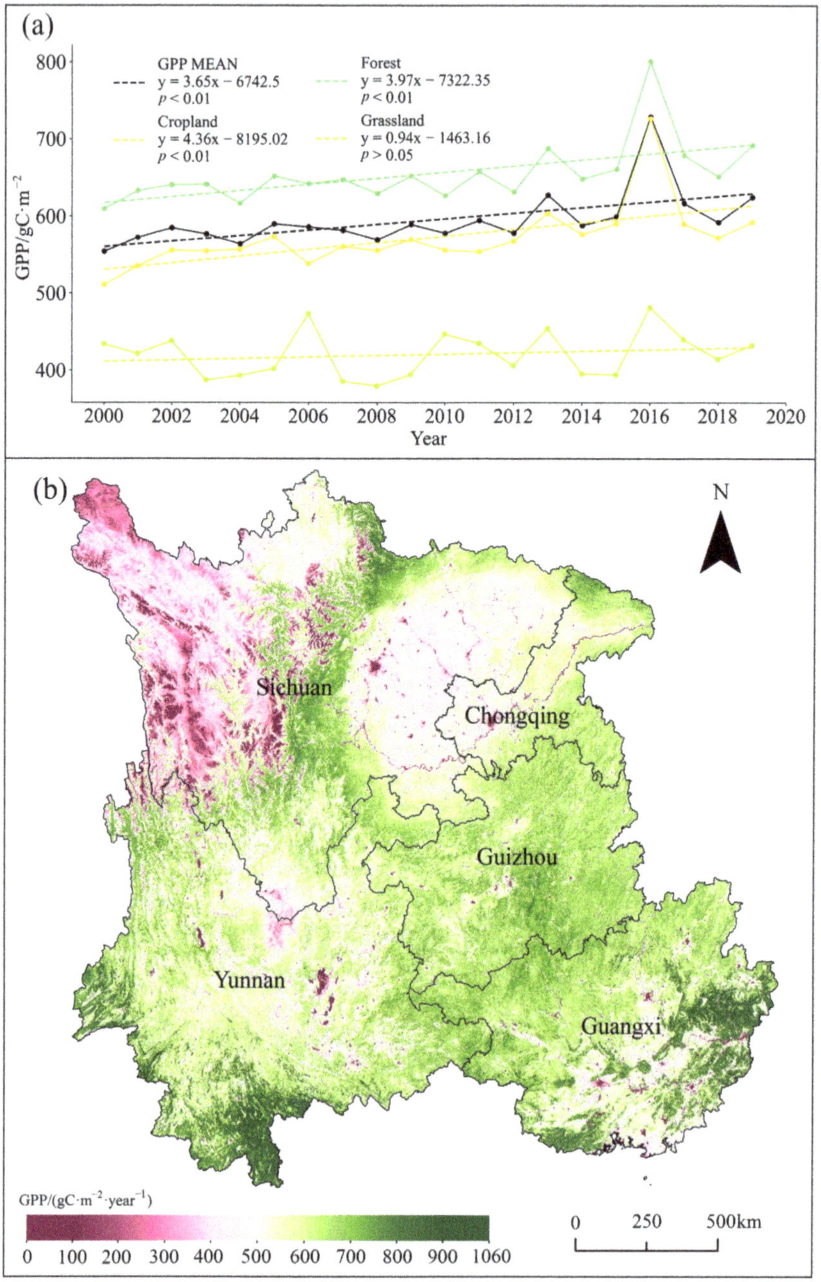

Figure 4. (**a**) Interannual GPP changes in different land use types; the dashed line shows the variation trend of GPP in different land use types, where the black is GPP mean, the formula is $y = 3.65x - 6742.5$; the yellow is cropland's GPP, the formula is $y = 4.36x - 8195.02$; the cyan is forest's GPP, the formula is $y = 3.97x - 7322.35$; the light green is grassland's GPP, the formula is $y = 0.94x - 1463.16$ and (**b**) spatial distribution pattern of GPP values of different land cover types in Southwest China from 2000 to 2019.

3.4. Changes in the Center of Gravity of NDVI and GPP

From 2000 to 2019, the migration direction of the NDVI center of gravity of different land cover types also showed varying differences. The center of gravity of the NDVI value of cropland moved to the southwest by 0.12° in the meridian and 0.12° in latitude. Among them, the center of gravity of cropland moved in the southeast direction, moving 0.037° in the meridian and 0.12° in latitude from 2000 to 2005; the center of gravity moved in the west direction, moving 0.16° in the meridian and 0.01° in latitude from 2005 to 2019; and the center of gravity of the woodland moved to the northeast, moving 0.03° in the meridian and 0.06° in latitude. However, the center of gravity moved to the northwest from 2000 to 2005 and continued to move to the northeast from 2005 to 2019. The center of gravity of the grassland continued to move to the northwest from 2000 to 2019 by 0.06° in the meridian and 0.16° in latitude. Since 2000–2006 was the main implementation period of the ecological project, the composition of land cover changed greatly, and the change direction of the NDVI center of forest, cropland, and grassland changed (Figure 5).

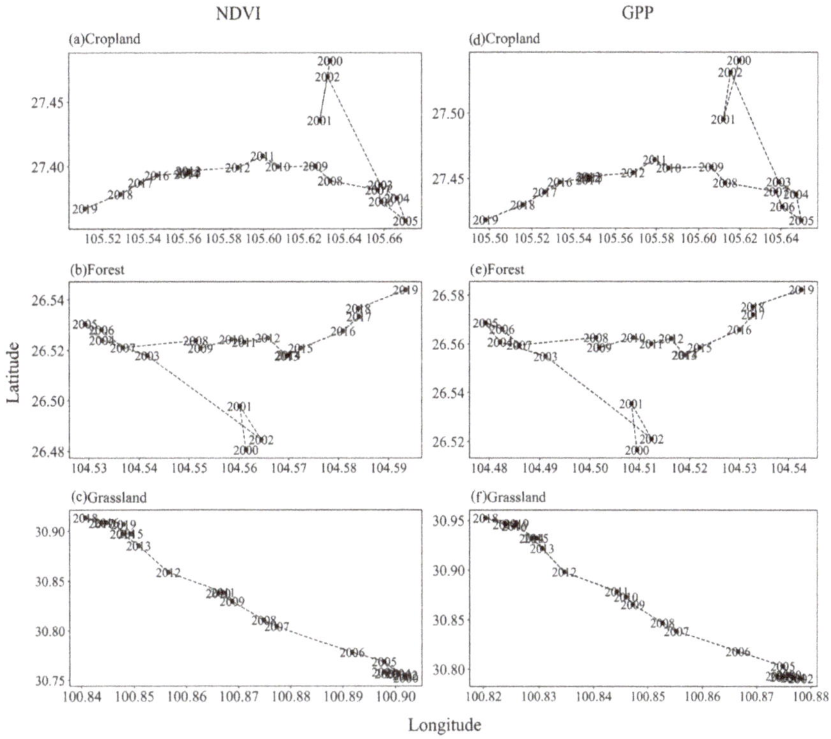

Figure 5. Changes in the center of gravity of NDVI and GPP. (Note: (**a–c**): Changes in the NDVI center of gravity of cropland, forest and grassland from 2000 to 2019; (**d–f**) Change in the GPP center of gravity of cropland, forest and grassland from 2000 to 2019).

The study found that the migration direction of the GPP gravity center of the same type of LULC type had been consistent with the gravity center migration direction of its NDVI (Figure 5). During the study period, the center of GPP of cropland moved to the southwest. The GPP center of gravity of cropland moved to the southeast from 2000 to 2005, by 0.03° in the meridian and 0.12° in latitude, while the GPP center of gravity of cropland moved to the southwest from 2005 to 2019, and the migration amplitude increased, moving 0.15° in the meridian, while there was almost no change in latitude. The GPP center of gravity of the forest moved to the northeast. From 2000 to 2005, it first moved to the northwest, moving 0.03° in the meridian and 0.052° in latitude, and then it moved to the northeast, moving 0.063° in the meridian and 0.014° in latitude. The GPP center of gravity of the grassland continued to move to the northwest in the study period, by 0.05° in the meridian and 0.15° in latitude.

From 2000 to 2019, the center of gravity of the lowest coverage shifted to the southeast, moving 0.36° in the meridian direction and 0.24° in the latitudinal direction, with a small but complex change in the inter-annual migration direction. The center of gravity of low coverage also moved to the southeast, 0.77° in the longitudinal and 0.24° in the latitudinal direction. During 2000–2008, the low coverage center of gravity first shifted back and forth to the northwest, and then folded to the southeast after 2008. The medium coverage center of gravity moved to the northwest, with a larger migration range, moving 1.24° in the meridian and 1.14° in the latitudinal direction. The center of gravity of the high coverage migrated to the northwest but migrated to the southeast first from 2000 to 2011, moving 0.157° in the meridian and 0.01° in latitude. After 2011, it migrated to the northwest, moving 0.26° in the meridian and 0.08° in latitude; the center of gravity of the highest coverage migrated to the southeast as a whole, moving 1.32°in the meridian and 0.53° in latitude (Figure A1).

Under different vegetation coverage levels, the migration direction of the gravity center of GPP was basically the same as that of vegetation NDVI (Figure A1). During the study period, the gravity center of GPP of lowest coverage and low coverage migrated to the southeast direction, with 0.36°and 0.77°in longitude, respectively, with a latitude migration of 0.24° for both. The center of gravity of the GPP with medium coverage moved to the northwest, moving 1.30° in the meridian and 1.19° in latitude. The center of gravity of the high-coverage GPP moved to the southeast from 2000 to 2008, moving 0.21° in the meridian and 0.22° in the latitudinal direction, then turned around and moved to the northwest direction after 2008, moving 0.31° in the meridian direction, and 0.30° in the latitudinal direction. The center of gravity of the highest-coverage GPP moved to the southeast direction, moving 1.37° in the meridian direction and 0.61° in the latitudinal direction.

4. Discussion

4.1. Spatiotemporal Variation in Land Types and NDVI and GPP

The results revealed that the rate of returning farmland to forest increased significantly in the study period (2000–2019), and cropland was the main land source for forest expansion. The area changes of LULC types were as follows: the area of cropland increased first and then decreased, while the area of forest decreased first and then increased, and the grassland continued to decrease in southwest China (Figure 2). This trend was consistent with the trend of land-use change in China [19]. In the study region, cropland was mainly distributed in the east of Sichuan, the center of Guangxi, the west of Guizhou, and the east of Yunnan (Figure 1). These regions were suitable for agricultural activities due to their flat terrain. The increase in agricultural intensification and productivity has led to the expansion of cropland in these regions, the main source of which was forest and grassland. Due to the development of ecological engineering, the phenomenon of "returning farmland to forest" appeared in the margin of Sichuan Basin, northwest and southwest Yunnan, southeast Guizhou, and south Guangxi, which effectively controlled the expansion of cropland. This indicates that the ecological restoration project in this region was the main driver of LULC change in most regions of southwest China and has achieved some results at this stage [55]. Second, rapid urbanization was also one of the most common causes of farmland loss [56]. At the beginning of the 21st century, the urban area of the Sichuan Basin increased by 66,000 hm^2, accounting for 68.31% of the decreased area of cultivated land [57]. The grassland was mainly concentrated in the northwestern part of Sichuan, where the altitude is higher, combined with a cold and dry climate. There are rivers flowing through this region, and the precipitation is relatively abundant. Under the background of climate warming, this is conducive to the growth of vegetation, and with the LULC change, in some regions, grassland has shifted to forest.

The study found that the NDVI and GPP of vegetation in the growing season had shown a significant upward trend as a whole, and the growth trends had been different for different LULC types in southwest China from 2000 to 2019 (Figures 3b and 4a). This is consistent with the conclusion that the main vegetation types in China show a dynamic greening trend, which is the result of the combined effects of climate change, LULC types of distribution, and human activities (such as ecological engineering and agricultural management), among other factors [58–60]. In general, the increase in the NDVI value in southwest China was mainly due to ecological restoration caused by ecological projects such as returning farmland to forest, natural forest protection plans, and closing mountains for afforestation [55,61]. The average NDVI and GPP of vegetation in the growing season in southwest China were high in the southeast and low in the northwest (Figures 3a and 4b), which was mainly related to the spatial distribution of LULC types. In southwest Yunnan, southeast Guangxi, and Sichuan Basin, high mean NDVI and GPP levels in the growing season were mainly distributed in forests, while northwest Sichuan and Sichuan Basin were mainly distributed in grasslands and croplands, which were lower than those of forests on the whole. Compared with 2000, most regions became greener and more productive in 2019. The conversion of a large amount of cropland or grassland to forest and the ecological restoration of most regions (low coverage shift to high coverage) resulted in significant greening of vegetation. However, at the periphery of most cities, vegetation degraded, and productivity declined, indicating that urbanization has led to the loss of vegetation [62]. In addition, studies have shown that the interaction of temperature, precipitation, and solar radiation has different effects on vegetation greening [63–65], and sustained warming and decreased precipitation are key factors in limiting vegetation growth [66]. In the past 20 years, the overall climate has been dry and warm, with a significant upward trend in temperature (0.42 °C/10 year) but no significant change in precipitation in southwest China [36,48]. The occurrence of extreme weather events and natural disasters hinders the growth of vegetation, resulting in a decline in regional vegetation coverage [13,67]. In recent decades, severe droughts have occurred frequently in southwest China [68–71], which have had a significant impact on grassland and cropland. The average NDVI in the vegetation growing season has shown a decreasing trend (Figure 3b). However, the area of forest in this study region was about 2.4 times that of cropland, and the increase in the greening effect on the whole region compensated for the decreasing trend of NDVI.

The productivity levels of different LULC types had different responses to influencing factors. For example, cropland and grassland were vulnerable to extreme climate disasters, but the positive effects of human activities on the effective management of cropland weakened the negative effects of drought. From 2000 to 2019, the area of cropland decreased, but the average GPP value in the growing season showed an increasing trend in southwest China ($p < 0.01$), which may be related to the improvement in the productivity of cropland by agricultural management measures in recent years [24]. The forest ecosystem was relatively stable, and human activities, such as deforestation and afforestation, had a greater impact on the productivity of the forest ecosystem than the impact of extreme climates [72]. The average GPP of forest in the growing season showed a significant increasing trend, which was related to the increase in forest area and vegetation restoration in this region [73]. Studies have shown that vegetation productivity exhibits different growth patterns at different stages of forest age [74,75]. Due to the conversion of grassland to other land types, the amount of grassland has been decreasing continuously for 20 years. However, with the influence of ecosystem protection policies in recent years, the GPP of grassland still showed a fluctuating growth trend (Figure 4a). These results indicate that the overall carbon sequestration capacity of the southwest region is gradually increasing, and the environmental quality of the ecosystem is gradually improving, which corresponds to the significant increase in the vegetation coverage in southwest China in recent years.

4.2. Migration Changes in NDVI and GPP Centroid

Our study indicated that the migration direction of the NDVI center of gravity of cropland, forest, and grassland was the same as that of GPP during 2000–2019 in southwest China (Figure 5), and the spatial distribution pattern of vegetation cover and productivity gravity center changed with a change in LULC distribution. In southwest China, the center of gravity of cropland moved westward, and there were inflection points of the center of gravity in 2001 and 2005. In 2001, the project of returning cropland to forest, ecological protection in the east, and the natural forest protection plan in the west were fully launched. Therefore, the area of cropland increased, and the area of forest decreased during 2000–2005. After 2005, ecological engineering began to achieve results, which showed that the cropland area decreased and the forest area increased, which was consistent with the results of previous studies [76]. These results indicated that the LULC change caused by the ecological restoration project was the direct cause of the change in the vegetation growth and productivity center of gravity in southwest China over the past 20 years. The inflection point times of the change in the center of gravity were consistent with the planning and implementation time of the restoration project. The grassland continued to migrate to the northwest, which may be related to the conversion of grassland to other land types in the southeast, while the northwest region had a higher altitude, less human activity, and fewer changes in land cover types [36]. The study also found that the change in GPP's gravity centers of different vegetation NDVI grades was highly similar to the migration direction of NDVI's gravity center changes of different grades (Figure A1). This indicated that the increase or decrease in vegetation productivity in southwest China was related to the restoration or degradation of vegetation; the productivity increased in the region of vegetation restoration, and the productivity decreased in the region of vegetation degradation. From 2000 to 2019, the change in the center of gravity of the lowest coverage was relatively stable, and the direction and distance of the center of gravity were not large. The center of gravity of low coverage and high coverage mainly shifted to the southeast, and the center of gravity of medium coverage and high coverage mainly shifted to the northwest. This was mainly because of the spatial distribution, composition, and climatic characteristics of LULC types of different vegetation grades at different stages, which affected the change in vegetation cover and the center of gravity of productivity. Therefore, reasonable planning of regional cover-type composition is of great significance to effectively improve regional vegetation NDVI and GPP.

The shift in vegetation types in the direction of the center of gravity indicates that the expansion of vegetation in the direction of migration or the degradation of vegetation in the opposite direction, and the long-term shift in the center of gravity in a single direction, lead to the imbalance of ecosystem structure and function. For example, the southern and southwestern regions of southwest China are the main regions of returning farmland to forest. Due to the favorable ecological conditions in this region, the vegetation coverage has shifted from medium coverage to high coverage. Recent studies have also confirmed that afforestation measures have achieved good results in improving vegetation cover and promoting carbon sequestration [58,77]. However, it was also found that the excessive growth of large regions of forest consumes surface water, resulting in a shortage of regional water resources [78,79]. In addition, local negative effects have also occurred due to the planting of unsuitable tree species [80]. To sum up, large-scale afforestation may rapidly improve the vegetation greening degree in the region in the short term, but it cannot guarantee the long-term stable development of the ecological environment. In addition, because of the heterogeneity of the growth of different vegetation types, it is understood that the composition structure of land cover types will affect the ecosystem balance and sustainable development in the southwest region of southwest China under different topographic and landform conditions.

4.3. Implications for Future

Our study found that the LULC change was beneficial to the improvement in regional greening and plant productivity in southwest China. Karst landforms are widely distributed in the study area. Due to its unique binary structure, precipitation is rapidly lost, which leads to low utilization of precipitation by vegetation [81]. Moreover, the karst regions have thin soil layers and poor water storage capacities, and the climate warming trend may have an inhibitory effect on the growth of forests [14]. Studies have shown that ecological restoration and LULC pattern change not only ameliorate land degradation [40] but also affect local and regional climate [82]. For example, land surface temperature decreases significantly when forest is converted to cropland [83]. The global warming trend has a greater impact on ecosystems [84,85], which has led to dramatic changes in land cover types and plant biomass in the high northern latitudes and promoted the expansion of woody shrubs and forest areas [86]. Meanwhile, the rise in temperature has removed local environmental boundaries, allowing alpine plant species to move to higher altitudes [87]. These results indicate that in future implementations of ecological projects, it is necessary to consider the structural composition and environmental carrying capacity of LULC types under the background of climate warming so that they can more effectively serve the maintenance of local, sustainable ecosystem balance.

5. Conclusions

We investigated NDVI and GPP in the growing season and changes in the spatiotemporal variation patterns of their centers of gravity in southwest China from 2000 to 2019. We found that the LULC structure of the study area changed greatly from 2000 to 2019. The area of forest increased while the area of cropland and grassland decreased, and cropland was the main contributor to the forest increase. For the entire study region, both NDVI and GPP in the growing season showed a generally increasing trend. However, there were differences in the increasing trend of NDVI and GPP among different LULC types. From the perspective of spatial distribution, NDVI and GPP showed a distribution pattern of high in the southeast and low in the northwest, corresponding to the fact that forest is mainly distributed in the southeast and grassland is concentrated in the northwest; the ecological restoration project has greatly improved the vegetation coverage, 45.6% of the region showed a greening trend. In 2000, it was dominated by medium coverage and high coverage (32.32%, 54.73%), while high coverage and highest coverage predominated in 2019 (49.29%, 34.37%). Under the same vegetation coverage, the spatiotemporal variation distribution of the center of gravity of NDVI and GPP was basically the same. From 2000 to 2019, the spatial variation in the center of gravity of NDVI and GPP at the lowest coverage was relatively stable; the center of gravity of low coverage shifted to the east, the center of gravity of medium coverage and the high coverage both shifted to the northwest, and the center of gravity of the highest coverage moved to the southeast. This indicates that although the southwest region is greening, there are differences in vegetation growth and carbon sequestration capacity among different land cover types, resulting in changes in the vegetation center of gravity. Therefore, the combination structure of vegetation cover and land cover type should be prioritized in the future to ensure the balance of the ecosystem in the southwest and maintain sustainable development.

Author Contributions: Conceptualization, L.R. and X.L.; methodology, X.L.; software, M.Z. and C.Y.; formal analysis, X.L. and W.Y.; investigation, M.Z., Q.W., Z.Z., and C.Y.; data curation, X.L.; writing—original draft preparation, X.L.; writing—review and editing, L.R. and X.L.; visualization, X.L.; supervision, X.L.; project administration, X.L.; funding acquisition, L.R. All authors have read and agreed to the published version of the manuscript.

Funding: This research was funded by the key project of the "14th Five-Year" National Key R&D Program "Typical Fragile Ecosystem Protection and Restoration", grant number "2022YFC080900" and the Karst plateau canyon rocky desertification comprehensive control and scale management technology and demonstration of ecological industry, grant number "2016YFC0502603", Project of National "Thirteenth Five-year" Key Research and Development Programme.

Institutional Review Board Statement: Not applicable.

Informed Consent Statement: Informed consent was obtained from all subjects involved in the study.

Data Availability Statement: Not applicable.

Acknowledgments: We thank the anonymous reviewers for their valuable comments. We gratefully acknowledge the design of L.R. and the contribution of co-authors.

Conflicts of Interest: The authors declare no conflict of interest.

Appendix A

Table A1. Transformation matrix of NDVI grades from 2000 to 2019 (km^2).

2000–2019	0~0.35	0.35~0.55	0.55~0.75	0.75~0.85	0.85~1.0
0~0.35	4397.62	2772.06	630.53	95.78	45.96
0.35~0.55	1089.16	11,384.48	11,465.65	1269.25	169.92
0.55~0.75	418.50	8424.23	136,924.63	259,673.48	39,192.96
0.75~0.85	57.80	1560.78	44,215.18	395,140.43	311,870.61
0.85~1.0	5.03	48.18	1050.40	22,088.73	121,421.98

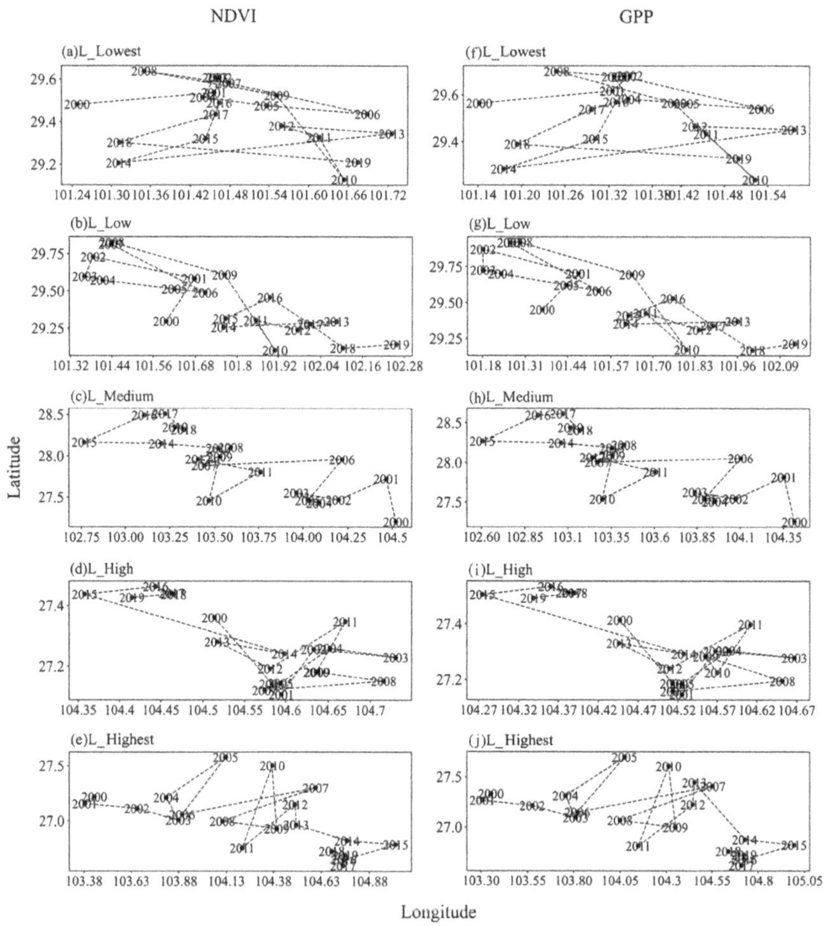

Figure A1. Variation in the center of gravity of different NDVI grades and its GPP. (Note: (**a**–**e**): Changes in the center of gravity of the five NDVI grades from 2000 to 2019; (**f**–**j**): Changes in the center of gravity of GPP for five NDVI grades from 2000 to 2019).

References

1. Liu, H.; Zhang, M.; Lin, Z.; Xu, X. Spatial heterogeneity of the relationship between vegetation dynamics and climate change and their driving forces at multiple time scales in Southwest China. *Agric. For. Meteorol.* **2018**, *256*, 10–21. [CrossRef]
2. Yan, W.; Wang, H.; Jiang, C.; Jin, S.; Ai, J.; Sun, O.J. Satellite view of vegetation dynamics and drivers over southwestern China. *Ecol. Indic.* **2021**, *130*, 108074. [CrossRef]
3. Huang, J.; Ge, Z.; Huang, Y.; Tang, X.; Shi, Z.; Lai, P.; Song, Z.; Hao, B.; Yang, H.; Ma, M. Climate change and ecological engineering jointly induced vegetation greening in global karst regions from 2001 to 2020. *Plant Soil* **2022**, *475*, 193–212. [CrossRef]
4. Yue, Y.; Liao, C.; Tong, X.; Wu, Z.; Fensholt, R.; Prishchepov, A.; Jepsen, M.R.; Wang, K.; Brandt, M. Large scale reforestation of farmlands on sloping hills in South China karst. *Landsc. Ecol.* **2020**, *35*, 1445–1458. [CrossRef]
5. Li, Y.; Zhang, Y.; Lv, J. Interannual variations in GPP in forest ecosystems in Southwest China and regional differences in the climatic contributions. *Ecol. Inform.* **2022**, *69*, 101591. [CrossRef]
6. Luo, X.; Jia, B.; Lai, X. Contributions of climate change, land use change and CO_2 to changes in the gross primary productivity of the Tibetan Plateau. *Atmos. Ocean. Sci. Lett.* **2020**, *13*, 8–15. [CrossRef]
7. Tian, S.; Wang, S.; Bai, X.; Luo, G.; Li, Q.; Yang, Y.; Hu, Z.; Li, C.; Deng, Y. Global patterns and changes of carbon emissions from land use during 1992–2015. *Environ. Sci. Ecotechnol.* **2021**, *7*, 100108. [CrossRef]
8. Zhu, L.; Song, R.; Sun, S.; Li, Y.; Hu, K. Land use/land cover change and its impact on ecosystem carbon storage in coastal areas of China from 1980 to 2050. *Ecol. Indic.* **2022**, *142*, 109178. [CrossRef]
9. Fang, Z.; Ding, T.; Chen, J.; Xue, S.; Zhou, Q.; Wang, Y.; Wang, Y.; Huang, Z.; Yang, S. Impacts of land use/land cover changes on ecosystem services in ecologically fragile regions. *Sci. Total Environ.* **2022**, *831*, 154967. [CrossRef]
10. Jiang, Z.; Lian, Y.; Qin, X. Rocky desertification in Southwest China: Impacts, causes, and restoration. *Earth Sci. Rev.* **2014**, *132*, 1–12. [CrossRef]
11. Dai, Q.; Peng, X.; Yang, Z.; Zhao, L. Runoff and erosion processes on bare slopes in the karst rocky desertification area. *CATENA* **2017**, *152*, 218–226. [CrossRef]
12. Mokhtar, A.; He, H.; Alsafadi, K.; Li, Y.; Zhao, H.; Keo, S.; Bai, C.; Abuarab, M.; Zhang, C.; Elbagoury, K.; et al. Evapotranspiration as a response to climate variability and ecosystem changes in southwest, China. *Environ. Earth Sci.* **2020**, *79*, 312. [CrossRef]
13. Ali, S.; Haixing, Z.; Qi, M.; Liang, S.; Ning, J.; Jia, Q.; Hou, F. Monitoring drought events and vegetation dynamics in relation to climate change over mainland China from 1983 to 2016. *Environ. Sci. Pollut. Res.* **2021**, *28*, 21910–21925. [CrossRef] [PubMed]
14. Sun, H.; Wang, X.; Fan, D.; Sun, O.J. Contrasting vegetation response to climate change between two monsoon regions in Southwest China: The roles of climate condition and vegetation height. *Sci. Total Environ.* **2022**, *802*, 149643. [CrossRef] [PubMed]
15. Bai, X.Y.; Wang, S.-J.; Xiong, K.-N. Assessing spatial-temporal evolution processes of karst rocky desertification land: Indications for restoration strategies. *Land Degrad. Dev.* **2013**, *24*, 47–56. [CrossRef]
16. Sweeting, M.M. *Karst in China: Its Geomorphology and Environment*; Springer: Berlin/Heidelberg, Germany, 2012; Volume 15.
17. Zhu, Z.; Piao, S.; Myneni, R.B.; Huang, M.; Zeng, Z.; Canadell, J.G.; Ciais, P.; Sitch, S.; Friedlingstein, P.; Arneth, A.; et al. Greening of the Earth and its drivers. *Nat. Clim. Chang.* **2016**, *6*, 791–795. [CrossRef]
18. Zheng, K.; Wei, J.-Z.; Pei, J.-Y.; Cheng, H.; Zhang, X.-L.; Huang, F.-Q.; Li, F.-M.; Ye, J.-S. Impacts of climate change and human activities on grassland vegetation variation in the Chinese Loess Plateau. *Sci. Total Environ.* **2019**, *660*, 236–244. [CrossRef]
19. Liu, J.; Kuang, W.; Zhang, Z.; Xu, X.; Qin, Y.; Ning, J.; Zhou, W.; Zhang, S.; Li, R.; Yan, C.; et al. Spatiotemporal characteristics, patterns, and causes of land-use changes in China since the late 1980s. *J. Geogr. Sci.* **2014**, *24*, 195–210. [CrossRef]
20. Fang, J.; Guo, Z.; Hu, H.; Kato, T.; Muraoka, H.; Son, Y. Forest biomass carbon sinks in E ast A sia, with special reference to the relative contributions of forest expansion and forest growth. *Glob. Chang. Biol.* **2014**, *20*, 2019–2030. [CrossRef]
21. Ouyang, Z.; Zheng, H.; Xiao, Y.; Polasky, S.; Liu, J.; Xu, W.; Wang, Q.; Zhang, L.; Xiao, Y.; Rao, E.; et al. Improvements in ecosystem services from investments in natural capital. *Science* **2016**, *352*, 1455–1459. [CrossRef]
22. Li, C.; Bai, X.; Tan, Q.; Luo, G.; Wu, L.; Chen, F.; Xi, H.; Luo, X.; Ran, C.; Chen, H.; et al. High-resolution mapping of the global silicate weathering carbon sink and its long-term changes. *Glob. Chang. Biol.* **2022**, *28*, 4377–4394. [CrossRef] [PubMed]
23. Tian, Y.; Wang, S.; Bai, X.; Luo, G.; Xu, Y. Trade-offs among ecosystem services in a typical Karst watershed, SW China. *Sci. Total Environ.* **2016**, *566*, 1297–1308. [CrossRef] [PubMed]
24. Chen, C.; Park, T.; Wang, X.; Piao, S.; Xu, B.; Chaturvedi, R.K.; Fuchs, R.; Brovkin, V.; Ciais, P.; Fensholt, R.; et al. China and India lead in greening of the world through land-use management. *Nat. Sustain.* **2019**, *2*, 122–129. [CrossRef] [PubMed]
25. Xiong, L.; Bai, X.; Zhao, C.; Li, Y.; Tan, Q.; Luo, G.; Wu, L.; Chen, F.; Li, C.; Ran, C.; et al. High-resolution datasets for global carbonate and silicate rock weathering carbon sinks and their change trends. *Earth's Future* **2022**, *10*, e2022EF002746. [CrossRef]
26. Brandt, M.; Yue, Y.; Wigneron, J.-P.; Tong, X.; Tian, F.; Jepsen, M.R.; Xiao, X.; Verger, A.; Mialon, A.; Al-Yaari, A.; et al. Satellite-observed major greening and biomass increase in south China karst during recent decade. *Earths Future* **2018**, *6*, 1017–1028. [CrossRef]
27. Zhang, S.; Bai, X.; Zhao, C.; Tan, Q.; Luo, G.; Wu, L.; Xi, H.; Li, C.; Chen, F.; Ran, C.; et al. China's carbon budget inventory from 1997 to 2017 and its challenges to achieving carbon neutral strategies. *J. Clean. Prod.* **2022**, *347*, 130966. [CrossRef]
28. Pettorelli, N.; Vik, J.O.; Mysterud, A.; Gaillard, J.-M.; Tucker, C.J.; Stenseth, N.C. Using the satellite-derived NDVI to assess ecological responses to environmental change. *Trends Ecol. Evol.* **2005**, *20*, 503–510. [CrossRef]
29. Fensholt, R.; Proud, S.R. Evaluation of earth observation based global long term vegetation trends—Comparing GIMMS and MODIS global NDVI time series. *Remote Sens. Environ.* **2012**, *119*, 131–147. [CrossRef]

30. Zhang, X.; Liu, K.; Li, X.; Wang, S.; Wang, J. Vulnerability assessment and its driving forces in terms of NDVI and GPP over the Loess Plateau, China. *Phys. Chem. Earth, Parts A/B/C* **2022**, *125*, 103106. [CrossRef]
31. Wang, Z.; Liu, S.; Wang, Y.; Valbuena, R.; Wu, Y.; Kutia, M.; Zheng, Y.; Lu, W.; Zhu, Y.; Zhao, M.; et al. Tighten the Bolts and Nuts on GPP Estimations from Sites to the Globe: An Assessment of Remote Sensing Based LUE Models and Supporting Data Fields. *Remote Sens.* **2021**, *13*, 168. [CrossRef]
32. Ito, A.; Inatomi, M.; Huntzinger, D.N.; Schwalm, C.; Michalak, A.M.; Cook, R.; King, A.W.; Mao, J.; Wei, Y.; Mac Post, W.; et al. Decadal trends in the seasonal-cycle amplitude of terrestrial CO_2 exchange resulting from the ensemble of terrestrial biosphere models. *Tellus B Chem. Phys. Meteorol.* **2016**, *68*, 28968. [CrossRef]
33. Piao, S.; Sitch, S.; Ciais, P.; Friedlingstein, P.; Peylin, P.; Wang, X.; Ahlström, A.; Anav, A.; Canadell, J.G.; Cong, N.; et al. Evaluation of terrestrial carbon cycle models for their response to climate variability and to CO_2 trends. *Glob. Chang. Biol.* **2013**, *19*, 2117–2132. [CrossRef] [PubMed]
34. Wu, L.; Wang, S.; Bai, X.; Tian, Y.; Luo, G.; Wang, J.; Li, Q.; Chen, F.; Deng, Y.; Yang, Y.; et al. Climate change weakens the positive effect of human activities on karst vegetation productivity restoration in southern China. *Ecol. Indic.* **2020**, *115*, 106392. [CrossRef]
35. Chen, Y.; Wang, S.; Wang, Y. Spatiotemporal Evolution of Cultivated Land Non-Agriculturalization and Its Drivers in Typical Areas of Southwest China from 2000 to 2020. *Remote Sens.* **2022**, *14*, 3211. [CrossRef]
36. Jiang, S.; Chen, X.; Smettem, K.; Wang, T. Climate and land use influences on changing spatiotemporal patterns of mountain vegetation cover in southwest China. *Ecol. Indic.* **2021**, *121*, 107193. [CrossRef]
37. Ding, Z.; Zheng, H.; Liu, Y.; Zeng, S.; Yu, P.; Shi, W.; Tang, X. Spatiotemporal Patterns of Ecosystem Restoration Activities and Their Effects on Changes in Terrestrial Gross Primary Production in Southwest China. *Remote Sens.* **2021**, *13*, 1209. [CrossRef]
38. Tan, Z.; Tao, H.; Jiang, J.; Zhang, Q. Influences of climate extremes on NDVI (normalized difference vegetation index) in the Poyang Lake Basin, China. *Wetlands* **2015**, *35*, 1033–1042. [CrossRef]
39. Wang, W.; Wang, W.J.; Li, J.S.; Wu, H.; Xu, C.; Liu, T. The impact of sustained drought on vegetation ecosystem in Southwest China based on remote sensing. *Procedia Environ. Sci.* **2010**, *2*, 1679–1691. [CrossRef]
40. Tong, X.; Brandt, M.; Yue, Y.; Horion, S.; Wang, K.; De Keersmaecker, W.; Tian, F.; Schurgers, G.; Xiao, X.; Luo, Y.; et al. Increased vegetation growth and carbon stock in China karst via ecological engineering. *Nat. Sustain.* **2018**, *1*, 44–50. [CrossRef]
41. Yue, X.; Zhang, T.; Shao, C. Afforestation increases ecosystem productivity and carbon storage in China during the 2000s. *Agric. For. Meteorol.* **2021**, *296*, 108227. [CrossRef]
42. Xiao, J.; Zhou, Y.; Zhang, L. Contributions of natural and human factors to increases in vegetation productivity in China. *Ecosphere* **2015**, *6*, art233–20. [CrossRef]
43. Hurtt, G.C.; Chini, L.P.; Frolking, S.; Betts, R.A.; Feddema, J.; Fischer, G.; Fisk, J.P.; Hibbard, K.; Houghton, R.A.; Janetos, A.; et al. Harmonization of land-use scenarios for the period 1500–2100: 600 years of global gridded annual land-use transitions, wood harvest, and resulting secondary lands. *Clim. Chang.* **2011**, *109*, 117–161. [CrossRef]
44. Zhao, A.; Zhang, A.; Liu, X.; Cao, S. Spatiotemporal changes of normalized difference vegetation index (NDVI) and response to climate extremes and ecological restoration in the Loess Plateau, China. *Arch. Meteorol. Geophys. Bioclimatol. Ser. B* **2018**, *132*, 555–567. [CrossRef]
45. Piao, S.; Fang, J.; Ji, W.; Guo, Q.; Ke, J.; Tao, S. Variation in a satellite-based vegetation index in relation to climate in China. *J. Veg. Sci.* **2004**, *15*, 219–226. [CrossRef]
46. Liu, H.; Gong, P.; Wang, J.; Clinton, N.; Bai, Y.; Liang, S. Annual dynamics of global land cover and its long-term changes from 1982 to 2015. *Earth Syst. Sci. Data* **2020**, *12*, 1217–1243. [CrossRef]
47. Rundqvist, S.; Hedenås, H.; Sandström, A.; Emanuelsson, U.; Eriksson, H.; Jonasson, C.; Callaghan, T.V. Tree and shrub expansion over the past 34 years at the tree-line near Abisko, Sweden. *Ambio* **2011**, *40*, 683–692. [CrossRef]
48. Lian, Y.; You, G.J.-Y.; Lin, K.; Jiang, Z.; Zhang, C.; Qin, X. Characteristics of climate change in southwest China karst region and their potential environmental impacts. *Environ. Earth Sci.* **2015**, *74*, 937–944. [CrossRef]
49. Yang, J.; Huang, X. 30 m annual land cover and its dynamics in China from 1990 to 2019. *Earth Syst. Sci. Data Discuss.* **2021**, 1–29.
50. Piao, S.; Fang, J.; Zhou, L.; Guo, Q.; Henderson, M.; Ji, W.; Li, Y.; Tao, S. Interannual variations of monthly and seasonal normalized difference vegetation index (NDVI) in China from 1982 to 1999. *J. Geophys. Res. Earth Surf.* **2003**, *108*, 4401. [CrossRef]
51. Zhang, Y.; Zhu, Z.; Liu, Z.; Zeng, Z.; Ciais, P.; Huang, M.; Liu, Y.; Piao, S. Seasonal and interannual changes in vegetation activity of tropical forests in Southeast Asia. *Agric. For. Meteorol.* **2016**, *224*, 1–10. [CrossRef]
52. Hou, W.; Gao, J.; Wu, S.; Dai, E. Interannual variations in growing-season NDVI and its correlation with climate variables in the southwestern karst region of China. *Remote Sens.* **2015**, *7*, 11105–11124. [CrossRef]
53. Holben, B.N. Characteristics of maximum-value composite images from temporal AVHRR data. *Int. J. Remote Sens.* **1986**, *7*, 1417–1434. [CrossRef]
54. Yuan, L.; Chen, X.; Wang, X.; Xiong, Z.; Song, C. Spatial associations between NDVI and environmental factors in the Heihe River Basin. *J. Geogr. Sci.* **2019**, *29*, 1548–1564. [CrossRef]
55. Pei, J.; Wang, L.; Wang, X.; Niu, Z.; Kelly, M.; Song, X.-P.; Huang, N.; Geng, J.; Tian, H.; Yu, Y.; et al. Time series of Landsat imagery shows vegetation recovery in two fragile karst watersheds in southwest china from 1988 to 2016. *Remote Sens.* **2019**, *11*, 2044. [CrossRef]
56. López, E.; Bocco, G.; Mendoza, M.; Duhau, E. Predicting land-cover and land-use change in the urban fringe: A case in Morelia city, Mexico. *Landsc. Urban Plan.* **2001**, *55*, 271–285. [CrossRef]

57. Liu, J.; Zhang, Z.; Xu, X.; Kuang, W.; Zhou, W.; Zhang, S.; Li, R.; Yan, C.; Yu, D.; Wu, S.; et al. Spatial patterns and driving forces of land use change in China during the early 21st century. *J. Geogr. Sci.* **2010**, *20*, 483–494. [CrossRef]
58. Lu, F.; Hu, H.; Sun, W.; Zhu, J.; Liu, G.; Zhou, W.; Zhang, Q.; Shi, P.; Liu, X.; Wu, X.; et al. Effects of national ecological restoration projects on carbon sequestration in China from 2001 to 2010. *Proc. Natl. Acad. Sci. USA* **2018**, *115*, 4039–4044. [CrossRef]
59. Chen, Y.; Chen, L.; Cheng, Y.; Ju, W.; Chen, H.Y.H.; Ruan, H. Afforestation promotes the enhancement of forest LAI and NPP in China. *For. Ecol. Manag.* **2020**, *462*, 117990. [CrossRef]
60. Zhang, S.; Bai, X.; Zhao, C.; Tan, Q.; Luo, G.; Cao, Y.; Deng, Y.; Li, Q.; Li, C.; Wu, L.; et al. Limitations of soil moisture and formation rate on vegetation growth in karst areas. *Sci. Total Environ.* **2022**, *810*, 151209. [CrossRef]
61. Pan, Y.; Birdsey, R.A.; Fang, J.; Houghton, R.; Kauppi, P.E.; Kurz, W.A.; Phillips, O.L.; Shvidenko, A.; Lewis, S.L.; Canadell, J.G.; et al. A large and persistent carbon sink in the world's forests. *Science* **2011**, *333*, 988–993. [CrossRef]
62. Deng, J.S.; Wang, K.; Hong, Y.; Qi, J.G. Spatio-temporal dynamics and evolution of land use change and landscape pattern in response to rapid urbanization. *Landsc. Urban Plan.* **2009**, *92*, 187–198. [CrossRef]
63. Nemani, R.R.; Keeling, C.D.; Hashimoto, H.; Jolly, W.M.; Piper, S.C.; Tucker, C.J.; Myneni, R.B.; Running, S.W. Climate-driven increases in global terrestrial net primary production from 1982 to 1999. *Science* **2003**, *300*, 1560–1563. [CrossRef] [PubMed]
64. Forzieri, G.; Alkama, R.; Miralles, D.G.; Cescatti, A. Satellites reveal contrasting responses of regional climate to the widespread greening of Earth. *Science* **2017**, *356*, 1180–1184. [CrossRef] [PubMed]
65. Yang, H.; Hu, J.; Zhang, S.; Xiong, L.; Xu, Y. Climate Variations vs. Human Activities: Distinguishing the Relative Roles on Vegetation Dynamics in the Three Karst Provinces of Southwest China. *Front. Earth Sci.* **2022**, *10*, 799493. [CrossRef]
66. Piao, S.; Wang, X.; Ciais, P.; Zhu, B.; Wang, T.; Liu, J. Changes in satellite-derived vegetation growth trend in temperate and boreal Eurasia from 1982 to 2006. *Glob. Chang. Biol.* **2011**, *17*, 3228–3239. [CrossRef]
67. Guan, Y.; Zheng, F.; Zhang, P.; Qin, C. Spatial and temporal changes of meteorological disasters in China during 1950–2013. *Nat. Hazards* **2015**, *75*, 2607–2623. [CrossRef]
68. Peng, J.; Zhang, Q.Y.; Bueh, C. On the characteristics and possible causes of a severe drought and heat wave in the Sichuan-Chongqing region in 2006. *Clim. Environ. Res.* **2007**, *12*, 464–474.
69. Huang, R.; Liu, Y.; Wang, L.; Wang, L. Analyses of the causes of severe drought occurring in Southwest China from the fall of 2009 to the spring of 2010. *Chin. J. Atmos. Sci.* **2012**, *36*, 443–457.
70. Sun, L.; Ren, F.M.; Wang, Z.Y.; Liu, Y.Y.; Liu, Y.J.; Wang, P.L.; Wang, D.Q. Analysis of climate anomaly and causation in August 2011. *Meteorol. Mon.* **2012**, *38*, 615–622.
71. Huang, Y.; Xu, C.; Yang, H.; Wang, J.; Jiang, D.; Zhao, C. Temporal and spatial variability of droughts in southwest china from 1961 to 2012. *Sustainability* **2015**, *7*, 13597–13609. [CrossRef]
72. Qiao, Y.; Jiang, Y.; Zhang, C. Contribution of karst ecological restoration engineering to vegetation greening in southwest China during recent decade. *Ecol. Indic.* **2021**, *121*, 107081. [CrossRef]
73. Tong, X.; Brandt, M.; Yue, Y.; Ciais, P.; Rudbeck Jepsen, M.; Penuelas, J.; Wigneron, J.-P.; Xiao, X.-P.; Song, X.-P.; Horion, S.; et al. Forest management in southern China generates short term extensive carbon sequestration. *Nat. Commun.* **2020**, *11*, 129. [CrossRef] [PubMed]
74. Wang, S.; Zhou, L.; Chen, J.; Ju, W.; Feng, X.; Wu, W. Relationships between net primary productivity and stand age for several forest types and their influence on China's carbon balance. *J. Environ. Manag.* **2011**, *92*, 1651–1662. [CrossRef] [PubMed]
75. He, L.; Chen, J.M.; Pan, Y.; Birdsey, R.A.; Kattge, J. Relationships between net primary productivity and forest stand age in US forests. *Glob. Biogeochem. Cycles* **2012**, *26*, 394–415. [CrossRef]
76. Xu, X.; Liu, H.; Lin, Z.; Jiao, F.; Gong, H. Relationship of abrupt vegetation change to climate change and ecological engineering with multi-timescale analysis in the karst region, Southwest China. *Remote Sens.* **2019**, *11*, 1564. [CrossRef]
77. Zhang, C. Carbonate rock dissolution rates in different landuses and their carbon sink effect. *Chin. Sci. Bull.* **2011**, *56*, 3759–3765. [CrossRef]
78. Škerlep, M.; Steiner, E.; Axelsson, A.L.; Kritzberg, E.S. Afforestation driving long-term surface water browning. *Glob. Chang. Biol.* **2020**, *26*, 1390–1399. [CrossRef]
79. Feng, X.; Fu, B.; Piao, S.; Wang, S.; Ciais, P.; Zeng, Z.; Lü, Y.; Zeng, Y.; Li, Y.; Jiang, X.; et al. Revegetation in China's Loess Plateau is approaching sustainable water resource limits. *Nat. Clim. Chang.* **2016**, *6*, 1019–1022. [CrossRef]
80. Zhang, L.; Sun, P.; Huettmann, F.; Liu, S. Where should China practice forestry in a warming world? *Glob. Chang. Biol.* **2022**, *28*, 2461–2475. [CrossRef]
81. Zhou, Q.; Luo, Y.; Zhou, X.; Cai, M.; Zhao, C. Response of vegetation to water balance conditions at different time scales across the karst area of southwestern China—A remote sensing approach. *Sci. Total Environ.* **2018**, *645*, 460–470. [CrossRef]
82. Peng, S.-S.; Piao, S.; Zeng, Z.; Ciais, P.; Zhou, L.; Li, L.Z.X.; Myneni, R.B.; Yin, Y.; Zeng, H. Afforestation in China cools local land surface temperature. *Proc. Natl. Acad. Sci. USA* **2014**, *111*, 2915–2919. [CrossRef] [PubMed]
83. Jiang, L.; Zhang, H.; Zhao, F.; Zhang, L.; Wang, X. Warming/cooling effect of cropland expansion during the 1900s~ 2010s in the Heilongjiang Province, Northeast of China. *Int. J. Biometeorol.* **2022**, *66*, 1379–1390. [CrossRef]
84. Heyder, U.; Schaphoff, S.; Gerten, D.; Lucht, W. Risk of severe climate change impact on the terrestrial biosphere. *Environ. Res. Lett.* **2011**, *6*, 034036. [CrossRef]
85. Xi, H.; Wang, S.; Bai, X.; Tang, H.; Luo, G.; Li, H.; Wu, L.; Li, C.; Chen, H.; Ran, C.; et al. The responses of weathering carbon sink to eco-hydrological processes in global rocks. *Sci. Total Environ.* **2021**, *788*, 147706. [CrossRef] [PubMed]

86. Xue, S.-Y.; Xu, H.-Y.; Mu, C.-C.; Wu, T.-H.; Li, W.-P.; Zhang, W.-X.; Streletskaya, I.; Grebenets, V.; Sokratov, S.; Kizyakov, A.; et al. Changes in different land cover areas and NDVI values in northern latitudes from 1982 to 2015. *Adv. Clim. Chang. Res.* **2021**, *12*, 456–465. [CrossRef]
87. Vanneste, T.J.M. Climate Change and Alpine Vegetation Shifts on Mountaintop Summits in Norway. Ph.D. Thesis, Universiteit Gent, Gent, Belgium, 2016.

land

MDPI

Article

Analysis Long-Term and Spatial Changes of Forest Cover in Typical Karst Areas of China

Fei Chen [1,2,3], Xiaoyong Bai [2,3,*], Fang Liu [1], Guangjie Luo [4], Yichao Tian [5], Luoyi Qin [2], Yue Li [6], Yan Xu [2], Jinfeng Wang [1,2,3], Luhua Wu [2,3], Chaojun Li [2,3,7], Sirui Zhang [2,3,7] and Chen Ran [2,3,7]

1 College of Resources and Environmental Engineering, Guizhou University, Guiyang 550025, China
2 State Key Laboratory of Environmental Geochemistry, Institute of Geochemistry, Chinese Academy of Sciences, Guiyang 550081, China
3 CAS Center for Excellence in Quaternary Science and Global Change, Xi'an 710061, China
4 Guizhou Provincial Key Laboratory of Geographic State Monitoring of Watershed, Guizhou Education University, Guiyang 550018, China
5 College of Resources and Environmental, Beibu Gulf University, Qinzhou 535011, China
6 College of Public Management, GuiZhou University of Finance and Economics, Guiyang 550025, China
7 University of Chinese Academy of Sciences, Beijing 100049, China
* Correspondence: baixiaoyong@vip.skleg.cn

Abstract: In recent decades, China has exhibited the fastest and most remarkable social-economic development in the world. As a result of such development, the forest cover of the country has undergone radical changes. This paper aims to develop a method for analyzing long-term and spatial changes in forest cover based on historical maps and remote sensing images. Moreover, we will focus on the reduction or restoration of forests distributed at different altitudes, slopes, soils, and lithologic types in different periods, to reveal the problems that should be paid attention to in forest restoration in karst areas. A typical county of China was selected as the study area. A historical military operation map was considered the principal source of basic data. These data were then combined with Landsat satellite images to conduct quantitative analysis on changes in the spatial area and location of forest cover with a long time series. The findings are as follows: in terms of time series, the forest area in the study area showed a trend of decreasing at first and then increasing, with the year 1986 as the turning point. In terms of spatial patterns, a considerable difference is observed in regions with changes in forest cover under different historical periods. Changes are obvious in limestone areas, rock soil areas, and areas with an elevation of 2000–2500 m and a slope gradient of 6°–15°. Spatial–temporal changes in forest cover reflect the effects of the war, national policy, and economic development to some extent. All these results indicate that, despite its limitations, a historical map is a valuable document for studying an ecological environment.

Keywords: forest cover; old maps; spatial–temporal changes; forest landscape; transfer matrix

Citation: Chen, F.; Bai, X.; Liu, F.; Luo, G.; Tian, Y.; Qin, L.; Li, Y.; Xu, Y.; Wang, J.; Wu, L.; et al. Analysis Long-Term and Spatial Changes of Forest Cover in Typical Karst Areas of China. *Land* **2022**, *11*, 1349. https://doi.org/10.3390/land11081349

Academic Editor: Francisco Manzano Agugliaro

Received: 7 June 2022
Accepted: 15 August 2022
Published: 18 August 2022

1. Introduction

Forests play an important role in maintaining the ecological balance of the earth [1,2]. The forest is a significant symbol of the ecological environment [3], and it is the material base for forestry production [4,5]. Forest can not only regulate atmospheric circulation and water cycle, but also affect climate change, and play an important role in protecting water and soil resources and preventing wind and sand [6–8]. The rapid change in forest cover is causing the loss of habitat, biodiversity, and climate change [9,10]. Social and economic conditions in different periods have different impacts on land vegetation cover [11]. The spatial reconstruction of historical forests is helpful for a better understanding of the changes human beings have made to the surface and their impacts on the environment [12]. Therefore, long-term and spatial dynamic changes in forest cover have been an important concern for global ecologists, environmentalists, and so on [13–16].

Remote sensing data can provide the continuous change in surface elements in time and space, and it plays an irreplaceable role in regional ecological environment monitoring [17]. However, during the period when satellites were not yet launched, satellite images were unavailable, and the surface of the Earth and its features could not be located accurately [18]. Thus, studies on the spatial framework of an ecological environment are limited to the years after 1972 [19]. This issue has always been a challenge for some ecologists [20]. With openness and sharing of data, an increasing number of available historical maps, especially historical military maps, can be obtained easily on the Internet or in libraries [21,22]. Thus, historical maps are an important data source for scientists to reveal the changes in the surface landscape before remote sensing images. Some internationally well-known scholars have conducted a considerable amount of valuable research with the use of these maps [23–26].

Because of the important ecological and economic functions of forests, it is very important to monitor long-term changes by using old maps and written sources, because this method enables people to monitor trends from different time dimensions and find the reasons for the current situation [16,27]. For example, Skaloš et al. [28] carried out a comparison of landscape developments in Sweden and the Czech Republic by using old maps. Pelorosso et al. [29] used historical maps and recent remote sensing-derived maps to reduce misleading changes and to assess spatial aggregation errors based on a data integration procedure using landscape metrics in Italy. Skaloš et al. [16] analyzed long-term land-cover changes in central Bohemia and contributed to a better understanding of the dynamics of forest land using old military survey maps and orthophotograph maps, covering more than 250 years. Furthermore, researchers of some countries have used military maps to study past land cover changes in Slovenia [30,31], Germany [32], Sweden [33], Norway [34,35], and the Czech Republic [28,36,37].

On the other hand, many researchers have studied the long-term land cover in China. He et al. [11] reconstructed the forest cover in China from 1700–2000, and found that the deforestation mainly occurred in southwest China, the hilly regions of south China, the southeast of Gansu province, and northeast China from 1700 to the 1960s. Yang et al. [38] evaluated the reliability of global historical land-use scenarios for forest data in China and pointed out that these global historical land-use scenarios could not accurately reveal the spatial and temporal pattern of China's forests due to differences in data sources, reconstruction methods and spatial scales. Li et al. [39] used the historical forest area allocation model to reconstruct forest cover between 1780 and 1940 in Northeast China. Liu et al. [40] synthesized historical maps and aerial images to describe long-term land-use change and landscape dynamics for a region near Chancellorsville, USA, from 1867 to 2014. In a word, many researchers use historical maps combined with remote sensing images to reveal long-term changes in forest/land cover and proved that climate change and human activities are the main influencing factors, but mainly on the national and regional scales. Aiming at the change in long-time series forest cover in karst areas, Tanacs et al. [41] used an integrated GIS of historical data (18th–19th century military maps, old forest management plans, aerial imagery, etc.) to describe the example of the Haragistya karst plateau and how the forests of the Aggtelek karst region were used in the last few centuries and to what extent they were affected by anthropogenic activity. However, there is still a lack of research on long-time series forest cover in typical karst areas of China.

The southwest China karst area is located in the center of East Asia karst area, and it is one of the three largest karst areas in the world [42]. Karst ecosystem is a fragile ecosystem, which is affected and restricted by the special geological background [43,44]. Because of the special geological and climatic conditions, the bedrock in this area is exposed, which has the basic characteristics of little soil reserves, discontinuous soil distribution, complex and diverse micro-landforms, etc. [45,46]. These characteristics have caused some problems in karst areas, such as high rock exposure rate, slow soil formation speed, easy loss, and weak soil water and fertilizer retention capacity [47,48]. Among them, karst rocky desertification with a large scale and high frequency of change is the most serious ecological disaster in Southwest China [49,50]. In the past, intense human activities have significantly changed the structure of karst ecosystems, especially the change in land cover [51–53]. The whole karst ecosystem is more sensitive to changes in environmental conditions [54]. Taking the vegetation in the karst area as an example, the research shows that the geochemistry of karst bedrock can affect the growth of vegetation by adjusting the water-holding capacity of the weathered layer, which makes the vegetation productivity in the karst area more susceptible to drought [55]. The fragility and stability of karst ecology and the fragile natural recovery ability have brought challenges to local economic development and environmental protection [56,57]. It is of great significance for karst ecosystems to cope with global changes and achieve sustainable development by exploring the long-time sequence of forest cover changes in karst areas [58].

This study mainly reveals the dynamic changes in forest cover in a long time series from 1944 to 2013 in typical karst areas of Southwest China through historical maps and remote sensing image data and analyzes the different spatial distribution patterns of forest cover from four aspects: altitude, slope, soil type, and lithology. Furthermore, we try to explore the possible influencing factors of forest cover increase or decrease in different change periods. It is hoped that our research results can fully understand the dynamics and temporal and spatial differences in the evolution process and trend of forest cover in recent decades, provide theoretical reference for comprehensive control of rocky desertification in karst mountain areas and rural revitalization, and provide the certain theoretical basis for promoting the sustainable development of karst areas in southwest China.

2. Materials and Methods

2.1. Study Area

Xuanwei County in Yunnan Province, Western China was selected as the study area. Xuanwei has a land area of 6502 km^2, with a low-latitude highland monsoonal climate, average annual precipitation of 986 mm, and an average annual temperature of 13.3 °C. The area is located in the watershed zone between Yangtze River and Pearl River, which has a maximum elevation of 2868 m, a minimum elevation of 920 m, a relative altitude difference of 1948 m, numerous sloping fields, and few mountainous areas, and a relatively large slope gradient. The karst landform in the area spreads widely, with the outcrop reaching 3300 km^2 and accounting for 52.74% of the entire land area of the county. The soil in the study area mainly consists of red earth and limestone. Xuanwei County is located in the "Hump Course" of World War II. The critical section of Yunnan–Burma, and Yunnan–Guizhou Highway played a significant role in the victory of the Allies during World War II (Figure 1). To guarantee smooth passage in this important transport corridor during the war, the U.S. army conducted several detailed works to study the geographical elements of the forests in this region, and the resulting historical military maps reflect the actual and ecological landscapes in the area during that period.

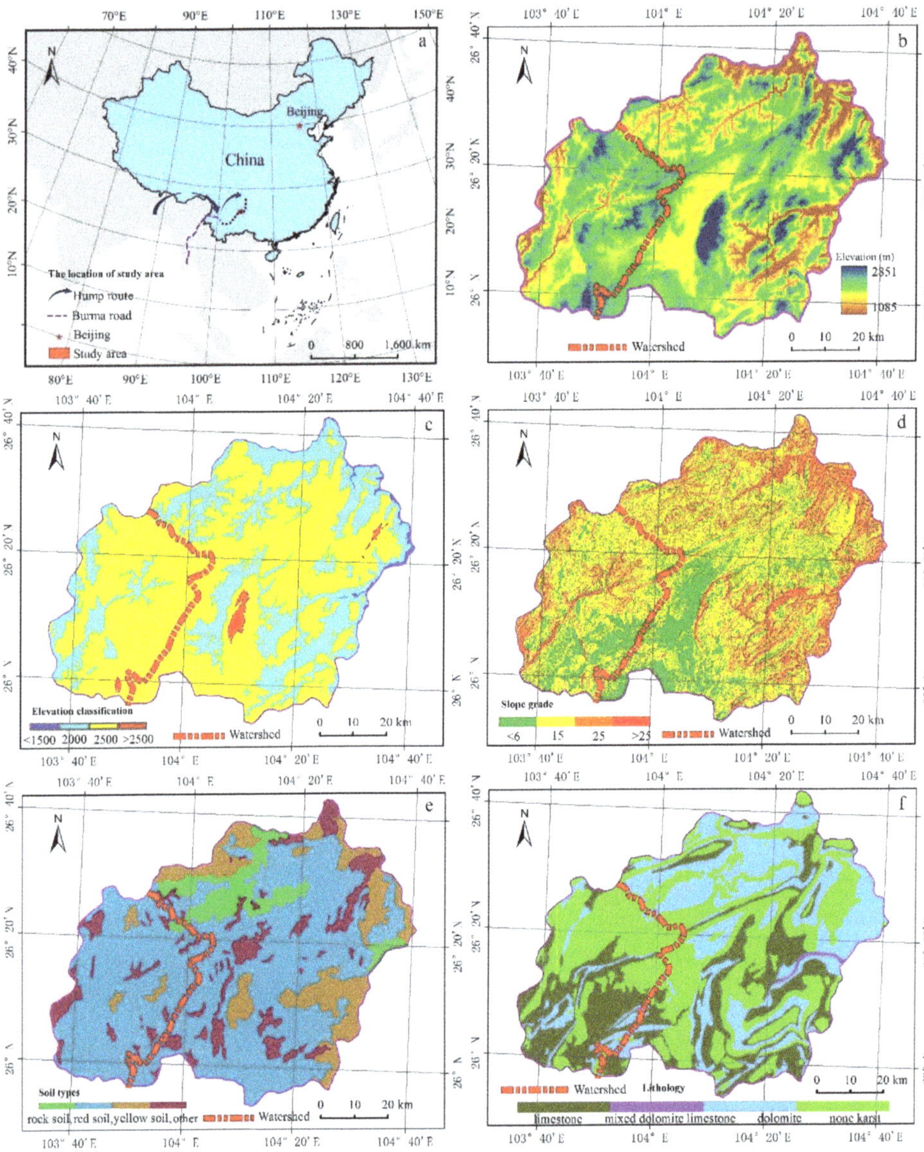

Figure 1. The maps of the location and main natural background of the study area. Note the following: (**a**) location; (**b**) elevation; (**c**) elevation classification; (**d**) slope grade; (**e**) soil type; (**f**) lithology. (We constructed this map using ArcGIS9.3 (http://www.esri.com/arcgis/about-arcgis) (accessed on 12 November 2021)).

2.2. Data Collection and Processing

2.2.1. Historical Map

The horizontal axis Mercator projection and the U.S. Hayford ellipsoid were applied in 1990 to the military topographic map from China drawn by the Army Map Service of the U.S. Army Corps of Engineers during World War II. The map was drawn based on the supergraph of the topographic map that was investigated and drawn by Japan during the height of its military aggression in 1942. This map reflects the situation in 1944. It was revised in 1945 and published in 1954. This map can be freely obtained from the website of the University of Texas libraries (http://www.lib.utexas.edu/maps/ams/) (accessed on 12 November 2021).

2.2.2. Remote Sensing Images

The period was from 1977 to 2013, with a total of five periods. Data from each period were obtained from the University of Maryland (http://glcf.umiacs.umd.edu/data/landsat/) (accessed on 18 December 2020), and the data sharing plan for Earth observation was obtained from the website of the Chinese Academy of Sciences (http://ids.ceode.ac.cn/query.html) (accessed on 18 December 2020). The latter included the following images: Landsat Multispectral Scanner in 1977; Landsat with Thematic Mapper (TM) in 1986, 2000, and 2007; Landsat with Enhanced TM Plus in 2013. To guarantee data accuracy, the seasons of the images that are used are consistent.

Other data.

Digital elevation model (DEM) data with a resolution of 90 m were obtained from the International Scientific Data Service platform (http://datamirror.csdb.cn/) (accessed on 18 December 2020). Lithological spatial distribution and soil type data were gathered from the preliminary study materials of the project group [8,28].

2.2.3. For the Historical Military Map

The original historical military map was scanned, the boundaries were cut, and a seamless splice was created using Adobe Photoshop. The resulting map was then combined with a civilian topographic map to conduct overall scanning or accuracy check of the original historical military map (Figure 2). The contents for examination include the karst cave and its size and location, the morphological characteristics of the mountains, the trend of rivers, and the quantity and form of depressions. If the aforementioned features on the historical military map correspond with those on the civilian topographic map, then the historical military map is accurate and the preliminary scanning is acceptable. The location and distribution of the forest were then examined when the preliminary scanning was acceptable. If the preliminary scanning was not acceptable, then the causes of inconsistencies should be examined. If the scanning failed again, then the historical military map might be inaccurate and should thus be abandoned. During the examination of forest location and distribution, local chronicles, local history, agricultural literature, medical materials, military climate graphs, military hydrographs, military environmental diagrams, military installation diagrams, and analysis diagrams of the combat capacity of war zones should be used as references to achieve a comprehensive analysis. If the findings passed the examination, then the historical military map satisfied the research requirements. The historical military map was then cut according to the size of the study area, and digitization and vectorization were implemented to generate a preliminary diagram of forest distribution. In the statistical calculation of the area and sample inspection, an accuracy rate not lower than 95% should be guaranteed. A diagram of forest distribution during wartime could then be generated. A vector diagram of the spatial distribution of forest cover was obtained to establish the spatial and attribute databases of the forest pattern in the study area in 1944 (Figure 3a). The procedure should be repeated if the accuracy rate could not be guaranteed because of problems during digitization and vectorization.

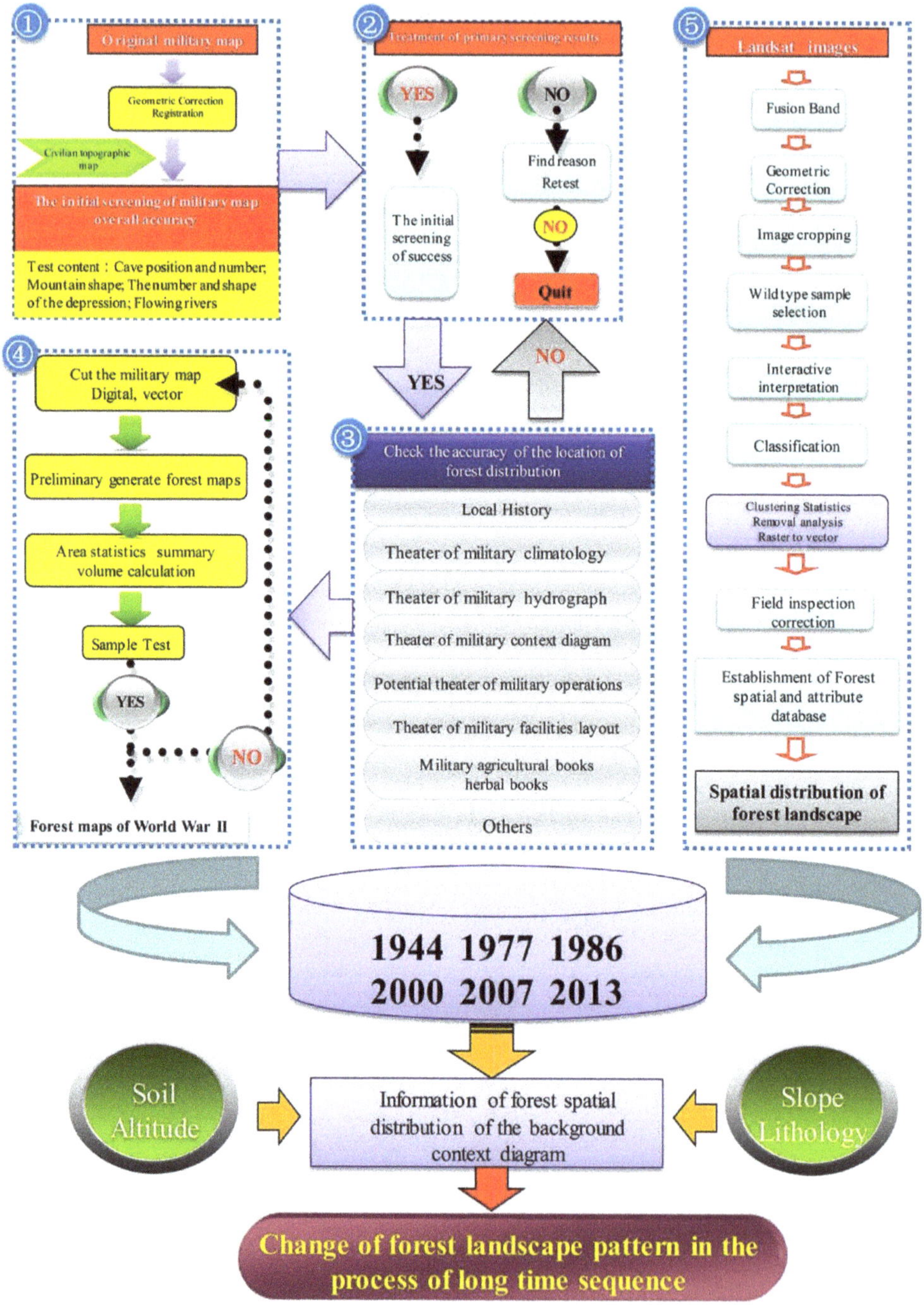

Figure 2. The technique flowchart showing the study of spatial patterns of forest landscape based on the military map and remote sensing images. (We constructed this figure using WPS office (https://platform.wps.cn/) (accessed on 29 November 2021)).

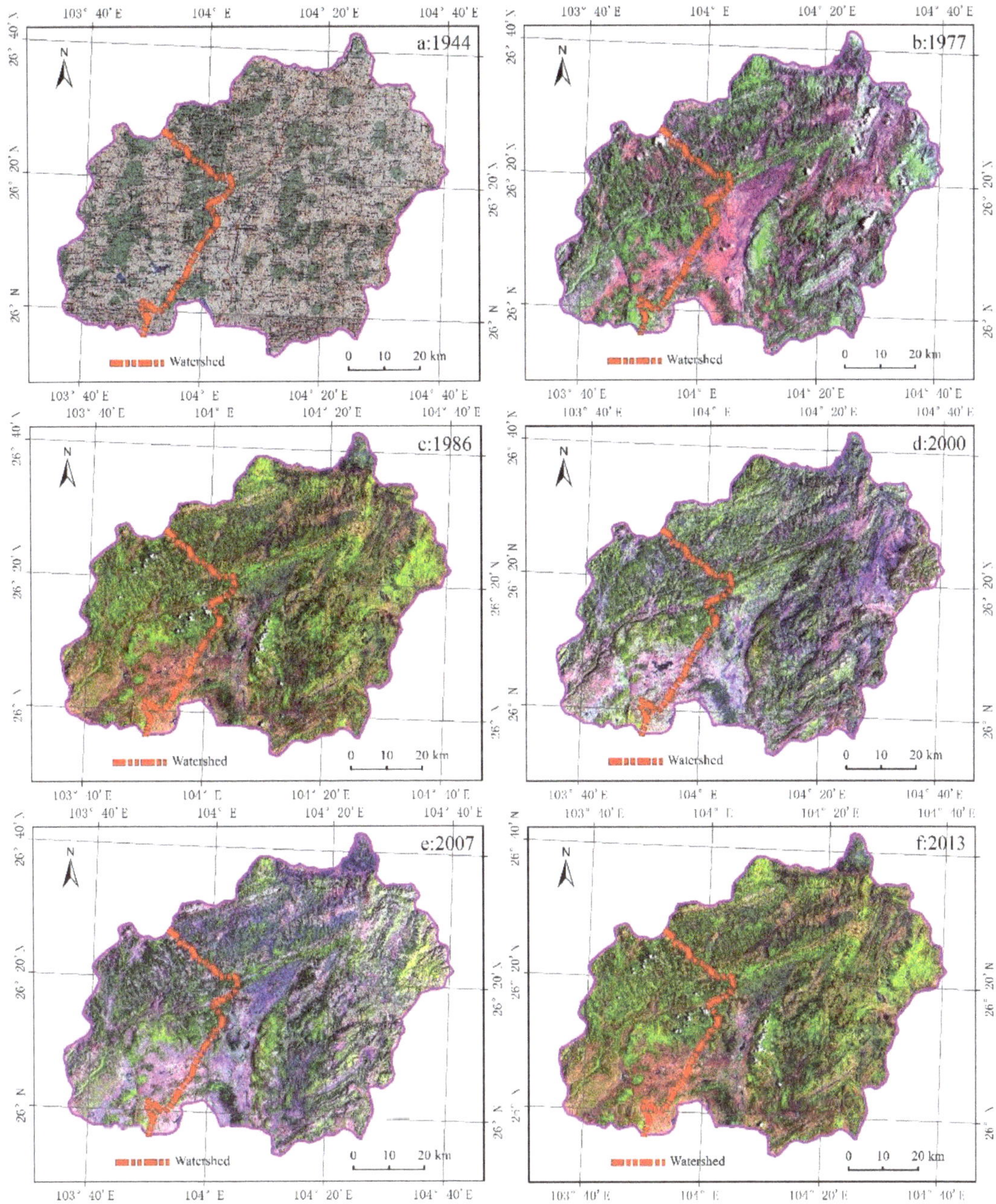

Figure 3. The military map and remote sensing images in different times. (We constructed this map using ArcGIS9.3 (http://www.esri.com/arcgis/about-arcgis) (accessed on 19 December 2021)).

2.2.4. For the Remote Sensing Images

All Landsat images applied near-infrared, red light, and green light wave bands to perform standard false color composition. Radiation correction was performed to achieve spectrum enhancement, radiation enhancement, and geometric precision correction for five-period images under ERDAS IMAGINE environment. The geometric precision correction was performed to scan and input a 1:50,000 topographic map into the computer to conduct projection disposal. This map was used as the main control data source to correct remote sensing images, and the average position error was controlled within a pixel. Field type samples were selected to perform visual interpretation under man–machine interaction, and then the supervised classification was applied. The obtained data were combined with DEM, weather, hydrological, vegetation, soil, land-use change, forest monitoring, and the corresponding social statistics data of the study area to check and amend the forest distribution diagram, establish the forest spatial and attribute databases in the study area beginning in 1977, and generate a vector diagram of spatial distribution information of the forest cover for the following periods: 1977, 1986, 2000, 2007, and 2013 (Figure 3b–f, respectively).

2.2.5. Analyzing Spatial–Temporal Evolvement of the Forest Cover

The dissolve tool was used to integrate spatial distribution data of forest cover for six periods into ArcGIS, which were classified into two codes: forest land and non-forest land. The intersect tool was used to calculate data intersection between two periods as a group. The area field was added to the attribute table and calculated. Next, the attribute table was converted into shapefile attribute format (.dbf) and opened in Microsoft Excel. The commands "Pivot Table" and "Pivot Diagram" in the data menu were implemented, which generated a transfer matrix of the forest cover with two periods after appropriate changes were made. A transfer matrix with different periods was generated when the preceding process was repeated.

Data on soil type, elevation, soil, and lithology were generated from the distribution layer of soil type, sea level elevation, slope gradient, and lithology, respectively. Overlay analysis of the layers of spatial distribution information on forest cover under different historical periods was performed with the aforementioned layers in OVERLAY EVENTS. Under different classification conditions, the spatial distribution information of the forest cover was extracted under various periods concerning different soil types, elevation, gradient, and lithology. The corresponding distribution diagrams were drawn, and the area calculation function of ARC/INFO was used to establish area statistics. The transfer matrix method was used to calculate the area and direction of spatial transfer for the forest cover under different periods and natural backgrounds.

3. Results

3.1. Overall Changed Process and Characteristics in Time

From Figures 4 and 5, it can be seen that the change in forest area and scale showed a trend of first decreasing and then increasing in the study area from 1944 to 2013, and an obvious turning point appeared in 1986. These changes can be roughly divided into two stages: forest degradation before 1986 and forest restoration after 1986. It can be seen intuitively from Figure 4 that the forest coverage of the six years studied is quite different. The forest area was at its maximum in 1944 (up to 24.28% of the study area) and its minimum in 1986 (only 8.5% of the study area), with a difference of 2.84 times. The forest area decreased annually before 1986 and gradually increased after 1986, reaching 19.97% of the study area in 2013. However, even though ecological restoration and improvement were carried out from 2000 to 2013 (Figure 5), the forest coverage rate still did not reach the scale at the beginning of the research period.

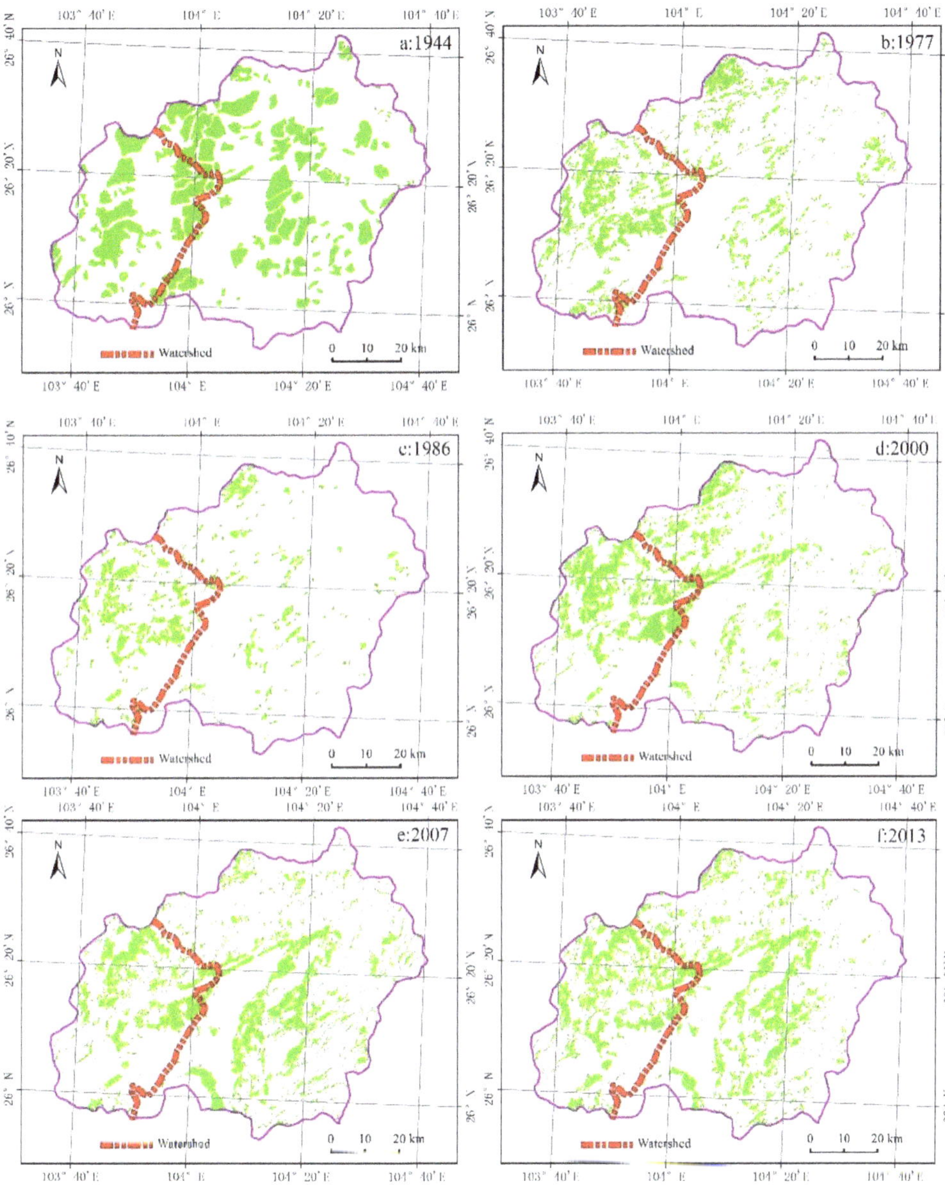

Figure 4. The spatial distribution information maps of forest landscape in the different historical times. (We constructed this map using ArcGIS9.3 (http://www.esri.com/arcgis/about-arcgis) (accessed on 25 December 2021)).

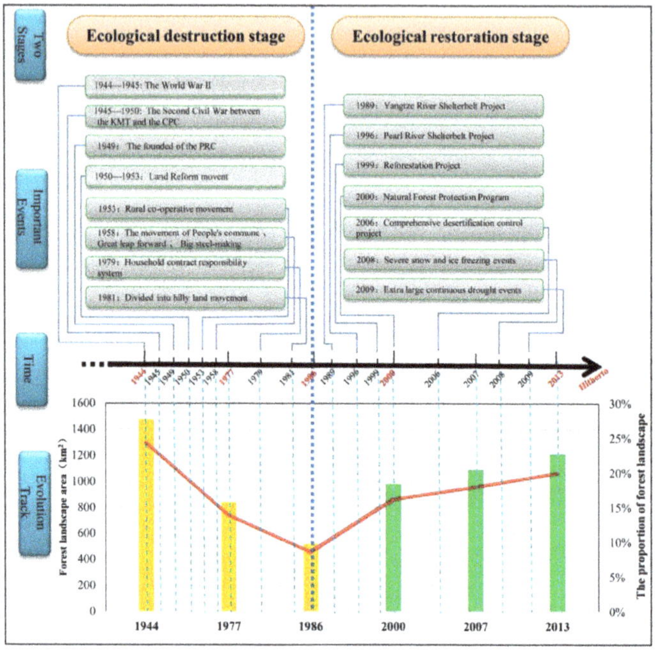

Figure 5. The evolution process of forest landscape and important historical events related to it. (We constructed this figure using WPS office (https://platform.wps.cn/) (accessed on 30 December 2021)).

3.1.1. Rate of Change in the Forest Cover under Different Historical Periods

In the analysis of the previous part, it has been found that the evolution trend of forest cover area in the whole study period is first decreasing and then increasing. However, it remains to be analyzed which period changes more quickly. It can be seen from Table 1 that before 1986, the evolution frequency was negative (referring to the decrease in forest area), and after 1986, it was positive (the increase in forest area).

Table 1. The annual changing rates of forest landscape in different historical periods. (We constructed this table using WPS office (https://platform.wps.cn/) (accessed on 19 December 2021)).

Time	The Annual Changing Rates	Changing Rates
1944–1977	−0.32%	−10.43%
1977–1986	−0.48%	−5.29%
1986–2000	0.51%	7.61%
2000–2007	0.26%	1.82%
2007–2013	0.33%	1.97%

Specifically, from 1944 to 1977, although the overall change rate was −10.43%, the average annual change frequency was only −0.32% because the two years were separated by 33 years. The minimum frequency of change was from 2000 to 2007, with a change rate of only 1.82%. The difference between the highest and lowest frequency of change is nearly 5.73 times. However, a big change does not necessarily mean a quick change rate. This variable is also related to the number of years. From 1986 to 2000, the annual average rate of change was the largest, about 0.52%. Although the overall frequency of change in this period is 7.61%, which is lower than 10.43% in 1944–1977, the average annual frequency of change was higher. It shows that in this period from 1986 to 2000, a series of national ecological protection projects and policies have obvious effects on the restoration of forest cover, such as the Yangtze River Shelter-belt Project (1989), Pearl river Shelter-belt Project (1996), and Reforestation Project (1999) (Figure 5).

3.1.2. Transfer Direction of the Forest Cover under Different Time Series

It can be seen from Figure 6 that there are obvious differences in changes between forest and other land types in various historical periods. Specifically, from 1944 to 1977, the conversion of forest to other land types was 1165 km ^{2}, which was significantly larger than the conversion area of other land types to forest in this period (Figure 6a). From 1977 to 1986, although the changes from forest to other land types and the changes from other land types to forested land were scattered in space, it was also obvious that the changes from the forest to other land types were more (Figure 6b). From 1986 to 2000, the change in other land types to the forests was the most obvious, which was the fastest period of ecological restoration, mainly in the northwest of the study area (Figure 6c). From 2000 to 2007, other land types still changed to forest (467 km^2), which mainly happened in the southeast of the study area (Figure 6d). From 2007 to 2013, the forest was converted into other land types (255 km^2), mainly occurring in the middle of the study area, which may be related to the increase in temporary forest in construction land caused by urbanization during this period [59]. At the same time, during this period, other land types of 366 km^2 were also converted into forest (Figure 6e). From the whole research period from 1944 to 2013, the conversion of forests to other land types is higher than that of other land types (Figure 6f), which is also consistent with the overall analysis results of the previous part of forest coverage.

3.2. *Changes in Forest Cover under Different Influencing Factors*

The changes in forest cover under the different elements of factors such as elevation, slope, soil types and lithology in different time periods are shown in Table 2 and Figure 7. Table 2 shows the change area of forest under various factor levels in different time periods. Figure 7 shows the percentage of forest change area in different factor levels in the total forest area in different historical periods. In Figure 7, the positive and negative values before the abscissa percentage indicate the changing direction of forest increase and decrease, respectively. The negative percentage indicates the proportion of decreased area in the total forest area at the beginning of this period, and the positive percentage indicates the proportion of increased forest area in the total forest area.

Table 2. The distribution and change information of forest landscape under various natural background conditions during different historical stages (unit: km^2). (We constructed this table using WPS office (https://platform.wps.cn/) (accessed on 19 December 2021)).

Background Conditions	Classification	Time Sequence					
		1944	1977	1986	2000	2007	2013
Elevations	1000–1500 m	2.80	0.09	0.00	1.72	1.88	0.00
	1500–2000 m	358.17	119.16	52.02	169.28	114.50	160.79
	2000–2500 m	1103.26	712.82	458.77	800.70	951.19	1042.31
	2500–3000 m	11.78	14.74	12.49	16.91	31.76	10.27
Slope	≤6°	200.60	40.50	29.86	66.50	93.88	105.95
	6–15°	813.51	467.45	302.35	552.54	616.94	681.42
	15–25°	356.83	273.19	157.31	281.28	277.53	313.62
	>25°	96.87	61.11	31.00	82.96	104.82	105.68
Soil Tapes	Other soil	116.01	47.83	22.14	59.29	43.84	58.08
	Rock soil	168.49	50.77	22.86	91.01	39.01	59.19
	Red soil	966.61	610.70	386.99	677.95	816.22	943.19
	Yellow soil	217.27	133.13	88.62	155.32	194.30	146.60
Lithologies	Dolomite Mixed	375.64	200.78	115.28	172.80	270.57	305.81
	dolomite-limestone	2.27	1.31	0.34	1.80	4.59	4.99
	Limestone	427.23	183.21	86.07	221.99	210.99	239.53
	Non-karst	663.23	457.12	318.91	586.97	607.22	656.72

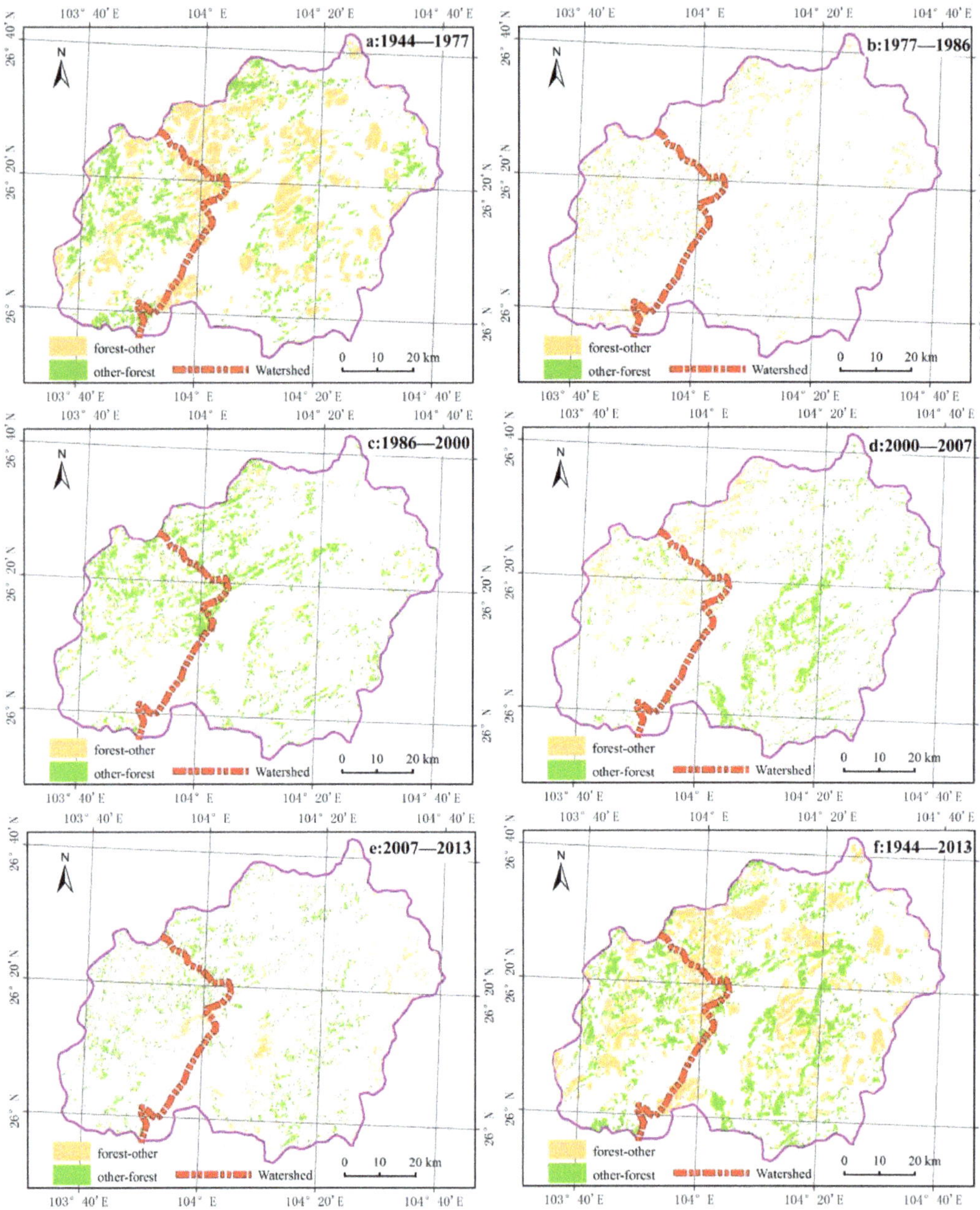

Figure 6. The space transfer matrix maps of forest landscape in different historical periods. (We constructed this map using ArcGIS9.3 (http://www.esri.com/arcgis/about-arcgis) (accessed on 27 December 2021)).

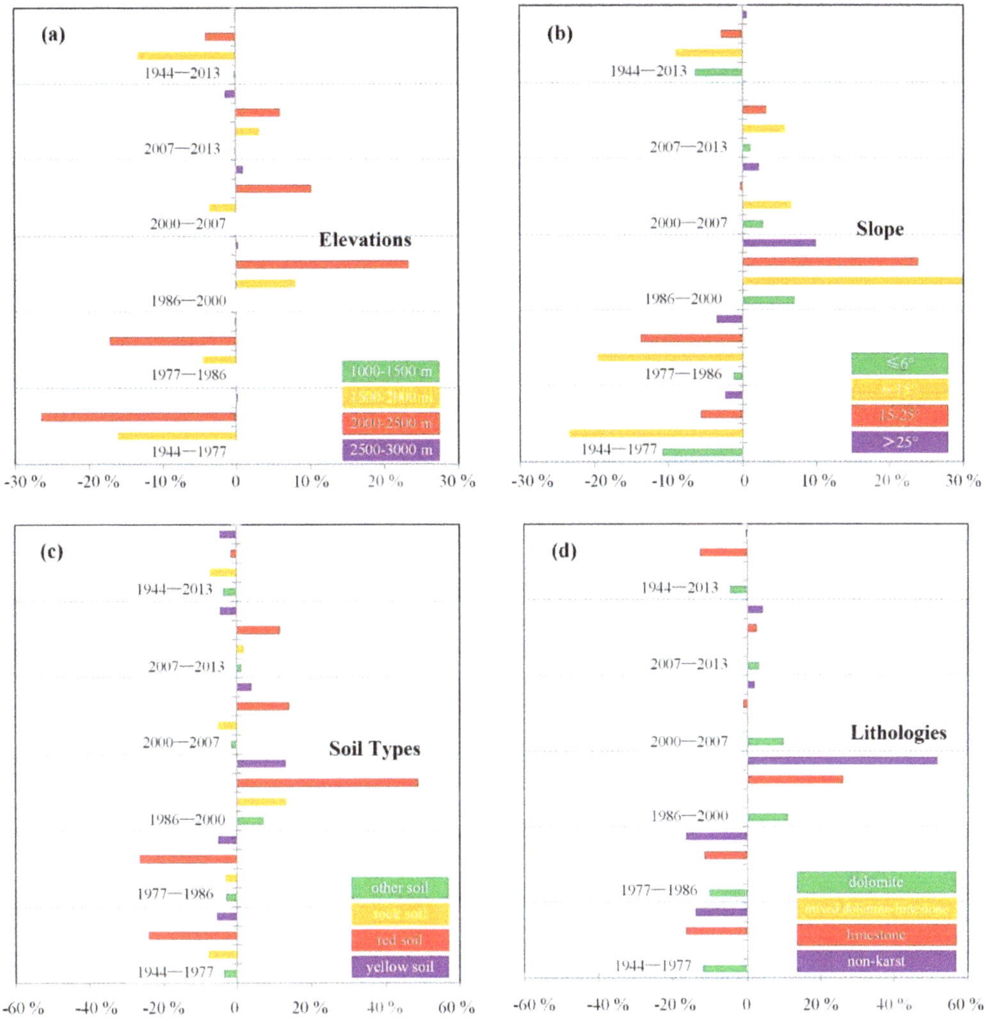

Figure 7. The percentage of forests changing area with different elements grades in the total forest area in different historical periods. (We constructed this figure using WPS office (https://platform.wps.cn/) (accessed on 27 December 2021)).

3.2.1. Changes under Different Elevations

Under different levels of elevation, the forest area in three altitude ranges, 1000–1500, 1500–2000 and 2000–2500 m, decreased from 1944 to 2013 (Table 2). Specifically, the forest distributed in 2000–2500 m reached the highest value of 1103.26 km^2 at the beginning of the study, with the largest decrease of 458.77 km^2 in 1986, and then gradually recovered. Only the forest area in the range of 2500–3000 m above sea level shows an increasing trend, which may be related to fewer human activities in high altitude areas [60].

The percentage of forest change area in different elevations in the total forest area can be seen more intuitively from Figure 7a. Among them, the biggest change area percentage was from 1944 to 1977, and the change area percentage of the forest with 2000–2500 m was −26.45%. Secondly, from 1986 to 2000, the change area percentage of 2000–2500 m forest was 23.26%.

3.2.2. Changes under Different Slopes

Table 2 also shows the forest area under different slopes, from 1944 to 2013, only the forest area with a slope >25° increased, while the other three slopes included a decrease in the forest areas with a slope ≤6°, 6–15° and 15–25°. Among them, the reduced area is mostly distributed in the range of slope 6–15°, which decreased from 813.51 km^2 in 1944 to only 681.41 km^2 in 2013, indicating that forests with gentle slope in this range is more likely to be occupied by cultivated land or other land types. Forests with steep slopes, such as those with a slope of more than 25°, are more likely to be preserved, because they are less damaged [61,62].

The difference in forest changes area percentage between different periods and slope is shown in Figure 7b. The period 1986 to 2000 presented a relatively large change area percentage, accounting for 48.26% at 6°–15° and 23.90% at 15°–25°, with a total of more than 72%. This result is significantly associated with the "grain for green" policy during this period. For the periods of 1944–1977 and 1977–1986, the changes were mainly negative, and the gradients mainly focused at 6°–15° and 15°–25°, respectively. Therefore, the region with the gradient 6°–25° was frequently interrupted by human activities [63].

3.2.3. Changes under Different Soil Types

The forest area of four different soil types (other soil, rock soil, red soil and yellow soil) in the study area showed a decreasing trend from 1944 to 2013 (Table 2). Among them, the biggest decrease in forest was in rock soil, which dropped sharply from 168.49 km^2 in 1944 to 22.86 km^2 in 1986, and the ecological restoration after that only increased to 59.19 km^2 in 2013. The largest forest distribution area is red soil, which was 966.61 km^2 in 1944, and decreased sharply to 386.99 km^2 in 1986. After that, the forest coverage gradually recovered to 943.19 km^2 in 2013. Therefore, once the vegetation of karst rock soil type is destroyed, it is much more difficult to restore it than other soil types [64].

The changes in forests area percentage in different soil types in different periods are shown in Figure 7c, among which the biggest change area percentage is that the positive growth rate of forest distributed on red soil was 55.89%, from 1986–2000. However, before this, the negative percentage of change in this soil type were almost offset. Specifically, during the periods of 1944–1977 and 1977–1986, the reduction area percentage of forests distributed on red soil was −24.11% and −26.57%, respectively. However, the change area percentage of forests distributed in karst soil is −7.40%, which means that forests distributed in laterite is easier to recover after being destroyed than that in karst.

3.2.4. Changes under Different Lithologies

The distribution and change characteristics of forests under different lithology generally show that the reduced area of forests in karst lithology (including dolomite mixed with limestone) is obviously larger than that of non-karst lithology (Table 2). Among them, limestone is the largest forest decrease, which decreased sharply from 427.23 km^2 in 1944 to 86.07 km^2 in 1986, and then recovered to only 239.53 km^2 in 2013, only recovering to nearly half of the forest area at the early stage. However, the forests distributed in non-karst lithology decreased by half from 1944 to 1986, and gradually increased to 656.72 km^2 in 2013 in the process of restoration, which is similar to the forest area in 1944.

As shown in Figure 7d, the rapid growth of forest area percentage, that is, from 2000 to 2007, it was about 51.78% in non-karst areas and 26.11% for limestone areas, which may have a great relationship with the implementation of natural forest protection projects during this period. On the whole, the limestone area had the biggest negative area percentage from 1944 to 2013, which was −12.72%, and the dolomite area had −4.73%. It also shows that vegetation restoration in karst areas is difficult for non-karst areas [65,66].

4. Discussion

4.1. Comparison with Other Studies

This part of the discussion is mainly aimed at the comparative analysis of other studies that use historical maps combined with remote sensing images to reveal the temporal and spatial changes in long-time series forest cover. At the same time, the comparison is mainly carried out from two aspects: one is to cover all or part of the study period (1944–2013), and the other is to include or relate to the study area (typical karst area in southwest China).

He et al. [67] revealed the trend and main process of forest dynamics from 1700 to 1998 by using historical documents, modern surveys and statistical data, and the results of previous studies. Among them, during the rapid decline from 1700 to 1949, the northeast, southwest and southeast regions suffered the most serious decline, and the coverage rate of most provinces fell by more than 20%. During the recovery period from 1949 to 1998, the western provinces (including Yunnan) increased by over 5%. In addition, another article by the researcher [11] shows that from 1700 to the 1960s, deforestation mainly occurred in southwest China. Judging from the changing trend and the general turning point, the trend of first worsening and then recovering is consistent. In addition, other studies using historical maps and remote sensing images to reveal the long-term changes in forests do not include or involve the study area of this paper, such as Taiwan Province Province [68], Hainan Island [69], Heilongjiang Province [70], etc.

Although there is a lack of research on forest evolution in southwest karst area by using historical maps and remote sensing images, much research that only uses remote sensing data to reveal forest or vegetation cover changes in karst area can also provide a reference for the second period of this study (the forest restoration stage after 1986). For example, Tong et al. [71] used the gimms-3 g Normalized Difference Vegetation Index (NDVI) from the period 1982–2012 to evaluate the effect of ecological engineering vegetation restoration in Yunnan, Guizhou and Guangxi. It was found that although the whole vegetation area was afforested, the restoration rates were different in different areas. On this foundation, Zhang et al. [72] also used the gimms-3 g NDVI from 1982 to 2016 to study the trend of vegetation change in Guizhou, Guangxi and Yunnan, and they found that the trend of vegetation greening in karst areas was strengthened from 1982 to 2016, and ecological engineering was the main reason for the increase in vegetation in karst areas, while the climate was the main driving factor for the decrease in vegetation in non-karst areas. This is consistent with the trend of forest restoration after 1986 in this research. Similarly, using NDVI data, Xu et al. [73] examined the vegetation mutation in Southwest China from 1982 to 2015, and found that the mutation point appeared in 2001, and the trend of NDVI changed from no significant increase to significant increase after the mutation point. For the above researches on forest or vegetation cover in karst areas of southwest China, the time scale mainly concentrated after 1982, and the forest cover showed a consistent increasing trend.

In the last 30 years, there is still a consensus that vegetation will turn green, whether in China [74–76] or in the region [77,78]. The research at the China national level before the 1970s shows that southwest China is a region with significant reduction in vegetation cover [11,67], but the research on forests in southwest karst area before 1970s is very scarce. It may be limited by the difficulty in obtaining remote sensing image data, and it also highlights the advantages of this research in combining historical maps with remote sensing images to deal with this problem.

4.2. Events and Factors That Dominate Forest Cover Changes in Different Periods

Forest deterioration from 1944 to 1986: Many incidents occurred during this period, including World War II (also called the "Anti-Japanese War" in China), the civil war between the Chinese Nationalist Party and the Chinese Communist Party from 1945 to 1950 (the Liberation War), the founding of the People's Republic of China in 1949, the shifting in the national system from capitalism to socialism, the land reform movement in 1950 (i.e., the transformation from feudal land ownership to private land ownership for peasants), the rural cooperative movement in 1953 (i.e., the transformation from private land ownership for peasants to collectivization and socialization of agricultural land), the population policy (i.e., "many hands make work easy") in 1958, the "Great Leap Forward," the smelting of steel, and the movement to establish communes for rural residents in 1958. World War II, the Chinese Civil War, the change in the national system, the change in land ownership, or the smelting of steel might have resulted in the sharp deterioration and even the loss of forest vegetation. The policy on family contract business was implemented, particularly the transfer of the collective operation of lands and forests to families or individual corporations, the implementation of a "system of fixed output for households, work contracted to households, and mountain contracted to households," and the policy to divide privately farmed hilly lands and forest lands among individuals. Farmers were afraid of the change in ownership of such lands, Thus, they engaged in large-scale firewood gathering and logging, which might have damaged forest vegetation.

Forest recovery from 1986 to 2013: Construction projects were conducted to protect the forest system of the Yangtze River Basin in 1989 and the Pearl River Basin in 1996. A project to return grain plots to forests was also implemented in 1999. This project involved ecological construction engineering with the strongest policy, largest investment, widest coverage, and the highest extent of public engagement in China. The project was also the largest one that supports and benefits farmers, with funds of more than CNY4.3 trillion provided by the central government, thereby becoming the largest ecological construction project in the world. The implementation of various projects effectively promoted the increase in forest coverage rate and the reduction in soil erosion incidents. China formally launched conservation programs for natural forest resources in 2000 to strictly manage and protect ecological public welfare forests, strongly develop forestation, and adjust and optimize the ecological structure of forest zones. These programs greatly improved the regional ecological environment and reduced water and soil erosion areas. The comprehensive termination of the stony desertification project was implemented in 2006. In this project, the drainage basin was considered as a unit, the damaged natural ecological system was gradually recovered by increasing the vegetation land cover and conserving water and soil, and the extent of karst rocky desertification was effectively reduced.

The change in forest cover is affected by both natural and human factors [79,80], but the dominant factors are different in different time periods [81]. Although the past forest destruction has brought about the deterioration of the ecological environment, fortunately, a series of ecological projects have made great contributions to the restoration of forest vegetation [82,83].

4.3. Limitations and Future Research Prospects

First of all, due to the limitation of data sources, there was a long period between 1944 and 1977, and only the historical map of 1944 was used, which caused uncertainty of forest change trend analysis at present. In future research, we can increase the number of historical maps obtained as reasonably as possible before the 1970s, or update the latest year to the latest year. Secondly, in the correlation analysis of influencing factors, this study only considered four factors: elevation, slope, soil types and lithology. Then, on the time scale of several decades, these four factors will not change much. In future research, climate factors such as temperature and precipitation can be considered for analysis.

5. Conclusions

Based on the historical map of 1944 and Landsat remote sensing satellite images, this paper quantitatively analyzed the spatial distribution and change in forests from 1944 to 2013, aiming at the evolution law of long-time series spatial distribution characteristics of forest cover in karst areas and the events and factors that may affect the forest changes in the unsustainable stage. The main conclusions are as follows: (1) the forest area in the study area showed a trend of decreasing at first and then increasing. From 1944 to 1986, the deterioration phase of forest area decreased, and from 1986 to 2013, the restoration phase of forest gradually increased. (2) The forest with an altitude of 2000~2500 m changes most frequently in the damage recovery stage, and the forest with a slope of 6~15° faces the greatest risk of damage. (3) The changing characteristics of forests with different soil types and lithology in different stages show that it is more difficult to restore forests in karst areas after they are destroyed. To sum up, the use of historical maps can better solve the temporal and spatial evolution of long-time series of forest cover before there is no remote sensing image due to the limitation of satellite launch time. At the same time, it is necessary to pay attention to forest protection in the subsequent social and economic development of karst areas, to avoid damage and increase the cost of rehabilitation.

Author Contributions: Conceptualization, X.B.; methodology, Y.T.; software, L.Q.; validation, Y.L. and Y.X.; formal analysis, J.W. and L.W.; investigation, C.L., C.R. and S.Z.; resources, F.L. and G.L.; data curation, X.B.; writing—original draft preparation, F.C.; supervision, X.B.; project administration, X.B.; funding acquisition, X.B. All authors have read and agreed to the published version of the manuscript.

Funding: This research work was supported jointly by the Western Light Cross-team Program of Chinese Academy of Sciences (No. xbzg-zdsys-202101), National Natural Science Foundation of China (No. 42077455 & No.42167032), Strategic Priority Research Program of the Chinese Academy of Sciences (No. XDB40000000 & No. XDA23060100), Guizhou Provincial Science and Technology Projects (No. 2022-198), High-level innovative talents in Guizhou Province (No. GCC [2022]015-1 & No. 2016-5648), Guizhou Provincial 2020 Science and Technology Subsidies (No. GZ2020SIG), Opening Fund of the State Key Laboratory of Environmental Geochemistry (No. SKLEG2022206 & No. SKLEG2022208).

Institutional Review Board Statement: Not applicable.

Informed Consent Statement: Not applicable.

Data Availability Statement: Not applicable.

Acknowledgments: We would like to thank all the authors and reviewers for their great guidance and help in writing this manuscript.

Conflicts of Interest: The authors declare no conflict of interest.

References

1. Yang, J.P. The important role of forest ecological system in economic development. *For. Econ.* **1999**, *4*, 1–5.
2. Bonan, G.B. Forests and climate change: Forcings, feedbacks, and the climate benefits of forests. *Science* **2008**, *320*, 1444–1449. [CrossRef] [PubMed]
3. Hansen, M.C.; Potapov, P.V.; Moore, R.; Hancher, M.; Turubanova, S.A.; Tyukavina, A.; Thau, D.; Stehman, S.V.; Goetz, S.J.; Loveland, T.R.; et al. High-resolution global maps of 21st-century forest cover change. *Science* **2013**, *342*, 850–853. [CrossRef]
4. Blonska, E.; Malek, S.; Januszek, K.; Józef, B.; Wanic, T. Changes in forest soil properties and spruce stands characteristics after dolomite, magnesite and serpentinite fertilization. *Eur. J. For. Res.* **2015**, *134*, 981–990. [CrossRef]
5. Hansen, M.C.; Stehman, S.V.; Potapov, P.V. Quantification of global gross forest cover loss. *Proc. Natl. Acad. Sci. USA* **2010**, *107*, 8650–8655. [CrossRef]
6. Mayer, A.L.; Kauppi, P.E.; Angelstam, P.K.; Zhang, Y.; Tikka, P.M. Importing timber, exporting ecological impact. *Science* **2005**, *308*, 359–360. [CrossRef] [PubMed]
7. Elbakidze, M.; Andersson, K.; Angelstam, P.; Armstrong, G.W.; Axelsson, R.; Doyon, F.; Hermansson, M.; Jacobsson, J.; Pautov, Y. Sustained yield forestry in Sweden and Russia: How does it correspond to sustainable forest management policy? *Ambio* **2013**, *42*, 160–173. [CrossRef]
8. Xu, Y.Q.; Xiao, F.J.; Yu, L. Review of spatio-temporal distribution of net primary productity in forest ecosystem and its responses to climate change in China. *Acta Ecol. Sin.* **2020**, *40*, 4710–4723.
9. Hu, X.S.; Wu, C.Z.; Hong, W.; Qiu, R.Z.; Li, J.; Hong, T. Forest cover change and its drivers in the upstream area of the Minjiang River, China. *Ecol. Indic.* **2014**, *46*, 121–128. [CrossRef]
10. Lin, Y.Y.; Hu, X.S.; Zheng, X.X.; Hou, X.Y.; Zhang, Z.X.; Zhou, X.N.; Qiu, R.Z.; Lin, J.G. Spatial variations in the relationships between road network and landscape ecological risks in the highest forest coverage region of China. *Ecol. Indic.* **2019**, *96*, 392–403. [CrossRef]
11. He, F.N.; Li, S.C.; Zhang, X.Z. A spatially explicit reconstruction of forest cover in China over 1700–2000. *Glob. Planet. Chang.* **2015**, *131*, 73–81. [CrossRef]

12. Hansen, M.C.; Potapov, P.V.; Pickens, A.H.; Tyukavina, A.; Hernandez-Serna, A.; Zalles, V.; Turubanova, S.; Kommareddy, I.; Stehman, S.V.; Song, X.P.; et al. Global land use extent and dispersion within natural land cover using Landsat data. *Environ. Res. Lett.* **2022**, *17*, 034050. [CrossRef]
13. Blanco, V.; Brown, C.; Rounsevel, M. Characterising forest owners through their objectives, attributes and management strategies. *Eur. J. For. Res.* **2015**, *134*, 1027–1041. [CrossRef]
14. Luo, G.J.; Wang, S.J.; Bai, X.Y.; Liu, X.M.; Liu, X.M.; Cheng, A.Y. Delineating small karst watersheds based on digital elevation model and eco-hydrogeological principles. *Solid Earth* **2016**, *7*, 457–468. [CrossRef]
15. Skaloš, J.; Engstová, B. Methodology for mapping non-forest wood elements using historic cadastral maps and orthophoto maps as a basis for management. *J. Environ. Manag.* **2010**, *91*, 831–843. [CrossRef]
16. Skaloš, J.; Engstová, B.; Trpáková, I.; Šantrůčková, M.; Podrázský, V. Long-term changes in forest cover 1780–2007 in central Bohemia, Czech Republic. *Eur. J. For. Res.* **2012**, *131*, 871–884. [CrossRef]
17. Zhang, J. Extraction of Rocky Desertification Information in Karst Area of Southwest China Based on Time-Series Remote Sensing Data. Master's Thesis, China University of Geosciences, Beijing, China, 2021.
18. Bai, X.Y.; Wang, S.J.; Xiong, K.N. Assessing spatial-temporal evolution processes of karst rocky desertification land: Indications for restoration strategies. *Land Degrad. Dev.* **2013**, *24*, 47–56. [CrossRef]
19. Li, Y.B.; Shao, J.A.; Yang, H.; Bai, X.Y. The relations between land use and karst rocky desertification in a typical karst area, China. *Environ. Geol.* **2009**, *57*, 621–627. [CrossRef]
20. Li, Y.B.; Luo, G.J.; Bai, X.Y.; Wang, Y.Y.; Wang, S.J.; Xie, J.; Yang, G.B. The correlations among arable land, settlement and karst rocky desertification-cases study based on typical peak-cluster depression. *Acta Ecol. Sin.* **2014**, *34*, 2195–2207.
21. Andraž, Č.; Marjan, J.; Krištof, O.S. Past and Present Forest Vegetation in NE Slovenia Derived from Old Maps. *Appl. Veg. Sci.* **1998**, *1*, 253–258.
22. Alfonso, T.; Dina, S.; Pietro, P. Rural landscape planning through spatial modelling and image processing of historical maps. *Land Use Policy* **2015**, *42*, 71–82.
23. Zámolyi, A.; Szekely, B.; Draganits, E.; Timar, G. Neotectonic control on river sinuosity at the western margin of the Little Hungarian Plain. *Geomorphology* **2009**, *122*, 231–243. [CrossRef]
24. Lieskovsky, J.; Kaim, D.; Balazs, P.; Boltiziar, M.; Chmiel, M.; Grabska, E.; Kiraly, G.; Konkoly-Gyuro, E.; Kozak, J.; Antalova, K.; et al. Historical land use dataset of the Carpathian region (1819–1980). *J. Maps* **2018**, *14*, 644–651. [CrossRef]
25. Roccati, A.; Faccini, F.; Luino, F.; De Graff, J.V.; Turconi, L. Morphological changes and human impact in the Entella River floodplain (Northern Italy) from the 17th century. *Catena* **2019**, *182*, 104122. [CrossRef]
26. Fernandes, M.R.; Aguiar, F.C.; Martins, M.J.; Rivaes, R.; Ferreira, M.T. Long-term human-generated alterations of Tagus River: Effects of hydrological regulation and land-use changes in distinct river zones. *Catena* **2020**, *188*, 104466. [CrossRef]
27. Zdenek, K.; Marika, S.; Jan, S. Analyzing Land Cover Change—The Impact of the Motorway Construction and Their Operation on Landscape Structure. *J. Geogr. Inf. Syst.* **2014**, *6*, 559–571.
28. Skaloš, J.; Weber, M.; Lipský, Z.; Trpakova, I.; Santruckova, M.; Uhlirova, L.; Kukla, P. Using old military survey maps and orthophotograph maps to analyse long-term land cover changes—Case study (Czech Republic). *Appl. Geogr.* **2011**, *31*, 426–438. [CrossRef]
29. Pelorosso, R.; Leone, A.; Boccia, L. Land cover and land use change in the Italián central Appeninnes: A comparison of assessment methods. *Appl. Geogr.* **2009**, *29*, 35–48. [CrossRef]
30. Petek, F.; Urbanc, M. The Franziscean Land Cadaster as a key to understanding the 19th-Century cultural landscape in Slovenia—Franciscejski kataster kot ključ za razumevanje kulturne pokrajinev Slovenijiv 19. *Acta Geogr. Slov.* **2004**, *44*, 89–112. [CrossRef]
31. Rajšp, V.; Ticho, M.; Grabnar, M.; Kološa, V.; Serše, A.; Trpin, D. Joseske mapovani Slovenija na vojaškem zemljevidu 1763–1787 opisi. (Josephinische Landesaufnahme 1763–1787 für das Gebiet der Republik Slowenien Landesbeschreibung). *Ljublana* **1995**, *345*, 19.
32. Bender, O.; Boehmerb, H.J.; Jens, D.; Schumacher, K.P. Using GIS to analyse long-term cultural landscape change in Southern Germany. *Landsc. Urban Plan* **2005**, *70*, 111–125. [CrossRef]
33. Cousins, S.A.O.; Eriksson, Å.; Franzén, D. Reconstructing past land use and vegetation patterns using palaeogeographical and archaeological data. A focus on grasslands in Nynäs by the Baltic Sea in south-eastern Sweden. *Landsc. Urban Plan* **2002**, *61*, 1–18. [CrossRef]
34. Domaas, S.T. The reconstruction of past patterns of tilled fields from historical Cadastral Maps using GIS. *Landsc. Res* **2007**, *32*, 23–43. [CrossRef]
35. Hamre, L.N.; Domaas, S.T.; Austad, I.; Rydgren, K. Land-cover and structural changes in a western Norwegian cultural landscape since 1865, based on an old cadastral map and a field survey. *Landsc. Ecol.* **2007**, *22*, 1563–1574. [CrossRef]
36. Sklenička, P.; Molnarova, K.; Brabec, E.; Kumble, P.; Pittnerova, B.; Pixova, K.; Salek, M. Remnants of medieval field patterns in the Czech Republic: Analysis of driving forces behind their disappearance with special attention to the role of hedgerows. *Agric. Ecosyst. Environ.* **2009**, *129*, 465–473. [CrossRef]
37. Trpák, P.; Trpáková, I. Landscape function analysis based on evaluation of indicator maps and sketches of stable cadastre maps, Krajina 2002. In *Krajina 2002—Od Poznání k Integraci*; Ministerstvo Životního Prostředí: Ústí nad Labem, The Czech Republic, 2002; pp. 85–91. (In Czech)

38. Yang, F.; He, F.N.; Li, M.J.; Li, S.C. Evaluating the reliability of global historical land use scenarios for forest data in China. *J. Geogr. Sci.* **2020**, *30*, 1083–1094. [CrossRef]
39. Li, S.C.; He, F.N.; Zhang, X.Z. An approach to spatially explicit reconstruction of historical forest in Northeast China. *J. Geogr. Sci.* **2014**, *24*, 1022–1034. [CrossRef]
40. Liu, D.; Toman, E.; Zane Fuller, Z.; Chen, G.; Londo, A.; Zhang, X.S.; Zhao, K.G. Integration of historical map and aerial imagery to characterize long-term land-use change and landscape dynamics: An object-based analysis via Random Forests. *Ecol. Indic.* **2018**, *95*, 595–605. [CrossRef]
41. Tanacs, E.; Szmorad, F.; Barany-Kevei, I. A review of the forest management history and present state of the Haragistya karst plateau (Aggtelek Karst, Hungary). *Acta Carsolog.* **2007**, *36*, 441–451. [CrossRef]
42. Zhao, Z.Q.; Hou, L.S.; Cai, Y.L. The process and mechanism of soil degradation in karst area in Southwest China. *Earth Sci. Front.* **2006**, *13*, 185–189.
43. Yuan, D.X. Rock desertification in the sutbtropical karst of South China. *Z. Für Geomorphol. Neue Folge* **1997**, *108* (Suppl. Bd), 81–90.
44. Cao, J.H.; Yuan, D.X.; Pan, G.X. Some soil features in karst ecosystem. *Adv. Earth Sci.* **2003**, *18*, 37–44.
45. Wang, S.J.; Sun, C.X.; Feng, Z.G.; Liu, X.M. Mineralogical and geochemical characteristics of the limestone weathering profile. *Acta Mineral. Sin.* **2002**, *22*, 19–29.
46. Cao, J.H.; Yuan, D.X.; Zhang, C.; Pan, G.X. Karst Ecosystem constrained by geological conditions in southwest China. *Earth Environ.* **2004**, *32*, 1–8.
47. Wang, S.J.; Ji, H.B.; Ouyang, Z.Y.; Zhou, D.Q. The Preliminary Study of Carbonate Rock Weathering and Soil Formation. *Sci. Sin. (Terrae)* **1999**, *5*, 441–449.
48. Cao, J.H.; Jiang, Z.C.; Yang, D.S.; Pei, J.G.; Yang, H.; Luo, W.Q. Soil loss tolerance and prevention and measurement of Karst area in southwest China. *Soil Water Conserv. China* **2008**, *12*, 40–45+72.
49. Wang, S.J.; Li, R.L.; Sun, C.X.; Zhang, D.F.; Li, F.Q.; Zhou, D.Q.; Xiong, K.N.; Zhou, Z.F. How types of carbonate rock assemblages constrain the distribution of Karst rocky desertified land in Guizhou Province, PR China. *Land Degrad. Dev.* **2004**, *15*, 123–131. [CrossRef]
50. Li, Y.B.; Bai, X.Y.; Qiu, X.C.; Zhou, G.F.; Lan, A.J.; Zhou, X.; Xiong, K.N. Karst rocky desertification and land use of correlation studies. *Resour. Sci.* **2006**, *28*, 67–73.
51. Li, Y.B.; Bai, X.Y.; Wang, S.J.; Qin, L.Y.; Tian, Y.C.; Luo, G.J. Evaluating of the spatial heterogeneity of soil loss tolerance and its effects on erosion risk in the carbonate areas of southern China. *Solid Earth* **2017**, *8*, 661–669. [CrossRef]
52. Chen, F.; Zhou, D.Q.; Bai, X.Y.; Xiao, J.Y.; Qian, Q.H. Spatial-temporal evolution of karst rocky desertification and future trends based on CA-Markov methods in typical karst valley. *J. Agric. Resour. Environ.* **2018**, *35*, 174–180.
53. Li, H.W.; Wang, S.J.; Bai, X.Y.; Cao, Y.; Tian, Y.C.; Luo, G.J.; Chen, F.; Li, Q.; Wu, L.H.; Wang, J.F.; et al. Effects of climate change and ecological restoration on carbonate rock weathering carbon sequestration in the karst valley of Southwest China. *Acta Ecol. Sin.* **2019**, *39*, 6158–6172.
54. Huang, Q.; Cai, Y.; Xing, X. Rocky desertification, anti desertification, and sustainable development in the karst mountain region of Southwest China. *Ambio* **2008**, *37*, 390–392. [CrossRef] [PubMed]
55. Jiang, Z.H.; Liu, H.Y.; Wang, H.Y.; Peng, J.; Meersmans, J.; Green, S.M.; Quine, T.A.; Wu, X.C.; Song, Z.L. Bedrock geochemistry influences vegetation growth by regulating the regolith water holding capacity. *Nat. Commun.* **2020**, *11*, 2392. [CrossRef] [PubMed]
56. Yuan, D.X. Global view on Karst rock desertification and integrating control measures and experiences of China. *Pratacult. Sci.* **2008**, *25*, 19–25.
57. Jiang, Y.J.; Liu, X.M.; He, S.Y.; He, B.H.; Xie, J.P.; Luo, W.J.; Bai, X.Y.; Xiao, Q. Research and development of comprehensive rehabilitation measures for land rocky desertification in Karst trough valley area. *Acta Ecol. Sin.* **2016**, *36*, 7092–7097.
58. Yu, M.; Li, Y.B.; Luo, G.J. Evolution trend of rocky desertification in karst mountain areas of southwest China. *Acta Ecol. Sin.* **2022**, *42*, 4267–4283.
59. Liu, Y.S.; Wang, J.Y.; Deng, X.Z. Rocky land desertification and its driving forces in the karst areas of rural Guangxi, Southwest China. *J. Mt. Sci.* **2008**, *5*, 350–357. [CrossRef]
60. Liu, H.Y.; Jiao, F.S.; Yin, J.Q.; Li, T.Y.; Gong, H.B.; Wang, Z.Y.; Lin, Z.S. Nonlinear relationship of vegetation greening with nature and human factors and its forecast—A case study of Southwest China. *Ecol. Indic.* **2020**, *111*, 106009. [CrossRef]
61. Jiang, Z.C.; Lian, Y.Q.; Qin, X.Q. Rocky desertification in Southwest China: Impacts, causes, and restoration. *Earth-Sci. Rev.* **2014**, *132*, 1–12. [CrossRef]
62. Chen, F.; Wang, S.J.; Bai, X.Y.; Liu, F.; Zhou, D.Q.; Tian, Y.C.; Luo, G.J.; Li, Q.; Wu, L.H.; Zheng, C.; et al. Assessing spatial-temporal evolution processes and driving forces of karst rocky desertification. *Geocarto Int.* **2021**, *36*, 262–280. [CrossRef]
63. Zhang, Z.H.; Hu, G.; Zhu, J.D.; Luo, D.H.; Ni, J. Spatial patterns and interspecific associations of dominant tree species in two old-growth karst forests, SW China. *Ecol. Res.* **2010**, *25*, 1151–1160. [CrossRef]
64. Xu, E.Q.; Zhang, H.Q.; Li, M.X. Mining spatial information to investigate the evolution of karst rocky desertification and its human driving forces in Chang Shun, China. *Sci. Total Environ.* **2013**, *458–460*, 419–426. [CrossRef]
65. Chen, H.S.; Fu, Z.Y.; Zhang, W. Soil water processes and vegetation restoration in karst regions of southwest China. *Chin. J. Nat.* **2018**, *40*, 41–46.

66. Zhu, X.C.; Ma, M.G.; Tateno, R.; He, X.H.; Shi, W.Y. Effects of vegetation restoration on soil carbon dynamics in Karst and non-karst regions in Southwest China: A synthesis of multi-source data. *Plant Soil* **2021**, 1–15. [CrossRef]
67. He, F.N.; Ge, Q.S.; Dai, J.H.; Rao, Y.J. Forest change of China in recent 300 years. *J. Geogr. Sci.* **2008**, *18*, 59–72. [CrossRef]
68. Yang, X.; Jin, X.; Yang, Y.; Yang, Y.K.; Song, J.N.; Zhang, T.; Zhou, Y.K. Spatially explicit changes of forestland in Taiwan Province from 1910 to 2010. *J. Geogr. Sci.* **2022**, *32*, 441–457. [CrossRef]
69. Lin, S.L.; Jiang, Y.Z.; He, J.K.; Ma, G.Z.; Xu, Y.; Jiang, H.S. Changes in the spatial and temporal pattern of natural forest cover on Hainan Island from the 1950s to the 2010s: Implications for natural forest conservation and management. *PeerJ* **2017**, *5*, e3320. [CrossRef] [PubMed]
70. Zhang, L.; Liu, Z.; Liu, D.Y.; Xiong, Q.; Yang, N.; Ren, T.W.; Zhang, C.; Zhang, X.D.; Li, S.M. Crop Mapping Based on Historical Samples and New Training Samples Generation in Heilongjiang Province, China. *Sustainability* **2019**, *11*, 5052. [CrossRef]
71. Tong, X.W.; Wang, K.L.; Yue, Y.M.; Brandt, M.; Liu, B.; Zhang, C.H.; Liao, C.J.; Fensholt, R. Quantifying the effectiveness of ecological restoration projects on long-term vegetation dynamics in the Karst regions of Southwest China. *Int. J. Appl. Earth Obs. Geoinf.* **2017**, *54*, 105–113. [CrossRef]
72. Zhang, X.M.; Yue, Y.M.; Tong, X.W.; Wang, K.L.; Qi, X.K.; Deng, C.X.; Brandt, M. Eco-engineering controls vegetation trends in southwest China Karst. *Sci. Total Environ.* **2021**, *770*, 145160. [CrossRef]
73. Xu, X.J.; Liu, H.Y.; Lin, Z.S.; Jiao, F.S.; Gong, H.B. Relationship of Abrupt Vegetation Change to Climate Change and Ecological Engineering with Multi-Timescale Analysis in the Karst Region, Southwest China. *Remote Sens.* **2019**, *11*, 1564. [CrossRef]
74. Piao, S.L.; Yin, G.D.; Tan, J.G.; Cheng, L.; Huang, M.T.; Li, Y.; Liu, R.G.; Mao, J.F.; Myneni, R.B.; Peng, S.S.; et al. Detection and attribution of vegetation greening trend in China over the last 30 years. *Glob. Chang. Biol.* **2015**, *21*, 1601–1609. [CrossRef] [PubMed]
75. Li, Z.D.; Wang, S.; Li, C.J.; Ye, C.C.; Gao, D.X.; Chen, P. The trend shift caused by ecological restoration accelerates the vegetation greening of China's drylands since the 1980s. *Environ. Res. Lett.* **2022**, *17*, 044062. [CrossRef]
76. Zhao, H.W.; Wu, C.Y.; Wang, X.Y. Large-scale forest conservation and restoration programs significantly contributed to land surface greening in China. *Environ. Res. Lett.* **2022**, *17*, 024023. [CrossRef]
77. Wang, C.X.; Liang, W.; Yan, J.W.; Jin, Z.; Zhang, W.B.; Li, X.F. Effects of vegetation restoration on local microclimate on the Loess Plateau. *J. Geogr. Sci.* **2022**, *32*, 291–316. [CrossRef]
78. Brandt, M.; Yue, Y.M.; Wigneron, J.P.; Tong, X.W.; Tian, F.; Jepsen, M.R.; Xiao, X.M.; Verger, A.; Mialon, A.; Al-Yaari, A.; et al. Satellite-observed major greening and biomass increase in South China karst during recent decade. *Earth's Future* **2018**, *6*, 1017–1028. [CrossRef]
79. Qiao, Y.N.; Chen, H.; Jiang, Y.J. Quantifying the impacts of lithology on vegetation restoration using a random forest model in a karst trough valley, China. *Ecol. Eng.* **2020**, *156*, 105973. [CrossRef]
80. Li, Y.; Piao, S.L.; Laurent, Z.X.; Chen, A.P.; Wang, X.H.; Ciais, P.; Huang, L.; Lian, X.; Peng, S.S.; Zeng, Z.Z.; et al. Divergent hydrological response to large-scale afforestation and vegetation greening in China. *Sci. Adv.* **2018**, *4*, eaar4182. [CrossRef]
81. Wang, J.; Liu, P.; Gao, Z.; Bai, S.B.; Cao, G.J.; Qu, G.X. Temporal-spatial variation of the channel in Jiangsu reach of the Yangtze River during the last 44 years. *Dili Xuebao/Acta Geogr. Sin.* **2007**, *62*, 1185–1193.
82. Tong, X.W.; Brandt, M.; Yue, Y.M.; Horion, S.; Wang, K.L.; De, K.W.; Tian, F.; Schurgers, G.; Xiao, X.M.; Luo, Y.Q.; et al. Increased vegetation growth and carbon stock in China Karst via ecological engineering. *Nat. Sustain.* **2018**, *1*, 44–50. [CrossRef]
83. Tong, X.W.; Brandt, M.; Yue, Y.M.; Ciais, P.; Jepsen, M.R.; Penuelas, J.; Wigneron, J.P.; Xiao, X.M.; Song, X.P.; Horion, S.; et al. Forest management in southern China generates short term extensive carbon sequestration. *Nat. Commun.* **2020**, *11*, 129. [CrossRef] [PubMed]

land

MDPI

Article

Analysis of the Land Use Dynamics of Different Rural Settlement Types in the Karst Trough Valleys of Southwest China

Yiyi Zhang [1], Yangbing Li [1,*], Guangjie Luo [2], Xiaoyong Bai [1,3], Juan Huang [1], Fang Tang [1] and Meng Yu [1]

1 School of Geography and Environmental Sciences, Guizhou Normal University, Guiyang 550001, China
2 Guizhou Provincial Key Laboratory of Geographic State Monitoring of Watershed, Guizhou Normal College, Guiyang 550018, China
3 State Key Laboratory of Environmental Geochemistry, Institute of Geochemistry, Chinese Academy of Sciences, Guiyang 550081, China
* Correspondence: li-yabin@gznu.edu.cn

Abstract: Rural settlements are the basic spatial units of rural geography research, and it is essential to explore the dynamic changes in land use on a rural settlement scale to promote the development of the rural revitalization strategy. The study took different rural settlement types in karst trough valleys as examples and applied geographic information mapping trajectory models, buffer zone spatial analysis, the nearest neighbor index, and other research methods. We explored the land use dynamic change in the buffer zone of different settlement types in the karst trough valley from 1964 to 2021 in the long time series and micro-spatial dimensions. We analyzed the homogeneity, variability, and coupling characteristics of land use evolution in typical settlements. The results indicate the following: (1) From 1964 to 2021, the karst trough valley settlements as a whole showed an aggregation state, and the settlements could be classified into four categories: expanding settlements (ES), atrophic settlements (AS), balancing rural settlements (BS), and decreasing settlements (DS) according to the settlement life cycle theory and settlement development index measurement. (2) Different expansion and shrinkage of land use buffer changes exist for different settlement types. The closer the ES is to the location of the settlement center, the richer the land use type; the further the AS from the settlement center, the richer the land use type; the BS is not affected by the distance; and the DS settlement shows dynamic changes. (3) Land use dynamic change in settlements is driven by multiple integrated factors, and there is variability in the driving factors of different settlement types. (4) In this paper, through a case study, we propose the research idea that land use change (LUCC) reflects land use transformation (LUT) in different rural settlement types from a settlement-scale perspective, and land use transformation further causes the development of rural settlement transformation (RUT). Our study revealed the LUCC—LUT—RUT interaction feedback mechanism of karst trough valley settlements in Southwest China. This study aims to enrich the theoretical research framework of rural transformation at the settlement scale, on the one hand, and to provide case studies for developing countries with karstic mountain valley landscapes, such as China, on the other.

Keywords: rural settlement; land use; land use transition; rural transformation development

Citation: Zhang, Y.; Li, Y.; Luo, G.; Bai, X.; Huang, J.; Tang, F.; Yu, M. Analysis of the Land Use Dynamics of Different Rural Settlement Types in the Karst Trough Valleys of Southwest China. *Land* **2022**, *11*, 1572. https://doi.org/10.3390/land11091572

Academic Editor: Xuesong Kong

Received: 11 August 2022
Accepted: 9 September 2022
Published: 14 September 2022

1. Introduction

In the context of rapid global industrialization and urbanization, 45% of the world's population still lives in rural areas and is affected by globalization [1–3]. Meanwhile, the disparity and decline of rural development have become global issues in globalization [4,5]. Since the end of World War II, differential rural problems have gradually emerged in both developed and developing countries, with rural areas in the United States, the United Kingdom, and Canada, as well as developed countries in Europe and the United States, facing urban–rural disparities, concentrated poverty, cultural conflicts, and irrational land use [6–8]. Developing countries, such as Africa and India, face severe land degradation, widespread poverty, food insecurity, and other problems in rural areas [9,10]. With the prominence of global rural problems, it is easy to see that rural land development and land use are among the critical issues of rural development, so it has become a human consensus to solve rural problems and explore the long-term evolution trend and sustainable development of rural areas and their land use [11].

China has always been a major agricultural country, and its villages play a vital role in ensuring its agricultural and food security and sustainable development [12]. In recent years, along with the continuous promotion of urban–rural integration and rural revitalization strategy, rural-related research topics have gradually become a major socio-economic issue of concern to researchers and the Chinese government [13]. As one of the carriers of rural society, village settlements are widely used in rural studies [14]. A rural settlement is a production and living place formed by the interaction of rural residents with the surrounding natural environment, socio-economic environment, and cultural environment [15]. The integrated development and optimal spatial reorganization of all elements of rural settlements have become an academic hotspot driven by the urgent need and strategy for the sustainable development of national rural society [16].

Rural settlements are the basic spatial units of rural geography and an important part of the study of human–land relations [17]. Academics have conducted many studies on rural settlements, rural settlement land use, and rural settlement development and have achieved remarkable results. In studying the evolution of rural settlements, scholars initially focused on the formation and orientation of rural settlements [18,19]. With the development of geographic information technology, scholars have paid more attention to the types of rural settlements [20], spatial distribution, scale [21–24], density [25,26], driving mechanisms [27], transformational development, spatial reconfiguration, etc. [28,29]. In the studies related to land use change in rural settlements, scholars have focused on national [30,31], provincial [32], regional [25,33,34], and basin [35,36] scales of study, and used research methods such as the landscape pattern index method [37], neighborhood buffer analysis [38], spatial autocorrelation analysis [39], and sample zone analysis [40] to explore the core issues [41,42] of the spatial layout of rural settlement land use, driving mechanisms of land use evolution, and spatial optimization and reorganization. In the study of rural settlement development, scholars have devoted themselves to the study of rural settlement development types and patterns [12,17,43], development potential [44], multifunctionality [13,45,46], and sustainable development [47,48]. Several studies have analyzed rural settlements, land use, and rural settlement development separately. However, previous studies have only observed single-factor dynamic changes in rural settlements. There is a lack of in-depth research on the comprehensive development types of different rural settlements, the evolution of settlement development, and surrounding land use elements and there is still a lack of research on the dynamic changes in land use at the microscopic scale of rural settlements in long time series. Few studies have examined the integrated perspective of human–land interaction and rural development. The inadequacy of these studies make it difficult for us to accurately judge and grasp the regularity and stages of the evolution process of rural settlements and to scientifically optimize the planning related to rural settlements in rural revitalization. Therefore, it is necessary to study the dynamic changes in land use in rural settlements on a microscopic scale.

The karst trough valley is one of the typical karst landform types in China, as it is a large area with a flat topography at the bottom of the trough. The region is characterized by high population pressure, low land carrying capacity, relatively lagging socio-economics, considerable topographic relief, and significant differences in spatial patterns of land use. Based on this, clarification of the "people (settlement)—land (land use) relationship and its evolution" in the trough valley is vital for the territory. In the context of the current dualistic development of urban–rural territorial systems, the number, scale, and pattern of land use in and around different rural settlement types in the trough valley varies, and the differences in land use evolution of different rural settlement types reflect the differences in regional rural socio-economic development. Because of this, to explore the land use evolution pattern at the settlement scale in the karst trough valley, this study grasps the "type" of settlement with the help of "typical" settlements. The study uses "micro-scale long time series" to map "large scale short time series." The study selects different types of settlements in karst troughs and valleys as research objects and explores the spatial and temporal evolution characteristics of land use around different types of settlements on the settlement scale. At the same time, we analyze the land use transfer trajectories of different types of settlements; explore the land use rise and shrinkage patterns; and reveal the interactive feedback mechanisms of land use change (LUCC), land use transformation (LUT), and rural settlement transformation (RST) processes. We aim to explore the driving factors of the spatial differentiation of rural areas to provide a more scientific and reasonable reference basis for the land use of rural settlements under different geomorphological conditions in other karst trough and valley areas, to achieve a balance between the supply and demand of land use in rural settlements in karst troughs and valleys, to promote sustainable rural socio-economic development and ecological–environmental protection, and to provide a reference for enriching the research framework and typical cases of rural land use evolution and land use transformation in China.

Accordingly, we proposed the following research question: What is the development type of rural settlements in the karst trough valley of Southwest China? How does land use change across rural settlement types on the buffer scale? What natural or socio-economic factors influence land use change in rural settlements? Specifically, we tested two main hypotheses: (1) Land use change in rural settlements is driven by natural and socio-economic factors. (2) In the karst trough valley area of Southwest China, there is an interactive feedback mechanism of land use change–land use transformation–rural transformation development. To test these hypotheses, we selected the karst trough valley area in Southwest China as the study area. In Section 1, we briefly describe the study area and data sources. In Section 2, we list the appropriate research methods and the selection of typical clusters. Section 3 analyzes the results of different rural settlement types, land use buffer changes, land use transfer trajectories, and land use coupling states in karst trough valleys. In Section 4, we analyze the drivers of land use change and feedback mechanisms for different settlement types and highlight our research uncertainties and future research directions. Finally, the conclusions of our study are presented in Section 5.

2. Study Area and Data Sources

2.1. Study Area

A typical karst trough valley (LangXi trough valley) in the southwest karst mountainous region was selected as the research object. LangXi trough valley is located in YinJiang County of Tongren region in northeastern Guizhou Province, and the administrative area covers several townships in the territory, including BanXi town, TaiShui town, LangXi town, Eling town, LuoChang township, BaoXi township, and Tinzhai township, with a total area of 130.34 km^2. The geographical position is 108°24′36.43″–108°34′8.62″ E; 27°53′59.55″–28°6′22.16″ N. There are 35 administrative villages in LangXi karst trough valley, including SanCun village, XiBu village, HeXi village, etc. Based on the study's purpose and the area's topographical features, the study area was divided into five trough valley locations: the top of the trough valley, the trough slope, the dam, and the east and west slopes (Figure 1).

2.2. Data Sources

Spatial data and non-spatial data were selected for the study, in which spatial remote sensing image data were selected as the primary data for 1964, 1999, 2004, 2014, and 2021. Resolutions were in the order of 2.7 m, 2.7 m, 10 m, 2.5 m, and 2.5 m. After the field survey of the study area, the spatial remote sensing images were interpreted and the land use data were visually interpreted for each year.; According to the Land Use Status Classification (GTB-21010-2017) and the actual situation of the trough valley area, The land use types in the study area were further divided, where arable land was classified as steep slope, gentle slope, and flat dam (slope > 15°, 7° < slope < 15°, slope < 7°); grassland was classified as high cover, medium cover, and low cover, and the others were interpreted according to the images in turn (Table 1). The accuracy of the classified land use types was corrected using ENVI5.0 and verified by combining field research with sampling, and the accuracy of the land use vector map for each period reached 87%, meeting the needs of land use analysis. The non-spatial data were mainly extracted from field research and government statistics.

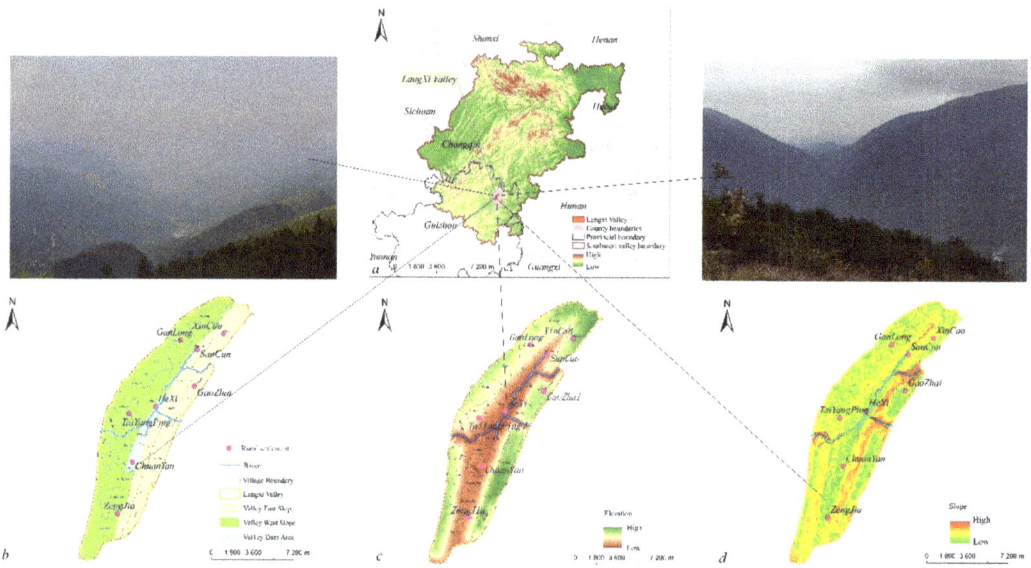

Figure 1. The study area ((**a**) Langxi trough location map; (**b**) Typical settlement distribution map; (**c**) Trough and valley elevation map; (**d**) Slope map of trough and valley).

Table 1. Trough land use classification.

Land Use Properties	Level 1 Land Use Grade	Level 2 Land Use Grade	Code
Production land	Arable land 01	Flat dam arable land	0101
		Gentle slope arable land	0102
		Steep-slope arable land	0103
	Garden 02	Orchards	0201
		Tea gardens	0202
		Vegetable Garden	0203
Living land	Construction Land 03	Industrial Land	0301
		Industrial and mining land	0302
		Rural residential area	0303
		Urban settlements	0304
		Road	0305
Ecological land	Forest 04	Forested land	0401
	Other lands 05	Abandoned land	0501
	Grassland 06	Low cover grassland	0601
		Medium cover grass	0602
		High cover grassland	0603
		Irrigated grassland	0604
	Water 07	River, Reservoir	0701

3. Methods

3.1. Settlement Selection and Classification

The difference in settlement types in the karst valley areas reflects the land-bearing capacity and the human–land relationship in the karst mountains. The classification of settlement types in the valley areas aims to reveal the characteristics of settlement differences in karst valley areas, evolution rules, and driving factors. The classification study helps us grasp the land use changes in different types of settlements. Research on the classification of settlement types usually follows the principles of wholeness, dominance, and the feasibility of development and classifies rural settlements into different types based on the geographical environment, location conditions, economic development, ecological environment, social culture, and farmers' wishes, and then formulates the corresponding optimization strategies [35,49–51]. In this study, based on the avoidance of administrative and large scattered villages, we selected eight typical settlement units in the trough valley region for the study and explored the land use buffer scale changes for individual settlements. Combining the previous research results, in order to grasp the land use evolution pattern guided by human activities in different settlement environments in the trough valley region, considering the geographical differences in the natural environments in which the settlements are located and the types of settlement evolution, drawing on the literature [52–54], we reintegrated the total rate of change and the average annual rate of change with modified formulas and calculated the total rate of change and the average annual net rate of change formulas for the analysis of rural settlement change. Finally, we used the settlement change index for settlement type classification and classified the typical settlement types in the karst trough valley area as follows (Figure 2): expanding settlements (ES: ZengJia, SanCun, and ChuanYan), atrophic settlements (AS: Ganlong and XinCao), disappearing settlements (DS: TaiYangPing), and balancing rural settlements (BS: HeXi and GaoZhai).

3.2. Buffer Analysis

Buffer analysis is used to establish a certain distance of a faceted area around a spatial object given a spatial object, the extent of which is determined by the radius R of the area to identify the radiation or influence of the analyzed object on the neighboring objects. Generally, the buffer area for a spatial object B is defined as follows:

$$B = \{X_i = \{\mathrm{X_i} | \mathrm{d}(X, O)\} \leq R\} \quad (1)$$

B is the target buffer influence range, X_i is the location of any point in the target field (rural settlement point), O is the analysis object, d is the minimum Euclidean distance, and R is the buffer radius [55]. Different settlement types in the karst trough valley area have different levels of settlement development, and the surrounding land use changes show differences. In this paper, we use the buffer zone analysis method to select different types of settlements as the center point and establish buffer zones (range 50 m, 100 m, 150 m, 200 m, 250 m, 300 m, 350 m, 400 m) according to the center of the settlement; we perform a study on the dynamic variation of surrounding land use under different types of settlements. The buffer zone analysis in this paper was used to establish a polygon land use layer with a certain radius around the settlement as the center point, overlay the established layer with the target layer, and then analyze the results to explore further the evolution and coupling relationship between different types of settlements and their surrounding land use in the trough region.

3.3. Geographic Information Mapping Trajectory and the Index Model of Comprehensive Land Use Intensity

ArcGIS10.2 software was used to overlay five phases of land use maps and analyze the spatial change process of land use using change mapping with the following equation:

$$Y = G_1 10^{n-1} + G_2 10^{n-2} + \ldots + G_n 10^0 \quad (2)$$

where Y is the n–digit number calculated synthesis of the land use code; n is the number of periods of land use; G_n is the nth period of land use unit.

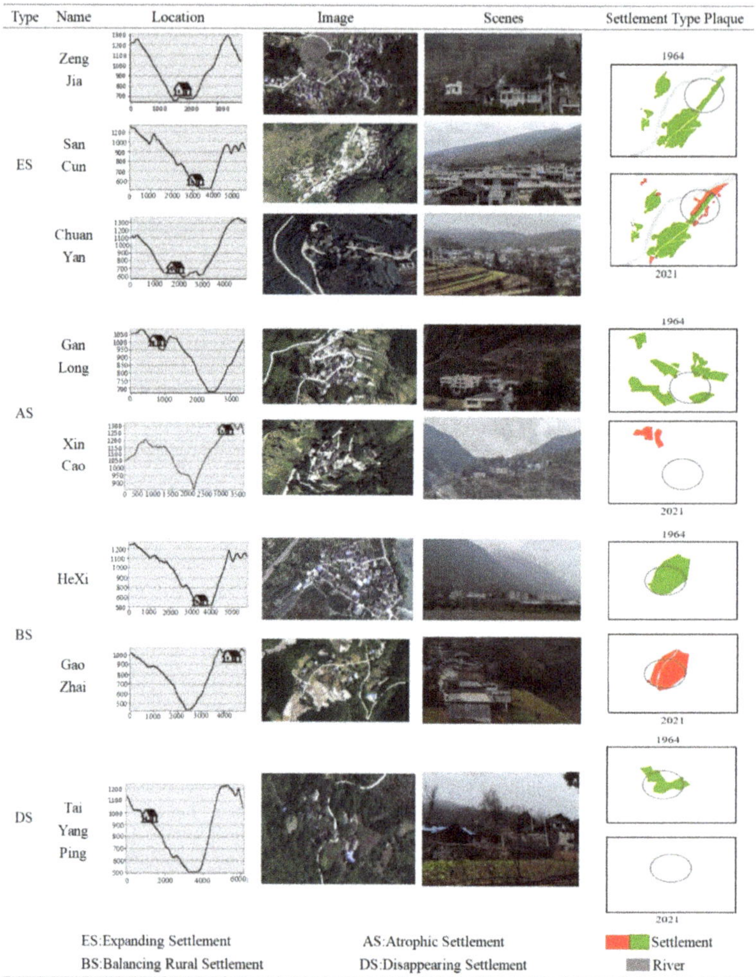

Figure 2. Typical settlement division and selection.

The index model of comprehensive land use intensity [56] was constructed to model the change in land use intensity around each settlement in the study area so that it could be implemented in the rural settlement spatial unit. The specific equation is as follows:

$$L = \sum_{i=1}^{n} A_i C_i = \sum_{i=1}^{n} A_i (S_i / S) \tag{3}$$

where L is the land use intensity of a single sample; A_i and C_i are the graded indices of land use intensity at level i and the percentage of area occupied in the sample; S_i is the area of land use type at level i in the sample; S is the total land area of the sample.

3.4. Average Nearest Neighbor

The Average Nearest Neighbor Index (ANN) is derived from the average distance between each rural settlement's center of mass and its nearest neighbor's center of mass and is one of the most common methods used to determine the spatial distribution pattern of rural settlements. The average Nearest Neighbor Index value is distributed between [−1, 1], and the closer the result is to 1, the more discrete the distribution is, and the opposite is the more clustered [31].

$$ANN = \gamma_\alpha \gamma_\beta = \frac{\sum \frac{d_{\min}}{n}}{\frac{\sqrt{n/A}}{2}} = \frac{2\sqrt{\lambda}}{N} \sum d_{\min} \tag{4}$$

ANN is the average nearest neighbor index. γ_α is the average distance of nearest neighbors of village settlement points; γ_β is the theoretical average under the random spatial distribution of village settlement points. $d_{\min}$ is the distance between a village settlement point and the nearest neighboring village settlement; n is the number of village settlements; A is the total area of spatial units; λ is the spatial distribution density of village settlements.

3.5. Standard Deviational Ellipse

Standard deviational ellipse (SDE) can accurately reveal the spatial distribution center, dispersion, and directional trends of geographical elements and is a spatial statistical method to quantitatively analyze the overall characteristics of the spatial distribution of geographical elements [56,57]. The rotation angle is the angle formed by clockwise rotation from due north to the central axis, reflecting the main trend direction of its distribution, and the long axis characterizes the dispersion of rural settlement sites in the main trend direction, whose mathematical expression is [56,57]:

$$(A) \approx \tan\theta = \left\{ \left[\left(\sum_{i=1}^{n} w_i^2 x'^2_i - \sum_{i=1}^{n} w_i^2 y'^2_j \right) \right] + \sqrt{\left(\left[\sum_{i=1}^{n} w_i^2 x'^2_i - \sum_{i=1}^{n} w_i^2 y'^2_j \right]^2 + 4 \left(\sum_{i=1}^{n} w_i^2 x'^2_i y'^2_j \right)^2 \right)} \right\} / 2 \left(\sum_{i=1}^{n} w_i^2 x'^2_i y'^2_j \right) \tag{5}$$

$$(B) \approx \delta_x = \sqrt{\left[\sum_{i=1}^{n} (w_i x'_i \cos\theta - w_i y'_i \sin\theta)^2 / \sum_{i=1}^{n} w_i^2 \right]} \tag{6}$$

$$(B) \approx \delta_y = \sqrt{\left[\sum_{i=1}^{n} (w_i x'_i \sin\theta - w_i y'_i \cos\theta)^2 / \sum_{i=1}^{n} w_i^2 \right]} \tag{7}$$

where (A) and (B), the azimuthal angle is derived from $\tan\theta$, δ_x, and δ_y are the standard deviations along the x and y axes, respectively, and x_i and y_i represent the coordinate deviations from the mean center, x_i, and y_i indicate the deviation of coordinates from the mean center. The center (center of gravity) is the average distribution center of the rural settlement land space in the trough valley area. The center uses the main trend direction of rural settlement distribution as the azimuth, the standard deviation in the x–direction and y–direction as the ellipse axis, and the spatial distribution ellipse of rural settlement land is constructed to explain the characteristics of centrality, directionality, and spatial distribution pattern of the evolution of rural settlement type land in the trough valley area. Meanwhile, the direction, intensity, and spatial dispersion trends of the development changes of rural settlements in karst trough valleys are identified by the standard deviation ellipse eigenvalues in different years. This paper calculated the standard deviation ellipse parameters of rural settlement sites in karst valleys with the help of the ArcGIS software spatial statistics module and visualized the results.

3.6. Kernel Density Estimation

Kernel density estimation (KDE) is a non–parametric density calculation method, which reveals the distribution characteristics of points through the spatial variation of the density of settlement points, and is suitable for measuring the spatial distribution density of rural settlement sites:

$$f(x, y) = \frac{1}{nh^2} \sum_{i=1}^{n} k\left(\frac{d_i}{n}\right) \tag{8}$$

where $f(x,y)$ is the density estimate at point (x,y); $k()$ is the kernel function; bandwidth $h > 0$; n is the number of observations; di is the distance of (x,y) location from the ith element. The higher the kernel density value, the higher the density of spatial distribution of rural settlement sites [58].

4. Results

4.1. Spatial Pattern Analysis of the Evolution of Rural Settlement Types in the Karst Trough

In order to clarify the spatial aggregation characteristics of the evolution of rural settlements in the LangXi trough valley and to classify the types of settlements, the regional rural settlements were analyzed using the nearest neighbor index. The nearest neighbor index analysis was conducted on the regional rural settlements, and the results of the analysis showed that, during the nearly 60 years from 1964 to 2021, the z–value of LangXi trough valley rural settlement was less than 1 in all four time periods, then the trough valley rural settlement showed a clustering trend. The significance level was less than 0.01, indicating that the spatial aggregation of rural settlement types within the trough valley territorial system rejects the null hypothesis of random distribution. From 1964 to 1999, the average observed distances in this period were all smaller than the expected average distances, and the nearest neighbor ratio was approximately 0.4 (0.37–0.39), with a significance level of $p < 0.01$, indicating that the karst valley rural settlements showed an overall clustering trend in this period. The number of rural settlement patches clusters significantly decreased, the average observed distance slightly increased, and the nearest neighbor ratio slightly decreased, from 0.396 to 0.372 (Table 2 and Figure 3). This shows that the spatial agglomeration of settlements tended to weaken with time evolution. From 2004 to 2021, the cluster z–value decreased sharply from −20.76 to −48.39, indicating that the spatial agglomeration of the clusters showed a sharp weakening trend over time.

Table 2. The nearest neighbor ratio of rural settlements in 1964 and 2021.

Year	Coverage Observation Distance(M)	Expected Average Distance(M)	Nearest Neighbor Ratio	Z-Value	p-Value
1964	133.9159	358.7075	0.3733	−19.1443	0.0000
1999	152.4767	384.9336	0.3961	−16.6215	0.0000
2004	123.5568	331.7834	0.3724	−20.7609	0.0000
2021	39.8482	185.4402	0.2150	−48.3908	0.0000

Drawing the standard deviation ellipse of the spatial distribution of rural settlement patches can explain the characteristics of centrality, direction, and spatial distribution patterns of rural settlement types in the karst trough valley area. Meanwhile, the direction and intensity of rural settlement development changes and their spatial dispersion trends can be identified by the standard deviation ellipse characteristic values in different periods. The average length of the x-axis from 1964 to 2021 was 1.4 km, the average length of the y-axis was 1.6 km, the rotation angle decreased from 25.59° to 25.31°, and the deviation range of the main parameters of the standard deviation ellipse for each year was approximately 2%, and the basic spatial pattern of the settlement was relatively stable and maintained its distribution in the W–N direction (Figure 4). This shows that the basic spatial pattern of the settlement in the study area is controlled by the topography of the trough valley and trough dam, as well as the topography of the trough dam, which is surrounded by mountains on both sides, with east-west trough slopes and narrow north-south slopes. The center of the standard deviation ellipse is the center of gravity of rural settlements in the corresponding year, and its migration changes can reflect the overall spatial process of the evolution of rural settlement types in the study area. The center of gravity of the settlement in 1964 was used as the coordinate origin to measure the rate and direction of the settlement center of gravity migration in each period and visualize it. The calculation results show that the average annual rate of gravity migration was 32.12 m/a. In directional change, the gravity of the settlement shifted southeast from 1964 to 2021 and pointed to the trough dam area. The spatial evolution of hotspot areas of rural settlement types in the trough valley differed in each period, and the spatial directionality was stronger from 2004 to 2021 than from 1964 to 2004. The main reason for this is the accelerated urbanization and industrialization of the trough valley area since 2004 and the significant changes in the spatial pattern of settlements.

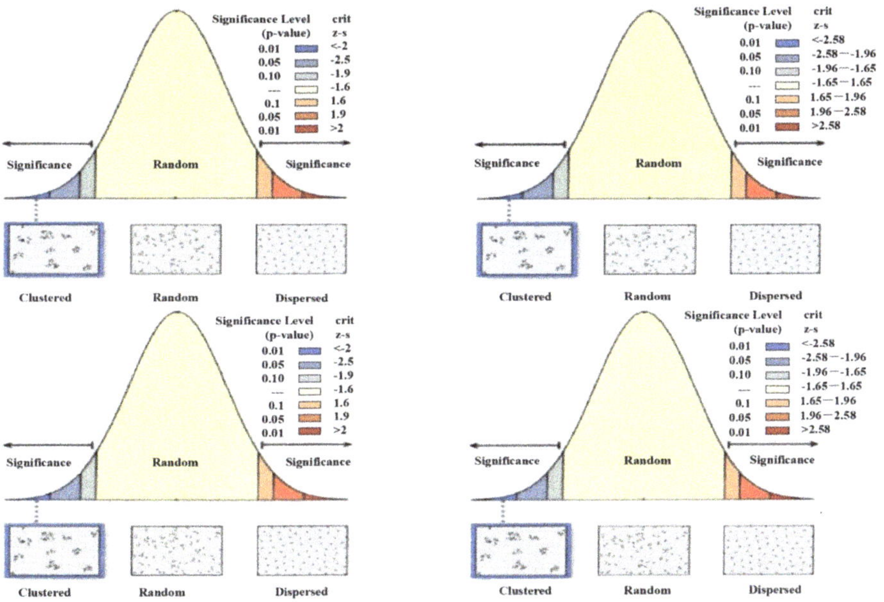

Figure 3. Distribution of Nearest Neighbor Index of Rural Settlements, 1964–2021.

The study extracted the center of gravity of each settlement patch, used the center of gravity to represent the settlement, and calculated the spatial distribution density of the settlement. Using the ArcGIS nuclear density module, a spatial analysis of nuclear density was conducted to classify the settlement nuclear density in each period into the background, low density, medium density, and high-density zones (Figure 4). The settlement nucleation density of each period was also classified into the background, low density, medium density, and high-density zones (Figure 4). The spatial heterogeneity of settlement density distribution in the trough and valley area is prominent, and the high-value area of settlement density distribution from 1964 to 2021 tended to be the trough and dam areas. The background area is mainly located on the slope and top of the valley, part of the geological environment in this area is not suitable for forming settlements, and the distribution of settlements is small. The medium and high-density areas are mainly located in the karst valley trough and dam area with flat terrain, convenient transportation, and good farming conditions and are primarily distributed in a band. The high-density areas are distributed along the traffic arteries and the Yinjiang River, while the low-density areas are scattered in the two wings of the troughs and valleys.

4.2. Analysis of the Buffer Zones in Land Use Change around Typical Rural Settlement Types

From 1964 to 2021, buffer changes, the types, and amounts of land use around the settlements showed differential change characteristics within the buffer area. The land use changes in the settlement's 0 to 400 m buffer zone showed the evolution characteristics of three buffer interval dimensions. The land use mapping of 0 to 50 m, 50 to 200 m, and 200 to 400 m buffers in four troughs and valleys in typical settlement classes had varying characteristics (Figure 5).

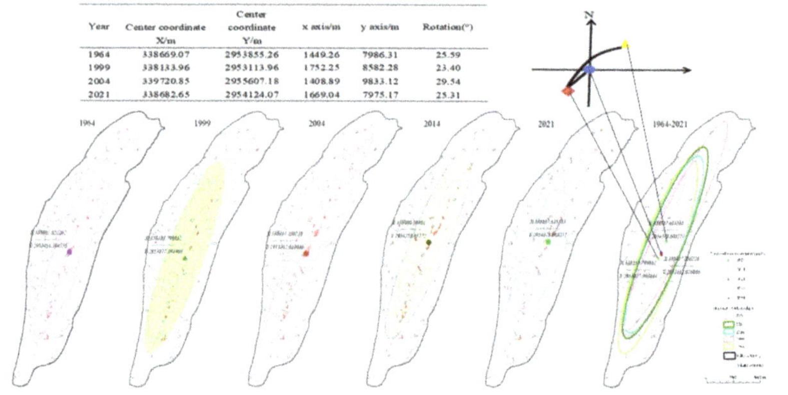

Year	Center coordinate X/m	Center coordinate Y/m	x axis/m	y axis/m	Rotation(°)
1964	338669.07	2953855.26	1449.26	7986.31	25.59
1999	338133.96	2953113.96	1752.25	8582.28	23.40
2004	339720.85	2955607.18	1408.89	9833.12	29.54
2021	338682.65	2954124.07	1669.04	7975.17	25.31

(**a**)

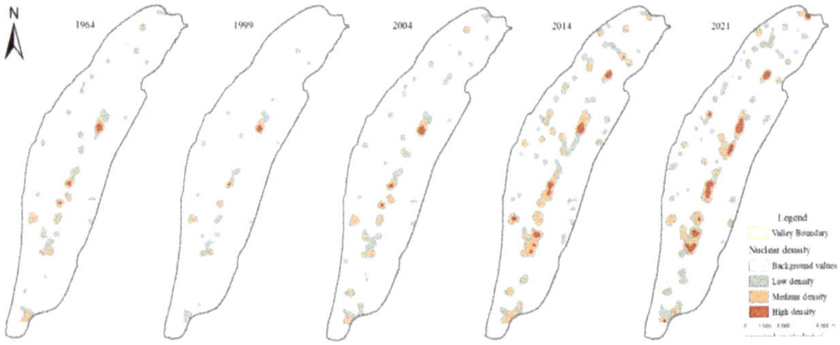

(**b**)

Figure 4. (**a**) Kernel density analysis of rural settlements from1964 to 2021. (**b**) Gravity shift and standard deviational ellipse of rural settlement distribution from 1964 to 2021.

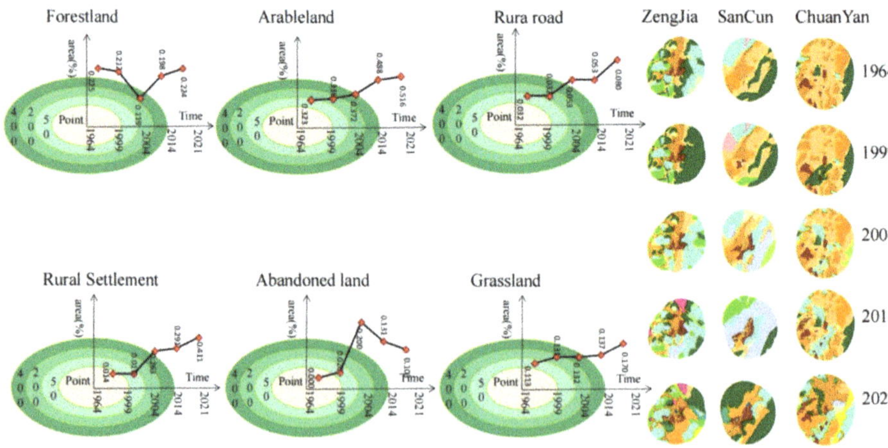

Figure 5. *Cont.*

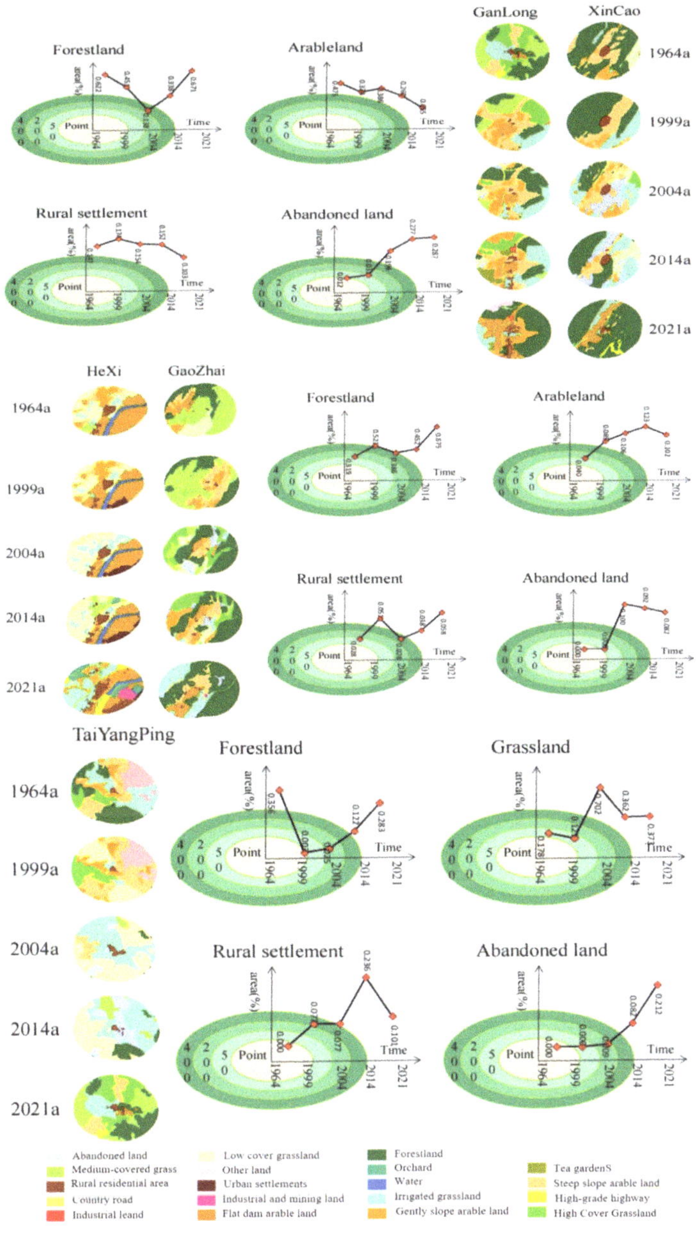

Figure 5. Evolution of land use in the typical settlement type's 0–400 m buffer zone.

In general, the change in land type in the buffer zone of the ES of the dam of karst trough valley is mainly concentrated in the buffer range of 0 to 50 m, and the land use types around the settlement are mainly steep-slope arable land, gentle slope arable land, flat dam arable land, and rural settlement, and the number of land types accounts for 30.84%, 13.21%, 12.67%, and 8.23% of the buffer area, respectively. In the buffer zone, the overall trend of land use change shows the expansion of forest land and abandoned land, while rural residential areas and arable land maintain a balance. However, within the 0 to 50m buffer zone, the land use land types around the BS are mainly rural residential areas, orchards, flat dam arable land, and steep-slope arable land, with the numbers accounting for 12.13%, 10.36%, 28.31%, and 21.62%, respectively. The rate of change is mainly based on the expansion of rural settlements, flat dams of arable land over time, and the increase in abandoned land and orchards. The changes in the land use buffer zone around the AS of Xinchao and Ganlong are mainly manifested in four types of land: forest land, arable land, and abandoned land (Figure 5). Within the 0 to 50 m buffer zone, the overall land use types are mainly steep slope arable land, gently sloping arable land, rural residential areas, and abandoned land, accounting for 36.12%, 23.03%, 12.35%, and 9.16% of the buffer zone area, respectively. The land use around the DS Taiyangping has apparent differences in land use types, quantity, and structural changes in the buffer zone dimension; the number of land use types has increased, and the change in land use quantity is mainly in forestland, cropland, grassland, and abandoned land. In the 0 to 50 m buffer zone range, the buffer zone land types of rural settlements, low-cover grassland, and flat dam cropland show an increase and decrease with time.

4.3. Mapping and Trajectory Analysis of Land Use Changes around Typical Rural Settlement Types

The study introduces a geographic information mapping trajectory model to analyze the spatial and temporal trajectories of land use change around typical settlements in the LangXi trough valley (Figure 6). From 1964 to 1999, the land use around the trough dam of ES showed an expansion of steep-slope arable land, flat-dam arable land, grassland, and rural residential area, and a contraction of forestland and low-cover grassland in the time-series change characteristics. The trajectory mapping of land use change is as follows: forestland→steep-slope arable land, forestland→flat dam arable land, forestland→irrigated grassland, gently sloping arable land→irrigated grassland, low-cover grassland→steep-slope arable land, low-cover grassland→rural residential areas. The land use at the trough valley top and the trough slope of the AS is mainly steep-slope arable land, flat dam arable land, irrigated grassland expansion, and forestland shrinkage. The trajectory of land use change around the settlement is mainly forestland→steep-slope arable land, forest land→flat dam arable land, forest land→gently sloping arable land. BS at the top of the trough valley is dominated by the expansion of steep-slope arable land, flat dam arable land, grassland, and rural residential area, and the contraction of forestland and low-cover grassland. The trajectory mapping of land use change is mainly: forestland→steep-slope arable land, forestland→flat dam arable land, forestland→irrigated grassland; gently sloping arable land→irrigated grassland, low cover grassland→steep-slope arable land, low cover grassland→rural residential area. The trough slope of the DS is mainly dominated by the expansion of arable land and the shrinkage of forestland grassland, and the land use change trajectory is mainly: forestland→steep-slope arable land, forestland→flat dam arable land, forestland→irrigated grassland, steep-slope arable land→gently sloping arable land; the intensity of arable land use is higher in the trough slope in this period.

From 2004 to 2014, the land use types of the ES were mainly forestland, abandoned land, grassland expansion, steep slope arable land, irrigated grassland, flat-dam arable land, and gently sloping arable land contraction. The land use change trajectory is mainly steep-slope arable land→abandoned land, gentle slope arable land→abandoned land, flat dam arable land→abandoned land, low cover grassland, and irrigated grassland→abandoned land. The land use types of the AS are mainly an expansion of grassland and abandoned land, and contraction of forestland, irrigation grass, gently sloping arable land, and steep sloping arable land, and the trajectory mapping of the surrounding land use changes are mainly forestland→steep sloping arable land, forestland→flat dammed arable land, and forestland→gently slope arable land. The types of land use around the BS are mainly gently sloping arable land, abandoned land, irrigated grassland, low cover grassland, high cover grassland, rural road expansion, steep-slope arable land, forest land, and flat dam arable land contraction. The trajectory of land use change is as follows: steep-slope arable land to fallow land, steep-slope arable land to grassland, forest land to irrigated grassland, forest land to gently sloping arable land, forest land→irrigated grassland, forest land→country roads, gently sloping arable land→abandoned land. The land use types around the DS are mainly arable land, grassland expansion, and forest land contraction, and the land use change trajectory mainly shows forestland→arable land, forest land→irrigated grass, forestland→high cover grass, and irrigated grass→high cover grass.

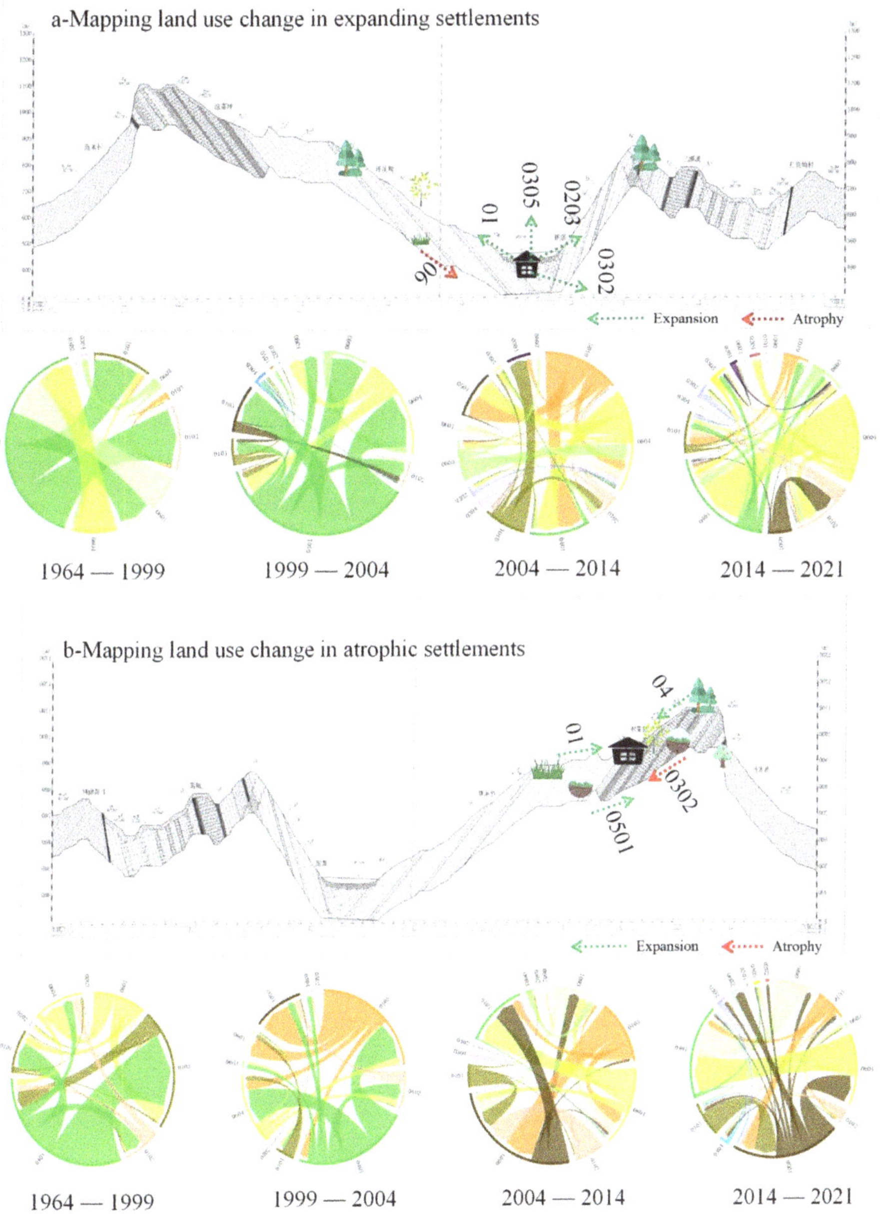

Figure 6. *Cont.*

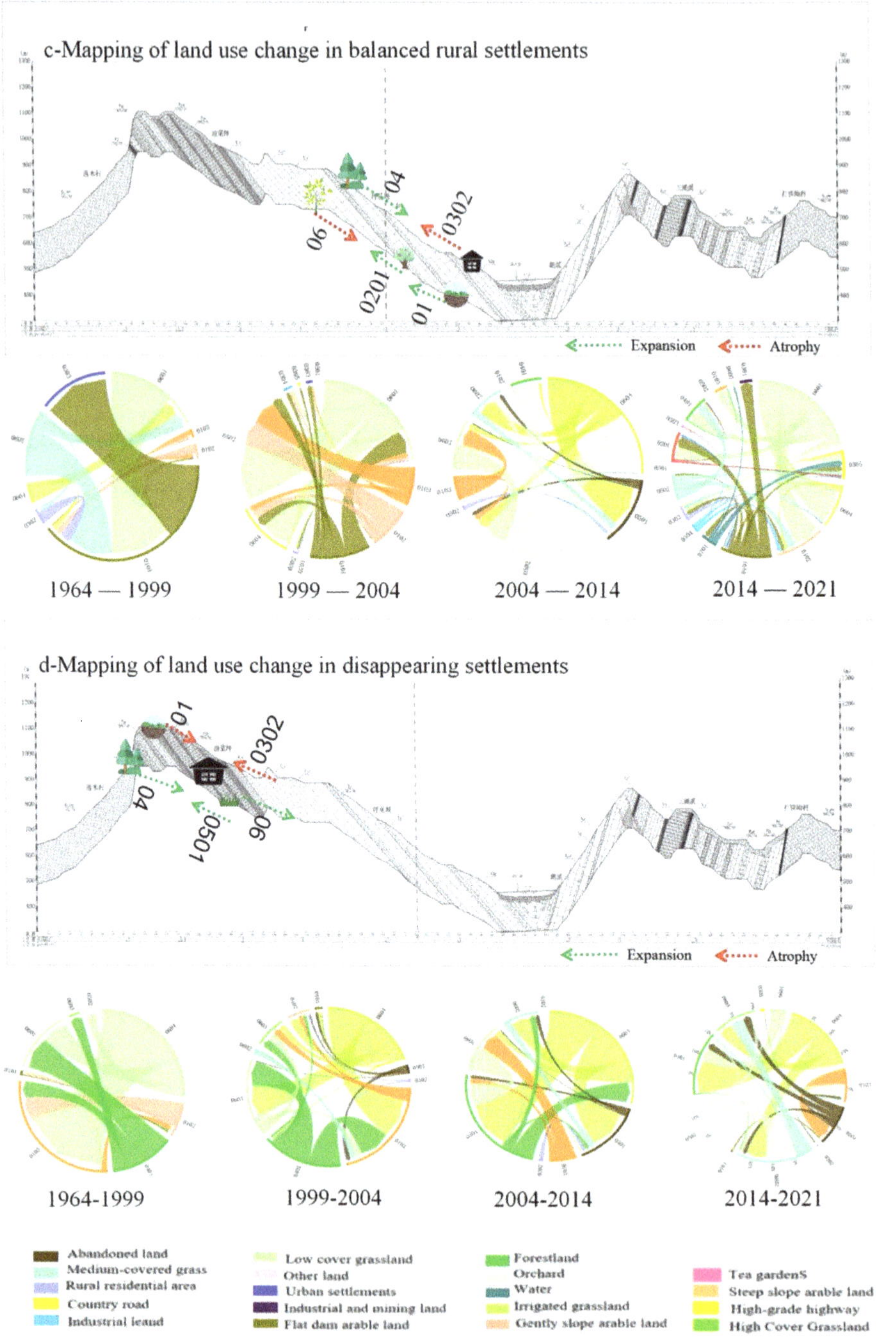

Figure 6. Mapping of land use transfer trajectories in typical rural settlement types.

From 2014 to 2021, the land use types of the ES were forestland, flat dam arable land, rural road expansion, abandoned land, and steep-slope arable land, and slow slope arable land contraction was dominant, and the land use change trajectory around the settlement is mainly: abandoned land→flat dam arable land, irrigated grass→forestland, low cover grass→forestland, abandoned land→forestland. The AS mainly focuses on expanding forestland, abandoned land and grassland, flat dam arable land, gently sloping arable land, and steep sloping arable land shrink. The BS is dominated by the expansion of forest land, rural roads, tea gardens, flat dam arable land, and the shrinkage of arable land and grassland. The land use types of the DS are mainly forestland, abandoned land, grassland expansion, steep-slope arable land, irrigated grassland, flat dam arable land, and slow slope arable land contraction. The trajectory of land use change is mainly steep-slope arable land→abandoned land, gentle slope arable land→abandoned land, flat dam arable land→abandoned land, low cover grassland, and irrigated grassland→abandoned land; the overall trend of land types around the settlement showed shrinkage at this stage.

4.4. Coupled Analysis of Land Use Evolution around Typical Rural Settlement Types

The evolution of land use around different settlement types in the karst trough valley has a mutual influence relationship, and different settlement types in the trough valley were divided into three time periods according to the land use change patterns around different settlements: 1964–2004, 2004–2014, and after 2014. There are dynamic evolutionary coupling processes and dynamic coupling strengths between different settlement types and their surrounding land use changes in the three time periods (Figure 7). Among them, the ES at the bottom of the trough, such as ZengJia, SanCun, Chuangyan, and the surrounding land use coupling process, show a complete coupling situation. The settlement area showed a local expansion from 1964 to 2004, a balanced expansion from 2004 to 2014, and a significant core expansion after 2014. The land use change around the settlement showed different trends with the settlement expansion, the land use area of steep-slope arable land and gentle slope arable land expanded, and the flat dam arable land decreased. The clusters and their surrounding land use showed an increasing linear trend in coupling intensity.

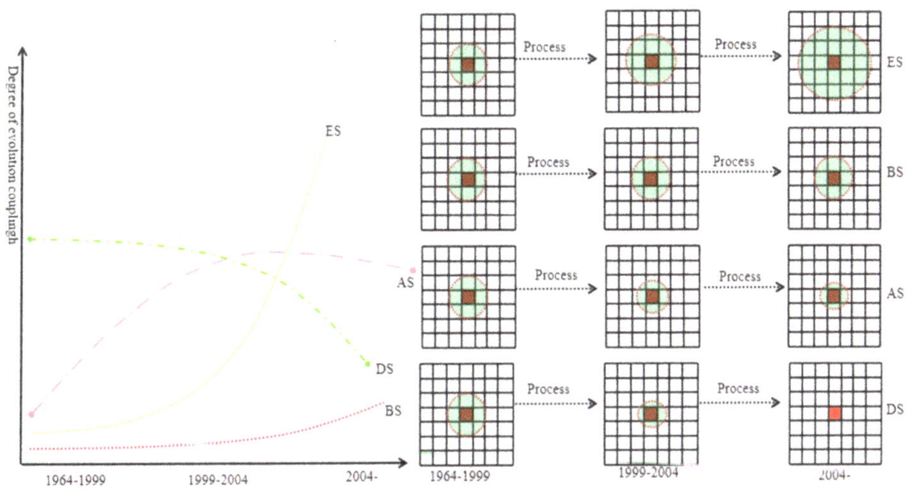

Figure 7. Schematic diagram of the coupling evolution of different settlement types and land use.

Regarding the BS of HeXi and GaoZhai in 1964–2021, the coupling process between the settlement and its surrounding land use showed a remote continuous coupling state, and the core of the settlement and the land use around the settlement its coupling intensity showed an inverse linear growth trend. The AS and the surrounding land use changed in this period, and the coupling process of surrounding land use with the settlement as the core showed a local coupling, which exhibited a local expansion from 1964 to 2004. From 2004 to 2014, it showed local shrinkage; after 2014, it showed rapid shrinkage. The land use changes around shrinking settlements showed a gradual decrease in cultivated land on gentle slopes, steep slopes, and flat dams; minor changes in the surrounding forest land; and an increase in the irrigation and grassland areas, and overall, the coupling intensity of shrinking settlements and their surrounding land use changes showed a trend of increasing first and then decreasing. The area of DS showed local expansion from 1964 to 2004; balanced shrinkage from 2004 to 2014; and after 2014, the settlement center transitioned into a significant extinction phenomenon, the settlement and the surrounding land use showed a fundamental coupling disorder, and the coupling intensity of the settlement core and the surrounding land use of the settlement decreased linearly.

5. Discussion

5.1. Analysis of the Homogeneity and Heterogeneity of Land Use Changes around Typical Rural Settlement Types

5.1.1. Homogeneity Analysis of Land Use Changes around Different Settlement Types

The analysis of land use changes in the buffer zone of the ES, AS, DS, and BS found that the homogeneity of land use changes around different settlement types in karst trough valleys mainly manifested in the spatial dimension and the temporal pattern dimension. Homogeneity was manifested as follows: Firstly, spatially, the increment in trough dam settlement showed an inverted U-shaped variation with land class (Figure 8), the ES, and an increase in land types. The number of settlements and land types of DS on the trough slopes showed a linear change of "\". A single land type appeared when the settlement on trough slopes died out. At the top of the trough AS with BS, the number of communities and land types showed an "L" change; the development of settlements at the top of the troughs appeared to be flat, and the number of land types decreased. Second, the land types around the ES at the top and bottom of the trough were more and less influenced by the settlement within 200 m. At the same time, within the 200 m buffer zone and outside the 200 m buffer zone, the land use around the early and late settlement retained a particular slope of sloping arable land. Third, the abandonment phenomenon existed inside and outside different settlement types' 200 m buffer zone. In addition, the homogeneity of the land use of settlements in terms of temporal characteristics was shown by the annual enrichment of land use land types around different settlement types from 1964 to 2021. Land use structure and function around different settlement types in karst trough valleys showed dynamic changes.

5.1.2. Analysis of the Variability of Land Use Change around Different Settlements

Different settlement types in karst trough valleys have differences in the buffer zone and temporal characteristics of land use changes around them. From the analysis of the overall land use changes of different settlement types, it was found that the ES land use evolution at the bottom of the trough valley showed that the closer the location to the center of the settlement, the richer the land use type. Moreover, its land use buffer type and land type shift showed that it was dominated by arable land, orchards, and tea gardens (Figure 8). The farther it is away from the settlement center, the more homogeneous the land use type is, and the land use type showed the ecological restoration type of use, such as forest land, abandoned land, and grassland. The land use evolution pattern of the BS at the top of the trough showed that the land use types were richer regardless of the distance from the settlement, and the land use types in the buffer zone and around the settlement were mainly arable land. The land use changes in the AS showed a closer distance to the settlements on the slopes of the troughs and valleys. The primary land use type for more homogeneous land use type was mainly arable land, abandoned land, and other valuable methods. The more distant the settlement, the richer the land use type was. The primary land use type is steep-slope arable land, grassland, irrigation grass, and forestland. The DS land use changes on the slopes of the trough valley mainly show that the rural residential areas in the center of the settlement are abandoned, and their buffer zone land class and the land class transfer around the settlement are mainly abandoned land, grassland, and forestland, while in the buffer zone dimension, the closer to the settlement center, the lower the land class richness, and vice versa. Regarding the characteristics of the temporal pattern of land use change in the settlement, the abandoned land, grassland, and forestland around the settlement expanded, and the arable land and residential areas shrank from 1964 to 2021.

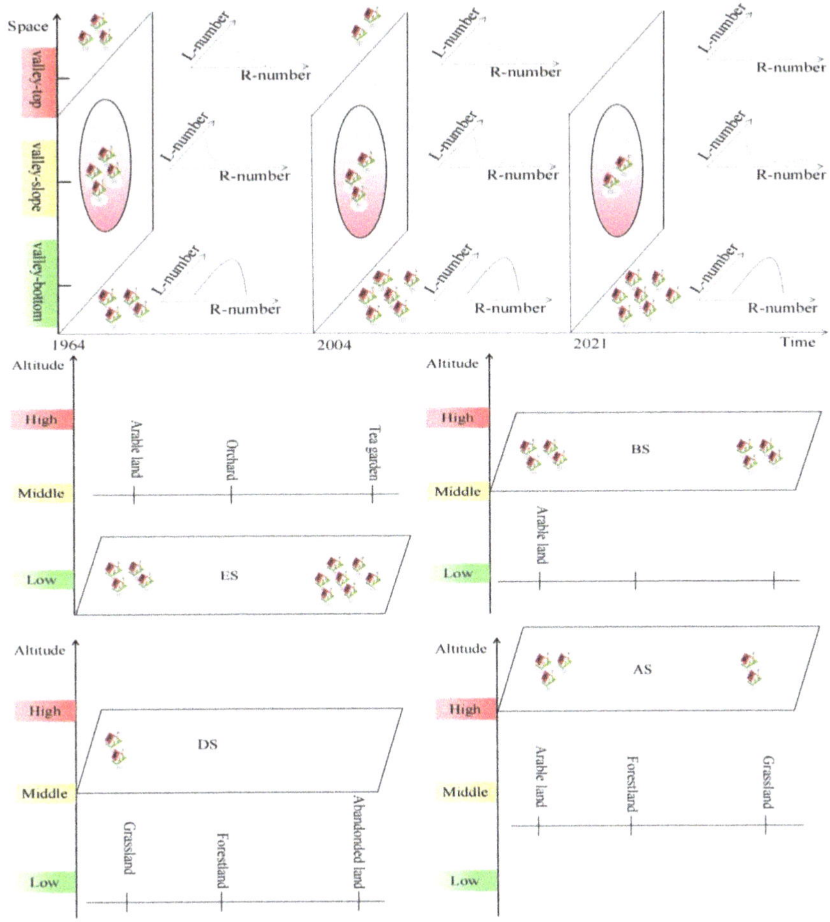

Figure 8. Homogeneity and heterogeneity analysis of land use change in buffer zones of settlement types.

5.2. Driving Mechanisms of Land Use Change in Typical Rural Settlement Types

Considering the characteristics of typical settlement types in karst trough valleys, the analysis of land use changes around settlements revealed that land use changes around rural settlements in karst trough valleys are driven by multiple factors and are the result of the interaction between the natural environment and human activities. Natural, human, socio-economic, and environmental factors influence land use changes around different settlement types. Among them, topography, climate, hydrology, geology, soil, and other physical, geographic, and environmental factors directly influence regional differences in rural settlements, especially in the driving mechanism of land use change in rural settlements.

In this study, the influence of regional geological composition and soil texture on the land type change around rural settlements was relatively weak, so the mechanism driving factors were not explored here. In the unique environment of regional karst geomorphology, in addition to geological features partially influencing factors, the changes in cultivation conditions and radius caused by the dual factors of elevation and slope become the main limiting drivers of land use changes around rural settlements. Therefore, land use changes in rural settlements in the LangXi karst trough valley are driven by natural geographic and environmental factors (Figure 9). The land use change in the settlement is driven by the positive double feedback mechanism of topography and slope, with the dynamic change pattern of "low-low expansion." The average slope of the typical settlements in SanCun, Chuangyan, and ZengJia is 5–15°, the average elevation is 580 to 795 m, and the land use around the settlements is forest land and arable land expansion (Figure 9). The driving pattern of land use change in and around a shrinking slotted slope settlement showed "low-low shrinkage" dynamics. The average slope of typical settlements in Ganlong and Xinchu is 15–20°, and the elevation is between 920 and 1170 m. Balanced rural settlements form "medium-medium average" and "high-high extinction" dynamic change driving patterns, respectively, with the medium-high slopes and land use changes showing dynamic equilibrium and extinction.

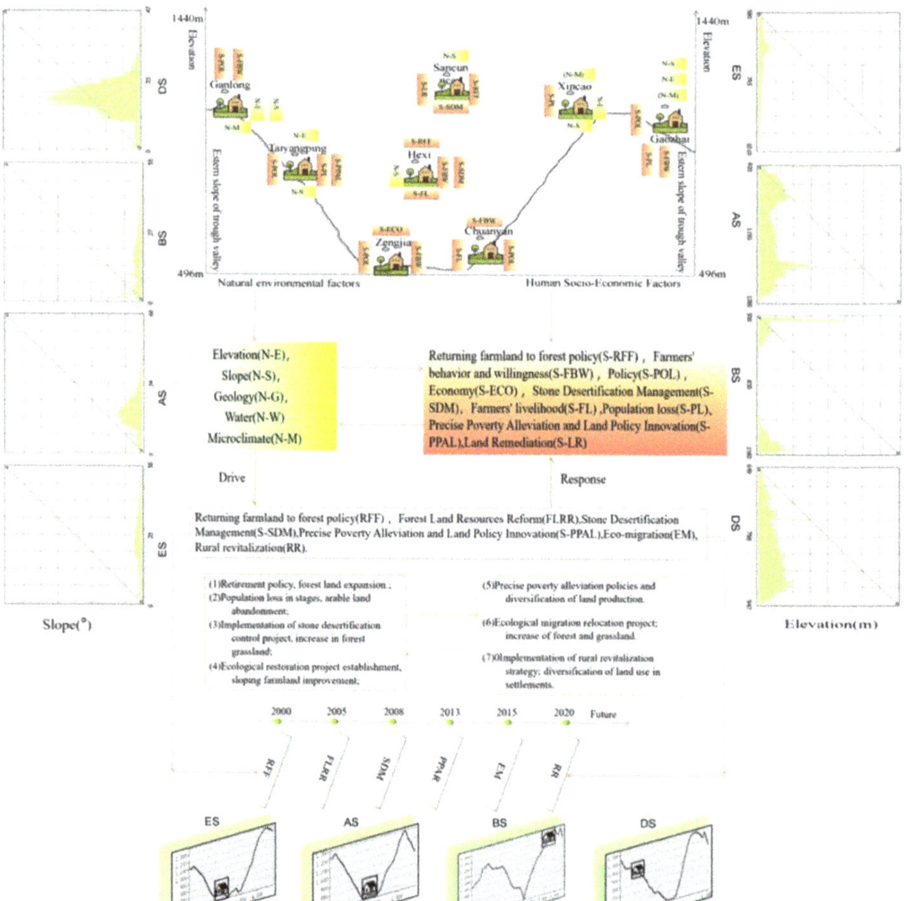

Figure 9. Driving mechanisms of land use change around typical settlements.

In summary, the ES showed a trend of spatial expansion to lower elevations and elevation zones over time. The AS showed the trend of atrophy of high elevation and high-altitude settlement centers over time, while the BS showed the trend of balanced rural development in the middle elevation of troughs and valleys. The DS showed the development trend of high altitude and high elevation to restrict the land use of clusters.

The change of land use in and around rural settlements is a process of selective regional development under the combined influence of natural resource conditions and human and socio-economic infrastructure conditions; it can be seen as the result of competition between rural settlement land and other types of surrounding land. Natural factors determine the basic structure of rural settlements and their surrounding land pattern and constitute the substrate for the growth and development of rural settlements. Moreover, they have a significant influence on the origin, change, and extinction of rural settlements in karst troughs; human and socio-economic factors directly or indirectly cause the activity state of rural settlements in the trough and constitute the main drivers of dynamic changes in land use in and around rural settlements on short and medium time scales. The land use changes around the rural settlements in typical karst trough valleys are mainly driven by policy orientation, population change orientation, farmer willingness orientation, farmer livelihood orientation, and economic development regarding human and socio-economic factors, specifically including returning farmland to forest policy (S-RFF), farmers' behavior and willingness (S-FBW), policy (S-POL), economy(S-ECO), Stone Desertification Management (S-SDM), farmers' livelihood (S-FL), population loss (S-PL), Precise Poverty Alleviation and Land Policy Innovation (S-PPAL), and Land Remediation (S-LR) (Figure 9). Under the policy-driven guidance, land use changes around typical rural settlements in the trough valley are influenced by the policy of returning farmland to forest, stone desertification control projects, ecological restoration project construction, and sloping land improvement policy. Precise poverty alleviation and the policy of stone desertification management are significant. Driven by population orientation, the population's age structure in the trough area is dominated by young children and older adults, and the labor capacity of rural settlements is weak. The labor capacity determines their labor distance and intensity, and the labor distance and intensity affect the planting and tending of rural settlement land, thus affecting the land use pattern of rural settlements in the karst trough area. For example, the population loss of TaiYangPing in the trough valley slope of DS gradually increased from 2000 to 2020 (Table 3), and the loss rate reached 50%, and a large number of young people in the settlement went out or moved out, and the land use pattern around the settlement changed from comprehensive mixed-use land type to single abandoned land type, and the settlement gradually showed the extinction trend. The labor force of Ganlong and XinCao has left, and middle-aged and young people have been lost, among which the loss of people in Ganlong increased from 21% in 2000 to 46% in 2020. There was dynamic stability in the rate of human flow loss in the trough dam settlements of HeXi and GaoZhai from 2000 to 2020, and the land use changes around the settlements showed a balanced state. The population of the expanding settlement is returning to the land, and the settlement and land use are expanding. Driven by the livelihood orientation of farmers, typical settlements in karst trough valleys diversify with agrochemical livelihoods and show diversity in surrounding land use changes. In general, the evolution of rural settlements at the top and slopes of trough valleys with higher elevation is dominated by more decisive geographical factors; rural settlements in trough dam areas show a stronger correlation with socio-economic factors and are more strongly influenced by population, policies, and economic development levels; rural settlements and their surrounding land use changes have obvious clustering effects toward transportation, rivers, and cultural centers.

5.3. Discussion of the LUCC–LUT–RST Interaction Feedback Mechanism for Typical Rural Settlement Types

Various socio-economic issues mapped out in the development of rural settlements are reflected in their land use. Generally, drawing from the settlement land use changes can reflect the trend changes in regional land use patterns and formulate regular summaries. Land use morphological change is the core element of land use transition research. Land use transformation is the external expression of the transformation of rural settlement development, so the relationship between land use change (LUCC), land use transition (LUT), and rural settlements transformation (RST) are inseparable.

Table 3. The population loss rate in typical rural settlements, 2000–2020.

Settlement	2000		2010		2020	
	Typical Settlement Population Loss Rate	**The Average Regional Population Loss Rate**	**Typical Settlement Population Loss Rate**	**The Average Regional Population Loss Rate**	**Typical Settlement Population Loss Rate**	**The Average Regional Population Loss Rate**
Ganlong	0.211	0.066	0.402	0.074	0.460	0.074
SanCun	0.211	0.104	0.303	0.123	0.386	0.094
HeXi	0.211	0.201	0.403	0.242	0.409	0.243
ZengJia	0.166	0.224	0.275	0.151	0.225	0.144
ChuanYan	0.166	0.103	0.275	0.065	0.275	0.062
GaoZhai	0.277	0.132	0.474	0.146	0.436	0.150
XinCao	0.276	0.111	0.475	0.133	0.437	0.137
TaiYangPing	0.211	0.329	0.413	0.415	0.492	0.496

Typical settlements and land use changes in karst trough valleys show the characteristics of 0 to 50 m, 50 to 200 m, and 200 to 400 m buffer zone circle changes (Figure 10). The process, intensity, and pattern of land use changes around different settlement types are different, thus forming the different processes of settlement land use evolution. From the analysis of the land use transfer matrix and change mapping of settlements, it was found that the ES presents the land use transition process of production–life–ecology (LUT-1); and the AS presents the land use life–production–ecology transition process (LUT-2); the BS presents the production–ecology–ecology land use transition process (LUT-3), and the DS forms the ecology–ecology–ecology land use transition process (LUT-4).

Within the karst trough valley territorial system, driven by external factors of urbanization and agricultural modernization, the land use and the dynamic change of rural settlements in the trough valley are reflected in the rural development level (RSD-level: R_a, R_b, R_c, R_d) and land use (LUD: LUD–a LUD–b, LUD–c, LUD–d) changes and are driven by both to form a regional RST process curve T_a~T_d. Meanwhile, land use change development LUD-a~LUD-d (land use morphology change) and rural settlement development level (RSD-level) change together to promote the typical settlement type of the buffer area land use for the transformation interaction process (transformation process formed by ecology–production–life interaction), thus forming the LangXi trough valley rural settlement development and interaction process curve for land use and dividing the typical settlement development in the region into four stages. In the first stage, through the valley settlement subsistence–forest arable land competition stage R_a, the settlement is in the subsistence stage of maintaining basic livelihoods. Food production is the main purpose. Agricultural development is in the primary rough expansion stage, including the land use pattern LUD, which is a manifestation of agricultural land and ecological land competition; land use change, which is a manifestation of forest land; grassland reduction; arable land expansion, such as deforestation; trough slope reclamation; grass reclamation; and other behavioral activities, in the trough valley settlement. In the second stage of the trough valley settlement production, the agricultural land use development stage R_b, the trough valley settlement development is in the stage of sizeable agricultural development, the initial rise of agricultural modernization, and agricultural development from the initial rough expansion gradually transitioned to intensive production, manifested by agricultural chemical planting technology to improve crop yields in the trough valley area. Land use pattern LUD-b shows that the expansion of arable land area is gradually slowing down, while with the change of ecological and land policies, ecological land such as woodland and grassland is being restored. In the third stage of settlement labor, the land abandonment stage Rc, with the continuous promotion of industrialization, urbanization, and agricultural modernization, the population of rural settlements in the trough valley migrated to the cities and towns; the further transformation of settlements occurred, and the development of settlements shifted to a stage dominated by outworking. Land use changes were mainly manifested in the expansion of urban construction land, rural settlements, forestland, and other habited and ecological land, as well as the reduction in arable land and other productive land. In the fourth stage, the land use diversification stage Rd, which focuses on the diversified transformation of the settlement, the development of the trough valley settlement entered the post-industrial stage, the proportion of non-agricultural output value increased rapidly, and the settlement started to become multifunctional and diversified. In this stage, the LUD-d land use change was mainly manifested in the further expansion of habitable and ecological land use and a slight decrease in production land use.

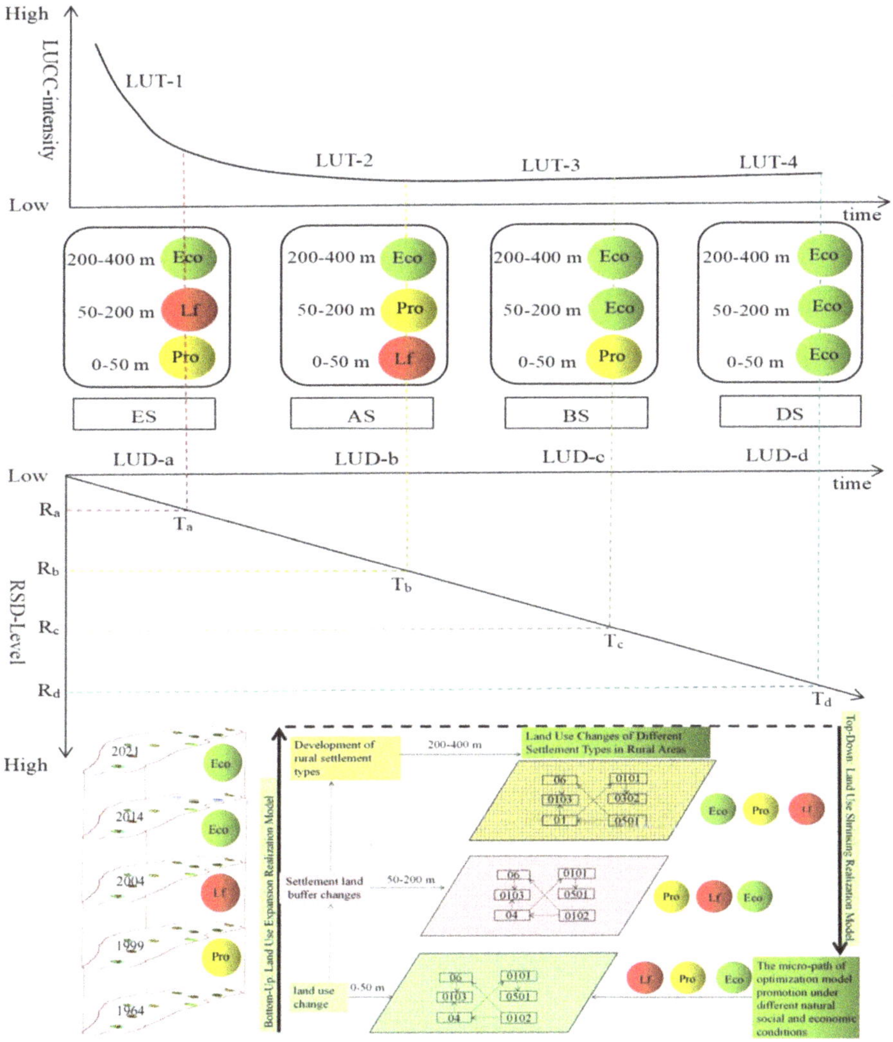

Figure 10. The interactive feedback mechanism of LUCC-LUT-RST valley rural settlements.

LUT and rural settlement development succession will form differentiated RSD geographical types. Different rural settlement types present different levels of rural development. In summary, the land use change analysis of typical rural settlements in karst trough valleys from 1964 to 2021 under the characteristics of the long time series dimension revealed that the development of typical rural settlements in trough valleys presents four development stages: the traditional production function stage, the traditional industrial development stage (life function stage), the ecological restoration stage (ecological function restoration–development stage), and the ecological function enhancement stage. Within the karst trough valley, spatial reorganization optimization patterns and promotion micro-paths under different natural socio-economic conditions are formed through typical settlement land class changes and shifts, i.e., top-down land use contraction patterns and bottom-up expansion realization patterns of land use. In general, the transformation of land use function around the settlements in the trough valley is relatively apparent, with the land use around the expanded settlements changing from an agricultural production function to a living production function from 1964 to 2021; and the land use around the atrophied settlements changing from agricultural production function to ecological production function, and the land use around the balanced rural settlements changing from an agricultural production function to production and living function. The land use around the disappearing settlements showed changes from production and living functions to ecological functions.

5.4. Contribution to Research, Limitations, and Future Work

In our study, we revealed the mechanisms of settlement-scale land use change, land use transformation, and rural settlement transformation and development reciprocal feedbacks based on settlement micro-scale and case studies in southwest China's karst trough valley area. However, for the study data, we used image data with different resolutions (2.7 m, 2.7 m, 10 m, 2.5 m, and 2.5 m), Thus when we performed the spatial transformation of the settlement data, there were deviations of about 0.01%, and these deviations can lead to highly slight changes in the rural settlement data.

Based on our analysis of the research on the dynamics of land use buffer zone changes in different settlement types, we synthesized the results and reflected on the limitations of this paper, and we believe that future research could also include the following aspects. (1) To analyze the dynamic changes of land use in the buffer zone of rural settlements in other geomorphic regions and to reveal the homogeneity and differences in land use dynamics in the buffer zone of rural settlements in different geomorphic regions. (2) In other geomorphic areas, the influence of other factors, such as spatial accessibility of the test settlement and watershed, on the dynamic changes of land use in rural settlements are fully considered. (3) Based on various big data models, simulate and predict the future dynamic land use changes in rural settlements. (4) Validate the land use change–land use transformation–rural transformation development model in other geomorphic areas through empirical research.

6. Conclusions

The study selected the LangXi trough valley, a typical karst trough valley, as the research object and analyzed the buffer zone evolution of land use of typical settlement types within the karst trough valley at the long time series evolutionary sequence and settlement unit scale and the conclusions of the study showed that:

(1) In the evolution and development of rural settlements in the karst trough valley, there are differences in the evolution pattern of different settlement types and surrounding land use. According to the geographical differences between the natural environment of settlements and the theory of life cycle, the types of rural settlements in the LangXi trough valley are classified as following: expanding settlements in the trough dam area, atrophying settlements on trough slopes, disappearing settlements on trough slopes, and balancing permanent settlements on trough slopes and trough valley tops, taking into account the increment in and decrement in an offset of land use changes in rural settlements;

(2) The homogeneity and heterogeneity of land use changes around different rural settlement types in karst trough valleys are concentrated in spatial and temporal characteristics. The spatial homogeneity is concentrated in the different settlement types, and the amount of settlement development and its surrounding land types shows inverted "U", "\," and "L" type change characteristics, respectively. The spatial heterogeneity is reflected in the different settlement types, different distances from the center of the settlement, and the different richness of their land types. The homogeneity of the chronological evolution was shown from 1964 to 2021, with different settlement types and a year-by-year enrichment of the surrounding land use land types. The heterogeneity was shown by the different maps of change around the rural settlements in the trough valley from 1964 to 2021;

(3) The spatial and temporal patterns of socio-economic development of a typical rural settlement are reflected in its land use change, which is one of the manifestations of land use transformation. The land use transformation results counteracted the development of the rural transformation. Therefore, the results of land use evolution analysis of typical settlement types in karst trough valleys revealed the interactive feedback process and mechanism of land use change, rural transformation development, and land use transformation. This interactive process reflected the spatial reorganization optimization pattern and promotion micro-path driven by natural socio-economic conditions in the karst trough valley region, i.e., top-down land use contraction pattern and bottom-up land use expansion pattern. At the same time, the LUCC—LUT—RST interactive feedback mechanism aims to enrich the research framework of land use transformation and rural transformation at the settlement scale on the one hand and to provide case studies for developing countries with karstic mountainous trough and valley landscapes such as China in the global rural problem-solving process on the other hand. Meanwhile, the LUCC—LUT—RST interactive feedback mechanism aims to enrich the research framework of land use transition and rural transformation at the settlement scale on the one hand. On the other hand, the global rural problem-solving process provides case studies for developing countries with karst mountainous troughs and valley landscapes, such as China.

Author Contributions: Conceptualization, writing original draft preparation, writing review, and editing, Y.Z.; methodology, project administration, supervision, funding acquisition, Y.L.; supervision, Validation, Project administration. G.L.; Validation, Projected ministration. X.B.; Validation, Project administration. J.H.; Visualization, Formal analysis, F.T.; data curation, Visualization, M.Y. All authors have read and agreed to the published version of the manuscript.

Funding: This research work was supported jointly by the National Natural Science Foundation of China (No. 4206010008) and the National Key R&D Program Project (2016YFC0502300).

Data Availability Statement: Not applicable.

Conflicts of Interest: The authors declare no conflict of interest.

References

1. Abegaz, G.; Rahmato, D. Rural land use issues and policy: An overview. In *Land Tenure & Land Policy in Ethiopia after the Derg: Second Workshop of the Land Tenure Project*; University of Trondheim: Trondheim, Norway, 1994.
2. Ashley, C.; Maxwell, S. Rethinking Rural Development. *Dev. Policy Rev.* **2001**, *19*, 395–425. [CrossRef]
3. Hoggart, K.; Paniagua, A. What rural restructuring? *J. Rural Stud.* **2001**, *17*, 41–62. [CrossRef]
4. Mihai, F.; Iatu, C. Sustainable Rural Development under Agenda 2030. *HAL* **2020**, *4*, 9–18.
5. Song, X.-P.; Hansen, M.C.; Stehman, S.V.; Potapov, P.V.; Tyukavina, A.; Vermote, E.F.; Townshend, J.R. Global land change from 1982 to 2016. *Nature* **2018**, *560*, 639–643. [CrossRef]
6. Mawunyo, D.F. Food security in rural sub-Saharan Africa: Exploring the nexus between gender, geography and off-farm employment. *World Dev.* **2019**, *113*, 26–43.
7. Jayne, T.S.; Snapp, S.; Place, F.; Sitko, N. Sustainable agricultural intensification in an era of rural transformation in Africa. *Glob. Food Secur.* **2019**, *20*, 105–113. [CrossRef]

8. Sun, P.-L.; Xu, Y.-Q.; Wang, C. The topographic gradient effect of land use change in the Beijing-Tianjin poverty belt. *Trans. Chin. Soc. Agric. Eng.* **2014**, *30*, 12.
9. Govindan, K.; Loisi, R.V.; Roma, R. Greenways for sustainable rural development: Integrating geographic information systems and group analytic hierarchy process. *Land Use Policy* **2016**, *50*, 429–440.
10. Zhan, Q.; Lu, F. *Rural Social Change and Rural Governance*; China Agriculture Press: Beijing, China, 2006.
11. Bański, J.; Wesołowska, M. Transformations in housing construction in rural areas of Poland's Lublin region—Influence on the spatial settlement structure and landscape aesthetics. *Landsc. Urban Plan.* **2010**, *94*, 116–126. [CrossRef]
12. Li, Y.; Fan, P.; Liu, Y. What makes better village development in traditional agricultural areas of China? Evidence from long-term observation of typical villages. *Habitat Int.* **2019**, *83*, 111–124. [CrossRef]
13. Jia, K.; Qiao, W.; Chai, Y.; Feng, T.; Ge, D. Spatial distribution characteristics of rural settlements under diversified rural production functions: A case of Taizhou, China. *Habitat Int.* **2020**, *102*, 102–201. [CrossRef]
14. Wen, Y.; Zhang, Z.; Liang, D.; Xu, Z. Rural Residential Land Transition in the Beijing-Tianjin-Hebei Region: Spatial-Temporal Patterns and Policy Implications. *Land Use Policy* **2020**, *96*, 104700. [CrossRef]
15. Feng, Y.; Long, H. Progress and Prospects of Spatial Reconfiguration of Rural Settlements in Mountainous Areas of China. *Prog. Geogr.* **2020**, *39*, 14.
16. Chen, C.-L.; Xu, M.-J.; Zhu, P.; Dong, L. Spatial and temporal pattern evolution and intensive use changes in rural settlements in Jiangsu Province over the past 30 years. *Trans. Chin. Soc. Agric. Eng.* **2021**, *29*, 2124–2135.
17. Zhang, B.L.; Zhang, F.R.; Zhou, J.; Yan-Bo, Q.U. Functional Evolution of Rural Settlement Based on Micro-perspective: A Case Study of Hetaoyuan Village in Yishui County, Shandong Province. *Sci. Geogr. Sin.* **2015**, *35*, 1272–1279.
18. Erturk, S.A. The settlement characteristics of Bursa plain and its environs. *Procedia Soc. Behav. Sci.* **2011**, *19*, 371–380. [CrossRef]
19. Jin, Q. The history and recent trends in the study of rural settlement geography in China. *Acta Geogr. Sin.* **1988**, *4*, 311–317.
20. Tan, X.; Liu, Z.; Yanhua, H.E.; Tan, J.; Zhang, Y.; Zhou, G. Regional differentiation and type division of rural settlements to South of Yangtse River: A case study of Changsha. *Geogr. Res.* **2015**, *34*, 2144–2154.
21. Yang, R.; Liu, Y.; Long, H.; Wang, Y.; Zhang, Y. Spatial Distribution Characteristics and Optimized Reconstructing Analysis of Rural Settlement in China. *Sci. Geogr. Sin.* **2016**, *36*, 170–179.
22. Chen, Y.; Ge, Y. Spatial Point Pattern Analysis on the Villages in China's Poverty-stricken Areas. *Procedia Environ. Sci.* **2015**, *27*, 98–105. [CrossRef]
23. Zhang, S.; Tong, B.; Hao, J. An Analysis of the Temporal and Spatial Evolution Characteristics of Settlement in Zhenglan Banner of Inner Mongolia (1933–1983). *Econ. Geogr.* **2018**, *3*, 147–156.
24. Yang, K.; Song, Y.; Xue, D. Spatial and temporal evolution and influencing factors of rural settlement land use on the Loess Plateau. *Resour. Sci.* **2020**, *42*, 14.
25. Gong, J.; Jian, Y.; Chen, W.; Liu, Y.; Hu, Y. Transitions in rural settlements and implications for rural revitalization in Guangdong Province. *J. Rural Stud.* **2019**, *93*, 359–366. [CrossRef]
26. Ma, X.; Qiu, F.; Li, Q.; Shan, Y.; Cao, Y. Spatial Pattern and Regional Types of Rural Settlements in Xuzhou City, Jiangsu Province, China. *Chin. Geogr. Sci.* **2013**, *23*, 482–491. [CrossRef]
27. Chen, C.; Jin, Z. Spatio-temporal change of land use pattern of rural settlements in developed area: A case study of Huishan district in Wuxi city. *Geogr. Res.* **2015**, *34*, 2155–2164.
28. Chen, Y.; Xie, B. The spatial evolution and restructuring of rural settlements in Jiangnan hilly region: A case study in South Jiangxi. *Geogr. Res.* **2016**, *35*, 184–194.
29. Lu, M.; Wei, L.; Ge, D. Spatial optimization of rural settlements based on the perspective of appropriateness–domination: A case of Xinyi City. *Habitat Int.* **2020**, *98*, 102148. [CrossRef]
30. He, R.; Chen, G.; Liu, S.; Guo, S.; Liu, Y. Progress and Trends of Research on the Geography of Rural Settlements in China. *Prog. Geogr.* **2012**, *31*, 1055–1062.
31. Chen, Y.; Liu, Y.; Li, Y. The state of agricultural development and industrial prosperity in rural revitalization in China. *Geogr. Res.* **2019**, *38*, 11.
32. Li, Y.; Li, R.; Luo, G.; Xie, J.; Xu, Q. Village evolution patterns and driving mechanisms in Guizhou's distinct crest depression region over the past 50 years. *Acta Ecol. Sin.* **2018**, *38*, 13.
33. Ran, C.; Li, Y.; Liang, X. Land use effects of settlement change in a typical Three Gorges reservoir area watershed. *Acta Ecol. Sin.* **2020**, *40*, 12.
34. Wang, Q.; Tang, F.; Li, Y.; Huang, J.; Bai, X. Spatial and temporal variation of landscape pattern evolution and its ecological security in karst areas: An example from the trough valley of northeastern Guizhou Province. *Acta Ecol. Sin.* **2021**, *41*, 19.
35. Luo, G.; Li, Y.; Wang, S. Analysis of the distribution pattern and evolution of karst mountain settlements: The example of Houzhaihe area in Putting County. *Resour. Environ. Yangtze Basin* **2010**, *19*, 6.
36. Liu, C.; Zhang, J. Evaluation of land suitability of rural settlements in typical counties in the upper reaches of Minjiang River based on ecological niche model. *Trans. Chin. Soc. Agric. Eng.* **2021**, *37*, 8.
37. Wei, Y. Fractal Characteristics of Rural Settlements in Mountainous, Hilly Areas and Their Optimization. Ph.D. Thesis, Southwestern University, Georgetown, TX, USA, 2019.
38. Shi, M.; Jie, Y.; Cao, Q. Landscape pattern evolution and mechanism analysis of rural settlements in an arid oasis. *Geogr. Res.* **2016**, *35*, 692–702.

39. Tang, G. Study on the spatial distribution pattern of rural settlements based on GIS: An example of Yulin area in northern Shaanxi. *Econ. Geogr.* **2000**, *20*, 4.
40. Ren, P.; Hong, B.T.; Zhou, J.M. Research of Spatio-temporal pattern and characteristics for the evolution of rural settlements based on spatial autocorrelation model. *Resour. Environ. Yangtze Basin* **2015**, *24*, 1993–2002.
41. Quanlin, L.I.; Xiaodong, M.A.; Yi, S. Analysis of the spatial pattern of rural settlements in northern Jiangsu. *Geogr. Res.* **2012**, *31*, 144–154.
42. Long, H.; Liu, Y. Rural restructuring in China. *J. Rural Stud.* **2016**, *47*, 387–391. [CrossRef]
43. Long, H.; Ma, L.; Zhang, Y.; Qu, L. Multifunctional rural development in China: Pattern, process, and mechanism. *Habitat Int.* **2022**, *121*, 102530. [CrossRef]
44. López-Penabad, M.C.; Iglesias-Casal, A.; Rey-Ares, L. Proposal for a sustainable development index for rural municipalities. *J. Clean. Prod.* **2022**, *357*, 131–876. [CrossRef]
45. Gu, X.; Xie, B.; Zhang, Z.; Guo, H. Rural multifunction in Shanghai suburbs: Evaluation and spatial characteristics based on villages. *Habitat Int.* **2019**, *92*, 102041. [CrossRef]
46. Qu, Y.; Jiang, G.; Zhao, Q.; Ma, W.; Zhang, R.; Yang, Y. Geographic identification, spatial differentiation, and formation mechanism of multifunction of rural settlements: A case study of 804 typical villages in Shandong Province, China. *J. Clean. Prod.* **2017**, *166*, 1202–1215. [CrossRef]
47. He, Y.; Wu, J.; Zhou, G.; Zhou, B.B. Discussion on rural sustainability and rural sustainability science. *Acta Geogr. Sin.* **2020**, *75*, 736–752.
48. Li, X.; Yang, H.; Jia, J.; Shen, Y.; Liu, J. Index system of sustainable rural development based on ecological livability. *Environ. Impact Assess. Rev.* **2021**, *86*, 106478. [CrossRef]
49. Chen, Z.; Li, Y.; Liu, Y. Distribution pattern characteristics and types of rural settlements in the Loess Hills and Gullies. *Trans. Chin. Soc. Agric. Eng.* **2017**, 33, 9.
50. Liu, M.; Guan, X.; Li, Q. Spatial distribution characteristics of rural settlements in mountainous, hilly areas. *Trans. Chin. Soc. Agric. Eng.* **2014**, *21*, 119–122.
51. Qhab, C.; Cheng, W. Quality evaluation and division of regional types of rural human settlements in China—ScienceDirect. *Habitat Int.* **2020**, *105*, 102278.
52. Qi, W.; Zhao, M.; Liu, S. Accounting for the mobile population at county and city scales and the evolution of geographical types in China, 1982–2010. *Acta Geogr. Sin.* **2017**, *72*, 16.
53. Liu, J.; Ning, J.; Kuang, W. Spatial and temporal patterns and new characteristics of land use change in China, 2010–2015. *Acta Geogr. Sin.* **2018**, *73*, 14.
54. Qiao, L.; Liu, Y.; Yang, R. Types of land use changes in rural settlements in China and strategies for regulation. *Trans. Chin. Soc. Agric. Eng.* **2015**, *31*, 8.
55. Ling, D.-Q.; Bi, S.-B.; Zuo, Y.; Li, J.-H.; Jiang, L. Study on the construction of an integrated model for buffer zone analysis. *Sci. Surv. Mapp.* **2019**, *44*, 7.
56. Jin, S.; Li, B.; Yang, Y.; Shi, P.; Shi, K.; Da, F. Urban distribution characteristics and their influencing factors in China. *Geogr. Res.* **2015**, *34*, 1352–1366.
57. Li, Q.; Ma, X.; Shen, Y. The spatial pattern of rural settlements in northern Jiangsu. *Geogr. Res.* **2012**, *31*, 11.
58. Song, W.; Li, H. Spatial pattern evolution of rural settlements from 1961 to 2030 in Tongzhou District, China. *Land Use Policy* **2020**, *99*, 105044. [CrossRef]

land

MDPI

Article

The Spatial and Temporal Evolution Pattern and Transformation of Urban–Rural Construction Land in Karst Mountainous Areas: Qixingguan District of Guizhou, Southwest China

Yaopeng Sun [1,2,3], Zhongfa Zhou [1,2,4,*], Denghong Huang [1,2,4], Quan Chen [1,2,4] and Min Fang [5]

1 School of Karst Science, Guizhou Normal University, Guiyang 550001, China
2 School of Geography & Environmental Science, Guizhou Normal University, Guiyang 550001, China
3 The Third Surveying and Mapping Institute of Guizhou Province, Guiyang 550004, China
4 State Engineering Technology Institute for Karst Desertification Control, Guiyang 550001, China
5 Guizhou Vocational and Technical Collage of Water Resources and Hydropower, Guiyang 551416, China
* Correspondence: fa6897@gznu.edu.cn

Abstract: Studying the temporal and spatial evolution pattern and transformation rule of urban–rural construction land in karst mountainous areas has important guiding significance for urban development boundary planning, red lines for ecological protection, and cultivated land protection. The present study took 46 townships (streets) in Qixingguan District of Guizhou Province, southwest China, as the research area; collected the current status of four-phase land use data in 2009, 2013, 2017, and 2020; and used GIS spatial analysis models and geographical detectors to analyze the temporal and spatial evolution pattern characteristics and influencing factors of urban–rural construction. The results showed the following: (1) Since 2009, the total area of urban–rural construction land has continued to increase; the largest area is rural residential land, followed by urban land and transportation land, with relatively little urban industrial and mining land, scenic spots, and special land. The growth rate of land used for transport increased rapidly, and urban land grew faster than rural residential land. (2) More than 57.72% of the newly increased urban–rural construction land came from cultivated land, but the transformation of cultivated land for construction gradually slowed down; 57.48% of urban–rural construction land was transferred for reclamation as cultivated land. During the study period, the transformation of cultivated land to construction land was more intense (the transfer out of cultivated land was greater than the transfer in by 9541.94 hm^2). (3) There are strong spatial differences in the density of urban–rural construction land, showing scattered agglomeration distribution, and the degree of aggregation in medium-high- and high-density areas is further strengthened, expanding to the east and southwest. (4) The growth of urban–rural construction land has been controlled by a variety of complex factors, the most influential of which are the completion of fixed asset investment in society as a whole and the total fiscal revenue, with explanatory power (PD) values of 0.819 and 0.607, respectively. Interactions between detection factors have a greater impact on the spatial differentiation of urban–rural construction land than single factors. The results of this study can provide basic research data and support the control and high-quality development of urban–rural construction land in Qixingguan District and karst mountain areas.

Keywords: karst mountain area; urban–rural construction land; land transformation; temporal and spatial evolution; geographical detector

Citation: Sun, Y.; Zhou, Z.; Huang, D.; Chen, Q.; Fang, M. The Spatial and Temporal Evolution Pattern and Transformation of Urban–Rural Construction Land in Karst Mountainous Areas: Qixingguan District of Guizhou, Southwest China. *Land* **2022**, *11*, 1734. https://doi.org/10.3390/land11101734

Academic Editor: Xiaoyong Bai

Received: 28 August 2022
Accepted: 1 October 2022
Published: 7 October 2022

1. Introduction

Land use transformation was first proposed by the British geographer Alan Grainger, inspired by the concept of forest transformation proposed by the Finnish scholar A.S. Mather [1]. The transforming land types mainly include cultivated land [2–4], forest land, urban land [5,6], rural land, and other land use types, and they also include the overall regional land use type [7,8]. With the acceleration of urbanization and increasing human activities, the transformation of urban–rural construction land in land use transformation has become an important socioeconomic phenomenon worldwide [9,10]. Urban–rural construction land is the core component of the land use system. It is not only an important indicator to control urban expansion and urban–rural construction, but also the main component of urban spatial planning and the spatial carrier of human non-agricultural economic production activities [11–13]. At present, under circumstances in which global warming is causing glaciers to melt and inundate some coastal areas, forests are sharply reduced, the environmental carrying capacity is fragile, and human activities affect vegetation restoration [14–16], how to realize scientific and sustainable utilization of limited land resources has become a key issue for the future development of the world. Whether urban–rural construction land can be reasonably controlled and allocated will directly affect the protection of cultivated land and the development of urbanization.

In recent years, many scholars have studied the evolution and driving factors of urban–rural construction land by applying current land use data and using different methods [17–20]. Weber and Puissant extracted land cover from 1986 to 1996 with time series satellite images (SPOT XS), used a prediction model to carry out an empirical analysis of the expansion characteristics of local construction land in Tunisia, and explained the development trend of the city in the future [21]. Based on Landsat satellite observation data, Masek et al. used the band reflection of NDVI to distinguish urban and agricultural land, and dynamically monitored the expansion and evolution process of urban construction land in Washington, DC, from 1973 to 1996. In conjunction with census data, they found that urban expansion was strongly correlated with the regional economic development level [22]. Saizen et al. quantitatively analyzed land use changes in the Osaka metropolitan area of Japan from 1979 to 1996 through GIS raster data, and the results showed that the main reason for the continuous increase in idle land in the suburbs was urban sprawl [23]. Ustaoglu and Aydmoglu used an integrated geographic information system (GIS) and multicriteria decision analysis (MCA) approach to assess the suitability of land use in the Pendikc area of eastern Istanbul, Turkey, for residential, industrial, commercial, and recreational development in the city [24].

Mann used a regression analysis model to study the main driving factors of changes in per capita construction land in rural areas and suggested that the implementation of incentive measures for construction land management by local governments could limit the expansion of construction land to a certain extent [25]. Colsaet et al. analyzed the scientific literature on the occupation of agricultural land by urban construction land and the determinants of urban expansion from 1990 to 2017, and suggested that population and income growth, transportation infrastructure, and car use were the main driving factors [26]. Bittner et al. studied the spatial evolution of land use in peri-urban areas of Israel based on time-series land use data supplemented by summary statistical analysis, and indicated that rural reorganization would have an important impact on the economic society and ecological environment [27]. Diogo and Koomen studied the process of land use change in Portugal between 1990 and 2006 and analyzed the impact of different driving forces on the formation of land use patterns during this period. They concluded that land expansion was positively correlated with economic development, and the driving influence of economic factors on land use change remained stable in a certain period.

The deployment of new infrastructure and the gradual implementation of territorial space planning policies will also affect land expansion [28]. However, the results of previous works illustrate that the main driving factors for the evolution of urban–rural construction land are the economic and social development level, social living conditions, and the policy and institutional environment, etc. [29,30]. Overall, most studies on the influencing factors of spatiotemporal changes of urban–rural construction land are based on linear analysis, trend analysis, and correlation analysis; however, it is still difficult to quantitatively decompose the influencing factors of such land changes.

The above studies are of great significance for understanding the driving factors of urban–rural construction land changes, but the disadvantage of the methods used is that they assume there is a constant and significant linear relationship between the driving forces and land changes across the entire time series. An idealized linear model or qualitative description can reveal the complex driving forces; in fact, there is no strict linear relationship between the transformation of urban–rural construction land and socioeconomic development, population density [31], or urban residential development [32]. Although studies have applied algorithms such as K-means [33] for classification and partitioning, statistical methods for spatial differentiation need to be further developed.

Geographical detector comprises a group of statistical methods that detect spatial differentiation and reveal driving forces [34]. Therefore, this study applied the geographic detector model to analyze the characteristics and driving mechanisms of the transformation of urban–rural construction land in typical karst mountain areas to explain the interactions between social and economic activities and changes in such land. The results of the present study are intended to help local governments explore whether any unreasonable, unsustainable land use resulted from unsustainable human activities and development practices, and thus adapt to current village planning and urbanization strategies and optimize the efficiency of land resource allocation. Moreover, this study can also provide reference for the urban–rural development of other karst areas and promote the integration of urban and rural transformation and land use transformation.

2. Materials and Methods

2.1. Study Area

The present study area is located within Qixingguan District (27°03′–27°46′ N, 104°51′–105°55′ E), Guizhou Province, southwest China, on the slopes of the Yunnan–Guizhou Plateau, sloping toward the eastern low mountains (Figure 1) and covering an area of 3411.14 km^2. The study area is adjacent to Yunnan Province in the west and Sichuan Province in the north and is the transportation and logistics hub of southwest China [35]. The elevation ranges from 456 to 2210 m. The terrain is high in the west and low in the east. The landform type is mainly mid-size mountains and hills. The western area has a concentration of high mountains, mid-size mountains, and valleys, and the terrain is steep, with ravines and mountains intertwined in the northeast. Most of the central area consists of river valley flats and middle mountains [36]. The type of soil in this area is mainly yellow soil (43.27%), which develops from limestone, followed by lime soil, yellow brown soil, paddy soil, and coarse bone soil [37]. This area is located in the hinterlands of the Wumeng Mountains. It is a typical karst mountain area with completely developed karst landforms and severe rocky desertification, resulting in a low ecological environment capacity and a fragile ecologic system. In terms of economic development, in 2020, the gross regional product of Qixingguan District was RMB 50.006 billion, a year-over-year increase of 4.4%; the fixed asset investment of the region increased by 3.7% over the previous year; the total fiscal revenue of the region was RMB 11.749 billion, an increase of −2.4% over the previous year; and the general public budget revenue was RMB 2.572 billion, an increase of −9.1% over the previous year.

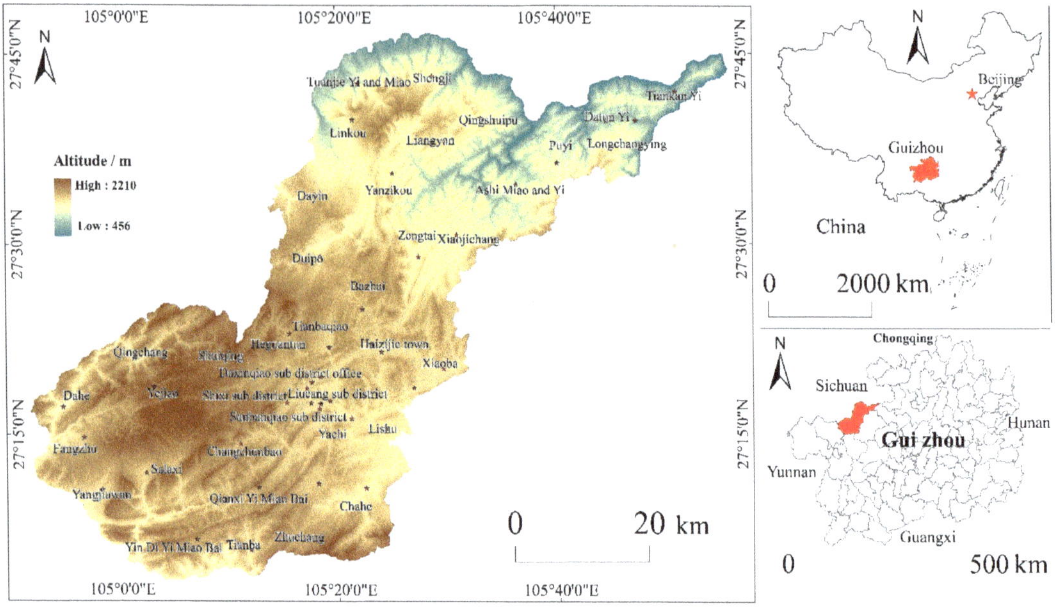

Figure 1. Topographic and location map of study area.

2.2. Data Resources

The data of the present study mainly come from the National Remote Sensing Center, Guizhou Branch (Guizhou Provincial Remote Sensing Center) and the State Engineering Technology Institute for Karst Desertification Control, Guiyang, China, including land use data of 2009, 2013, 2017, and 2020. Administrative division data from Qixingguan District, road vector data, statistical yearbook data for Qixingguan District from 2009 to 2020, and Seventh National Census Data were also used. This part of the data takes the 46 townships or streets as the basic unit for attribute assignment. Digital elevation model (DEM) data with 30 m resolution were acquired from the Geospatial Data Cloud (http://www.gscloud.cn/). In this study, the first land use included cultivated land, garden land, forest land, grassland, water area, construction land, and other land types. According to the Land Use Status Classification System (GB/T21010-2017) [38] and previous research [39], the second class of construction land included urban land, urban industrial and mining land, rural residential land, transportation land, scenic spots, and special land.

2.3. Analysis

2.3.1. GIS Spatial Analysis Method

We used ArcGIS software for basic data processing, topological analysis, summary statistics, spatial display, thematic map production, etc. [40], and the GIS spatial overlay tool to carry out overlay analysis on land types in two different periods. The final land use transition matrix was formed through statistical analysis. The land use transfer matrix can quantitatively describe the quantity and direction of mutual conversion between land types in a specific area in a certain period of time. By analyzing the transfer matrix of urban–rural construction land area, the total change between land types in two phases can be obtained. In this study, the land use transfer matrix [41] was used to describe the structural transformation trend between urban–rural construction land and other land types, and its mathematical expression is as follows:

$$S_{ij} = \begin{bmatrix} S_{11} & S_{12} & \cdots & S_{1m} \\ S_{21} & S_{22} & \cdots & S_{2m} \\ \vdots & \vdots & \vdots & \vdots \\ S_{m1} & S_{m2} & \cdots & S_{mm} \end{bmatrix} \quad (1)$$

where S is the area of land use, m is the type of land use, and i and j are the land use types used in the initial and final stages, respectively. The land use transfer matrix is mainly used to study the transfer of land use types between two adjacent periods to clarify variations in each type at the beginning of the study and the source and composition of each type at the end of the study [42].

2.3.2. Selection of Influencing Factors

Combined with the actual social economy and natural environment conditions in Qixingguan District, and based on the principle of data accessibility, the following 3 categories and 12 detection factors were selected as explanatory variables to explore the driving factors of urban–rural construction land transformation (Table 1). The influencing factors mainly include the economic development level [43,44], social living conditions, and basic natural conditions. The geomorphology of the present study area is of the mid-size mountain type, and it is located in the sloping zone of the transition from the eastern Yunnan Plateau to the original hills of the central Guizhou Mountains. The altitudinal variation is 1754 m. Altitude, terrain slope, and road network density are the natural constraints of urban spatial layout. Urban–rural construction land is the main spatial carrier in the process of regional economic and social development. The improvement of the overall economic development level (total social investment in fixed assets, total fiscal revenue, per capita GDP, per land GDP, and total industrial output value) continuously promotes the speed of urban–rural construction land expansion. In addition, the urbanization rate, population density, year-end resident population, and year-end salary of employees were used to represent the effects of regional social living conditions on the expansion of urban–rural construction land.

On the basis of index construction, SPSS 19.0 software was used to conduct the Kai-ser-Meyer-Olkin (KMO) test to check correlations and partial correlations between variables. The resulting values are between 0 and 1; the closer the KMO statistic is to 1, the stronger the correlation between variables, and the weaker the partial correlation, the better the effect of factor analysis. Bartlett's sphericity test judges whether the correlation matrix is a unit matrix, and if the independent factor analysis method of each variable is invalid. When the test results of the 12 indicators in 4 monitoring periods by SPSS showed a p-value < 0.05, this meant that the standard was met, the data were spherically distributed, and the variables had a spherical distribution independent of each other to a certain extent. The calculated KMO values were 0.665, 0.713, 0.785, and 0.692, which were all greater than the threshold of 0.5. Bartlett's test results were all significant at the 0.01 level, indicating a correlation between the variables of each index, and factor analysis could be carried out.

Table 1. Detection indicators of influencing factors of urban–rural construction land changes.

Influencing Factor	Number	Detection Factor	Access	Definition of Indicator
Basic natural conditions	X_{10}	Terrain slope	Based on DEM, obtained using ArcGIS slope calculation tool	Basis of natural conditions
	X_{11}	Average elevation	Based on DEM data, using ArcGIS software partition statistics	Basis of natural conditions
	X_{12}	Road network density	Total road area/total area	Degree of traffic development
Level of economic development	X_1	GDP per capita	Total GDP/total population of region	Level of economic development
	X_2	Average GDP	Total GDP/total area of region	Level of economic development
	X_9	Gross industrial output	Bijie Seven Star Customs District Statistical Yearbook	Level of industrial development
	X_7	Total fiscal revenue	Outline of National Economic and Social Development Plan	Level of economic strength
	X7	Completed investment in fixed assets of whole society	Total investment in social fixed assets/area of urban and rural construction land	Land use investment intensity
Social life conditions	X_3	Urbanization rate	Urban permanent population/total regional population	Population agglomeration level
	X_4	Population density	Total population of region/total area of region	Population agglomeration level
	X_5	Population density	Directly obtained from Bijie Qixingguan District Statistical Yearbook	Population agglomeration level
	X_8	Total salary of employees of unit at end of year	Directly obtained from Bijie Qixingguan District Statistical Yearbook	Social life conditions

2.3.3. Geographical Detector

The spatial distribution patterns of the geography or phenomena in a region are driven by both natural and human factors. By analyzing the relationship between the dependent and independent variables, the geographic detector can better describe the spatial heterogeneity of the dependent variable, and it is an effective spatial analysis method for revealing mechanisms [45]. It has been widely used [46–48]. If the independent variable has a significant effect on the dependent variable, then the spatial distributions of the two variables are similar. The formula is as follows:

$$PD = 1 - \frac{1}{n\sigma^2}\sum_{h=1}^{L} n_h\sigma_h^2 = 1 - \frac{SSW}{SST} \quad (2)$$

$$SSW = \sum_{h=1}^{L} N_h\alpha_h^2,\ SST = N\alpha^2 \quad (3)$$

In this formula, PD is the explanatory power, with a value ranging from 0 to 1; n and n_h are the numbers of samples in the entire area and in layer h, respectively; σ^2 and ${\sigma_h}^2$ are the dispersion variance of the entire area and layer h, respectively; L is the number of subareas; and SST and SSW are the total variance of the study area and the sum of the variance of the subregions, respectively. The larger the PD value, the stronger the driving effect of the detection factor on the evolution of urban–rural construction land.

The steps of the geographical detection operation are as follows:

(1) Extract information. In ArcGIS 10.5, villages and towns or streets are taken as the basic research units, and then the data of urban–rural construction land and influencing factors of each town or street are correlated according to the spatial location to generate

an attribute table and obtain the quantitative relationship between the corresponding urban–rural construction land and each selected indicator.

(2) Classify impact factors. Using the Reclassify tool in ArcGIS, each impact factor is classified according to the natural breakpoint method [49], and the classification value of each variable is extracted. Then, the per capita GDP, land average GDP, urbanization rate, total industrial output value, completion of fixed asset investment in the whole society, and total salary of employees at the end of the year are divided into 5 grades, and the average slope, average elevation, year-end total population, and population density are divided into 6 grades. In addition, total fiscal revenue and road network density are divided into 7 and 9 categories, respectively.

(3) Input the dependent variable Y (statistical value of urban–rural construction land area) and the independent variable X (gradual value of each influencing factor) into Excel Geodetector software (http://www.geodetector.cn) to detect the influence of factors and their interactions.

3. Results and Discussion

3.1. Spatiotemporal Pattern Analysis of Urban–Rural Construction Land

3.1.1. Structure and Spatial Pattern of Urban–Rural Construction Land

As seen from the change in the time-series law of scale structure, the area of urban–rural construction land in Qixingguan District increased from 10,034.97 hm^2 in 2009 to 22,879.86 hm^2 in 2020, thus more than doubling in those 10 years (Table 2). Using the natural breakpoint method, the quantity change of urban–rural construction land was divided into five zones: low, medium low, medium, medium high, and high density (Figure 2). The spatial distribution of urban–rural construction land changes in Qixingguan District during the four monitoring periods had similar uniformity, expanding outward from the central urban area and gradually increasing. Overall, the rural residential land in the study area is mainly characterized by the largest amount of land and broken map spots, which is consistent with the surface features of a broken karst mountainous area. In terms of map spots, in 2020, the number of map spots for urban use was 8155, while the number for rural residential land was up to 75,178. The medium-high-density and high-density areas of urban–rural construction land are mainly distributed in non-central urban areas and towns, indicating that rural residential land in the study area has the characteristics of large land occupation, scattered layout, and low land use efficiency. During the monitoring period, the area of urban land continued to increase, and its proportion reached a maximum of 28.86% in 2017.

As shown in Figure 2, the central urban area has always been in a low-density area, while the conclusions of previous works were mostly related to urban expansion and showed higher exponential growth than other regions [50]. A previous study suggested that high-density construction land areas, such as industry and transportation, should be mainly distributed in relatively good township areas, which disagrees with our study. There are two possible reasons for this: (1) The present study area is located in a karst mountain area, where the ecological environment is relatively fragile, the location is relatively remote, there is insufficient motivation for urban development motivation, economic and social development are seriously lagging behind, the planned urban expansion area is limited, and the administrative division of several streets in the central urban area is small. (2) The flow of farmers to more developed cities to work and the return of funds will increase the rural residential land area or improve the residential functions, resulting in a spreading of rural space, which will continue to increase the rural residential land use.

Table 2. Quantity of change in urban–rural construction land in Qixingguan District (2009–2020) (hm^2, %).

Classification of Land Use	Statistical Indicator	2009	2013	2017	2020
Urban land	Number of polygons	606	1060	2570	8155
	Area	2193.14	3234.55	4177.09	5587.29
	Proportion	21.85	25.34	28.86	24.42
Rural residential land	Number of polygons	22,648	22,863	24,382	75,178
	Area	7536.82	7614.13	7793.91	9552.45
	Proportion	75.11	59.65	53.85	41.75
Urban industrial and mining land	Number of polygons	436	655	679	1974
	area	195.82	668.46	671.82	2086.95
	proportion	1.95	5.24	4.64	9.12
Traffic land	Number of polygons	6	740	1569	18,251
	Area	7.53	1144.65	1721.40	5467.53
	Proportion	0.08	8.97	11.89	23.90
Scenic spots and special land	Number of polygons	256	256	262	888
	Area	101.66	103.83	108.36	185.64
	Proportion	1.01	0.81	0.75	0.81
Aggregate statistics	Total area	10,034.97	12,765.62	14,472.59	22,879.86

Note: In land use classification, proportion refers to the proportion of each land use type in urban–rural construction land.

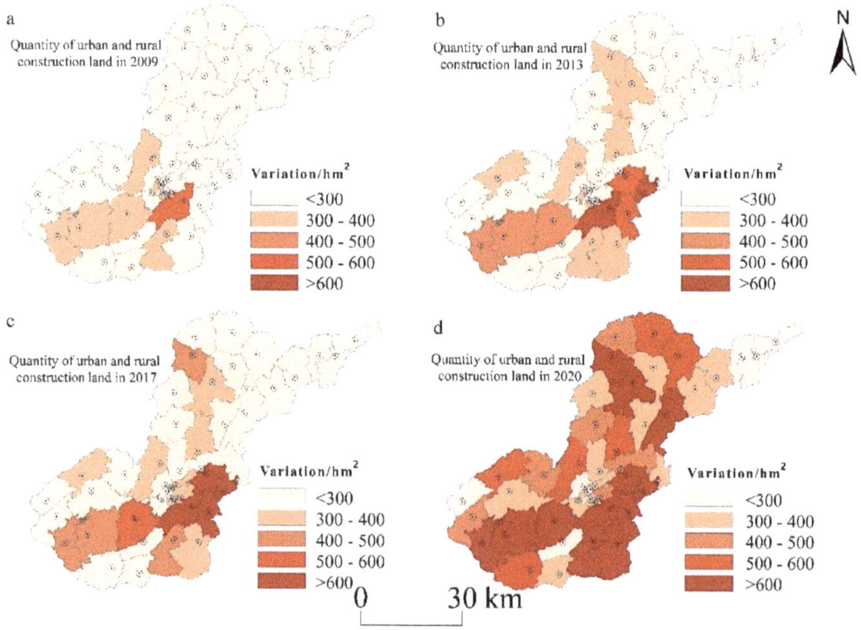

Figure 2. Spatial distribution of urban–rural construction land characteristics in Qixingguan District from 2009 to 2020.

It is worth noting that, during the study period, the quantity and scale of transportation land changed the most significantly; its area increased by 5460.00 hm^2, and the average annual growth rate in each monitoring period was higher than that of other land types, with an average annual increase of 546 hm^2 (Table 2). During the study period, the average annual growth rate was 82.01%. The reason for this is that, during the research period,

Guizhou Province achieved high-speed access between counties, highways to villages, and hardened roads between groups, establishing a main transportation network extending all over the study area. At the same time, with the development of urbanization, road facilities also continuously increased. The quantity and scale of urban industrial and mining land, scenic spots, and special land had little variation during the study period.

3.1.2. Sources and Trends of Urban–Rural Construction Land

1. Sources of urban–rural construction land transformation

A land use transfer matrix can show the transformation law of urban–rural construction land. As shown in Figure 3, the sources of urban–rural construction land in Qixingguan District greatly increased from 2009 to 2020, with the newly increased area being the most significant when comparing rural residential land, urban land, and transportation land. During that period, 89.36% of the newly increased area of urban–rural construction land came from three areas: urban, industrial, and mining land. The size of the newly added area encompassing scenic spots and special land is small. It was found that most of the newly increased urban–rural construction land (more than 89%) consists of agricultural land types, including arable land and forest land, with proportions larger than 57.72% and not less than 18.92%, respectively. Among the newly increased area of urban–rural construction land, 42.25% is rural residential land, mostly cultivated land (71.54%), and forest land (26%). A total of 94.43% of newly increased urban land consists of agricultural land types, including cultivated land (75.51%) and forest land (18.92%). The areas of newly increased urban industrial and mining land, transportation land, scenic spots, and special land differ slightly, with areas of 1240.65, 3127.94, and 89.65 hm^2, accounting for 64.91, 60.3, and 57.72%, respectively. During the study period, the proportions of grassland, gardens, water, and other areas occupied by various types of urban–rural construction land were small. Garden land had the largest area occupied by urban land (72.82 hm^2). This was mainly due to the large proportion of agricultural land in Qixingguan District (agricultural land accounted for more than 88.22% of the total area and cultivated land was approximately 44.46% during the monitoring period) and its wide distribution. It is inevitable that it will be occupied to a large extent during the rapid development of urbanization and industrialization.

2. Trend of urban–rural construction land trans formation

With regard to the direction of the reduction in urban–rural construction land in Qixingguan District during the monitoring period, as shown in Figure 4, in terms of absolute quantity, the main reason for the reduction in land from 2009 to 2020 was the reclamation of rural residential land. This represented an area of 5828.00 hm^2, accounting for 88.46% of the reduced urban–rural construction land area. A total of 92.19% of the transferred rural residential land was reclaimed as farmland and woodland, which accounted for 57.86% of the total. The second largest proportion was urban land, which had a reduced area of 544.80 hm^2, accounting for 8.27% of the reduced area of urban–rural construction land. This reduction was mainly due to conversion to arable land and forest land (together accounting for 82.17%). The reduction in area of urban industrial and mining land, scenic spots, and special land were relatively small: 141.73 and 71.84 hm^2, respectively. A total of 3.3% of the transferred urban industrial and mining land was reclaimed as arable land, and more than 52.67% was reclaimed as forest land and other nonarable agricultural land types. However, the transformation of transportation land showed a trend of continuous increase and did not decrease during the monitoring period. During the study period, transportation land only decreased by 1.90 hm^2, which again confirmed the remarkable results of road construction in Qixingguan District during the study period, changing it into an important land transportation hub from a regressive area in southwest China. In terms of absolute area, land converted from urban–rural construction land to grassland, garden, water, and other land types was far smaller than the land converted to cultivated land and woodland, which in recent years has been closely related to the many land consolidation works aimed at replenishing cultivated land.

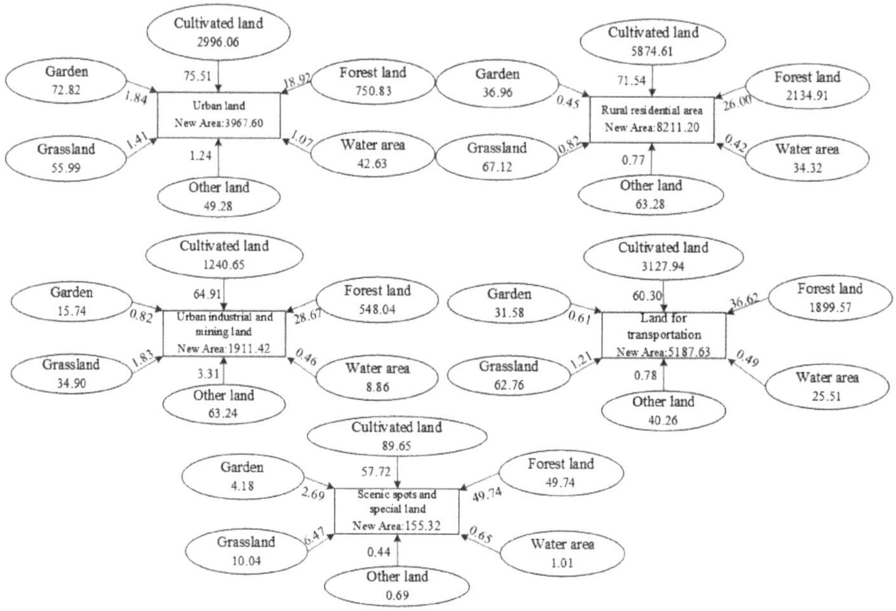

Figure 3. New sources of urban–rural construction land in Qixingguan District from 2009 to 2020 (hm², %).

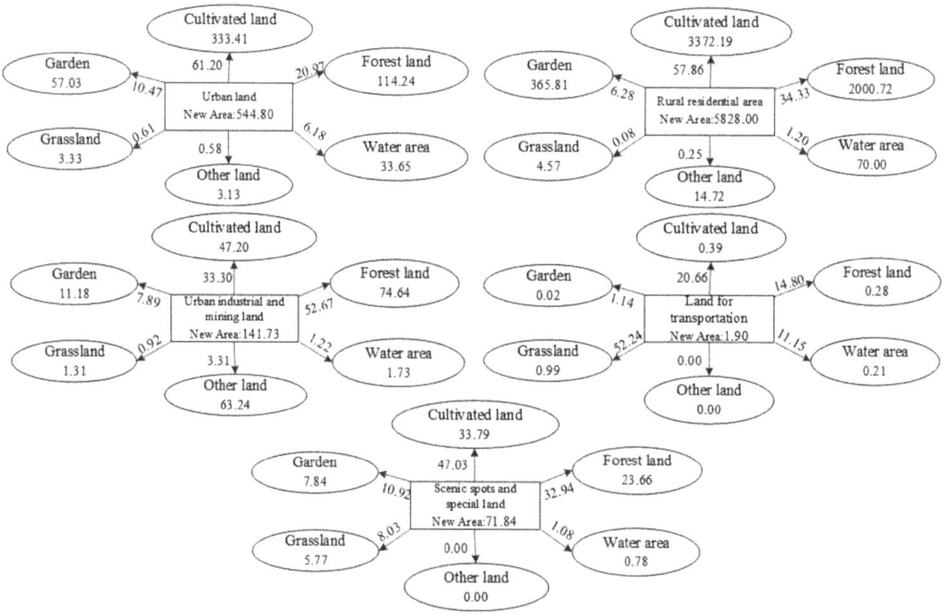

Figure 4. End points for the reduction in urban–rural construction land in Qixingguan District from 2009 to 2020 (hm²,%).

3.2. Analysis of Driving Factors of Urban–Rural Construction Land Transformation

3.2.1. Analysis of Detection Factor Influence

By testing the driving factors of urban–rural construction land, the influence of testing factor on driving factors was revealed. The PD value of each driving factor was calculated by the geographic detector (Table 3), and the influence of each factor on urban–rural construction land was determined. Table 3 shows the order of impact of various driving factors on urban–rural construction land (from high to low): total fixed asset investment in society as a whole > total fiscal revenue > road network density > total population at the end of the year > urbanization rate > per capita GDP > total annual wages of employees per unit > average GDP > total industrial output > population density > average slope > average elevation. According to the detection results for the driving factors, the total fixed assets investment for the whole society had the greatest impact on the growth of urban–rural land with an explanatory power of 0.819, which is similar to the results of studies conducted in China and internationally [46,47]. It is believed that total fixed assets investment improves people's living standards, stimulates the development of urban real estate, and promotes the expansion of urban construction land.

Figure 5 shows that the PD values of population density and total industrial output from 2009 to 2020 generally show increasing trends, and the PD values of other driving factors have increases and decreases during the monitoring period. From 2009 to 2013, except for increased PD values, average slope, average elevation, road network density, population density, and total industrial output value, the PD values of other driving factors showed decreasing trends to varying degrees. From 2013 to 2017, the PD values of total fixed asset investment of the whole society, average slope, average elevation, road network density, etc., showed a decreasing trend. From 2017 to 2020, the PD values of other driving factors showed an increasing trend, except for values for per capita GDP, total population at the end of the year, total fiscal revenue, and road network density. To summarize, the explanatory power of the driving factors during each monitoring period for Qixingguan District regarding the expansion of urban–rural construction land varies. Therefore, time series monitoring of driving factors would have important guiding significance for predicting the expansion of urban–rural construction land.

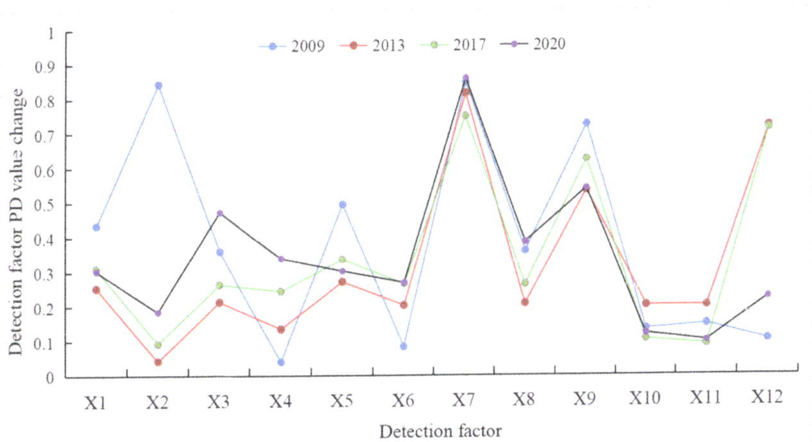

Figure 5. Change in PD value of detection factor in Seven Star Pass area from 2009 to 2020. X1, per capita GDP; X2, average GDP; X3, urbanization rate; X4, population density; X5, total population at end of year; X6, total industrial output; X7, total fixed assets investment of whole society; X8, total annual wages of employees per unit; X9, total fiscal revenue; X10, average slope; X11, average elevation; X12, road network density.

Table 3. PD values of detection factors.

Detection Factor	X_1	X_2	X_3	X_4	X_5	X_6	X_7	X_8	X_9	X_{10}	X_{11}	X_{12}
PD value	0.325	0.291	0.328	0.189	0.351	0.205	0.819	0.304	0.607	0.139	0.134	0.440
p value	0.073	0.414	0.251	0.553	0.096	0.434	0.000	0.148	0.000	0.386	0.566	0.193

3.2.2. Interaction Analysis of Detection Factors

The interaction detection in the geographic detector mainly identifies the degree of influence of interactions between driving factors on changes in urban–rural construction land in Qixingguan District. On this basis, interaction detection evaluates whether the influence of factors is independent or if it increases or decreases the explanatory power of the evolution of urban–rural construction land due to the interaction. Table 4 shows the relationships between driving factors affecting the evolution of urban–rural construction land by using the interaction detector. The results indicate that there are no factors that independently affect urban–rural construction land, and the interaction effect between driving factors is mainly manifested in the relationship of nonlinear enhancement and mutual enhancement, and the explanatory power of the interaction between all factors increases significantly: $X_9 \cap X_3$ (0.915) > $X_5 \cap X_3$ (0.891) > $X_7 \cap X_3$ (0.882) > $X_{12} \cap X_3$ (0.794) > $X_{10} \cap X_3$ (0.707) > $X_6 \cap X_3$ (0.642) > $X_4 \cap X_3$ (0.591) > $X_{11} \cap X_3$ (0.544). In particular, a significant interaction exists between other factors and the completion of fixed assets and total fiscal revenue in the whole society, and the interaction between total fiscal revenue and total industrial output has the strongest explanatory power. Overall, the influences of the driving factors on the evolution of urban–rural construction land are not independent, nor do they represent a simple superposition process. Rather, a mutual or nonlinear enhancement effect is shown, and the influence of the interactions between detecting factors on the spatial differentiation of urban–rural construction land is greater than that of a single factor.

Table 4. Interaction of detection factors.

	X_1	X_2	X_3	X_4	X_5	X_6	X_7	X_8	X_9	X_{10}	X_{11}	X_{12}
X_1	0.303											
X_2	0.870	0.185										
X_3	0.868	0.840	0.474									
X_4	0.863	0.840	0.591	0.339								
X_5	0.342	0.903	0.891	0.892	0.302							
X_6	0.868	0.814	0.642	0.410	0.891	0.268						
X_7	0.947	0.941	0.882	0.937	0.957	0.936	0.860					
X_8	0.868	0.817	0.542	0.646	0.896	0.557	0.878	0.386				
X_9	0.753	0.906	0.915	0.912	0.751	0.970	0.931	0.925	0.540			
X_{10}	0.717	0.720	0.707	0.513	0.724	0.514	0.902	0.538	0.721	0.120		
X_{11}	0.338	0.482	0.544	0.491	0.360	0.484	0.881	0.599	0.623	0.273	0.099	
X_{12}	0.819	0.529	0.794	0.851	0.848	0.829	0.916	0.764	0.913	0.625	0.676	0.225

3.3. Discussion

3.3.1. Unreasonable Structure of Urban and Rural Construction Land

Although the average annual growth rate of urban land in the study area (8.87%) is higher than that of rural residential land (2.18%), the total rural residential land area exceeds that of urban construction land. With regard to the proportion of urban–rural construction land in the study area, the proportion of rural residential land is the largest, and the map plot is broken. Although the proportion of the area has decreased since 2009, the proportion of rural residential land was still as high as 41.75% by 2020. Moreover, the number of map spots was almost twice that of urban construction land. For urban land, the area continued to increase in the four monitoring periods of the study; the area ratio increased from 21.85% in 2009 to 28.86% in 2017, it but decreased by 4.44% during the period 2017–2020. During the 13th Five-Year Plan period in China, the urbanization rate of the resident population in the study area increased to 54%, indicating that the proportion of land and population in rural areas is seriously imbalanced as a result of less population occupying more land. China's special long-term dual urban–rural structure is mainly manifested in a continuous increase in the demand for construction land in cities, and the continuously supply still occasioned an increasing shortage of construction land. However, in the countryside, it is manifested in the continuous increase in the population moving to cities, and the construction land does not decrease but increases. A large amount of rural construction land is thus idle [51].

The present study area is located in an underdeveloped area of karst mountains in southwest China, and the problem of unbalanced and insufficient urban and rural development is particularly prominent. Compared with China's economically developed regions (e.g., Yangtze River Delta) [52], the main features are that the structure of urban–rural construction land is unreasonable, the area of rural residential land is larger than the area of urban land, the population and residential areas are obviously scattered, and the driving forces of urban and rural development and urbanization are mainly external (e.g., state investment, project-driven development, administrative promotion), thus the development of urban–rural integration is limited by economic development, difficulty integrating urban and rural resources, etc. In other countries, such as the United States [53], Japan [54], the United Kingdom [55], and France [56], urban–rural integration has basically been achieved because it was developed earlier and their economies are more developed, resulting in fewer significant differences between urban and rural areas. However, China is still in the early stage of urban–rural integration; especially in the karst mountain area with fragile ecological environment, the promotion of urban–rural development is costly, difficult, and slow.

To address the problems in urban and rural land planning, Qixingguan District should formulate policies to control the expansion of urban construction land and the consolidation of rural land based on the experience of developed countries to deal with the relationship between urban and rural land to a certain extent. It is necessary to pay closer attention to the management of rural construction land, and to optimize the allocation of rural land and agricultural development through rural land consolidation, sorting out more land that can free up more land targets for the expansion of the central urban area. Moreover, urban land should be controlled within a certain range, in order to realize rational urban development and adjust the structure and layout of rural residential and urban land. This would satisfy the common development of urbanization and industrialization and improve rural development and urban–rural relations, which would improve the efficiency and effectiveness of land use, and at the same time avoid the continuous spread of the "hollowing out" phenomenon in rural areas.

3.3.2. Transformation of Cultivated Land into Construction Land

From 2009 to 2020, the main source of new urban–rural construction land in the study area was agricultural land (more than 89%, of which the minimum occupied cultivated land was 57.72%). Cultivated land converted to new urban–rural construction land amounted to 13,328.91 hm^2, but construction occupying cultivated land slowed down to a certain extent. From 2009 to 2020, the urban–rural construction land mainly transformed into reclaimed construction land. The area of reclaimed land transformed into cultivated land was 3786.97 hm^2; 61.20% of the transferred urban land was reclaimed as cultivated land, while 57.86% of the transferred rural residential land was reclaimed as cultivated land.

In terms of stages, the proportion of urban–rural construction land reclaimed as cultivated land shows a fluctuating trend, but the area of cultivated land transformed to urban–rural construction land continues to increase. In terms of area, the reduction of urban–rural construction land to grassland, garden, water, and other types of land is much smaller than the area of cultivated land and forest land, indicating that, in recent years, a large number of supplementary cultivated land projects were carried out in the study area, which will contribute to the protection of cultivated land to a certain extent. It is worth noting that under the condition that the land area remains stable, if the current growth trend is maintained and the cultivated land is recklessly constructed, it will lead to expanded urban development and the unreasonable and disorderly growth of rural residential land. The "red line" of cultivated land will be broken. At the same time, it will inevitably lead to wasted land resources and increased carrying capacity of land. Therefore, in the process of land development in the study area, more attention should be given to excavating the existing rural residential land, reforming and improving the countryside homestead management, and promoting the efficient and intensive use of urban–rural construction land by the use of clearing property rights, compensation systems for the use of resources, and standardized circulation. At the same time, it is also necessary to strengthen the supervision of the red line of cultivated land protection, resolutely implement the policy of linkage between urban land taking and rural land giving (LUTRG), and prohibit construction and development in non-construction areas, especially in karst areas with a fragile ecological environment.

3.3.3. Regular Evaluation of Planning and Dynamic Revision

From the detection results of driving factors (Section 3.3), the completion of fixed assets in the whole society has the greatest impact on the growth of urban and rural land use, and the PD value is 0.819, which agrees with previous works in other regions [57,58]. The completion of fixed asset investment in the whole society improved people's living standards, stimulated the development of urban real estate, and promoted the expansion of urban construction land. At the same time, our results imply that the PD value of each driving factor and the interaction between factors of urban and rural construction land are not constant in different monitoring periods and show dynamic changes with fluctuation. For example, the PD value of road network density as a single factor of urban–rural construction land was lower than 23% in 2009 and 2020, but it was greater than 70% in 2013 and 2017. However, China's current laws and regulations, such as those regarding land use planning or urban and rural planning, focus on vision planning, in an effort to develop an ideal planning scheme to achieve the "ultimate goal" throughout the planning period, while ignoring the suitability between the processes of "static" planning and "dynamic" implementation. In addition, there is limited ability to quickly and synchronously adjust the planning scheme when identifying changing factors, resulting in a greatly reduced implementation effect.

According to the changes in the influence of driving factors and the actual situation in the study area, the present study suggests that formulating land-related laws and regulations does not mean planning an ultimate blueprint, but a strategic and structural type of flexible and dynamic planning. Based on this, this study suggests that detecting the driving factors of urban–rural construction land in the study area should be included in the city-level urban physical examination category of "territorial space safety". The formation of the annual urban physical examination index system should be optimized, the key driving factors of urban–rural construction land in each monitoring period in the study area should be identified, and the implementation of the plan should be regularly scheduled, evaluated, and adjusted during planning. This not only could improve the utilization of urban–rural construction land resources and the efficiency of land resource utilization, but also more accurately predict the reasonable demand and spatial expansion direction of urban–rural construction land in the future.

3.3.4. Research Prospects

The following aspects should receive more attention in future work: (1) Research on the transformation of urban–rural construction land only considers the transfer between construction land and non-construction land, and mainly focuses on its dominant characteristics (quantity structure, source destination, spatial differentiation, etc.). Thus, it is necessary to consider urban–rural construction land in the next step in terms of the transformation between types and the analysis of the interaction mechanism of strengthening the hidden characteristics (input–output, utilization efficiency, economic density, function, etc.), to gain an in-depth understanding of the transformation characteristics and driving mechanism of urban–rural construction land. (2) The acquisition of basic data and indicators was limited to only four periods of land use data (2009, 2013, 2017 and 2020), which was not able to fully reflect the long-term and continuous process of urban–rural construction land transformation, resulting in insufficient research depth. It is extremely important for research to obtain multitemporal land use data and make comparisons with the natural environmental conditions, economic and social levels, policy and institutional environment, and other data in the same period. (3) The spatial distribution law, evolution process, and driving mechanism of urban–rural construction land reflect the interactive human–land relationship. Based on the limited availability and accessibility of data in this study, the selection of driving indicators of urban–rural construction land transformation needs to be further improved, and in the selection of indicators, the categories can be further increased, such as forest cover [59], technological progress, food security [60], karst basin water re-sources [61], etc., to clarify the mechanism of the driving factors of urban–rural construction land transformation more clearly. (4) Research results on carbon peaking and carbon neutralization surge since the double carbon target was proposed [62,63], the carbon emissions of urban–rural construction have attracted more attention [64]. In order to optimize carbon neutrality in new urbanization construction, more studies should be conducted that combine the processes of carbon emission reduction and carbon trading, relying on this huge carbon sink [65] to achieve the goal of carbon neutrality [66].

4. Conclusions

Using spatial analysis and geographical detectors, this study conducted a detailed analysis of the spatial and temporal evolution characteristics of urban–rural construction land in the study area and explored the effects of their driving factors. The main research conclusions are as follows:

(1) With regard to time series, the urban–rural construction land increased in Qixingguan District from 2009 to 2020, and most of it came from agricultural land. The proportion of cultivated land among the various land use types is not less than 57.72%. The fastest growing land types are rural residential, urban, and transportation land. The expansion of rural residential land is greater than that of urban land, and it continues to grow. With regard to the spatial distribution pattern, the spatial distribution expands from the surrounding area of the central city; until 2020, the distribution was medium high and high density in the northern, eastern, and southwestern townships of the study area, and medium low density in the central urban area.

(2) From 2009 to 2020, the ranking of various driving factors on urban and rural construction land in terms of impact was as follows (from high to low): total fixed asset investment in society as a whole > total fiscal revenue > road network density > total population at the end of the year > urbanization rate > per capita GDP > total annual wages of employees per unit > average GDP > total industrial output > population density > average slope > average elevation. The PD value of total fixed asset investment in the whole society and total fiscal revenue is above 60%; thus, they have become the main driving factors affecting the change of urban–rural construction land.

(3) The driving factors have interactive effects in terms of their impact on urban–rural construction land, and these effects show mutual and nonlinear enhancement. The present study reveals that the PD value of the driving factors in Qixingguan District in each monitoring period has changed. This study reveals the impact of these driving factors on urban–rural construction land and provides a foundation for studying the dynamics of the transformation of such land.

Author Contributions: Y.S.: Conceptualization, Methodology, Software, Writing—original draft. Z.Z.: Resources, Data curation, Formal analysis, Project administration, Funding acquisition. D.H.: Visualization. Q.C.: Supervision. M.F.: Investigation. All authors have read and agreed to the published version of the manuscript.

Funding: This research was funded by the State's Key Project of Research and Development Plan of China (2018YFB0505400), the National Natural Science Foundation of China (41661088), and the Guizhou Province High-level Innovative Talent Training Plan "Hundred" Level Talents (Qiankehe Platform Talents [2016]5674).

Institutional Review Board Statement: Not applicable.

Informed Consent Statement: Not applicable.

Data Availability Statement: Not applicable.

Acknowledgments: This study was supported financially by the State's Key Project of Research.

Conflicts of Interest: The authors declare no conflict of interest.

References

1. Grainger, A. National land use morphology: Patterns and possibilities. *Geography* **1995**, *80*, 235–245.
2. Ge, D.; Long, H.; Zhang, Y.; Ma, L.; Li, T. Farmland transition and its influences on grain production in China. *Land Use Policy* **2018**, *70*, 94–105. [CrossRef]
3. Jiang, G.; Zhang, R.; Ma, W.; Zhou, D.; Wang, X.; He, X. Cultivated land productivity potential improvement in land consolidation schemes in Shenyang, China: Assessment and policy implications. *Land Use Policy* **2017**, *68*, 80–88. [CrossRef]
4. Njoh, A.J. Municipal councils, international NGOs and citizen participation in public infrastructure development in rural settlements in Cameroon. *Habitat Int.* **2011**, *35*, 101–110. [CrossRef]
5. Gu, C.; Li, Y.; Han, S.S. Development and transition of small towns in rural China. *Habitat Int.* **2015**, *50*, 110–119. [CrossRef]
6. Li, T.; Long, H.; Liu, Y.; Tu, S. Multi-scale analysis of rural housing land transition under China's rapid urbanization: The case of Bohai Rim. *Habitat Int.* **2015**, *48*, 227–238. [CrossRef]
7. Song, X. Discussion on land use transition research framework. *Acta Geogr. Sin.* **2017**, *72*, 471–487, (In Chinese with English Abstract).
8. Lambin, E.F.; Meyfroidt, P. Land use transitions: Socio-ecological feedback versus socio-economic change. *Land Use Policy* **2010**, *27*, 108–118. [CrossRef]
9. Liu, Y.; Li, Y. Revitalize the world's countryside. *Nature* **2017**, *548*, 275–277. [CrossRef]
10. Tu, S.; Long, H. Rural restructuring in China: Theory, approaches and research prospect. *J. Geogr. Sci.* **2017**, *27*, 1169–1184, (In Chinese with English abstract). [CrossRef]
11. Ventura, S.J.; Niemann, B.J.; Moyer, D.D. A multipurpose land information system for rural resource planning. *J. Soil Water Conserv.* **1988**, *43*, 226–229.
12. Wasilewski, A.; Krukowski, K. Land conversion for suburban housing: A study of urbanization around Warsaw and Olsztyn, Poland. *Environ. Manag.* **2004**, *34*, 291–303. [CrossRef] [PubMed]
13. Pijanowski, B.C.; Robinson, K.D. Rates and patterns of land use change in the Upper Great Lakes States, USA: A framework for spatial temporal analysis. *Landsc. Urban Plan.* **2011**, *102*, 102–116. [CrossRef]
14. Livingstone, S.J.; Li, Y.; Rutishauser, A.; Sanderson, R.J.; Winter, K.; Mikucki, J.A.; Björnsson, H.; Bowling, J.S.; Chu, W.; Dow, C.F.; et al. Subglacial lakes and their changing role in a warming climate. *Nat. Rev. Earth Environ.* **2022**, *3*, 106–124. [CrossRef]
15. Marsh, S. Plummeting morale in the forest service: Why it should matter to Americans who love nature. *Mt. J.* **2018**, *20*, 2020.
16. Liu, M.; Bai, X.; Tan, Q.; Luo, G.; Zhao, C.; Wu, L.; Luo, X.; Ran, C.; Zhang, S. Climate Change Enhances the Positive Contribution of Human Activities to Vegetation restoration in China. *Geocarto Int.* **2022**, 1–24. [CrossRef]
17. Bourne, L.S. Reurbanization, uneven urban development, and the debate on new urban forms. *Urban Geogr.* **1996**, *17*, 690–713. [CrossRef]
18. Jensen, J.R.; Toll, D.L. Detecting residential land-use development at the urban fringe. *Photogramm. Eng. Remote Sens.* **1982**, *48*, 629–643.
19. López, E.; Bocco, G.; Mendoza, M.; Duhau, E. Predicting land-cover and land-use change in the urban fringe: A case in Morelia city, Mexico. *Landsc. Urban Plan.* **2001**, *55*, 271–285. [CrossRef]
20. Braimoh, A.K.; Onishi, T. Spatial determinants of urban land use change in Lagos, Nigeria. *Land Use Policy* **2007**, *24*, 502–515. [CrossRef]
21. Weber, C.; Puissant, A. Urbanization pressure and modeling of urban growth: Example of the Tunis Metropolitan Area. *Remote Sens. Environ.* **2003**, *86*, 341–352. [CrossRef]
22. Masek, J.G.; Lindsay, F.E.; Goward, S.N. Dynamics of urban growth in the Washington DC metropolitan area, 1973–1996, from Landsat observations. *Int. J. Remote Sens.* **2000**, *21*, 3473–3486. [CrossRef]
23. Saizen, I.; Mizuno, K.; Kobayashi, S. Effects of land-use master plans in the metropolitan fringe of Japan. *Landsc. Urban Plan.* **2006**, *78*, 411–421. [CrossRef]
24. Ustaoglu, E.; Aydınoglu, A.C. Suitability evaluation of urban construction land in Pendik district of Istanbul, Turkey. *Land Use Policy* **2020**, *99*, 104783. [CrossRef]
25. Mann, S. Institutional causes of urban and rural sprawl in Switzerland. *Land Use Policy* **2009**, *26*, 919–924. [CrossRef]

26. Colsaet, A.; Laurans, Y.; Levrel, H. What drives land take and urban land expansion? A systematic review. *Land Use Policy* **2018**, *79*, 339–349. [CrossRef]
27. Bittner, C.; Sofer, M. Land use changes in the rural-urban fringe: An Israeli case study. *Land Use Policy* **2013**, *33*, 11–19. [CrossRef]
28. Diogo, V.; Koomen, E. Land-use change in Portugal, 1990–2006: Main processes and underlying factors. *Cartogr. Int. J. Geogr. Inf. Geovis.* **2012**, *47*, 237–249. [CrossRef]
29. Duram, L.A. A pragmatic study of conventional and alternative farmers in Colorado. *Prof. Geogr.* **1997**, *49*, 202–213. [CrossRef]
30. Luo, P.; Du, Q.; Lei, Y.; Wang, T. Cellular Automata Based on Geographic Feature and Urban Land Use Evolvement. *Geomat. Inf. Sci. Wuhan Univ.* **2004**, *29*, 504–507, (In Chinese with English Abstract).
31. Thamodaran, R.; English, B.; Heady, E.O. *A Statewide Projection of Agricultural Land Losses to Non-Agricultural Land Uses*; Working Paper; Center for Agricultural and Rural Development, Iowa State University: Ames, IA, USA, 1981.
32. Dewi, D.I.K.; Ratnasari, R.A. Land use change in sub district Mranggen because of residential development. *Procedia-Soc. Behav. Sci.* **2016**, *227*, 210–215. [CrossRef]
33. Wang, J.F.; Li, X.H.; Christakos, G.; Liao, Y.L.; Zhang, T.; Gu, X.; Zheng, X.Y. Geographical detectors-based health risk assessment and its application in the neural tube defects study of the Heshun Region, China. *Int. J. Geogr. Inf. Sci.* **2010**, *24*, 107–127. [CrossRef]
34. Bai, H.; Li, D.; Ge, Y.; Wang, J.; Cao, F. Spatial rough set-based geographical detectors for nominal target variables. *Inf. Sci.* **2022**, *586*, 525–539. [CrossRef]
35. Wu, D.; Yang, L. Evaluation of Coordinated Development between Urbanization and Eco- environment in Wumeng Mountainous Areas of Guizhou—A Case Study of Qixingguan Region. *Anhui Agric. Sci. Bull.* **2016**, *22*, 97–102, (In Chinese with English Abstract).
36. Zhou, D.B.; Huang, T.L.; Xiang, T.G.; Xie, G.; Huang, X.Y. Current Situation Analysis and Suggestions of Forest Ecological Industry in Qixingguan District of Bijie City. *Mod. Agric. Sci. Technol.* **2020**, *759*, 144–146, (In Chinese with English Abstract).
37. Wang, X.H.; Zhang, Z.H.; Wang, H.B.; Xiao, B.; Liu, H.; Wang, T.; Wu, P.; Pan, Y.L. Geochemical Evaluation Results and significance of Cultivated Land Quality in Qixingguan District, Guizhou Province. *Guizhou Geol.* **2020**, *37*, 251–257, (In Chinese with English Abstract).
38. *GB/T 21010-2017*; Current Land Use Classification. Standardization Administration of China: Beijing, China, 2017. (In Chinese)
39. Zeng, Y.J. *Coupling and Coordination between Urban-Rural Construction Land Transition and Economic and Social Development in Banan District Land Resource Management*; Southwest University: Chongqing, China, 2019; (In Chinese with English Abstract).
40. Unwin, D.J. GIS, spatial analysis and spatial statistics. *Prog. Hum. Geogr.* **1996**, *20*, 540–551. [CrossRef]
41. Du, W.; Zhao, X.; Zhao, Z.; Chen, C.; Qian, D. Assessment and dynamic mechanisms of the land-use dominant morphology transition: A case study of Hainan Province, China. *Environ. Monit. Assess.* **2022**, *194*, 419. [CrossRef]
42. Liu, F.; Qin, T.; Girma, A.; Wang, H.; Weng, B.; Yu, Z.; Wang, Z. Dynamics of land-use and vegetation change using NDVI and transfer matrix: A case study of the Huaihe River Basin. *Pol. J. Environ. Stud.* **2018**, *28*, 213–223. [CrossRef]
43. Deng, X.; Huang, J.; Rozelle, S.; Uchida, E. Economic growth and the expansion of urban land in China. *Urban Stud.* **2010**, *47*, 813–843. [CrossRef]
44. Wu, K.; Zhang, H. Land use dynamics, built-up land expansion patterns, and driving forces analysis of the fast-growing Hangzhou metropolitan area, eastern China (1978–2008). *Appl. Geogr.* **2012**, *34*, 137–145. [CrossRef]
45. Wang, Y.; Wang, S.; Li, G.; Zhang, H.; Jin, L.; Su, Y.; Wu, K. Identifying the determinants of housing prices in China using spatial regression and the geographical detector technique. *Appl. Geogr.* **2017**, *79*, 26–36. [CrossRef]
46. Wang, J.F.; Zhang, T.L.; Fu, B.J. A measure of spatial stratified heterogeneity. *Ecol. Indic.* **2016**, *67*, 250–256. [CrossRef]
47. Ju, H.; Zhang, Z.; Zuo, L.; Wang, J.; Zhang, S.; Wang, X.; Zhao, X. Driving forces and their interactions of built-up land expansion based on the geographical detector–a case study of Beijing, China. *Int. J. Geogr. Inf. Sci.* **2016**, *30*, 2188–2207. [CrossRef]
48. Wang, J.F.; Xu, C.D. Geodetector: Principle and prospective. *J. Geogr.* **2017**, *72*, 116–134, (In Chinese with English Abstract).
49. Wu, Z.H.; Li, T. The comprehensive performance evaluation of the high-tech development zone: Analysis based on the natural breakpoint method. *Stat. Inf. Forum* **2013**, *28*, 82–88, (In Chinese with English Abstract).
50. García-Ayllón, S. Predictive diagnosis of agricultural periurban areas based on territorial indicators: Comparative landscape trends of the so-called "Orchard of Europe". *Sustainability* **2018**, *10*, 1820. [CrossRef]
51. Li, Y.; Li, Y.; Westlund, H.; Liu, Y. Urban–rural transformation in relation to cultivated land conversion in China: Implications for optimizing land use and balanced regional development. *Land Use Policy* **2015**, *47*, 218–224. [CrossRef]
52. Niu, X.; Liao, F.; Liu, Z.; Wu, G. Spatial-Temporal Characteristics and Driving Mechanisms of Land-Use Transition from the Perspective of Urban–Rural Transformation Development: A Case Study of the Yangtze River Delta. *Land* **2022**, *11*, 631. [CrossRef]
53. García-Ayllón, S. Rapid development as a factor of imbalance in urban growth of cities in Latin America: A perspective based on territorial indicators. *Habitat Int.* **2016**, *58*, 127–142. [CrossRef]
54. Hebbert, M. Urban sprawl and urban planning in Japan. *Town Plan. Rev.* **1986**, *57*, 141–158. [CrossRef]
55. Alcock, I.; White, M.P.; Lovell, R.; Higgins, S.L.; Osborne, N.J.; Husk, K.; Wheeler, B.W. What accounts for 'England's green and pleasant land'? A panel data analysis of mental health and land cover types in rural England. *Landsc. Urban Plan.* **2015**, *142*, 38–46. [CrossRef]
56. Newsome, W.B. *French Urban Planning, 1940–1968: The Construction and Deconstruction of an Authoritarian System*; Peter Lang: Bern, Switzerland, 2009.

57. Zhang, T. Community features and urban sprawl: The case of the Chicago metropolitan region. *Land Use Policy* **2001**, *18*, 221–232. [CrossRef]
58. Schumacher, M.; Durán-Díaz, P.; Kurjenoja, A.K.; Gutiérrez-Juárez, E.; González-Rivas, D.A. Evolution and Collapse of Ejidos in Mexico—To What Extent Is Communal Land Used for Urban Development? *Land* **2019**, *8*, 146. [CrossRef]
59. Chen, F.; Bai, X.; Liu, F.; Luo, G.; Tian, Y.; Qin, L.; Li, Y.; Xu, Y.; Wang, J.; Wu, L.; et al. Analysis Long-Term and Spatial Changes of Forest Cover in Typical Karst Areas of China. *Land* **2022**, *11*, 1349. [CrossRef]
60. Song, F.; Wang, S.; Bai, X.; Wu, L.; Wang, J.; Li, C.; Chen, H.; Luo, X.; Xi, H.; Zhang, S.; et al. A new indicator for global food security assessment: Harvested area rather than cropland area. *Chin. Geogr. Sci.* **2022**, *32*, 204–217. [CrossRef]
61. Wu, L.; Wang, S.; Bai, X.; Chen, F.; Li, C.; Ran, C.; Zhang, S. Identifying the Multi-Scale Influences of Climate Factors on Runoff Changes in a Typical Karst Watershed Using Wavelet Analysis. *Land* **2022**, *11*, 1284. [CrossRef]
62. Du, C.; Bai, X.; Li, Y.; Tan, Q.; Zhao, C.; Luo, G.; Wu, L.; Chen, F.; Li, C.; Ran, C.; et al. Inventory of China's Net Biome Productivity since the 21st Century. *Land* **2022**, *11*, 1244. [CrossRef]
63. Xiong, L.; Bai, X.; Zhao, C.; Li, Y.; Tan, Q.; Luo, G.; Wu, L.; Chen, F.; Li, C.; Ran, C.; et al. High-Resolution Data Sets for Global Carbonate and Silicate Rock Weathering Carbon Sinks and Their Change Trends. *Earth's Future* **2022**, *10*, e2022EF002746. [CrossRef]
64. Hu, S.; Zhang, Y.; Yang, Z.; Yan, D.; Jiang, Y. Challenges and opportunities for carbon neutrality in China's building sector—Modelling and data. *Build. Simul.* **2022**, *15*, 1899–1921. [CrossRef]
65. Li, C.; Bai, X.; Tan, Q.; Luo, G.; Wu, L.; Chen, F.; Xi, H.; Luo, X.; Ran, C.; Chen, H.; et al. High-resolution mapping of the global silicate weathering carbon sink and its long-term changes. *Glob. Chang. Biol.* **2022**, *28*, 4377–4394. [CrossRef] [PubMed]
66. Zhang, S.; Bai, X.; Zhao, C.; Tan, Q.; Luo, G.; Wu, L.; Xi, H.; Li, C.; Chen, F.; Ran, C.; et al. China's carbon budget inventory from 1997 to 2017 and its challenges to achieving carbon neutral strategies. *J. Clean. Prod.* **2022**, *347*, 130966. [CrossRef]

land

MDPI

Article

Responses of NDVI to Climate Change and LUCC along Large-Scale Transportation Projects in Fragile Karst Areas, SW China

Yangyang Wu [1,2,3], Lei Gu [4], Siliang Li [2], Chunzi Guo [2,5], Xiaodong Yang [6,7], Yue Xu [8], Fujun Yue [2], Haijun Peng [8], Yinchuan Chen [9], Jinli Yang [7], Zhenghua Shi [7] and Guangjie Luo [1,3,*]

1 School of Geography and Resources, Guizhou Education University, Guiyang 550018, China
2 School of Earth System Science, Tianjin University, Tianjin 300072, China
3 Guizhou Provincial Key Laboratory of Geographic State Monitoring of Watershed, Guizhou Education University, Guiyang 550018, China
4 College of Geography and Remote Sensing Sciences, Xinjiang University, Urumqi 830017, China
5 Administration of Ecology and Environment of Haihe River Basin and Beihai Sea Area, Ministry of Ecology and Environment of People's Republic of China, Tianjin 300061, China
6 Department of Geography and Spatial Information Techniques, Center for Land and Marine Spatial Utilization and Governance Research, Ningbo University, Ningbo 315211, China
7 College of Ecological and Environmental Sciences, Xinjiang University, Urumqi 830017, China
8 State Key Laboratory of Environmental Geochemistry, Institute of Geochemistry, Chinese Academy of Sciences, Guiyang 550081, China
9 Shanghai Ecology and Environment Scientific Research Center, Yangtze River Basin Ecological Environment Supervision and Administration Bureau, Ministry of Ecology and Environment of People's Republic of China, Shanghai 200120, China
* Correspondence: luoguangjie@gznc.edu.cn

Citation: Wu, Y.; Gu, L.; Li, S.; Guo, C.; Yang, X.; Xu, Y.; Yue, F.; Peng, H.; Chen, Y.; Yang, J.; et al. Responses of NDVI to Climate Change and LUCC along Large-Scale Transportation Projects in Fragile Karst Areas, SW China. *Land* **2022**, *11*, 1771. https://doi.org/10.3390/land11101771

Academic Editor: Xiaoyong Bai

Received: 31 August 2022
Accepted: 9 October 2022
Published: 12 October 2022

Abstract: The fragile karst habitat is extremely sensitive to human activities such as large-scale engineering construction. To explore the influence of the construction and operation of the GH (Guiyang-Huangguoshu) highway on the vegetation within a certain range and the response of NDVI to climate factors, Landsat data were used to synthesize annual NDVI maps using the maximum value compositing method. Trend, correlation, and coefficient of variation analyses were performed. The results demonstrate that: (1) During the construction and operation periods, NDVI showed an overall upward trend, and the NDVI value and growth rate in the contrast area were greater than those in the core area; (2) the correlation between temperature and vegetation cover along the GH highway was stronger than that between precipitation and vegetation; (3) construction of the GH highway has had a significant impact on the surrounding vegetation, with the impact on vegetation ecology along the road mainly concentrated within the 2 km range. The increase of artificial surfaces along the road has had a great impact on the NDVI, and the vegetation cover change in the core area is more significant than that in the contrast area; and (4) the overall disturbance of the GH highway project to the surrounding ecology was mainly observed in the form of low and medium fluctuations. This study aims to provide a reference for environmental assessment and management in karst areas.

Keywords: vegetation dynamics; road construction; influencing factors

1. Introduction

The environment is the basis of human survival and development, comprising the sum of various natural factors within and around human society [1]. At present, China is committed to conserving natural ecosystems, focusing on strengthening the protection of the environment in large river basins [2]. Guizhou Province is located in the southwest karst area of China—the largest continuous karst landform region in the world [3]—which spans the Yangtze River and the Pearl River. It is an important ecological barrier in the upper reaches of the "two rivers". The entire ecological quality of this region is in good condition [4]; however, its ecosystems are vulnerable due to their significant sensitivity to external disturbances (including human activities and climate change) in this zone [5]. The Chinese government completed built numerous transportation infrastructure projects in Southwest China in recent years. As a consequence, the depth of highway access, road quality and network level in Guizhou been significantly improved. However, road construction projects are often in conflict with ecological protection. The rapid expansion of road traffic has brought unprecedented challenges to the local environment, and the construction or expansion of various types of roads may (directly or indirectly) leads to serious degradation of the natural environment, as well as increased local plant mortality. Therefore, a primary ecological problem to be addressed is to determine the current situation and change trend of vegetation along highways in karst region.

Highway construction and other large-scale construction activities can directly affect the environment by changing the surface vegetation cover. A construction project may greatly change the topography of the original slope conditions, geological conditions and natural stability, leading to increased vulnerability of the surrounding environment. Furthermore, with the promotion of large-scale projects, economic activities along the project will be activated, potentially including unreasonable human cultivation, excessive reclamation, overgrazing and urban expansion, all of which can lead to a decrease in vegetation coverage [6–8]. Therefore, large-scale artificial engineering activities have a significant impact on the growth and distribution of land surface vegetation [9,10], and can even change the distribution of vegetation coverage at the regional scale [11]. In contrast, reasonable project construction planning and ecological protection measures, such as afforestation, mountain closure afforestation and the improvement of agricultural technology, can facilitate the recovery vegetation [12,13].

Vegetation is one of the key components of the terrestrial ecosystem playing a fundamental role in regulating energy exchange and material cycling [14], especially in the process of karst rocky desertification control and ecological restoration [15,16]. Evidence has shown that, climate change is an important environmental factor having a significant impact on vegetation dynamics [17,18]. It influences the function and structure of the ecosystem by acting on the growth and adaptation characteristics of plants [19]. Temperature and precipitation are the most direct and important factors for vegetation growth and phenology [20,21]. At present, the normalized vegetation index (NDVI) is widely used to monitor vegetation and explore its response to climate change [22]. NDVI is an effective indicator of vegetation growth status and vegetation coverage, and it has a good linear relationship with surface vegetation. In a study on the correlation between global climate factors and NDVI changes, it has been found that NDVI presented an increasing trend with the increase in temperature in the middle high latitudes of the Northern Hemisphere [23]. In a regional study, it has been found that the seasonal variation of NDVI in different years were also responsive to land processes [24]. A spatial–temporal variation trend has been observed to vegetation degradation along with its response to climate change and anthropogenic stress [25]. Temperature may be an important driving force limiting forest greening in mountainous areas due to recent climate warming [26]. The NDVI of Guizhou karst area has been found to be more affected by temperature than precipitation and was one of the provinces with the most obvious environmental improvement [27,28]. In karst areas, a significant increase in vegetation NDVI is has been closely related to climate warming, but weakly related to precipitation [27,29]. Climate change in karst regions typically presents a cold-dry trend, while vegetation NDVI presents a recovery trend [30].

Although many scholars have studied the response of vegetation to human activities and climate factors, few have studied the impact of large engineering construction on the environment in karst areas. Relevant studies have revealed the spatial and temporal response relationship between NDVI and climate factors along the Qinghai–Tibet (QT) railway, as well as human activities, indicating that the influence of construction and operation of the QT railway on NDVI tended to weaken outward from the QT railway, while temperature and precipitation were positively correlated with NDVI [31,32]. Human activities have contributed to the response relationship between regional vegetation change and climate change [33,34]. In karst areas, human activities tend to have a stronger role in vegetation improvement and degradation than climate change [35].

The Guiyang–Huangguoshu highway (GH highway) is the first highway built in the karst area. Over a long times scale (35 a), the construction and operation activities of this large project and the subsequent enhancement of human activities along the line were sufficient to change the original land-cover and affect the surrounding environment, resulting in the destruction of habitats along the line. Therefore, GH highway is an ideal research area. In this study, we take the earliest GH highway in the karst region as the research object. High-resolution NDVI data from 1986 to 2020 are used. The trend of NDVI in the area within 8 km of the GH highway route is analyzed. The impacts of temperature, climate, land use and land-cover change (LUCC) on the vegetation along the highway are comprehensively considered. We explored the long-term impact of road traffic engineering on vegetation, in order to provide guidance for future road traffic route planning and industrial layout in fragile karst regions.

2. Materials and Methods

2.1. Study Area

The GH highway is the first high-grade highway in the karst region [36]. It started construction in August 1986 and opened to traffic in May 1991. GH highway starts from Guiyang and finally reaches Huangguoshu, with a total length of 137 km (Figure 1). The average elevation of the road is about 1200 to 1300 m, low in the middle and high around the ends. The climate is subtropical monsoon with a mean annual temperature (MAT) of 15.3 °C. The mean annual precipitation (MAP) is about 1100 mm. The GH highway is located in a typical karst landform area, within karst landforms accounting for 76.5% [37].

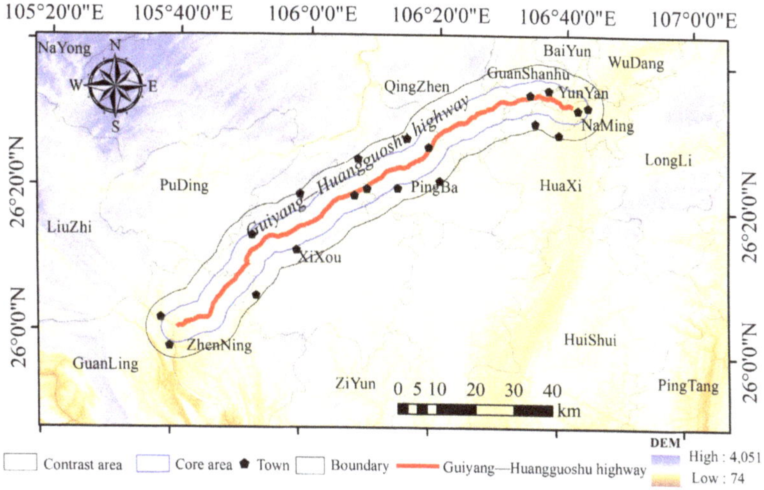

Figure 1. Location and elevation of the study area.

2.2. Data Collection and Processing

2.2.1. NDVI Data

Landsat 5, Landsat 7, Landsat 8 TM image data has been collected by the Google Earth Engine. The time resolution is 16 days and the spatial resolution is 30 m. The Maximum Value Composite (MSV) method [38] was used to generate the annual Maximum NDVI Value from 1986 to 2020. This study taken NDVI as an indicator and used the current situation vector data of GH highway to generate a 4 km buffer zone (hereafter referred to as the core area) in ArcGIS to evaluate the direct impact of human activities on the environment of GH highway. To evaluate the indirect effects of human activities on the environment of GH highway. We generated a 4 km buffer zone (hereafter referred to as the contrast area) in the periphery of the core area. The NDVI of the core area was divided into four buffer zones according to 1 km, and the direct impact of GH highway on the NDVI of human activities along the highway was evaluated.

2.2.2. Meteorological Data

Raster data of temperature and precipitation at 1 km resolution were used. The temperature and precipitation from 1986 to 2015 were collected from the Data Center of Institute of Geographic Sciences and Natural Resources Research, Chinese Academy of Sciences (https://www.resdc.cn/, accessed on 1 October 2022). The data of annual precipitation and annual temperature from 2016 to 2020 are from the monthly scale data provided by China Meteorological Administration (http://data.cma.cn/, accessed on 1 October 2022). The meteorological interpolation software ANUSPLIN was used to interpolate the temperature and precipitation data with a spatial resolution of 30 m. The digital elevation map (DEM) was introduced in the interpolation process to reduce the effect of topography on climate, thus minimizing the interpolation error and greatly improving the accuracy compared with other interpolation methods [39], which is more suitable for the analysis of meteorological elements of time series [40]. The above monthly synthesized temperature and precipitation data are extracted by ArcGIS to synthesize the annual average precipitation and annual average temperature data, and resampling into 30 m in ArcGIS.

2.2.3. Terrestrial Surface Data

The DEM digital elevation model is derived from the Geospatial Data Cloud (http://www.gscloud.cn/, accessed on 1 October 2022) with a spatial resolution of 30 m. LUCC data from 2000 to 2020 are obtained from the Globalland30 (http://www.globallandcover.com/, accessed on 1 October 2022), the spatial resolution is 30 m, the dataset includes ten types of land cover [41]. Wetlands are reclassified as water bodies due to its small area in the study region.

2.3. *Methods*

2.3.1. Trend Analysis Method

Trend analysis is a linear regression analysis of the changes of variables over timescales. It can not only track and analyze the change trend of variables, but also predict the change trend of variables. In the analysis of the change trend of inter-annual NDVI, the slope is the minimum power of the raster value of the time series, and the change value of spatial pixel on the time scale can be calculated by traversing pixel by pixel, and the change trend can be obtained [42]. The calculation method is as follows:

$$\varnothing_{Slope} = \frac{n \times \sum_{i=1}^{n}(i \times NDVI_i) - \sum_{i=1}^{n} i \sum_{i=1}^{n} NDVI_i}{n \times \sum_{i=1}^{n} i^2 - (\sum_{i=1}^{n} i)^2} \quad (1)$$

where, $\varnothing_{Slope}$ is pixel regression Slope, $NDVI_i$ is NDV value in the n year, and n is time length. When $\varnothing_{Slope} > 0$, it indicates an increasing $NDVI_i$ trend, and when $\varnothing_{Slope} < 0$, it indicates a decreasing NDVI trend.

2.3.2. Analysis of Correlation

Correlation analysis is a statistical method to study the correlation between two or more variables. In data analysis, it is often used to analyze the relationship between continuous independent variables and continuous dependent variables. When there are many features, Pearson correlation analysis is used. Pearson's correlation coefficient is a statistic reflecting the degree of linear correlation between two variables. The calculation formula goes as follows:

$$r_{xy} = \frac{\sum_{i=1}^{n}(x_i - \overline{x})(y_i - \overline{y})}{\sqrt{\sum_{i=1}^{n}(x_i - \overline{x})^2}\sqrt{\sum_{i=1}^{n}(y_i - \overline{y})^2}} \quad (2)$$

where, r_{xy} is the correlation coefficient of NDVI precipitation or temperature, which is between −1 and 1. The larger the value, the greater the correlation, and the smaller the value, the smaller the correlation. $\overline{x}$, $\overline{y}$ are the mean values of multi-year NDVI and precipitation or temperature, x_i, y_i are the NDVI values of the ith year and the temperature and precipitation values of the i year.

$$R_{12,3} = \frac{r_{12} - r_{13}r_{23}}{\sqrt{(1 - r_{13}^2)(1 - r_{23}^2)}} \quad (3)$$

where $R_{12,3}$, $R_{13,2}$, $R_{23,1}$ is the partial correlation coefficients among variables; $R_{12,3}$ is the partial correlation coefficient between r_1 and r_2 after fixing the variable r_3. $R_{12,3} > 0$ indicates positive correlation, that is, the two factors are correlated in the same direction; $R_{12,3} < 0$ indicate negative correlation, that is, the two elements of heterotrophy correlation; the larger the partial correlation coefficient is, the stronger the correlation between the two elements at the pixel is.

2.3.3. Coefficient of Variation

The coefficient of variation, also known as the "coefficient of dispersion", is a normalized measure of the degree of dispersion of a probability distribution. The calculation formula is shown below.

$$C_v = \frac{1}{\overline{x}}\sqrt{\frac{\sum_{i=1}^{n}(x_i - \overline{x})^2}{n-1}} \quad (4)$$

C_v stands for the coefficient of variation of NDVI; x_i stands for the NDVI value in the i-th year; $\overline{x}$ stands for the mean NDVI value in the n years. The higher the C_v value, the more discrete the data, the higher the variation degree of the corresponding NDVI value, and the greater the inter-annual variation. The smaller the C_v value is, the more the data is aggregated, the lower the variation degree of the corresponding NDVI value and the lower the inter-annual variation.

3. Results

3.1. *Trends in Time Scale and Spatial Change of NDVI*

In order to explore the influence of the construction and operation activities along the GH highway, the NDVI obtained along the GH highway was divided into the construction period (1986–1991) and the operation period (1992–2020) in the time scale. By piecewise fitting of time-series NDVI, the trend of NDVI change in each time period was obtained.

Within the construction period, the NDVI in both the core and contrast areas showed a clear upward trend (Figure 2a). The NDVI increased more significantly in the contrast area at 0.0170/a, while it increased in the core area at 0.0149/a (Figure 2a). The results revealed that the construction of the GH highway caused some damage to the surrounding vegetation, and the NDVI values decreased most significantly in the early stage of construction (1987).

During the operation period, the vegetation cover in the core and contrast areas of GH highway gradually improved and the NDVI showed a generally increasing trend. The growth rates in the core and contrast areas were 0.0024/a and 0.0027/a, respectively (Figure 2b). However, the change trend of NDVI obviously differed before and after 2000. The NDVI increased at a faster rate in the core area (0.0137/a) than in the contrast area (0.0130/a) during 1992 to 2000. After this period of growth, the NDVI reaching a relatively stable state, with little fluctuation around 2000. However, after this, the NDVI in both the core and the contrast areas decreased (at rate of −0.0030/a and −0.0002/a, respectively), showing two significant decreases (in 2010 and 2018) and one significant increase (2012) between 2001 and 2020.

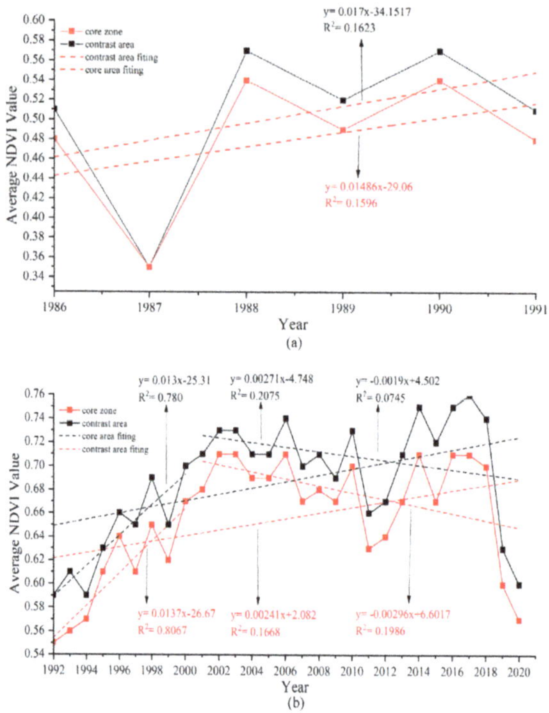

Figure 2. Inter annual dynamic change of average NDVI within construction (**a**) and operation (**b**) periods in the study area.

The overall NDVI level in the operation period was higher than that in the construction period, while the fluctuation of NDVI was smaller than that in the construction period. Within the construction period, the MAT decreased slightly, while the MAP and NDVI increased significantly (Figure 3a). The MAP increased at a rate of 12.04 mm/a, while the MAT decreased at a rate of −0.0180 °C/a.

The trend of MAT generally increased, while the MAP decreased (Figure 3b). In the fitting of temperature and precipitation from 1992 to 2020, the temperature increased at a rate of 0.0288 °C/a, while the precipitation decreased at a rate of −16.44 mm/a.

Considering that there are many other traffic routes, cities, towns and villages along the GH highway. The core area was divided into four buffer zones with a distance of 1 km in order to measure the spatial impact level of GH highway. In the temporal dimension, the NDVI showed an overall increase due to self-healing of the environment. In the spatial dimension, the construction and operation of the GH road negatively affected the vegetation within 2 km of the route. This impact was inversely proportional to the distance from the GH highway route. The NDVI values were higher within the 1 km buffer than the 2 km buffer, but did not increase outward at the 3 km and 4 km ranges (Figure 4a,b).

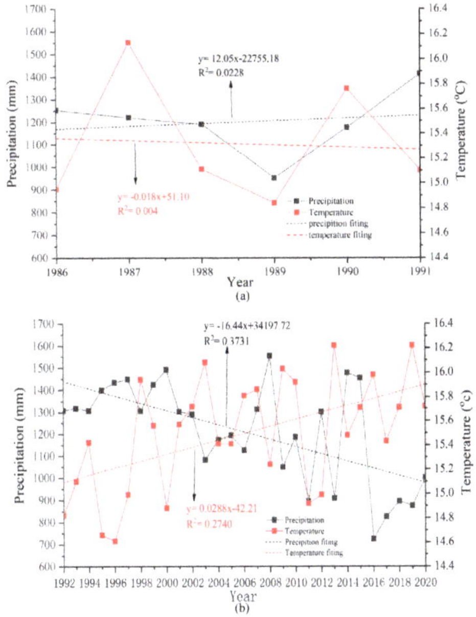

Figure 3. Trend fitting of MAT (**a**), MAP (**b**) and NDVI within the construction and operation periods.

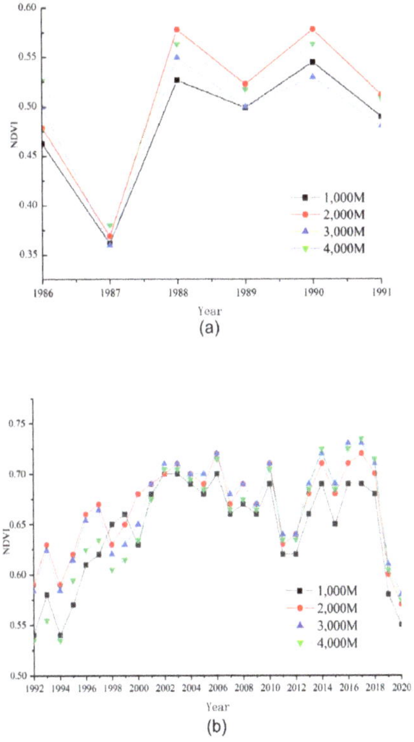

Figure 4. NDVI values at different distances within construction (**a**) and operation (**b**) periods.

Within the construction and operation period, the slopes of the regression equations in the core and contrast areas are concentrated between −0.05–0.15 and 0.002–0.015, respectively (Table 1).

Table 1. Classification statistics of different trends in core and contrast areas within construction and operation periods.

Period	Classification	Slope	Core Area			Contrast Area		
			Pixels	Area (km^2)	Percent	Pixels	Area (km^2)	Percent
Construction Period	Significantly Decrease	<−0.05	17526	14.08	0.17%	16562	13.31	0.13%
	Slightly Decrease	−0.05–0	219687	176.55	2.10%	204681	164.49	1.65%
	Slightly Increase	0–0.05	9691449	7788.51	92.42%	11551849	9283.62	92.99%
	Significant Increase	0.05–0.1	429426	345.11	4.10%	498663	400.75	4.01%
	More Significantly Increase	>0.1	127995	102.86	1.22%	151227	121.53	1.22%
Operation Period	Significantly Decrease	<−0.01	9413	7.56	0.09%	4997	4.02	0.04%
	Slightly Decrease	−0.01–0	140109	112.60	1.34%	85721	68.89	0.69%
	Slightly Increase	0–0.01	9867593	7930.07	94.10%	11781944	9468.54	94.84%
	Significant Increase	0.01–0.02	464370	373.19	4.43%	544514	437.60	4.38%
	More Significantly Increase	<0.02	4622	3.71	0.04%	5796	4.66	0.05%

Within the construction period, the extent of NDVI damage in the core area was greater than that in the contrast area. However, the total trend of NDVI change was still overall slightly increasing, with the proportion accounting for 92.42% and 92.99% in core and contrast areas, respectively. Within the operating period, the NDVI showed an increasing trend as before. However, the proportion showing a slight increase became larger, accounting for 94.10% and 94.84% in the core and contrast areas, respectively.

Spatial differences in the increase or decrease in NDVI were observed along the GH road. Slight and significant decreases were dominant near the road and in urban areas (Figure 5a), while slight increases were dominant elsewhere. In particular, the increase was more significant in the mountain forest area (Figure 5b).

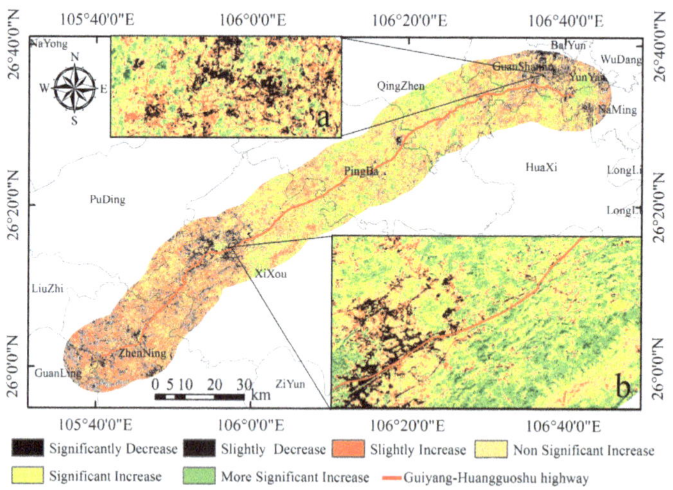

Figure 5. Spatial distribution of the trend of NDVI from 1986–2020. Trends were separated into the following classification: significantly decrease (<−0.005), slightly decrease (−0.005–0), slightly increase (0–0.005), non-significant increase (0.005–0.010), significantly increase (0.010–0.015), more significantly increase (>0.015).

3.2. Coefficient of Variation Analysis

The construction and operation of the GH freeway has had a negative impact on the stability of vegetation along the route. The fluctuations of NDVI in the construction period were greater than those in the operation period, and all fluctuations in the core area were greater than those in the contrast area. The NDVI, in terms of both periods and area were dominated by lower fluctuations, with lower fluctuations in the construction and operation periods of 845.12 km^2 and 993.80 km^2, respectively, and lower fluctuations in the core and contrast areas of 793.49 km^2 and 968.78 km^2, respectively. The high fluctuation of NDVI in the construction period were larger than those in the operation period, as well as were larger in the core than in the contrast area. The high fluctuation areas for the construction and operation periods were 280.64 km^2 and 121.98 km^2, respectively, while the high fluctuation sizes in the core and contrast areas were 225.54 km^2 and 177.08 km^2 respectively (Figure 6).

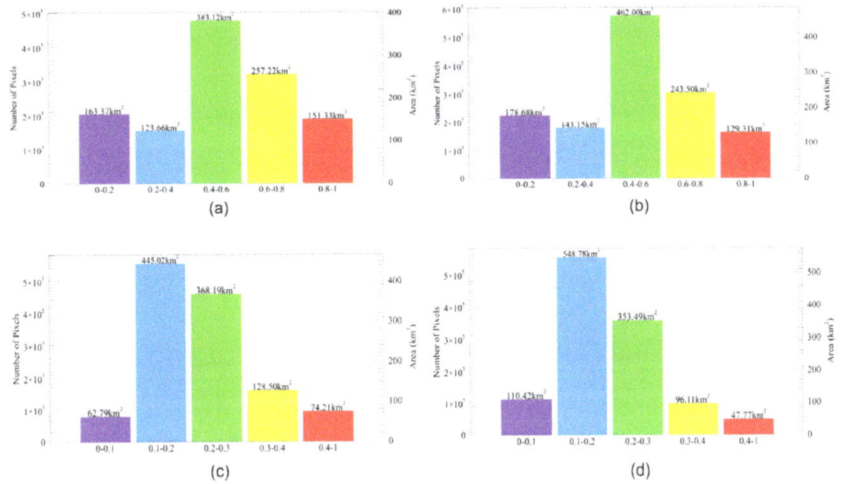

Figure 6. C_V for the core (**a**) and contrast (**b**) areas within the construction period divided into five levels; namely, low fluctuation (0–0.2), lower fluctuation (0.2–0.4), medium fluctuation (0.4–0.6), higher fluctuation (0.6–0.8) and high fluctuation (0.8–1). C_V in the core (**c**) and contrast (**d**) areas within the operating period was divided into five levels, namely, low fluctuation (0–0.1), lower fluctuation (0.1–0.2), medium fluctuation (0.2–0.3), higher fluctuation (0.3–0.4) and high fluctuation (0.4–1).

3.3. Correlation Analysis of NDVI with Temperature and Precipitation

The correlation coefficient between NDVI and precipitation from 1986 to 2020 was mainly concentrated between −0.45 and 0 (mean = −0.03), showing a low negative correlation, while, its correlation coefficient with air temperature mainly ranged from 0–0.5 (mean = 0.08), showing a low positive correlation. The mean partial correlation coefficients of NDVI with precipitation and temperature were 0.05 and 0.17, respectively, which were both less than 0.5 (Figure 7).

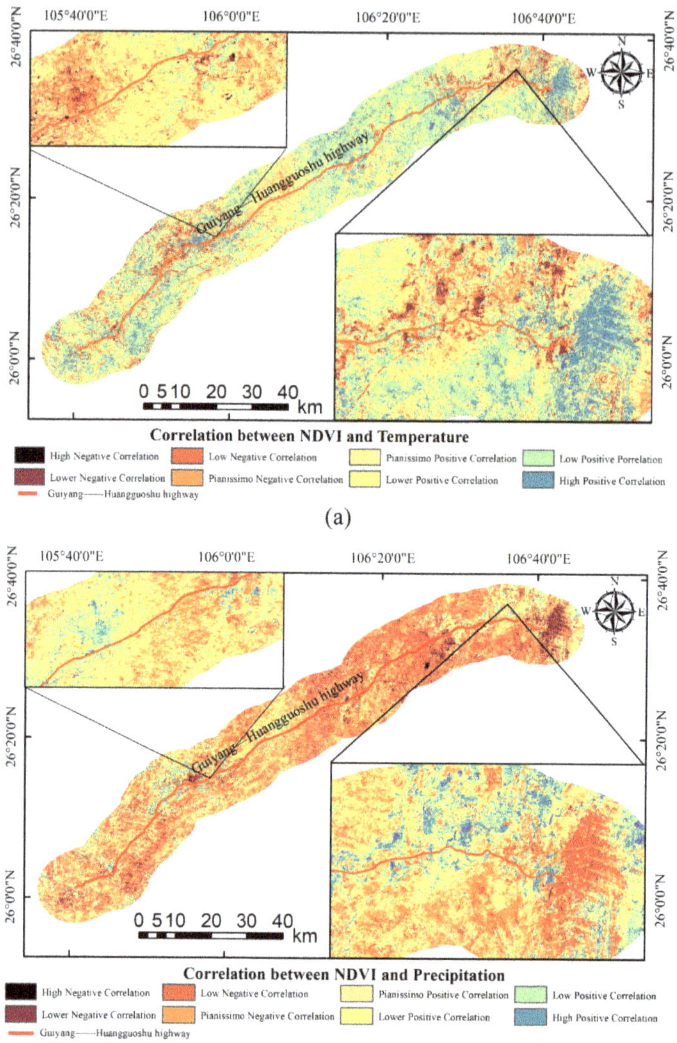

Figure 7. Correlation between temperature (**a**), precipitation (**b**), and NDVI from 1986 to 2020. The distribution range of correlation coefficients were divided into high negative correlation (<−0.6), lower negative correlation (−0.6–−0.4), low negative correlation (−0.4–−0.2), pianissimo negative correlation (−0.2–0), pianissimo positive correlation (0–0.2), lower positive correlation (0.2–0.4), low positive correlation (0.4–0.6) and high positive correlation (>0.6).

The correlation of NDVI with temperature was mainly low positive, while the correlation with precipitation is mainly low negative; however, the correlation between NDVI and precipitation was positive, while the correlation between NDVI and temperature was negative in where the underlying surface was artificial, especially in urban areas and along roads (Figure 7).

3.4. Analysis of Study Area LUCC

Wetlands were re-classified as water bodies before the calculations, due to their small size. The LUCC decreased by 0.39% for forest, 6.04% for cultivated land, 0.97% for water,

0.48% for shrub, and 0.97% for grass from 2000 to 2020, while the artificial cover increased by 8.67% (Table 2).

Table 2. LUCC Classification statistics from 2000 to 2020.

Classification	2000 LUCC		2020 LUCC	
	Area (km^2)	Percentage	Area (km^2)	Percentage
Cultivated Land	1070.71	45.82%	929.61	39.78%
Forest	510.85	21.86%	501.75	21.47%
Grass	338.67	14.49%	316.02	13.52%
Shrub	225.15	9.64%	214.16	9.16%
Water	64.46	2.76%	46.10	1.97%
Artificial Cover	126.94	5.43%	329.41	14.10%

The trend of NDVI decreased significantly along the GH highway (Figure 8), especially in urban areas along the route (Figure 8). The construction and operation of the GH highway and other later roads drove the development of towns along the route, leading to an expansion of artificial cover along the route and exacerbating the decline of NDVI (see Figure 5).

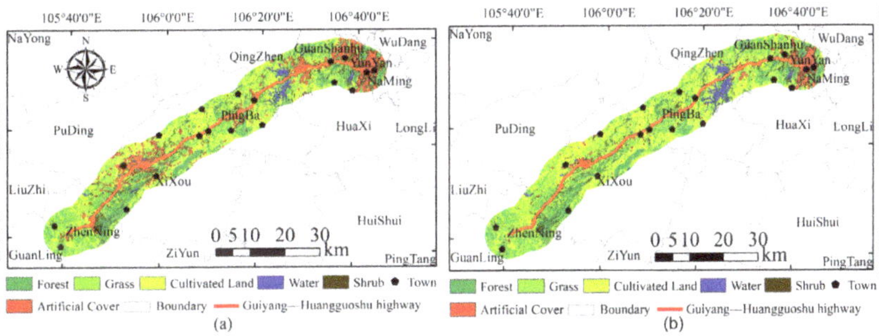

Figure 8. 2000 LUCC classification (**a**) and 2020 LUCC classification (**b**).

4. Discussion

4.1. Characteristics and Reasons for Change in NDVI during the Construction Period

Within the construction period, the annual average value and growth rate of the NDVI in the core area along the GH highway were smaller than those in the contrast area, and the coefficient of variation was larger than that in the contrast area, due to by the destruction of the original land-cover caused by the construction of the GH highway. The closer to the road, the greater the damage to the vegetation. This slowed the NDVI growth rate in the surrounding 4 km from the road, and breaks the environment for a certain distance along the line, thus increasing the variation in fluctuation of vegetation along the line. The fluctuation caused by highway construction activities on vegetation along the road were also larger closer to the road, indicating that the construction activities of GH highway had a negative effect on the stability of the surrounding environment. The construction and operation of the GH highway have increased the intensity of human activities in towns along the route, resulting in higher ecological fluctuations around the towns than other areas. The influence of the GH highway is mainly within 2 km, as the highway along the road are mainly within 2 km, making this the area with the strongest human activities. In the area far from the GH highway, the vegetation was weakly affected by the highway, and the heterogeneity of the surface was observed to have a greater impact on NDVI than the highway and human activities. In the early stage of construction (1987), the NDVI in both the area and the contrast areas declined sharply. This was due to the large-scale destruction of the surface vegetation in the early stage of the project construction, which led to a sharp decline in NDVI in this year. In addition, the rocky desertification was serious in this period, and there was a lack of relevant control work, leading to damage of the fragile environment [43]. The engineering construction in the core area can easily affect involve the vegetation cover in the contrast area, causing further damages.

4.2. Characteristics and Reasons for Change in NDVI during the Operation Period

Within the operation period, the overall NDVI in the core and contrast areas showed an upward trend, where the growth rate in the contrast area was greater than that in the core area; however the increase was small. Due to the implementation of environmental protection policy [44], the vegetation being slowly restored by regeneration, afforestation, and returning farmland to forest. The NDVI along the GH highway rose to a high value and the change is relatively stable in 2000. However, after 2000, the NDVI showed a downward trend. During the nearly 30 years of the operation period, human activities along the GH highway have substantially enhanced, as is reflected by an increase in artificial surfaces along the highway and the construction and widening of other roads along the highway. In addition, the NDVI showed an obvious downward trend in 2008 and 2009, related to the severe snow disaster in southwest China in the winter of 2008. A certain range of vegetation died due to freezing, which affected the maximum NDVI in the following year. The same significant decrease also occurred from 2010 to 2012, which was related to the destruction of vegetation on the original surface due to the construction of the Shanghai-Kunming highway in this section in 2010, and the severe drought in Guizhou in 2012, which resulted in a decline in and reduced growth of NDVI from 2010 to 2012. Furthermore, the NDVI decreased significantly in 2019–2020. Meteorological bureau data (http://gz.cma.gov.cn/ access on 1 October 2022) indicated that a spring drought occurred in the spring in Central Guizhou during 2019–2020. This spring drought caused the vegetation to be short of water during the growth period, leading to inhibition of vegetative growth throughout the year, thus significantly reducing the NDVI value over this period.

The construction and operation of GH highway has increased the intensity of human activities in towns along the line, these activities Include change of cropland area along the GH highway will also affect the vegetation [45], causing the fluctuation of NDVI along the GH highway to be higher than that in other regions. The construction and operation of the GH highway mainly affected the area within 2 km, where a large number of villages, towns, cities, and trunk roads are concentrated. These areas are characterized by strong human activities. In the area far from the GH highway, the vegetation is less affected by the highway, and the heterogeneity of the karst surface becomes the main factor affecting the NDVI. Although the impact of human engineering construction on the environment is not unique to karst, the heterogeneity, vulnerability, and sensitivity of the karst surface are strong [46]. Therefore, the impact of human activities tends to cause greater damage in karst areas. In recent decades, the frequency and intensity of extreme climate events have increased, and the impact of high temperatures and drought on the productivity of the mid-latitude ecosystem in the Northern Hemisphere has become greater and greater, leading to a more sensitively responsive of NDVI [47]. Over the past three decades, the reduction of precipitation and the increase in temperature have led to increased evaporation. The soil moisture in many areas has decreased [48], while the sensitivity of vegetation to soil moisture has generally increased [49]. In particular, soil moisture restricts karst ecological restoration [50]. The study area is a typical karst area, with thin soil layer, weak soil water holding capacity, many underground rivers, strong surface water infiltration, and easy soil water loss. Under the trend of long-term temperature rise and precipitation decrease, soil moisture evaporation will inevitably be intensified, leading to soil moisture reduction, which will further affect the growth and development of vegetation along the line. The enhancement of human activities along the GH highway will also lead to intensification of soil erosion, especially with expansion of cities along the highway. The construction of other highways will increase the sensitivity of vegetation in artificial surface areas to drought, resulting in water shortages during the peak growth season, ultimately leading to declining annual biomass.

4.3. Correlation between Climatic Factors and NDVI

Consistent with previous research, the correlation coefficient between NDVI and precipitation along the GH highway showed an insignificant negative correlation. During the nearly 30 years of the operation period, the slight decrease in total precipitation did not cause major a catastrophe for vegetation growth in the Southwest China; instead, the change in precipitation frequency made local rain recruitment more frequent, which partly compensated for the growth of southwestern vegetation being limited by the alternating time of dry and wet periods, rather than total precipitation [51]. At the same time, the partial correlation coefficients of NDVI with precipitation and temperature were 0.05 and 0.17, respectively; both were less than 0.5, which is basically consistent with previous research [29]. The proportion of karst landform in the study area is high (76.5%); together with the large change of karst underlying surface and the high degree of topographic relief, the spatial heterogeneity of temperature within a small range may be high. This results in temperature being a major factor controlling vegetative growth over a small range. A previous study has shown that terrain is generally a covariate of temperature, which is highly consistent with temperature change. Therefore, we did not consider terrain factors as covariates to participating in the partial correlation analysis [29].

Similar to previous studies, the correlation between NDVI and precipitation was weakly positive on the artificial surface, while the correlation between NDVI and temperature was low and negative on artificial surfaces [30]. Human activities are strong in artificial surface areas, and the environment is more fragile than in other areas. The low precipitation infiltration of hardened surface exacerbates water shortages and temperature increases, which may force the growth of vegetation to be slow. The surrounding ecology is fragile. The increase of precipitation makes the water supply needed for the growth of regional vegetation sufficient, thus reducing the vulnerability, while artificial surface vegetation is more sensitive to drought.

5. Conclusions

In this paper, NDVI and climate data were used to analyze the influence of the GH highway on the area within 8 km of the route. It was found that the annual mean and growth rate of NDVI in the core area within the construction and operation periods of the GH highway were smaller than those in the contrast area, the inter-annual variation fluctuated greatly, and the influence on the area was mainly within 2 km of the GH highway. Within the operation period, the NDVI reached a peak and then decreased slightly. Within the construction and operation period, the NDVI along the route increased overall, precipitation showed a downward trend, and temperature showed an upward trend. The correlation between NDVI and climate factors indicated that the correlation between NDVI and temperature is stronger than that between NDVI and precipitation. The influence of LUCC on NDVI was mainly manifested as an increase in artificial cover surface and the decline of other land-use types, resulting in the change of NDVI.

6. Limitations and Prospects

In this study, only the inter-annual variability of climate variables in response to NDVI was considered, seasonal variation of climate indicators was not considered; furthermore, only the annual maximum value of Landsat was used to synthesize the NDVI images, and multi-source remote sensing data fusion methods were not considered. In the future, multi-source remote sensing can be used to explore the corresponding relationship between NDVI and other climatic factors (e.g., surface soil humidity, evaporation, seasonal drought, and so on). Remote sensing data under nighttime lighting can also be adopted, in order to explore the correlation between vegetation cover and human activities.

Author Contributions: Conceptualization, Y.W. and S.L.; methodology, G.L.; software, L.G. and J.Y.; formal analysis, C.G. and F.Y.; investigation, L.G., J.Y., Z.S. and X.Y.; data curation, Y.C., H.P. and X.Y.; writing—original draft preparation, Y.W.; writing—review and editing, C.G. and S.L.; visualization, L.G.; supervision, Y.X.; project administration, G.L.; funding acquisition, Y.W. All authors have read and agreed to the published version of the manuscript.

Funding: This research was funded by Guizhou Provincial Science and Technology Projects (ZK [2022] YB334) and Doctoral Program of Guizhou Education University (X2021049).

Data Availability Statement: Not applicable.

Acknowledgments: We thank the anonymous reviewers for their valuable comments. We gratefully acknowledge the design of S.L. and the contribution of co-authors.

Conflicts of Interest: The authors declare no conflict of interest.

References

1. Zhang, Y.; Wang, X.; Li, C. NDVI dynamics under changing meteorological factors in a shallow lake in future metropolitan, semiarid area in North China. *Sci. Rep.* **2018**, *8*, 15971. [CrossRef] [PubMed]
2. Zhang, S.R.; Bai, X.Y.; Zhao, C.W.; Tan, Q.; Luo, G.J.; Wu, L.H.; Xi, H.P.; Li, C.J.; Chen, F.; Ran, C.; et al. China's carbon budget inventory from 1997 to 2017 and its challenges to achieving carbon neutral strategies. *J. Clean. Prod.* **2022**, *347*, 130966. [CrossRef]
3. Li, C.J.; Bai, X.Y.; Tan, Q.; Luo, G.J.; Wu, L.H.; Chen, F.; Xi, H.P.; Luo, X.L.; Ran, C.; Chen, H.; et al. High-resolution mapping of the global silicate weathering carbon sink and its long-term changes. *Glob. Chang. Biol.* **2022**, *28*, 4377–4394. [CrossRef]
4. Du, C.C.; Bai, X.Y.; Li, Y.B.; Tan, Q.; Zhao, C.W.; Luo, G.J.; Wu, L.H.; Chen, F.; Li, C.J.; Ran, C.; et al. Inventory of China's net biome productivity since the 21 st century. *Land* **2022**, *11*, 1244. [CrossRef]
5. Wu, Y.Y.; Liu, L.B.; Guo, C.Z.; Zhang, Z.H.; Hu, G.; Ni, J. Low carbon storage of woody debris in a karst forest in southwestern China. *Acta Geochim.* **2019**, *38*, 576–586. [CrossRef]
6. Shi, S.Y.; Yu, J.G.; Wang, F.; Wang, P.; Zhang, Y.C.; Jin, K. Quantitative contributions of climate change and human activities to vegetation changes over multiple timescales on the Loess Plateau. *Sci. Total Environ.* **2020**, *755*, 142419. [CrossRef]
7. Zhang, M.; Wu, X.Q. The rebound effects of recent vegetation restoration projects in Mu Us Sandy land of China. *Ecol. Indic.* **2020**, *113*, 106228. [CrossRef]
8. Liu, Y.; Li, Y.; Li, S.C.; Motesharrei, S. Spatial and temporal patterns of global NDVI trends: Correlations with climate and human factors. *Remote Sens.* **2015**, *7*, 13233–13250. [CrossRef]
9. Feng, K.; Wang, T.; Liu, S.; Yan, C.; Kang, W.; Chen, X.; Guo, Z. Path analysis model to identify and analyse the causes of aeolian desertification in Mu Us Sandy Land, China. *Ecol. Indic.* **2021**, *124*, 107386. [CrossRef]
10. Liu, X.; Pei, F.; Wen, Y. Global urban expansion offsets climate-driven increases in terrestrial net primary productivity. *Nat. Commun.* **2019**, *10*, 5558. [CrossRef] [PubMed]
11. Wang, C.G.; Zhao, H.R. The assessment of urban environment in watershed scale. *Procedia Environ. Sci.* **2016**, *36*, 169–175. [CrossRef]
12. Liu, D.; Chen, J.; Ou, Y.Z. Responses of landscape structure to the ecological restoration programs in the farming-pastoral ecotone of Northern China. *Sci. Total Environ.* **2020**, *710*, 136311. [CrossRef] [PubMed]
13. Sun, Z.; Mao, Z.; Yang, L.; Liu, Z.; Han, J.; Wang, H.; He, W. Impacts of climate change and afforestation on vegetation dynamic in the Mu U: Desert. *China. Ecol. Indic.* **2021**, *129*, 108020. [CrossRef]
14. Bégué, A.; Vintrou, E.; Ruelland, D.; Claden, M.; Dessay, N. Can a 25-year trend in Soudano-Sahelian vegetation dynamics be interpreted in terms of land use change? A remote sensing approach. *Glob. Environ. Chang.* **2011**, *21*, 413–420. [CrossRef]
15. Huang, Q.H.; Cai, Y.L. Spatial pattern of karst rock desertification in the middle of Guizhou Province, Southwestern China. *Environ. Geol.* **2007**, *52*, 1325–1330. [CrossRef]
16. Liu, M.; Bai, X.; Tan, Q.; Luo, G.; Zhao, C.; Wu, L.; Luo, X.; Ran, C.; Zhang, S. Climate Change Enhances the Positive Contribution of Human Activities to Vegetation restoration in China. *Geocarto Int.* **2022**, 1–24. [CrossRef]
17. Guo, B.; Zhou, Y.; Wang, S.X.; Tao, H.P. The relationship between normalized difference vegetation index (NDVI) and climate factors in the semiarid region: A case study in Yalu Tsangpo river basin of Qinghai-Tibet Plateau. *J. Mt. Sci.* **2014**, *11*, 926–940. [CrossRef]
18. Xia, C.F.; Li, J.; Liu, Q.H. Review of advances in vegetation phenology monitoring by remote sensing. *Nat. Remote Sens. Bull.* **2013**, *17*, 1–16. (In Chinese) [CrossRef]
19. Guo, K.M.; Qiu, T.; Zong, L.L.; Lu, H.X.; Yao, W. Change of NDVI in Hexi region and its response to temperature and precipitation. *Sci. Surv. Mapp.* **2021**, *46*, 83–89. (In Chinese) [CrossRef]
20. Nemani, R.R.; Keeling, C.D.; Hashimoto, H.; Jolly, W.M.; Piper, S.C.; Tucker, C.J.; Myneni, R.B.; Running, S.W. Climate-driven increases in global terrestrial net primary production from 1982 to 1999. *Science* **2003**, *300*, 1560–1563. [CrossRef]
21. Braswell, B.H.; Schimel, D.S.; Linder, E.; Moore, B. The response of global terrestrial ecosystems to interannual temperature variability. *Science* **1997**, *278*, 870–873. [CrossRef]

22. Justice, C.O.; Townshend, J.R.G.; Holben, B.N.; Tucker, C.J. Analysis of the phenology of global vegetation using meteorological satellite data. *Int. J. Remote Sens.* **1985**, *6*, 1271–1318. [CrossRef]
23. Myneni, R.B.; Keeling, C.D.; Tucker, C.J.; Asrar, G.; Nemani, R.R. Increased plant growth in the northern high latitudes from 1981 to 1991. *Nature* **1997**, *386*, 698–702. [CrossRef]
24. Ramírez-Cuesta, J.M.; Minacapilli, M.; Motisi, A.; Consoli, S.; Intrigliolo, D.S.; Vanella, D. Characterization of the main land processes occurring in Europe (2000–2018) through a MODIS NDVI seasonal parameter-based procedure. *Sci. Total Environ.* **2021**, *799*, 149346. [CrossRef]
25. Fokeng, R.M.; Fogwe, Z.N. Landsat NDVI-based vegetation degradation dynamics and its response to rainfall variability and anthropogenic stressors in Southern Bui Plateau, Cameroon. *Geosyst. Geoenviron.* **2022**, *1*, 100075. [CrossRef]
26. Prăvălie, R.; Sîrodoev, I.; Nita, I.A.; Patriche, C.; Dumitraşcu, M.; Roşca, B.; Tişcovschi, A.; Bandoc, G.; Săvulescu, I.; Mănoiu, V.; et al. NDVI-based ecological dynamics of forest vegetation and its relationship to climate change in Romania during 1987–2018. *Ecol. Indic.* **2022**, *136*, 108629. [CrossRef]
27. Ren, R.Y.; Hr, Z.H.; Liang, H.; Xia, C.H.; Zhang, L.; Yang, M.K. Spatiotemporal variation of NDVI and its response to changes in temperature and precipitation in Guizhou Province. *Bull. Soil Water Conserv.* **2021**, *28*, 118–129. (In Chinese) [CrossRef]
28. Jiang, L.G.; Liu, Y.; Wu, S.; Yang, C. Analyzing environment change and associated driving factors in China based on NDVI time series data. *Ecol. Indic.* **2021**, *129*, 107933. [CrossRef]
29. Liu, W.; Jiao, S.L.; Li, Y.G.; Mo, Y.S.; Zhao, Z.Q.; Zhang, J.; Zhao, M. Analysis on the correlation between vegetation cover of land surface and climatic factors in karst area. *Bull. Soil Water Conserv.* **2021**, *28*, 203–215. (In Chinese) [CrossRef]
30. Zhang, Y.R.; Zhou, Z.F.; Ma, S.B.; Zhang, Y.H. A study on response of vegetation to climate change based on NDVI in karst region-a case study at Liupanshui city in Guizhou Province. *Bull. Soil Water. Conserv.* **2014**, *34*, 114–117+2. (In Chinese) [CrossRef]
31. Yang, C.; Xing, Y.Q.; Ma, C. The response of NDVI along the QT railway to human activities and climate change. *Sci. Surv. Map.* **2022**, *47*, 137–145. (In Chinese) [CrossRef]
32. Hao, A.H.; Duan, H.C.; Wang, X.F.; Zhao, G.H.; You, Q.G.; Peng, F.; Du, H.Q.; Liu, F.Y.; Li, C.Y.; Lai, C.M.; et al. Different response of alpine meadow and alpine steppe to climatic and anthropogenic disturbance on the Qinghai-Tibetan Plateau. *Glob. Ecol. Conserv.* **2021**, *27*, e01512. [CrossRef]
33. He, P.; Xu, L.H.; Liu, Z.C.; Jing, Y.D.; Zhu, W.B. Dynamics of NDVI and its influencing factors in the Chinese Loess Plateau during 2002–2018. *Reg. Sustain.* **2021**, *2*, 36–46. [CrossRef]
34. Gao, W.D.; Zheng, C.; Liu, X.H.; Lu, Y.D.; Chen, Y.F.; Wei, Y.; Ma, Y.D. NDVI-based vegetation dynamics and their responses to climate change and human activities from 1982 to 2020: A case study in the Mu Us Sandy Land, China. *Ecol. Indic.* **2022**, *137*, 108745. [CrossRef]
35. Xu, Y.; Huang, W.T.; Lu, M.Y.; Ou, Y.X.; Zhang, Z.Y.; Li, M.J.; Guo, Z.D.; Ma, R.X. Vegetation Cover Change and the Relative Role of Climate Change and human Activities in Southwest Karst Areas. *Bull. Soil Water Conserv.* **2022**, *29*, 292–299. (In Chinese) [CrossRef]
36. Huang, Z. The first high-grade road in Guizhou, the construction process of Guiyang Huangguoshu highway. *Gui Lite Hist.* **2020**, *6*, 12–16. (In Chinese)
37. Chen, K.S.; Yin, Y.; Wu, J. Investigation and Analysis of Slope Protection on Guiyang-Huangguoshu Expressway. *J. Guizhou Univ. Nat. Sci.* **2011**, *45*, 120–123. (In Chinese) [CrossRef]
38. Holben, B.N. Characteristics of maximum-value composite images from temporal AVHRR data. *Int. J. Remote Sens.* **1986**, *7*, 1417–1434. [CrossRef]
39. Guo, B.; Zhang, J.; Meng, X.; Xu, T.; Song, Y. Long-term spatio-temporal precipitation variations in China with precipitation surface interpolated by ANUSPLIN. *Sci. Rep.* **2020**, *10*, 81. [CrossRef]
40. Price, D.T.; McKenney, D.W.; Nalder, I.A.; Hutchinson, M.F.; Kesteven, J.L. A comparison of two statistical methods for spatial interpolation of Canadian monthly mean climate data. *Agric. For. Meteorol.* **2000**, *101*, 81–94. [CrossRef]
41. Jun, C.; Ban, Y.; Li, S. Open access to Earth land-cover map. *Nature* **2014**, *514*, 434. [CrossRef]
42. Fensholt, R.; Proud, S.R. Evaluation of Earth Observation based global long term vegetation trends—Comparing GIMMS and MODIS global NDVI time series. *Remote Sens. Environ.* **2012**, *119*, 131–147. [CrossRef]
43. Wu, L.; Wang, S.; Bai, X.; Chen, F.; Li, C.; Ran, C.; Zhang, S. Identifying the Multi-Scale Influences of Climate Factors on Runoff Changes in a Typical Karst Watershed Using Wavelet Analysis. *Land* **2022**, *11*, 1284. [CrossRef]
44. Chen, F.; Bai, X.; Liu, F.; Luo, G.; Tian, Y.; Qin, L.; Li, Y.; Xu, Y.; Wang, J.; Wu, L.; et al. Analysis Long-Term and Spatial Changes of Forest Cover in Typical Karst Areas of China. *Land* **2022**, *11*, 1349. [CrossRef]
45. Song, F.; Wang, S.; Bai, X.; Wu, L.; Wang, J.; Li, C.; Chen, H.; Luo, X.; Xi, H.; Zhang, S. A New Indicator for Global Food Security Assessment: Harvested Area Rather Than Cropland Area. *Chin. Geogr. Sci.* **2022**, *32*, 204–217. [CrossRef]
46. Xiong, L.; Bai, X.Y.; Zhao, C.W.; Li, Y.B.; Tan, Q.; Luo, G.J.; Wu, L.H.; Chen, F.; Li, C.J.; Chen Ran, C.; et al. High-resolution datasets for global carbonate and silicate rock weathering carbon sinks and their change trends. *Earth's Future* **2022**, *10*, e2022EF002746. [CrossRef]
47. Jiao, W.; Wang, L.; Smith, W.K.; Chang, Q.; Wang, H.; D'Odorico, P. Observed increasing water constraint on vegetation growth over the last three decades. *Nat. Commun.* **2021**, *12*, 3777. [CrossRef] [PubMed]
48. Deng, Y.H.; Wang, S.J.; Bai, X.Y.; Luo, G.J.; Wu, L.H.; Chen, F.; Wang, J.F.; Li, Q.; Li, C.J.; Yang, Y.J.; et al. Characteristics of soil moisture storage from 1979 to 2017 in the karst area of China. *Geocarto Int.* **2021**, *36*, 903–917. [CrossRef]

49. Gampe, D.; Zscheischler, J.; Reichstein, M.; O'Sullivan, M.; Smith, W.K.; Sitch, S.; Buermann, W. Increasing impact of warm droughts on northern ecosystem productivity over recent decades. *Nat. Clim. Chang.* **2021**, *11*, 772–779. [CrossRef]
50. Li, W.; Migliavacca, M.; Forkel, M.; Denissen, J.M.C.; Reichstein, M.; Yang, H.; Duveiller, G.; Weber, U.; Orth, R. Widespread increasing vegetation sensitivity to soil moisture. *Nat. Commun.* **2022**, *13*, 3959. [CrossRef] [PubMed]
51. Luo, L.; Ma, W.; Zhuang, Y.; Zhang, Y.; Yi, S.; Xu, J.; Long, Y.; Ma, D.; Zhang, Z. The impacts of climate change and human activities on alpine vegetation and permafrost in the Qinghai-Tibet Engineering Corridor. *Ecol. Indic.* **2018**, *93*, 24–35. [CrossRef]

land

MDPI

Article

Evolution of Land Use Landscape Patterns in Karst Watersheds of Guizhou Plateau and Its Ecological Security Evaluation

Yue Li * and Huacai Geng

College of Public Administration, Guizhou University of Finance and Economics, Guizhou 550025, China
* Correspondence: liyue@mail.gufe.edu.cn

Abstract: Landscape ecological safety is of great significance in maintaining ecological balance, ecological protection, economic development, and promoting the sustainable use of regional land resources. This study collects three-phase remote sensing (RS) image data of 2000, 2010, and 2020 to elucidate the spatial and temporal changes in land use of the Guizhou Plateau Karst Watershed. We construct a landscape ecological security index using the ArcGIS and landscape pattern index method. With the spatial autocorrelation theory, we analyze the evolution of watershed landscape patterns and changes in characteristics of ecological security. The results show that the cultivated land is being converted to construction land in the watershed from 2000 to 2020. The percentage of the patch area of the constructed land is increasing. However, the patch cohesion of the cultivated land, grassland, and watershed is decreasing. The ecological safety of the Nanming River Basin landscape is in a positive trend, clustering in the central urban areas. The growth rate of the landscape ecological safety index increased by 5.80% from 2000 to 2020, and the aggregation of the ecological safety index was dominated by high–high aggregation and low–low aggregation, which was spatially positively correlated with the spatial aggregation effect. The findings provide a scientific reference for managing ecological balance and optimizing the land resource allocation in karst watersheds.

Keywords: karst; land use; landscape patterns; ecological safety

Citation: Li, Y.; Geng, H. Evolution of Land Use Landscape Patterns in Karst Watersheds of Guizhou Plateau and Its Ecological Security Evaluation. *Land* **2022**, *11*, 2225. https://doi.org/10.3390/land11122225

Academic Editor: Rui Alexandre Castanho

Received: 26 October 2022
Accepted: 4 December 2022
Published: 7 December 2022

1. Introduction

Landscape patterns are the spatial characteristics and structural composition of landscape elements that determine the environment, distribution, and composition of resources [1]. Landscape patterns play a vital role in understanding the ecological processes of a region and in evaluating and optimizing ecological security [2,3]. For decades, a direct influence has been observed between the landscape pattern's evolution and the ecosystem [4], bringing significant changes to the quality of the landscape ecosystem and land use patterns of natural and anthropogenic activities [5,6]. Landscape ecological security is a subsystem of land resource security [7], which is crucial to national and regional development and construction. Landscape ecological safety has been a new issue facing human society's long-term development since 2001. The Southwest Karst Region, in the central part of the Guizhou Plateau, is the largest and most populous continuous karst ecologically vulnerable area in the world [8]. This region is a hotspot and a key area for global climate change research. It also serves as a scientific research paradigm in the world and enhances an understanding of the comprehensive governance of degraded ecosystems [9,10]. However, the non-agricultural population is increasingly being concentrated in the region owing to recent large-scale urbanization, the expansion of industrial and commercial space, rapid changes in urban land use, and the consequent evolution of landscape patterns. Thus, a change in the landscape pattern can improve the urban green landscape layout [11], since the landscape structure is gradually showing strong rapid urbanization [12,13]. Moreover, the anthropogenic pressure and the transformation of regional ecosystems are increasing. Unreasonable human activities are also causing serious ecological and environmental problems [14,15], making ecological security face great challenges [16,17]. Thus, adopting a scientific and reasonable research method is critical in conducting an in-depth analysis of the ecological safety of the karst landscape.

Presently, ecological safety has become a research hot spot globally [18–24]. Studies have evaluated domestic and international landscape ecological safety, the ecological risk of the hotspot regional landscape, and the safety of urban clusters [25], coastal zones [26], and plateau mountains [27]. Ghosh et al. used the DEMATHE-ANP model to evaluate the ecological safety of the Kolkata Metropolitan Area in India [28]. Jiang et al. developed landscape classification on remote sensing (RS) images of the core region of Lijiang City and calculated the landscape pattern index as a driving factor [29]. From the perspectives of landscape patterns at the domestic and international levels, previous studies have established a solid conceptual base and methodological reference for exploring the evolutionary characteristics of regional landscape patterns and changes in ecological security patterns. Moreover, studies have explored landscape patterns and ecological security at a large scale, focusing on lakes [30], cultivated lands [31], and wetlands [32] in karst areas. Ren et al. used geographic information systems (GIS) and RS technology to analyze the spatial granularity effect of landscape patterns and identify the suitable spatial granularity of karst mountainous urban landscapes. However, these authors only explored a single landscape pattern index, which could not detect the regional ecological security status [33]. Liu et al. used the InVEST (Integrated Valuation of Ecosystem Services and Tradeoffs) model to explore the spatial and temporal variation characteristics of habitat quality in the Chishui River Basin and its coupling relationship with the landscape pattern [34]. Wang et al. constructed a landscape security index using the ArcGIS software and a landscape pattern index to study the evolution of landscape patterns in trough valley areas. These authors also used this software to explore the spatial and temporal divergence patterns in ecological security [35]. Peng et al. used the landscape ecological security theory to develop an evaluation model and understand the ecological security of cultivated landscapes in the karst mountains. These authors also used this theory to understand the direction of ecological security transfer and driving factors of cultivated lands in the karst mountains [36]. However, most of these studies only based their investigation on short-term time series data, with a lag in data updating. Studies that used long time-series data to evaluate ecological safety are scanty. Furthermore, non-karst places have been the primary focus of research on patterns of landscape ecological safety at large-scale levels, such as regions, watersheds, and municipal territories. As a significant ecologically sensitive territory in China, Guizhou is also one of the most extensive karst landscapes in China, and overcoming its ecological and environmental problems is the key to solving earth system science, which can assist in the promotion of the construction of ecological civilization in China and even worldwide. The Nanming River basin is a tributary of the Wu River in the Yangtze River system, and more than 90% of the basin's total area is comprised of karst landscapes; it is a crucial ecological barrier in the upper reaches of the Yangtze River and an essential cornerstone of ecological civilization development. However, due to the relative fragility of the watershed ecosystem, it is vulnerable to the effects of urbanization. With the expansion of urbanization, high-density human activities, land development, water quality degradation, and other issues are altering the land use type, landscape pattern, and ecological mechanism of the watershed, posing a significant threat to the ecological security of the watershed landscape, and coordinating the relationship between ecological preservation and utilization in the study area is a significant issue at now. Therefore, to effectively sustain a well-functioning ecosystem in the study region in the future, a scientific evaluation of the evolution of the land use landscape pattern and its ecological safety is essential.

Based on 2000, 2010, and 2020 RS image data, this study used the theory of spatial autocorrelation, GIS spatial analysis, and landscape pattern index to elucidate the characteristics of spatial and temporal changes in land use in the watershed. The study also reveals the evolution in landscape pattern in the watershed since 2001, analyzes the spatial and temporal changes of ecological security, and makes a scientific evaluation and diagnosis of ecological security. It aims to realize quantitative analysis and visualization of the dynamic evaluation of landscape patterns and ecological security in the study area from the spatial and temporal scales, disclose the evolution characteristics of the landscape pattern and the logic of ecological security pattern in the karst watershed under the human–land relationship, and provide scientific basis and advice for the sustainable development, proper development planning, scientific ecological planning and construction of the karst watershed. It also provides data references for maintaining ecological balance and optimizing land resource allocation and control in other karst areas of the same resource type in China as well as scientific and practical references for expanding international research on karst landscapes and ecological security.

2. Materials and Methods

2.1. Study Area

The Nanming River Basin is a tributary of the Wu River in the Yangtze River system. This Nanming River Basin is an important part of the Yangtze River Economic Belt, which is located at the social, economic, and cultural center of Guizhou Province of 26°15′–26°54′ N and 106°26′–107°15′ E. The watershed covers approximately 2158 km^2, and it is characterized by a subtropical monsoonal humid climate with an annual mean precipitation of 1200 mm. Meanwhile, its topographic is high in the southwest and low in the northeast, with an average slope drop of approximately 3.44. The karst landscape in the watershed is extremely developed; its soil is dominated by rice soil, limestone soil, and loam. The karst landscape comprises 93.17% of the study area (Figure 1).

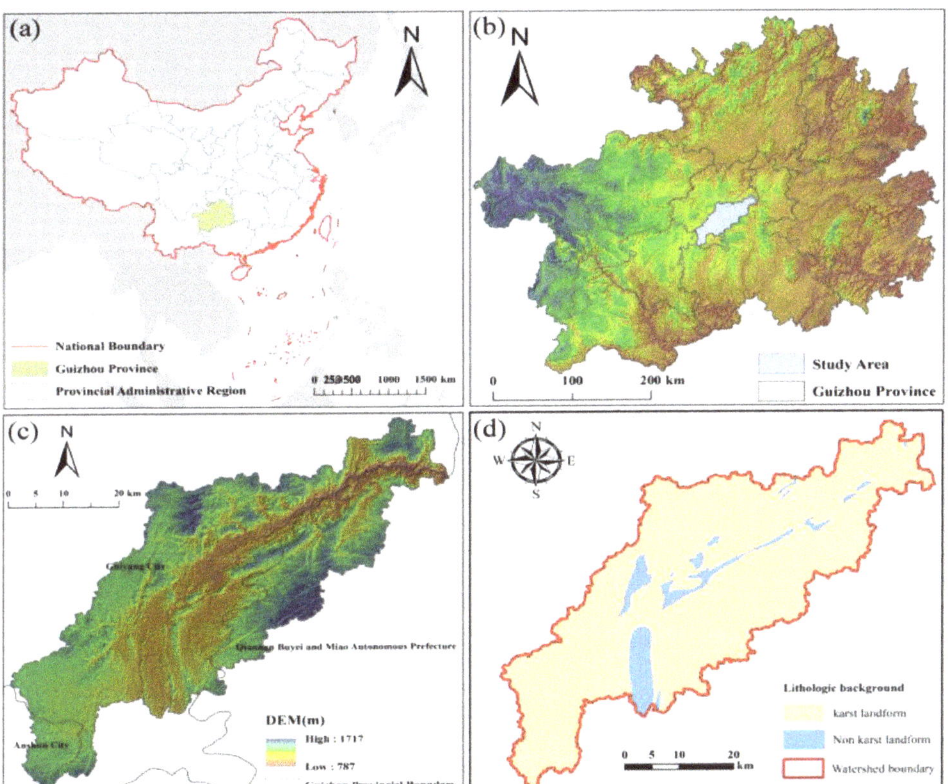

Figure 1. The basic information of the study area. Note: (**a**) shows location of Guizhou in China. (**b**) shows location of the study area in Guizhou. (**c**) shows the elevation of the study area. (**d**) shows the lithologic background.

2.2. Methods and Data Sources

2.2.1. Landscape Pattern Index Selection

The study combined previous studies that explored landscape characteristics of the basin [37–39], Patch Density (PD), and Patch Cohesion Index (COHESION) from the patch type level to reflect the degree of patch fragmentation in different landscape types. We also selected the Largest Patch Index (LPI) and Percentage of Landscape (PLAND) to identify dominant landscapes. The Patch Density (PD) and Contagion (CONTAG) were selected from the landscape level to reflect the degree of landscape fragmentation in the study area. Then, the Shannon Evenness Index (SHEI) and Shannon Diversity Index (SHDI) were selected to reflect the degree of landscape type diversification. The calculation formula and ecological significance of each index are detailed in the literature [40,41].

2.2.2. Moving Window Method

The moving window method in Fragstats 4.2 software was used to generate the landscape index raster map, and 1500 m was selected as the moving window radius after several calculations and comparisons [42].

2.2.3. Determination of Evaluation Unit

Considering the scope of the Nanming River watershed and sampling workload, referring to existing studies [43], the watershed was divided into 2 km × 2 km evaluation units as ecological safety evaluation plots, with a total of 640 sampling areas. Based on this, the landscape ecological safety index of each plot was calculated separately (Figure 2).

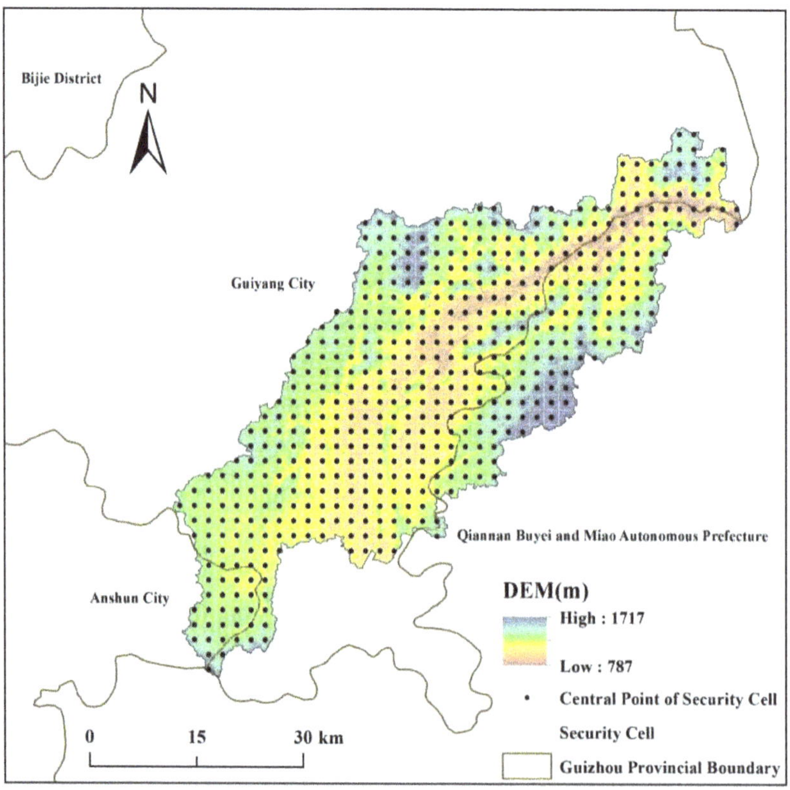

Figure 2. Sampling areas for ecological security assessment of landscape pattern in the study area.

2.2.4. Construction of the Landscape Ecological Security Index

The ecological safety index (ESI_k) of the land use landscape was calculated based on the landscape disturbance index and vulnerability index [44]. The equations are as follows:

$$ESI_k = \sum_{i=1}^{n} \frac{A_{ki}}{A_k} \times (1 - 10 \times LDI_I \times LVI_i) \quad (1)$$

$$LDI_i = aC_i + bH_i + cF_i \quad (2)$$

where ESI_k in (1) is the landscape ecological safety index of the k-th evaluation unit, and a larger ESI_k indicates a higher degree of ecological safety in the landscape and vice versa; LDI_i is the index of landscape disturbance; and LVI_i is the fragility index, based on the results of a previous study [45]. Each land use landscape fragility is specifically set to 5 levels: the value of constructed land is 1; a forest is 2; grassland is 3; cultivated land is 4, and water is 5. In this case, n is the number of landscape types; k is the number of evaluation units; A_{ki} is the area of the class i landscape of the k-th evaluated cell; and A_k is the total area of the k-th evaluation cell. In (2), C_i is landscape fragmentation, and H_i indicates the diversity index. F_i denotes the number of watershed landscape sub-dimensions; a, b, and c are the weights of C_i, H_i, and F_i, respectively, which are assigned to 0.5, 0.3, and 0.2, following the existing research results and the conditions in the study area [46].

2.2.5. Spatial Autocorrelation Analysis

We used spatial autocorrelation analysis to detect the spatial agglomeration of regional geographical phenomena [47]. Global Moran's I index was used to measure the overall spatial agglomeration characteristics of landscape ecological safety. The local spatial extent between regions was measured using the local Moran's I index [48,49]. The equations are as follows:

$$\text{Global Moran's I} = \frac{n \sum_{i=1}^{n} \sum_{j=1}^{m} W_{ij}(x_i - \overline{x})(x_j - \overline{x})}{\sum_{i=1}^{n} \sum_{j=1}^{m} W_{ij} \sum_{i=1}^{n} (x_i - \overline{x})^2} \quad (3)$$

$$\text{Local Moran's I}_i = \frac{(n-1)(x_i - \overline{x}) \sum_{j=1}^{n} W_{ij}(x_j - \overline{x})}{\sum_{j=1}^{n} W_{ij}(x_j - \overline{x})^2} \quad (4)$$

where n is the number of grids; $\overline{x}$ is the average vulnerability in the study area; x_i and x_j are the attribute values of the i-th, j-th raster, respectively ($i \neq j$), where i = 1, 2, 3, ..., n; j = 1, 2, 3, ..., m; W_{ij} is the weight value; W_{ij} = 1 when i and j are adjacent, and W_{ij} = 0 when they are far apart.

2.3. Data Sources

Landsat TM satellite 2000 images, 2010 images, and Landsat 8 satellite images of 2020 were used as the base data, which were obtained from the Geospatial Data Cloud platform (http://www.gscloud.cn/ accessed on 10 September 2022) with a spatial resolution of 30 m. First, the ENVI 5.3 software is used to pre-process the remote sensing images, such as geometric correction [50], atmospheric correction [51], and image enhancement [52], to complete the preparation and processing of fundamental geographic data. Second, a combination of supervised classification and human–computer interaction [53] was used to interpret and decipher the land use data into the waters, cultivated land, grassland, forest land, and constructed land according to the actual state of the watershed, which was not accounted for in the land use classification due to the tiny amount of unused land in the Nanming River watershed. Finally, the confusion matrix was utilized to rectify and validate the correctness of the results after land use classification [54], and the overall classification accuracy surpassed 85%, indicating that it could meet the analysis goals of this study. Furthermore, we obtained land use status maps for the study region in 2000, 2010, and 2020.

3. Analysis and Results

3.1. Research Framework

The changes in the characteristics of land use landscape were analyzed from 2000 to 2020 using GIS and RS technology. We also analyzed changes in the land use landscape pattern index from landscape patch scale and landscape scale. ArcGIS and landscape pattern index methods were used to construct the landscape ecological security index and to analyze the spatial and temporal variation mode of landscape pattern evolution and its ecological security (Figure 3).

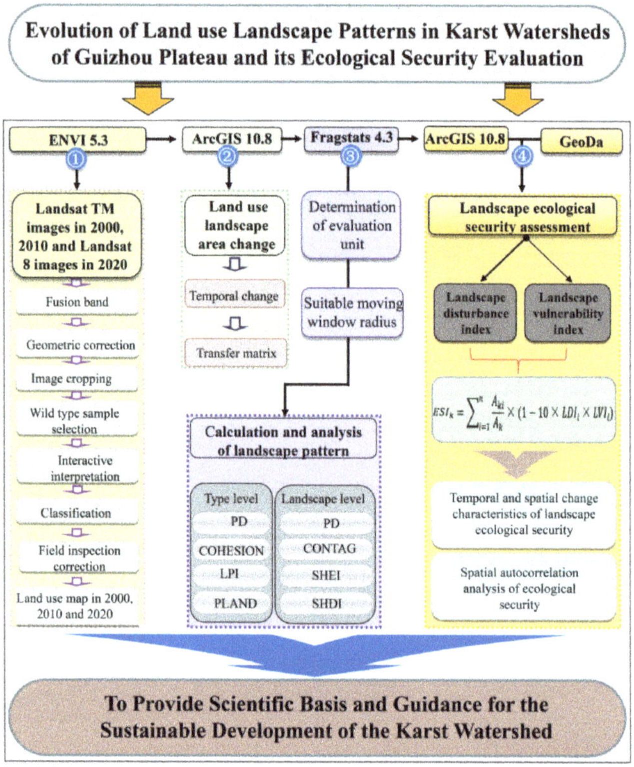

Figure 3. Research framework.

3.2. Analysis of the Characteristics of Land Use Landscape Area Change

Land use changes were influenced by the expansion of constructed land and loss of cultivated land in 2000–2020 (Figure 4) because of the influence of strong urban development activities. During this period, the cultivated land flowing to constructed land accounts for 36.48% of the landscape in the watershed. By 2020, constructed land accounted for approximately 20% of the watershed area, which was 2.31 times the shared area in 2000. After analyzing the changes in the land landscape in each period with the help of the land use transfer matrix, constructed land recorded the largest growth rate of 130.72% from 2000 to 2020. However, the proportion of grassland and cultivated land was decreasing, and the proportion of forest showed an increasing trend at first but later showed a decreasing trend. With the growth in urbanization, the Nanming River Basin has undergone a dynamic process of changing from natural to unnatural landscapes, indicating that the impact of human activities on the ecosystem of the watershed is more intense. This shows that the Nanming River Basin has undergone continuous economic development in the past 20 years since 2001.

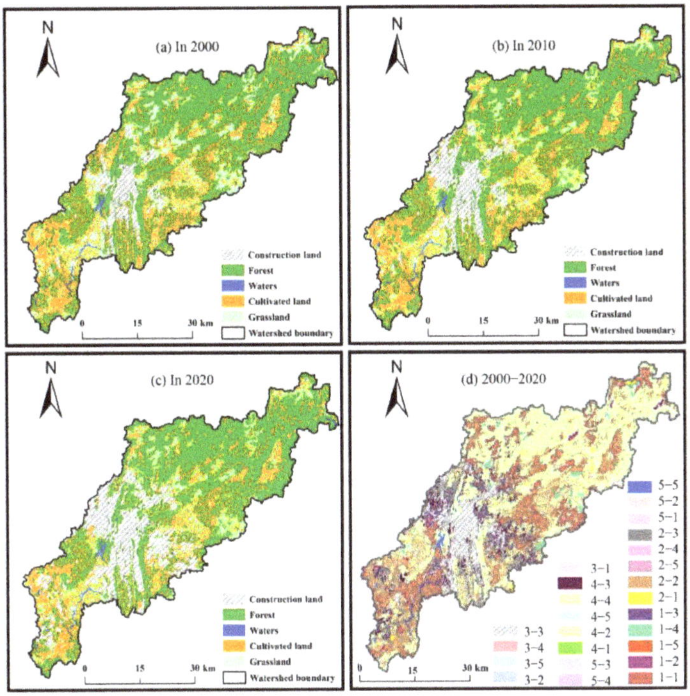

Figure 4. Land use status map of the study area in 2000 (**a**), 2010 (**b**), 2020 (**c**), and land use transfer map from 2000 to 2020 (**d**). Note: 1 indicates grassland; 2 indicates cultivated land; 3 indicates constructed land; 4 indicates forest; and 5 indicates waters.

3.3. Analysis of Spatial and Temporal Changes of Landscape Pattern Index

3.3.1. Analysis of Landscape Pattern Index Change on Patch Type Scale

As shown in Figure 5, the PD of the cultivated land is the largest among all land types, with a mean value of 0.24. This shows that the cultivated land has a profound influence on the landscape pattern of the Nanming River Basin. From 2000 to 2010, the PD of the forest, cultivated land, and constructed land increased. However, the PD of grassland showed a decreasing trend from 2010 to 2020. Except for the constructed land, the PD of the forest, constructed land, grassland, and water showed an increasing trend. The human ecological footprint has had a greater impact on the forest and cultivated landscapes over the past 20 years. Moreover, their ecological processes were more active, in which the expansion of the constructed landscapes was reasonably restrained during 2010–2020.

Among the landscapes in the watershed, the mean of the PLAND of the forest is the maximum of approximately 45.08%, which is about 28 times that of the smallest PLAND (waters), indicating that the forest offers more advantages in the landscape. This result agrees with the results of the spatial characterization of land use described above. From 2000 to 2010, the forest area and constructed land patches showed an increasing trend. From 2010 to 2020, the PLAND of the constructed land increased by 19.53%, which was 12% more than that of 2000. Nevertheless, the constructed land patches experienced continuously increasing trends during the study period. Since 2001, Nanming River Basin experienced comprehensive environmental improvement, which includes increasing projects on cultivated land, forest, and the construction of a comprehensive landscape on both sides of the river. With a rise in urbanization, a large construction continues to expand to the periphery of the city. However, patches of cultivated land and grassland continue to decline.

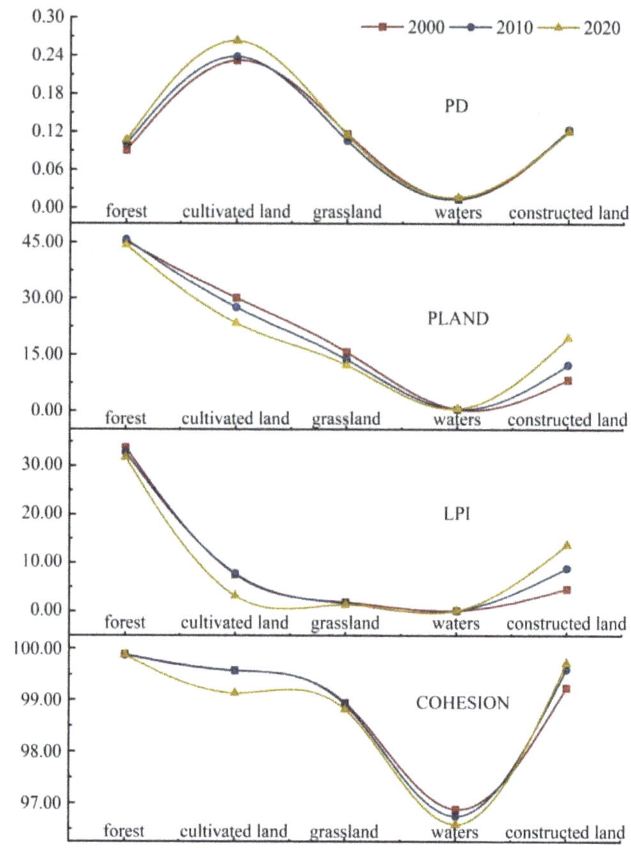

Figure 5. Change of landscape pattern index on patch type scale from 2000 to 2020.

From the LPI, the LPI of the forest land far exceeds that of other land types, indicating that forest is the major land substrate in the watershed. From 2000 to 2020, the LPI of constructed land continues to increase.

From the COHESION, the patch cohesiveness of the forest and constructed land is higher, indicating that the natural connectivity of these two types of land is efficient, and the distributions are patchy in spatial distribution. From 2000 to 2020, the COHESION of cultivated land, grassland, and waters generally declined, whereas the COHESION of constructed land continued to increase, reaching 35% from 2000 to 2010, which was three times that of 2010–2020. With large-scale land exploitation, the distribution of patches of cultivated land, grassland, and water fragments go through aggregation to fragmentation, with a decrease in the natural connectivity of the landscapes. However, the patches of constructed land gradually turn into aggregated blocks of spatial distribution.

3.3.2. Analysis of the Landscape Pattern Index Change on Landscape Scale

As shown in Figure 6, the PD and CONTAG have a more pronounced geographical variability. The CONTAG increased significantly in the southwestern part of the basin, and the PD was high in most of the northeastern part of the basin. Moreover, the spatial distribution of the CONTAG in these areas showed the opposite PD characteristics. We found that the high-value areas of CONTAG are concentrated in the urban center of Guiyang City, owing to its simple landscape structure, its constructed land as the matrix land, and a high degree of agglomeration, which greatly reduces the fragmentation of the landscape patches. Moreover, the low-value areas comprise the interspersed distribution of cultivated land, constructed land, and grassland, consisting of an intricate landscape structure with a high degree of landscape fragmentation in the watershed. The lower stream of the basin, at Wudang and Longli, has a lower level of landscape diversity, indicating that human activities in this area are infrequent, and the development is low. The SHEI and SHDI in the basin have similar spatial variations. In the southern part of Nanming District and the eastern part of Guanshan Lake District and Yunyan District upstream, the SHEI and SHDI have decreased significantly with a rise in urbanization and an increase in the project construction, which have become the main advantages of the landscape.

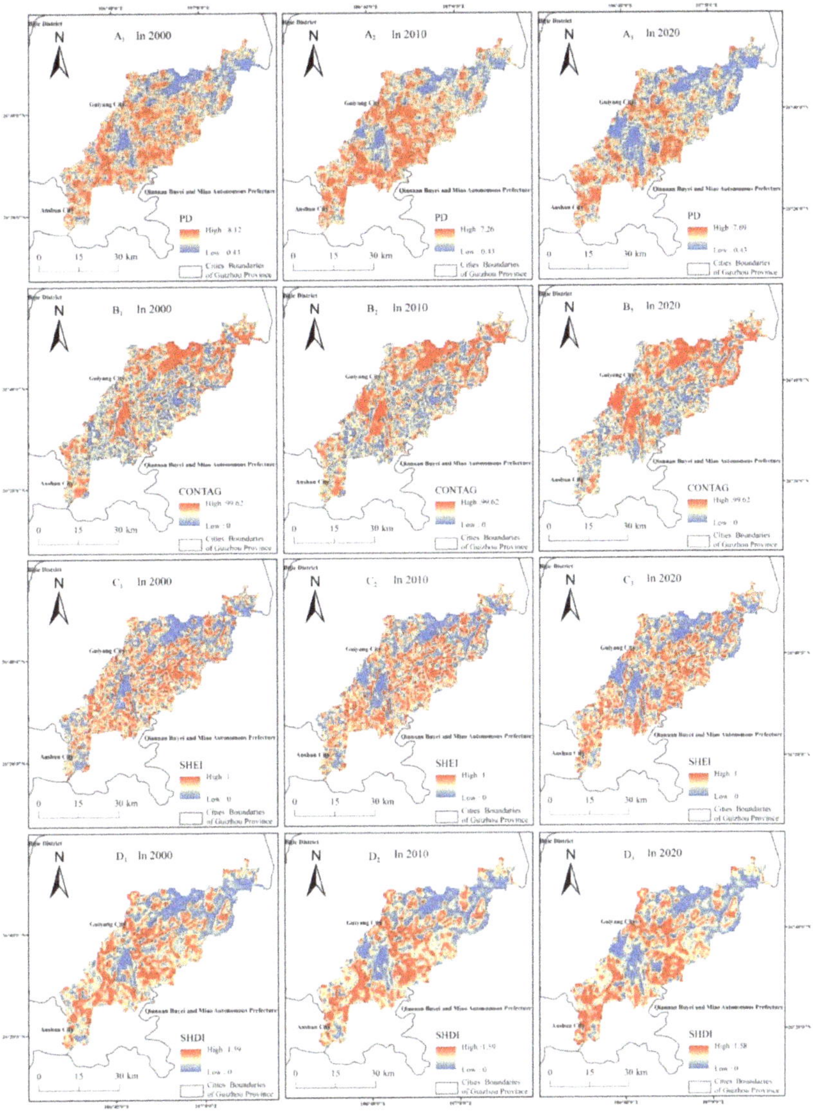

Figure 6. Spatial distribution of landscape pattern index at landscape level in the Nanming River Basin from 2000 to 2020. Note: (**A$_1$**) PD in 2000; (**A$_2$**) PD in 2010; (**A$_3$**) PD in 2020; (**B$_1$**) CONTAG in 2000; (**B$_2$**) CONTAG in 2010; (**B$_3$**) CONTAG in 2020; (**C$_1$**) SHEI in 2000; (**C$_2$**) SHEI in 2010; (**C$_3$**) SHEI in 2020; (**D$_1$**) SHDI in 2000; (**D$_2$**) SHDI in 2010; (**D$_3$**) SHDI in 2020.

3.4. *Landscape Ecological Safety Evaluation*

3.4.1. Landscape Ecological Safety Changes of Nanming River Watershed

The ecological safety index of the Nanming River Basin was divided into five classes (refer to related studies regarding each grade [55]). As shown in Figure 7 and Table 1, the overall landscape ecological safety shows a continuous positive trend in the last 20 years. The medium ecological safety zone is the largest area. The low-security area is concentrated in the southwest part of the watershed, which is characterized by a farming economy, underdeveloped economy, extensively cultivated land, and interspersed distribution of grassland, forest, and waters, which destroys the stability of the landscape. Driven by urbanization, a large patch of arable land is used for infrastructure construction to improve the standard of living. The construction of residential housing is increasing on the large arable land, leading to a rise in the landscape ecological risk. However, the low-security areas decline by about 25% in 2020. The shrinkage rate in the lower security zone was 11.13%. Spatially, the constructed land is connected to patches. The patches also converge and shift to the middle-security zone, revealing a gradual increase in ecological security. The higher security zone is increasing yearly, which is mainly concentrated in the northern part of the watershed, showing that the forest offers advantages to the landscape. Despite the recent government's policy of encouraging people to return to cultivated land, the forest has not yet been encroached upon by other landscape types, providing policy support to maintain a higher security-level state. The trend of a high-security zone is increasing, which is distributed in the center of Guiyang City and at a lower elevation at the watershed. With the obvious landscape advantages and contiguous urban housing, this area has become a stable landscape structure and has low landscape fragmentation.

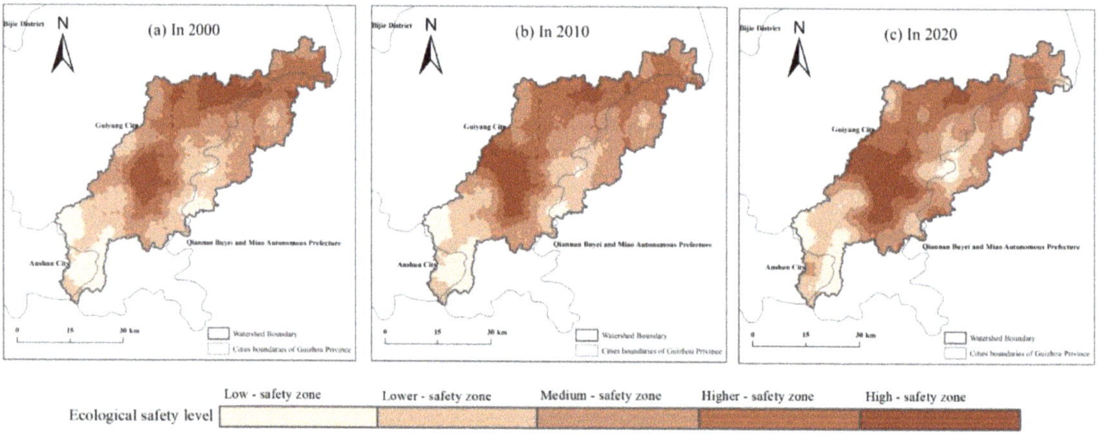

Figure 7. (**a**) Spatial distribution of landscape ecological security in 2000; (**b**) Spatial distribution of landscape ecological security in 2010; (**c**) Spatial distribution of landscape ecological security in 2020.

Table 1. Statistics of landscape ecological security area in the study area.

Ecological Security Level	2000		2010		2020	
	Area/km^2	Proportion	Area/km^2	Proportion	Area/km^2	Proportion
Low—safety zone	242	11.21%	219	8.14%	182	8.43%
Lower—safety zone	494	22.89%	467	28.76%	439	20.34%
Medium—safety zone	633	29.33%	463	34.93%	650	30.12%
Higher—safety zone	540	25.02%	693	32.11%	602	27.90%
High—safety zone	249	11.54%	256	7.63%	286	13.25%

3.4.2. Spatial Autocorrelation Analysis of Landscape Ecological Security Index

As shown in Figure 8, the Moran's I values of the landscape ecological safety index of the Nanming River Basin were 0.394, 0.464, and 0.488, greater than 0 for the study periods 2000, 2010, and 2020 at a significance level of $p < 0.05$. This indicates that the landscape ecological safety index in the study zone is correlated, and the spatial convergence is gradually increasing.

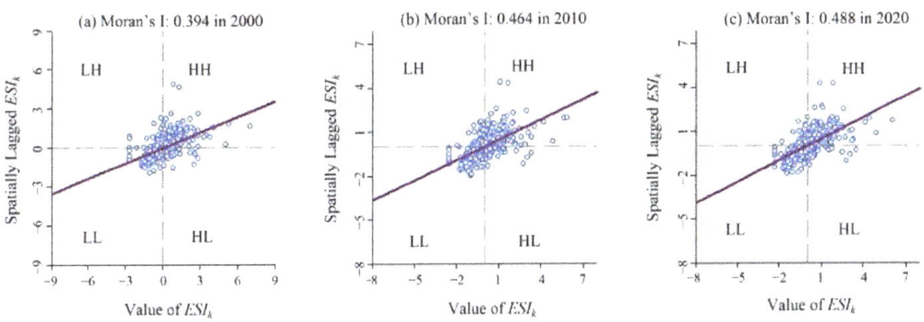

Figure 8. Scatter map of ecological security index of landscape pattern in the study area from 2000 to 2020.

As shown in Figure 9, the spatial clustering pattern of landscape ecological safety values in the Nanming River basin is characterized by high–high clustering and low–low clustering. The percentage of high–high concentration sample areas gradually increased over the three periods, ranging from 7.66% to 12.66%. From the local autocorrelation of the study area, the cluster structure of "high–high" values of the landscape ecosystem security index continues to extend outward from 2000. However, the range of "low–low" values continues to shrink. In terms of spatial distribution, the ecological safety high-value catchment area of the watershed is concentrated in the center of Guiyang City, with little distribution in the eastern part of the downstream Wuzhong. In this area, the terrain is relatively flat, and the topographic conditions are simple, with a single land-use landscape type as the main feature. Low-value ecological security catchment areas are focused on the central Huaxi and southern Pingba in the upper part of the watershed. At the same time, we found that the internal structure of the landscape in the adjacent areas of the region is finely fragmented. Additionally, each land use landscape type is disturbed by human activities and interspersed with each other; so the ecological safety is in a low-value state, and its stability may be difficult to maintain.

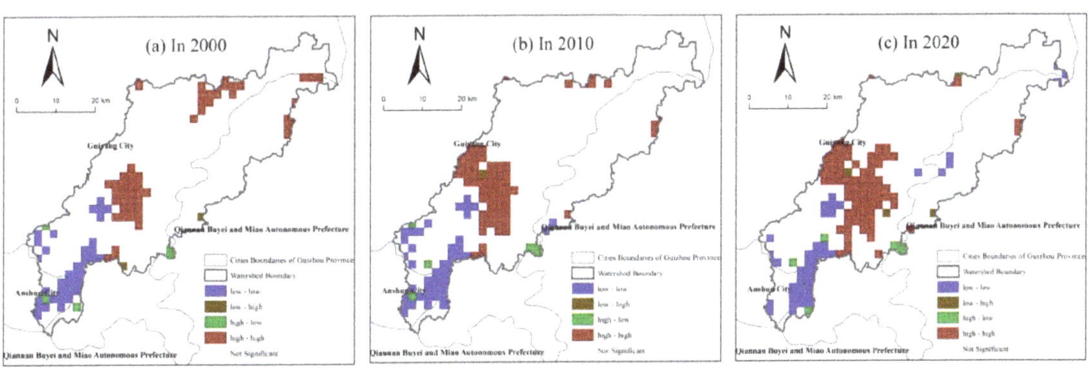

Figure 9. (**a**) LISA Map of Landscape Ecological Security in 2000; (**b**) LISA Map of Landscape Ecological Security in 2010; (**c**) LISA Map of Landscape Ecological Security in 2020.

4. Discussion

4.1. Landscape Ecological Safety Evolution Rules

(1) Change in the Landscape Ecological Safety Index

From 2000 to 2020, the landscape ecological security index showed an increasing trend, indicating that the ecological security of the Nanming River Basin gradually increased. Driven by the market economy, crop cultivation has increased in the upper watershed, making a certain amount of forest and grassland reclaimed as cropland. Moreover, a huge number of cultivated land has been converted to forest and constructed land, which was driven by the policy of returning farmland to forest and the construction of land expansion. This policy has increased the fragmentation of forest and cropland landscapes, decreased connectivity, and decreased the ecological security index. With urbanization and rapid economic development, the non-farm population has increased, and the disturbance to the watershed landscape from human activities has risen. This development enhances the distribution of the construction of landscapes favoring human ecological and living needs. Moreover, the contiguous distribution of constructed land decreases the fragmentation of landscape patches and increases the degree of stability because of the low vulnerability in the flat topography. Thus, the ecological safety index of the constructed land has maintained its maximum value in the last 20 years. Additionally, the implementation of artificial landscaping projects in the watershed has rationalized the layout of the grassland landscape and increased the landscape ecological safety index.

(2) Spatial Distribution of Landscape Ecological Safety Index

Table 2 shows a comparative analysis of the relevant literature [56–68]. In karst areas, the regional characteristics above the higher ecological safety level are manifested as follows: ① They are located at lower elevations and in urban centers with faster economic development. ② These areas are also the most concentrated belt of forest or the constructed land and forest of the landscape. Low-security and lower-security areas are characterized by cultivated land, grassland, and water. In non-karst areas, areas above medium security level are dominated by natural ecosystems such as forest and grassland, which are concentrated in agricultural areas or natural landscape protection zones far from urban centers. However, low-security level areas are distributed on constructed land, and natural landscape structures are fragmented by human interference.

Thus, the higher safety-level areas of the watershed are distributed in the central urban areas, whereas the constructed land is concentrated and contiguous. Moreover, the low and lower ecological safety areas are characterized by a distribution of interspersed and scattered forest, grassland, and cultivated land.

These results contradict results presented by previous studies conducted in non-karst areas. Compared with related studies in karst areas, the Panlong River Basin located in Kunming [60] has a social and geomorphic environment comparable to the study area. Lin et al. found that the ecological safety index of the constructed land within the Panlong River Basin was higher than that of other landscape types. Moreover, the landscape types formed by human behavioral activities have the characteristics of being the most resistant and stable to external disturbances. Taking the Dianchi watershed as a case study, Wu et al. analyzed ecological security and found that the high-security areas were located in the urban areas because urban housing was distributed in a row, with almost no other land use landscape. Moreover, the patches are highly connected and less fragmented, making them capable of resisting external disturbances [61]. These findings are in line with the results of this study.

From the above comparison of the landscape ecological security research in non-karst and karst areas, it is recommended to give more attention to the fallibility and fragility of natural ecosystems during the process of karst ecosystem restoration and reconstruction, such as forests and grassland. In non-karst areas, we should pay more attention to the uncontrolled expansion of constructed land and the quality of the surrounding natural ecosystem. In addition, in the follow-up study, the factors influencing the spatial distribution difference of ecological security of landscape patterns in non-karst and karst areas should be discussed in depth.

Table 2. Relevant study on ecological security of landscape pattern.

		High Security	Higher Security	Middle-Grade Security	Lower Security	Low Security
Karst area	Panlong River Basin, Kunming City [60]	construction land, forest land	Unused land, water area	/	Grassland, cultivated land	/
	Dianchi Watershed [61]	Most of the main urban area and part of Chenggong District	The main urban area of Kunming	Around Dianchi Lake, Most areas in Chenggong	The shore of Dianchi Lake	The water surface of Dianchi Lake
	Chishui River Basin [62]	Xishui County, Chishui City, and Tongzi County in Zunyi City		/	/	The junction of Guizhou and Sichuan where the Chishui River flows
	Wanzhou District, Chongqing [63]	Town center, suburb, and mountainous area with high altitude		/	Mainly low mountains and hills	Distributed along the Yangtze River basin
	Caohai Wetland, Guizhou [45]	Forest land in the northeast and southwest mountains	The central region is dominated by cultivated land		Around Caohai Lake District	
	This paper	Provincial capital city center with low altitude	North of the study area (mainly forest land)	Southeast of the study area (mainly forest land)	Southwest of the study area (mainly cultivated land)	In the southwest of the study area, mainly cultivated land, grassland, forest land, and water area are interlaced
Non-karst area	Yinchuan City [64]	/	/	Forestland and grassland		Cultivated land, construction land, and unused land
	Xinjiang Uygur Autonomous Region [65]	Oasis agricultural area		/	Taklimakan Desert, Turpan Basin Desert and the hinterland of Gurbantunggut Desert	
	State-operated Friendship Farm in Shuangyashan City, Heilongjiang Province [66]	Natural wetlands and woodlands in the north and east of the study area (1984)		/	Cultivated land or degraded grassland in the north and east of the study area (originally a natural wetland in 1984)	
	Huailai County, Zhangjiakou City, Hebei Province [67]	Mountains in the north and south of the study area and near Miyun Reservoir			Plain area and around the county seat	County town
	Haitan Island, Pingtan County, Fujian Province [44]	Hilly forest area and coastal protection forest			Sea reclamation area	

4.2. Limitations and Shortcomings

In interpreting land use type data through RS images and analyzing land use changes and landscape patterns in the Nanming River Basin, this study evaluated the spatiotemporal variation characteristics of ecological security in the basin. However, some shortcomings are found that require improvement. The results inevitably have certain errors when interpreting images through RS technology because the results have been influenced by objective factors and human subjective factors, thereby affecting the accuracy verification. Thus, the ground-based field data surveys and historical record data must be rectified to improve the accuracy of the interpretation results. The dynamic and landscape indicators of land use types only reflect the macro-structural changes influenced by topographic factors [68]. However, it is difficult to reveal the microstructural changes in the landscape. Thus, the microstructural changes in the landscape should be explored in future studies.

As a natural geographical unit, a watershed is the unification of multiple catchment areas within a natural environment [69]. However, watershed boundaries and administrative boundaries cannot completely overlap [70]. Thus, when exploring the influence of the natural environment and anthropogenic activities on watershed landscape patterns and ecological security, these influences cannot be fully quantified because of some constraints on the analysis of watershed change.

5. Conclusions

This study used 3S technology, the landscape pattern index method, and spatial autocorrelation theory to systematically analyze the landscape pattern evolution characteristics of the Nanming River watershed and evaluate the ecological security of the karst watershed in the Guizhou plateau. Several conclusions were obtained as follows. Forest was the leading landscape and mainland substrate in the watershed from 2000 to 2020. The cohesiveness of constructed land patches continues to increase. Moreover, the fragmentation and diversity of landscape pattern in the patchy distribution of forest and constructed land have declined. The growth rate of the Nanming River Basin Landscape Ecological Safety Index has increased by 5.80%, and the overall ecological safety has shown a continuous positive trend. The high-value ecological safety clusters are distributed in the central urban areas, where the constructed land is concentrated and contiguous. Moreover, the low-value clusters show the scattered distribution characteristics of forest, grassland, and cultivated land. Thus, the spatial clustering effect of the ecological security index is obvious, which is dominated by high–high clustering and low–low clustering types. The study reveals the landscape pattern evolution rules of the typical karst watershed in the Guizhou plateau since 2001 by systematically evaluating the spatiotemporal distribution characteristics of ecological security in the watershed. The findings provide scientific reference for maintaining the ecological balance in the watershed, optimizing land resource allocation and regulation, and improving the ecological environment of typical karst watershed geomorphic units.

Author Contributions: Y.L.: Methodology, Investigation, Formal Analysis, Visualization, Writing—Original Draft, Writing—Review and Editing. H.G.: Data Curation, Resources, Formal Analysis. All authors have read and agreed to the published version of the manuscript.

Funding: This study was jointly supported by the Youth Talent Growth Project of Guizhou Provincial Department of Education (Qian Jiao He KY [2022] No. 202), the Water Conservancy Science and Technology Funding Projects in Guizhou Province (KT202114), the Water Conservancy Science and Technology Funding Projects in Guizhou Province (KT202223), National Natural Science Foundation of China (42261044).

Institutional Review Board Statement: Not applicable.

Informed Consent Statement: Not applicable.

Data Availability Statement: Not applicable.

Conflicts of Interest: The authors declared that they have no conflict of interest to this work. We declare that we do not have any commercial or associative interest that represents a conflict of interest in connection with the work submitted.

References

1. O'Neill, R.V.; Krummel, J.R.; Gardner, R.H. Indices of landscape pattern. *Landsc. Ecol.* **1998**, *1*, 153–162. [CrossRef]
2. Wu, J.S.; Zhang, P.H. The effect of urban landscape pattern on urban waterlogging. *Acta Geogr. Sin.* **2017**, *72*, 444–456.
3. Wang, P.; Wang, Y.J.; Liu, X.P.; Wen, S.Q. Change of land use and landscape pattern in ecological resettlement area in central Ningxia. *J. Arid. Land Resour. Environ.* **2018**, *32*, 69–74.

4. Polasky, S.; Nelson, E.; Pennington, D.; Johnson, K.A. The impact of land-use change on ecosystem services, biodiversity and returns to landowners: A case study in the state of Minnesota. *Environ. Resour. Econ.* **2011**, *48*, 219–242. [CrossRef]
5. Macfarlane, R. Implementing agri-environment policy: A landscape ecology perspective. *J. Environ. Plan. Manag.* **1998**, *41*, 575–596. [CrossRef]
6. Yu, X.; Wu, K.N.; Yun, W.J.; Wei, H.B.; Song, Y.H.; Gao, X. Analysis on temporal and spatial variation of landscape ecological security in modern agricultural area. *Trans. Chin. Soc. Agric. Eng.* **2016**, *32*, 253–259.
7. Ou, D.H.; Xia, J.G. Landscape pattern optimization in peri-urban areas based on the particle swarm optimization method: A case study in Longquanyi District of Chengdu. *Geogr. Res.-Aust.* **2017**, *36*, 553–572.
8. Xiong, K.N.; Chen, Q.W. Discussion on karst rocky desert evolution trend based on ecologically comprehensive treatment. *Carsologica Sin.* **2010**, *29*, 267–273.
9. Wu, L.H.; Wang, S.J.; Bai, X.Y.; Chen, F.; Li, C.J.; Ran, C.; Zhang, S.R. Identifying the multi-scale influences of climate factors on runoff changes in a typical karst watershed using wavelet analysis. *Land* **2022**, *11*, 1284. [CrossRef]
10. Chen, F.; Bai, X.Y.; Liu, F.; Luo, G.J.; Tian, Y.C.; Qin, L.Y.; Li, Y.; Xu, Y.; Wang, J.F.; Wu, L.H.; et al. Analysis long-term and spatial changes of forest cover in typical karst areas of China. *Land* **2022**, *11*, 1349. [CrossRef]
11. Fu, F.J.; Liu, Z.H.; Huang, Q.D. Three-dimensional urban landscape pattern changes: A case study in the Central Business District of Futian, Shenzhen. *Acta Ecol. Sin.* **2019**, *39*, 4299–4308.
12. Hu, H.B.; Liu, H.Y.; He, J.F.; An, J. Influence of spatial difference on water quality in Jiuxiang River Watershed, Nanjing. *Environ. Sci.* **2012**, *33*, 794–801.
13. Hu, H.B.; Liu, H.Y.; He, J.F.; An, J. Effects of urbanization on the spatial heterogeneity of watershed ecosystem services value: A case study of Jiuxiang River Watershed in Nanjing City. *J. Nat. Resour.* **2011**, *26*, 1715–1725.
14. Song, F.J.; Wang, S.J.; Bai, X.Y.; Wu, L.H.; Wang, J.F.; Li, C.J.; Chen, H.; Luo, X.L.; Xi, H.P.; Zhang, R.R.; et al. A new indicator for global food security assessment: Harvested area rather than cropland area. *Chin. Geogr. Sci.* **2022**, *32*, 204–217. [CrossRef]
15. Du, C.C.; Bai, X.Y.; Li, Y.B.; Tan, Q.; Zhao, C.W.; Luo, G.J.; Wu, L.H.; Chen, F.; Li, C.L.; Ran, C.; et al. Inventory of China's net biome productivity since the 21st century. *Land* **2022**, *11*, 1244. [CrossRef]
16. Zhang, S.R.; Bai, X.Y.; Zhao, C.W.; Tan, Q.; Luo, G.J.; Wu, L.H.; Xi, H.P.; Li, C.J.; Chen, F.; Ran, C.; et al. China's carbon budget inventory from 1997 to 2017 and its challenges to achieving carbon neutral strategies. *J. Clean. Prod.* **2022**, *347*, 130966. [CrossRef]
17. Besser, H.; Mokadem, N.; Redhouania, B.; Rhimi, N.; Khlifi, F.; Ayadi, Y.; Omar, Z.; Bouajila, A.; Hamed, Y. GIS-based evaluation of groundwater quality and estimation of soil salinization and land degradation risks in an arid Mediterranean site (SW Tunisia). *Arab. J. Geosci.* **2017**, *10*, 350–370. [CrossRef]
18. Harvey, E.; Gounand, I.; Ward, C.L.; Altermatt, F. Bridging ecology and conservation: From ecological networks to ecosystem function. *J. Appl. Ecol.* **2017**, *54*, 371–379. [CrossRef]
19. Peng, J.; Pan, Y.J.; Liu, Y.X.; Zhao, H.J.; Wang, Y.L. Linking ecological degradation risk to identify ecological security patterns in a rapidly urbanizing landscape. *Habitat Int.* **2018**, *71*, 110–124. [CrossRef]
20. Xiong, L.; Bai, X.Y.; Zhao, C.W.; Li, Y.B.; Tan, Q.; Luo, G.J.; Wu, L.H.; Chen, F.; Li, C.J.; Ran, C.; et al. High-resolution datasets for global carbonate and silicate rock weathering carbon sinks and their change trends. *Earth's Future* **2022**, *10*, e2022EF002746. [CrossRef]
21. Yu, C.L.; Liu, D.; Feng, R.; Tang, Q.; Guo, C.L. Construction of ecological security pattern in Northeast China based on MCR model. *Acta Ecol. Sin.* **2021**, *41*, 290–301.
22. Liu, M.; Bai, X.Y.; Tan, Q.; Luo, G.J.; Zhao, G.W.; Wu, L.H.; Luo, X.L.; Ran, C.; Zhang, S.R. Climate change enhances the positive contribution of human activities to vegetation restoration in China. *Geocarto Int.* **2022**, *37*, 1–21. [CrossRef]
23. Cui, X.Y.; Fang, L.; Wang, X.R.; Kang, J.F. Urban eco-security assessment in the urban agglomerations based on DPSIR model:A case study of Yangtze River Delta, China. *Acta Ecol. Sin.* **2021**, *41*, 302–319.
24. Li, C.J.; Bai, X.Y.; Tan, Q.; Luo, G.J.; Wu, L.H.; Chen, F.; Xi, H.P.; Luo, X.L.; Ran, C.; Chen, H.; et al. High-resolution mapping of the global silicate weathering carbon sink and its long-term changes. *Glob. Chang. Biol.* **2022**, *28*, 4233–4398. [CrossRef]
25. Tieskens, K.F.; Schulp, C.J.E.; Levers, C.; Lieskovský, J.; Kuemmerle, T.; Plieninger, T.; Verburg, P.H. Characterizing European cultural landscapes: Accounting for structure, management intensity and value of agricultural and forest landscapes. *Land Use Policy* **2017**, *62*, 29–39. [CrossRef]
26. Li, Q.H.; Yan, J.Y. An evaluation of ecological security of coastal zone of sanya with DPSIR model. *Trans. Oceanol. Limnol.* **2021**, *43*, 147–154.
27. Zhang, Y.Y.; Li, Y.B.; Luo, G.J.; Bai, X.Y.; Huang, J.; Tang, F.; Yu, M. Analysis of the land use dynamics of different rural settlement types in the Karst Trough Valleys of Southwest China. *Land* **2022**, *11*, 1572. [CrossRef]
28. Ghosh, S.; Das Chatterjee, N.; Dinda, S. Urban ecological security assessment and forecasting using integrated DEMATEL-ANP and CA-Markov models: A case study on Kolkata Metropolitan Area, India. *Sustain. Cities Soc.* **2021**, *68*, 102773. [CrossRef]
29. Jiang, W.K.; Xie, D.T.; Wang, S.; Yan, J.Z.; Liu, F.S. Study on landscape pattern change in a cultural heritage protection area and its driving factors—A case study of the core area of Lijiang City. *J. Southwest Univ.* **2018**, *40*, 133–142.
30. Cao, B.S.; Shan, N.; Gu, Y.Y.; Ao, W.; Pang, B.; Dou, H.S.; Wang, W.L.; Zou, C.X. Evaluation of ecological security in Hulun Lake Watershed and its spatio-temporal distribution trend. *Res. Environ. Sci.* **2021**, *34*, 801–811.
31. Liu, Y.F.; An, R.; Qu, S.Q.; Liu, Y.L. Ecological security evaluation and obstacle factor diagnosis of the cultivated land in Fujian Province. *Chin. J. Agric. Resour. Reg. Plan.* **2022**, *43*, 1–15.

32. Wei, Q.Q.; Zhong, Y.X.; He, J.; Luo, L.L.; Qi, Y.R. Land use changes and ecological water requirement of Xinghai Lake Wetland. *J. Hydroecol.* **2022**, *43*, 27–34.
33. Ren, M.; Wang, Z.J.; Wang, Z.T.; Zeng, Y.J.; He, L.Y. Grain size effect of karst mountainous urban landscape pattern indices in the central Guizhou: A case study of Anshun City. *Chin. J. Ecol.* **2018**, *37*, 3137–3145.
34. Liu, C.J.; Zhou, G.F.; Huang, Q.F.; Cai, X.F. Spatio-temporal evolution analysis on habitat quality in mountainous watershed based on InVEST Model: Case of Chishui River Basin in Guizhou Province. *Yangtze River* **2021**, *52*, 62–69.
35. Wang, Q.; Li, Y.B.; Huang, J.; Hu, X.P.; Zhong, S.N. Effect of land use transformation process on ecosystem service value in Karst Trough Valley Area. *Res. Soil Water Conserv.* **2019**, *26*, 192–198.
36. Peng, W.J.; Shu, Y.G. Analysis of landscape ecological security and cultivated land evolution in the Karst mountain. *Acta Ecol. Sin.* **2018**, *38*, 852–865.
37. Zhao, Y.J.; Zeng, L.; He, Z.; Liang, Q. Interpretation and landscape pattern of Karst Landform based on multi-source remote sensing images. *Res. Soil Water Conserv.* **2017**, *24*, 158–162.
38. Zhang, X.; Wang, Z.J. Evaluation and characteristic analysis of urban landscape stability in karst mountainous cities in the central Guizhou Province. *Acta Ecol. Sin.* **2022**, *42*, 5243–5254.
39. Shi, S.N.; Li, X.Q.; Xie, B.G.; Hu, B.Q.; Tang, C.Y.; Yan, Y. Change and comparison of agricultural landscape patterns and ecological service values in Karst and Non-Karst Areas: A case study of Quanzhou County. *Trop. Geogr.* **2018**, *38*, 487–497.
40. Bu, R.C.; Hu, Y.M.; Chang, Y.; Li, X.Z.; He, H.S. A correlation analysis on landscape metrics. *Acta Ecol. Sin.* **2005**, *25*, 2764–2775.
41. Zhou, D.M.; Chen, C.Y.; Wang, M.J.; Luo, Z.W.; Kang, L.T.; Wu, S. Gradient and directional differentiation in landscape Pattern characteristics of urban ecological space based on optimal spatial scale: A case study in Changsha City, China. *J. Ecol. Rural Environ.* **2022**, *38*, 566–577.
42. Mei, Z.R.; Li, Y.J.; Kang, X.; Wei, S.B.; Pan, J.J. Temporal and spatial evolution in landscape pattern of mining site area based on moving window method. *Remote Sens. Nat. Resour.* **2019**, *31*, 60–68.
43. Yang, K.; Xin, G.X.; Jiang, H.Y.; Yang, C.X. Study on spatiotemporal changes of landscape ecological risk based on the optimal spatial scale: A case study of Jiangjin District, Chongqing City. *J. Ecol. Rural Environ.* **2021**, *37*, 576–586.
44. Dai, W.Y.; Huang, H.F.; Huang, W.L.; Wu, X.H.; Huang, J.B. Features of spatiotemporal differentiation of landscape ecological security in island ecologically fragile zones in Haitan Island. *Ecol. Sci.* **2017**, *36*, 152–159.
45. Ren, J.T.; Yang, K.M.; Chen, Q.L.; Mo, S.J.; Wang, Z.H. Evaluation on ecological security of land use landscape in Guizhou Caohai Wetland. *Environ. Sci. Technol.* **2018**, *41*, 158–165.
46. Zhang, Y.; Zhang, F.; Zhou, M.; Li, X.H.; Ren, Y.; Wang, J. Landscape ecological risk assessment and its spatio-temporal variations in Ebinur Lake region of inland arid area. *Chin. J. Appl. Ecol.* **2016**, *27*, 233–242.
47. Xiong, Y.; Wang, M.; Yuan, H.P.; Du, C.Y.; Wu, H.P. Landscape ecological risk assessment and its spatio-temporal evolution in Dongting Lake Area. *Ecol. Environ. Sci.* **2020**, *29*, 1292–1301.
48. Chen, Y.G. Reconstructing the mathematical process of spatial autocorrelation based on Moran's statistics. *Geogr. Res.* **2009**, *28*, 1449–1463.
49. Wu, J.; Chen, P.; Wen, S.F.; Fu, S.F.; Chen, Q.H. Ecological risk assessment of land use based on exploratory spatial data analysis (ESDA): A case study of Haitan Island, Fujian Province. *Chin. J. Appl. Ecol.* **2014**, *25*, 2056–2062.
50. Sun, Y.W.; Liu, B.; Di, K.C.; Liu, Z.Q. Automatic accurate geo-rectification of planetary remote sensing image based on reference image. *Remote Sens. Nat. Resour.* **2015**, *27*, 23–28. [CrossRef]
51. Han, X.Q.; Su, Y.; Li, J.; Zhang, Y.; Liu, J.; Gao, W.M. Atmospheric correction and verification of the SPOT remote sensing image in coastal zones. *Geogr. Res.* **2012**, *31*, 2007–2016.
52. Huang, K.Q.; Wang, Q.; Wu, Z.Y. Natural color image enhancement and evaluation algorithm based on human visual system. *Comput. Vis. Image Underst.* **2006**, *103*, 52–63. [CrossRef]
53. Zhou, K.; Yang, Y.Q.; Zhang, Y.N.; Miao, R.; Yang, Y.; Liu, L. Review of land use classification methods based on optical remote sensing images. *Sci. Technol. Eng.* **2021**, *21*, 13603–13613.
54. Song, J.W.; Zhang, Y.J.; Li, X.C.; Yang, W.Z. Comparison between GF-1 and Landsat-8 images in land cover classification. *Prog. Geogr.* **2016**, *35*, 255–263.
55. Ou, Z.R.; Zhu, Q.K.; Sun, Y.Y. Temporal and spatial variation of landscape ecological security in Yuanmou Dry-hot Valley. *Sci. Soil Water Conserv.* **2018**, *16*, 131–140.
56. Guo, C.J.; Liu, Y.X.; Li, H.F.; Sun, Y.T.; Yu, Y.C. Landscape characteristics and construction of landscape ecological security pattern in West Open Pit of Fushun Mine. *J. Shenyang Agric. Univ.* **2021**, *52*, 442–450.
57. Wang, Y.S.; Zhang, F.; Chen, R.; Qi, Y.X.; Liu, C.J. Comprehensive ecological security assessment: A case study of Urumqi City. *Arid Land Geogr.* **2021**, *44*, 427–440.
58. Yuan, Y.; Luo, Z.J.; Zhao, J.; Qi, S. Landscape ecological security evaluation of Nanchang City based on landscape structure and spatial statistics. *Res. Soil Water Conserv.* **2020**, *27*, 247–255.
59. He, C.L.; Pu, J.W.; Shen, J.X. Spatial-temporal changes and driving mechanisms of landscape ecological security in lower reaches of Lancang River during 2005–2018. *Bull. Soil Water Conserv.* **2020**, *40*, 219–227.
60. Lin, J. Research on the Construction of Ecological Security Pattern of Panlong River Basin Based on the Evolution of Landscape Pattern. Master's Thesis, Kunming University of Science and Technoligy, Kunming, China, 2021; pp. 19–22.

61. Wu, Z.M.; Yu, Z.X.; Li, Y.; Liu, J.; Wang, J.S.; Zhang, C. The study of land use evolution and landscape ecological security evaluation in Dianchi Lake Basin. *J. Southwest For. Univ.* **2021**, *41*, 122–129.
62. Ren, J.T.; Wang, Z.H.; Zuo, T.A.; Feng, T.; Yang, J.; Zu, W.T.; Li, H. Spatial-temporal dynamic analysis on landscape ecological safety of Chishuihe River Basin in Guizhou Province. *Water Resour. Hydropower Eng.* **2021**, *52*, 96–104.
63. Ma, S.W.; Xie, D.T.; Zhang, X.C.; Peng, Z.T.; Hong, H.K.; Luo, Z.; Xiap, J.J. Measures of land ecological security early warning and its spatial-temporal evolution in the ecologically sensitive area of the Three Gorges reservoir area: A case study of Wanzhou District, Chongqing City. *Acta Ecol. Sin.* **2017**, *37*, 8227–8240.
64. Zhang, X.D.; Zhao, Z.P.; Zhao, Y.X.; Gao, X.H.; Ma, Y.X.; Liu, N.J.; Ji, W.B. Landscape ecological risk assessment and ecological security pattern optimization construction in Yinchuan City. *Arid Land Geogr.* **2022**, *45*, 1626–1636.
65. Ma, J.; Jiao, L. Landscape ecological security evaluation of Xinjiang Uygur Autonomous Region from the perspective of production-life-ecological space. *Hubei Agric. Sci.* **2022**, *61*, 100–105.
66. Li, Z.; Hou, H.T.; Huang, Y.M.; Xu, H. Land ecological security evaluation of Coal- Resource Cities in Heilongjiang Province: A case study of Shuangyashan City. *Sci. Technol. Manag. Land Resour.* **2011**, *28*, 46–51.
67. He, W.L.; Huo, J.; Meng, M.X.; Yin, S.Q. The land use change and landscape ecological security assessment in Huailai County. *Chin. J. Soil Sci.* **2016**, *47*, 15–20.
68. Zang, Y.Z.; Liu, Y.S.; Yang, Y.Y. Land use pattern change and its topographic gradient effect in the mountainous areas: A case study of Jinggangshan city. *J. Nat. Resour.* **2019**, *34*, 1391–1404.
69. Lu, Y.P.; Jiang, Y.B.; Gao, J.Y.; Cai, Z.X. The division of the Ordovician small watershed in the west of Tahe and its geological significance. *Fault-Block Oil Gas Field* **2021**, *28*, 440–445.
70. Wu, Y.; Li, C.Y.; Zhang, C.F.; Shi, X.H.; Zhao, S.N.; Lin, T.T. A watershed delineation method for mountains, plains complex landform area based on ArcGIS and SWAT. *Arid. Land Geogr.* **2016**, *39*, 413–419.

land

MDPI

Article

Applicability Analysis of GF-2PMS and PLANETSCOPE Data for Ground Object Recognition in Karst Region

Yu Zhang [1], Chaoyong Shen [1,2,3,*], Shaoqi Zhou [1,*], Ruidong Yang [1], Xuling Luo [3] and Guanglai Zhu [1]

1 College of Resources and Environmental Engineering, Guizhou University, Guiyang 550025, China
2 Guizhou Academy of Sciences, Guiyang 550009, China
3 The Third Surveying and Mapping Institute of Guizhou Province, Guiyang 550004, China
* Correspondence: shency@gzchsy.cn (C.S.); shaoqizhou@gzchsy.cn (S.Z.); Tel.: +86-13-985-536-169 (C.S.); +86-18-286-133-628 (S.Z.)

Abstract: Remote sensing image with high spatial and temporal resolution is very important for rational planning and scientific management of land resources. However, due to the influence of satellite resolution, revisit period, and cloud pollution, it is difficult to obtain high spatial and temporal resolution images. In order to effectively solve the "space–time contradiction" problem in remote sensing application, based on GF-2PMS (GF-2) and PlanetSope (PS) data, this paper compares and analyzes the applicability of FSDAF (flexible spatiotemporal data fusion), STDFA (the spatial temporal data fusion approach), and Fit_FC (regression model fitting, spatial filtering, and residual compensation) in different terrain conditions in karst area. The results show the following. (1) For the boundary area of water and land, the FSDAF model has the best fusion effect in land boundary recognition, and provides rich ground object information. The Fit_FC model is less effective, and the image is blurry. (2) For areas such as mountains, with large changes in vegetation coverage, the spatial resolution of the images fused by the three models is significantly improved. Among them, the STDFA model has the clearest and richest spatial structure information. The fused image of the Fit_FC model has the highest similarity with the verification image, which can better restore the coverage changes of crops and other vegetation, but the actual spatial resolution of the fused image is relatively poor, the image quality is fuzzy, and the land boundary area cannot be clearly identified. (3) For areas with dense buildings, such as cities, the fusion image of the FSDAF and STDFA models is clearer and the Fit_FC model can better reflect the changes in land use. In summary, compared with the Fit_FC model, the FSDAF model and the STDFA model have higher image prediction accuracy, especially in the recognition of building contours and other surface features, but they are not suitable for the dynamic monitoring of vegetation such as crops. At the same time, the image resolution of the Fit_FC model after fusion is slightly lower than that of the other two models. In particular, in the water–land boundary area, the fusion accuracy is poor, but the model of Fit_FC has unique advantages in vegetation dynamic monitoring. In this paper, three spatiotemporal fusion models are used to fuse GF-2 and PS images, which improves the recognition accuracy of surface objects and provides a new idea for fine classification of land use in karst areas.

Keywords: spatiotemporal fusion; land use; high resolution; FSDAF; STDFA; Fit_FC

Citation: Zhang, Y.; Shen, C.; Zhou, S.; Yang, R.; Luo, X.; Zhu, G. Applicability Analysis of GF-2PMS and PLANETSCOPE Data for Ground Object Recognition in Karst Region. *Land* **2023**, *12*, 33. https://doi.org/10.3390/land12010033

Academic Editor: Giuseppe Modica

Received: 7 November 2022
Revised: 7 December 2022
Accepted: 12 December 2022
Published: 22 December 2022

1. Introduction

Karst land landscape accounts for about 12% of the global land area, and the environment is very fragile [1,2]. Karst landform accounts for more than 1/3 of China's land area [3], with strong karstification, which has always been a research focus [4–6]. In view of the increasingly prominent ecological problems in karst areas, the pressure of land use change on ecology gradually emerged. Therefore, it is of great significance to obtain more efficient and accurate land use classification methods for optimizing the allocation of land resources and realizing ecological restoration in fragile karst mountainous areas.

High-spatial-resolution data can provide rich spatial information, geometric structure, and texture information of ground objects and other details. High-temporal-resolution data can provide continuous changes of surface elements in time and space and play an irreplaceable role in regional ecological environment monitoring [7]. However, due to the limitations of satellite launch cost, technical conditions, and satellite revisit cycle, the remote sensing image of a single satellite has the problem of mutual restriction between spatial resolution and temporal resolution [8–12]. As a result, the accuracy of land use classification is not high, which limits the practical application of remote sensing data [9,13,14]. At the same time, optical remote sensing images are easily affected by atmospheric conditions such as clouds, which reduces the availability of data and further hinders the acquisition of time-continuous high-spatial-resolution images [15]. A cost-effective way to solve this problem is to develop a data fusion model. High-temporal and low-resolution data are combined with high-resolution and low-temporal data to obtain remote sensing images with high spatial resolution and high temporal resolution [16–18].

There are five main categories of spatiotemporal fusion algorithms: decomposition-based methods, weight-function-based methods, Bayesian-based methods, learning-based methods, and hybrid-based methods [19]. Decomposition-based methods employ linear spectral mixing theory in analyzing the composition of coarse pixels and decomposing them to estimate the value of fine pixels, including algorithms such as STDFA. This type of algorithm is simple in principle and easy to operate, but it cannot obtain good decomposition results in mixed areas with many land cover types [20]. The method based on the weight function estimates the fine pixel value by combining the information of all input images with the weight function, mainly including STARFM, STAARCH, ESTARFM, and other algorithms. Most of these methods involve empirical functions, and the fusion accuracy is poor when there are too many types of land cover or when abnormal changes such as land cover mutation occur [21]. The Bayesian-based data fusion method combines the time-related information in the image time series to transform the fusion problem into an estimation problem, mainly including BME, unified, and other algorithms. These methods lead to lower prediction accuracy when the land cover type changes [22,23]. Learning-based methods use machine learning to model the relationships between observed image pairs and predict unobserved images, mainly including algorithms such as SPSTFM and EBSPTM. This type of method can capture the main features in the prediction, including land cover type changes, etc., but cannot accurately preserve the shape of the predicted objects, especially irregular-shaped ground objects [24]. There are also some spatiotemporal fusion methods that combine the advantages of decomposition methods, Bayesian theory, weight functions, and learning methods to pursue better fusion effects, such as FSDAF algorithms. This type of method can deal with different land cover type change problems through the combination of multiple methods, which improves the prediction accuracy of the model, but also increases the complexity of the algorithm [25]. In addition, Wang et al. proposed a method by combining regression model fitting, spatial filtering, and residual compensation [26]. This method has some shortcomings in capturing image structure and texture, but it has a good fusion effect when the terrain changes greatly. It has great application value for remote monitoring of environment, agriculture, and ecology [27].

At present, most of the spatiotemporal fusion algorithms use Landsat data and MODIS, MERIS, and other medium- and low-spatial-resolution data for fusion, meaning that the fused data are far from fulfilling the actual needs. With the development of satellite technology and the improvements in sensor technology, the demand for high spatial resolution is increasing. However, research on spatiotemporal fusion using high-resolution images is scarce; in particular, the accuracy of the high-resolution fusion images is unknown [28–32]. At the same time, there is no research on the fusion accuracy of different models in different land use types in the current spatiotemporal fusion research. Therefore, this study will fill the gap in the current field of spatiotemporal fusion to facilitate better use of satellite remote sensing data.

As the images of PS satellite constellation have high temporal resolution and high spatial resolution, GF-2 has the highest resolution among Chinese civil land observation satellites. Therefore, in this paper, FSDAF, STDFA, and Fit_FC models are used to fuse high-spatial-resolution GF-2 and high-temporal-resolution PS data, and the fusion accuracy of each model is analyzed at the same time. This provides a new idea for fine classification of land use in karst areas, and analyzes the applicability of GF-2 and PS data for feature recognition in the Karst region. This can provide a scientific basis for further application research based on high-spatial-resolution satellites such as time series GF-2.

2. Materials and Methods

2.1. Study Area

Caohai (104°10′–104°20′ E, 26°47′–27°52′ N), in Guizhou Province, is a typical karst plateau wetland lake. It is located on the south side of the county seat of Yi, Hui, and Miao Autonomous County, Weining County, northwest Guizhou Province, and it provides a habitat for rare birds such as *Grus nigricollis*, unique to China (Figure 1). It is a complete and typical karst plateau small watershed, which requires frequent monitoring using remote sensing images. The terrain of the study area is the highest in the east, higher in the southwest, and the lake area is situated in the middle. The water outlet of the watershed is in the northwest, with an average elevation of 2171.7 m and a watershed area of approximately 96 km^2 [33,34]. The land use types in the region are complex and diverse, mainly including construction land, forest land, cultivated land, rivers, and lakes. As it is located in the karst plateau area of Southwest China and belongs to the humid subtropical plateau monsoon climate, the study area has poor light conditions, heavy rainfall, and cloud cover all year round. These factors lead to a serious lack of optical remote sensing image data, especially high-resolution data, and there is an urgent need for high-spatiotemporal-resolution images in daily production and scientific research activities [35–37].

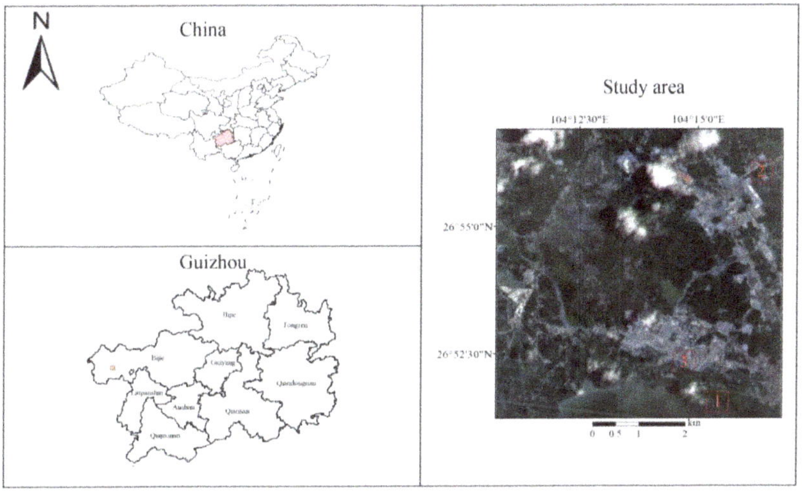

Figure 1. Location of the study region. (1) Land–water boundary area, (2) mountainous area, (3) urban area.

2.2. Data Sources

PS is the world's largest micro satellite group, consisting of hundreds of Dove (10 cm × 10 cm × 30 cm) satellites. PS data (https://www.planet.com/markets/education-and-research/, 15 March 2022) have a spatial resolution of 3 to 5 m, and the satellite can acquire data every day, with a short coverage period and fast update speed [38,39]. GF-2 is the first civil optical remote sensing satellite successfully launched by China in 2014, with a spatial resolution better than 1 m. It is the civil land observation satellite with the highest resolution in China. The GF-2 data come from the China Resources Satellite Application Center (http://www.cresda.com/CN/, 17 March 2022), the revisit period is 5 days, and the coverage period is 69 days [40–42].

This study selected the PS data of 2 scenes imaged on 15 April 2021 and 10 July 2021, and the GF-2 data of 2 scenes imaged on 15 April 2021 and 13 July 2021. Due to the limitation of revisit cycle, GF-2 does not have the image of 10 July 2021, so the scene with the closest time (13 July 2021) was selected. Among them, the GF-2 data for 15 April were used as the input known high-resolution low-temporal data. PS data for 15 April served as input known high-temporal low-resolution data. The PS data for 10 July were used as the high-temporal low-resolution data in the prediction period to simulate the high-resolution data in the corresponding period, and the GF-2 data for 13 July were used as the verification data for accuracy evaluation.

The data were radiometrically corrected using ENVI 5.3 software, and atmospheric correction was performed with the FLAASH Atmospheric correction module. Second, the PS data were converted to a UTM 50N/WGS84 projection and coordinate system and resampled to 1 m resolution using the nearest neighbor method. Finally, rectification was performed via the RPC Orthorectification Workflow tool to make the two images perfectly match. Finally, they were cropped to the same experimental area as the GF-2 data. In this study, four multispectral bands of GF-2 data and the corresponding PS band were selected as experimental bands. The specific band ranges are shown in Table 1.

Table 1. GF-2 PMS and PS image spectral ranges.

Band	Band Range of GF-2 (μm)	Band Range of PS (μm)
Blue	0.450~0.520	0.420~0.530
Green	0.520~0.590	0.500~0.590
Red	0.630~0.690	0.610~0.700
NIR	0.770~0.890	0.760~0.860

2.3. Methods

2.3.1. FSDAF Model

FSDAF integrates the method of mixed pixel decomposition and a weighting function, and provides better prediction results for changes in regional ground object types. The main steps are as follows: (1) classify the high-spatial-resolution images at time t_b; (2) use the reflectivity change of the PS image to estimate the time change of the corresponding ground object type from t_b to t_p; (3) use the category temporal change obtained in the previous step to predict the high-resolution image located in the t_p period and calculate the residual error of each pixel prediction of the PS image; (4) use the thin plate spline (TPS) function to predict the high-resolution image at time t_p; (5) calculate the residual distribution based on the thin-plate spline function; (6) use the neighborhood information to obtain the final prediction of the GF-2 image [25,27].

$$P_{high}(x_{ij}, y_{ij}) = B_{high}(x_{ij}, y_{ij}) + \sum_{k=1}^{n} [w_k \times \Delta R(x_k, y_k)] \tag{1}$$

$$\Delta P_{high}(x_{ij}, y_{ij}) = \varepsilon_{high}(x_{ij}, y_{ij}) + \Delta R_{high}(a) \tag{2}$$

In the formula, $\Delta P_{high}(x_{ij}, y_{ij})$ is the pixel change value at time t_b and time t_p; $\varepsilon_{high}(x_{ij}, y_{ij})$ is the residual of the high-spatial-resolution image assigned to the j-th pixel by the i-th pixel of high temporal resolution; $\Delta R_{high}(a)$ is the change value of surface cover type a of the high-spatial-resolution data between time t_b and time t_p; $P_{high}(x_{ij}, y_{ij})$ is the pixel value of the high-temporal-resolution image at time t_p; $B_{high}(x_{ij}, y_{ij})$ is the pixel value of the high-temporal-resolution image at time t_b; w_k is the weight value of the k-th similar pixel; $\Delta R(x_k, y_k)$ is the change value of pixel resolution at time t_b and time t_p. The residual value between the cell value of the base date and the cell value of the forecast date is calculated as follows:

$$\varepsilon_{high}(x_{ij}, y_{ij}) = m \times \varepsilon(x_i, y_i) \times W(x_{ij}, y_{ij}) \tag{3}$$

$$\varepsilon(x_i, y_i) = \Delta P_{low}(x_{ij}, y_{ij}) - \frac{1}{m}\left[\sum_{j=1}^{m} P_{t_p}^{TP}(x_{ij}, y_{ij}) - \sum_{j=1}^{m} B_{high}(x_{ij}, y_{ij})\right] \tag{4}$$

$$W(x_{ij}, y_{ij}) = CW(x_{ij}, y_{ij}) / \sum_{k=1}^{n} CW(x_{ij}, y_{ij}) \tag{5}$$

$$CW(x_{ij}, y_{ij}) = P_{t_p}^{SP}(x_{ij}, y_{ij}) - P_{t_p}^{TP}(x_{ij}, y_{ij}) + \varepsilon(x_i, y_i) \times [1 - HI \times \varepsilon(x_i, y_i)] \tag{6}$$

In the formula, m is the total number of high-spatial-resolution image pixels corresponding to the high-temporal-resolution image pixels (x_i, y_i); $\varepsilon(x_i, y_i)$ is the residual between the i-th pixel value predicted due to the time difference between the high spatial resolution at time t_b and time t_p; $CW(x_{ij}, y_{ij})$ is the weight of the assigned residual; $W(x_{ij}, y_{ij})$ is the weight of $CW(x_{ij}, y_{ij})$ normalized; $\Delta P_{low}(x_{ij}, y_{ij})$ is the pixel change value of the high-temporal-resolution image between time t_b and time t_p; $P_{t_p}^{TP}(x_{ij}, y_{ij})$ is the pixel value of the high-spatial-resolution image at time t_p predicted by the time difference; $P_{t_p}^{SP}(x_{ij}, y_{ij})$ is the pixel value of the high-spatial-resolution image at time t_p predicted after TPS optimization parameters; *HI* is the homogeneity coefficient, i.e., in the moving window, when the k-th high-spatial-resolution pixel (with a high temporal resolution) and the moving center pixel (x_{ij}, y_{ij}) have the same land cover type, *HI* is taken as 1; otherwise, *HI* takes a value of 0.

2.3.2. STDFA Model

The STDFA algorithm is a class of spectral unmixing methods. The algorithm first classifies high-resolution low-temporal images of known periods based on K-means, which is set to 5 categories in this paper, and uses Equation (7) to calculate the richness of each category in each high-temporal and low-resolution pixel. The corresponding subregion is determined by taking a high-temporal low-resolution pixel as the center and calculating the average reflectance value of each category in the subregion by Formula (8). Then, we assign this value to the corresponding class of high-resolution low-temporal pixels within the center pixel [36,37,43].

$$f(X, c) = N(X, c)/m \tag{7}$$

In the formula, $f(X, c)$ is the richness of category c in the high-temporal and low-resolution pixel X in the known period; $N(X, c)$ is the number of high-resolution and low-temporal pixels belonging to category c in pixel X; m is the number of high-temporal and low-temporal pixels contained in the high-temporal and low-temporal pixels X. We select the D high-temporal and low-resolution pixels with the highest abundance in each category, find the difference between these high-temporal and low-resolution pixels in the known period and the predicted period, and then use the least squares method to fit the high-resolution pixels of each category. Thus, we obtain the change in reflectivity of low-temporal pixels.

$$X(t) = \sum_{c=0}^{k} f(X, c) \times \overline{x}(c, t) \tag{8}$$

Limitation factor:

$$\sum_{c=0}^{k} f(X, c) = 1,\ f(X, c) \geq 0 \tag{9}$$

In the formula, t represents the prediction period and the known period t_0 and t_k; $\overline{x}(c, t)$ represents the average reflectance of category c in the high-temporal and low-resolution pixel X; k is the total number of categories. We calculate the average reflectivity of category c in the known period and the predicted period, respectively, and through an SRCM (surface reflectance calculation model), based on Equation (10) [19], the high-resolution low-temporal data of the final forecast period can be obtained.

$$x(c,t_0) = \overline{x}(c,t_0) - \overline{x}(c,t_k) + x(c,t_k) \tag{10}$$

In the formula, $x(c,t_0)$ and $x(c,t_k)$ represent the reflectivity of high-resolution low-temporal pixels belonging to category c in the prediction period and the known period, respectively.

2.3.3. Fit_FC Model

The Fit_FC algorithm is based on a linear model for data spatiotemporal fusion. It uses low-spatial-resolution and high-temporal-resolution data in the known period and the predicted period to fit linear coefficients, and then applies the coefficient to the known period of high-spatial-resolution and low-temporal-resolution data [36,44,45]. Taking the high-temporal-resolution pixel X as the center, we determine the neighborhood subregion, where the size of the subregion is 5 high-temporal and low-resolution pixels, and fit Formula (11) to the coefficients a and b.

$$X(t_0) = a \times X(t_k) + b \tag{11}$$

The predicted initial high-resolution low-temporal data can be obtained by applying the coefficients a and b to the high-resolution low-temporal pixels corresponding to the central pixel X in the known period. In addition, the residual value R can be obtained by Equation (12).

$$R = X(t_0) - (a \times X(t_k) + b) \tag{12}$$

In order to eliminate the "block effect" caused by the fusion of high- and low-resolution data, Formula (13) [13] is used to determine the similar neighborhood pixels centered on a high-resolution low-temporal pixel.

$$\sqrt{\sum_{b=1}^{nb} (x(t_k) - x_{neigh}(t_k))^2 / nb} \tag{13}$$

In the formula, nb represents the number of bands involved in the calculation; $x(t_k)$ and $x_{neigh}(t_k)$ represent the high-resolution low-temporal center pixel and its neighbors in the known period. The smallest D high-resolution low-temporal pixels are selected as similar pixels, and the corresponding weights are given according to the normalized distance from the central pixel. For the initially predicted high-resolution low-temporal data, firstly, based on similar pixels and their weight values, the initial correction of the central high-resolution and low-temporal pixels is obtained by means of weighted summation. The residual R is then linearly interpolated to ensure that it has the same resolution as the high-resolution low-temporal data, and, based on the obtained similar pixels and weights, the reflectance value of the central high-resolution low-temporal pixel is corrected again to obtain the final result.

2.3.4. Accuracy Evaluation

Using the real GF-2 band image acquired on 13 July 2021 as the verification image, visual interpretation and correlation analysis methods were used to evaluate the accuracy of the fusion image from both qualitative and quantitative aspects. The visual interpretation method can directly analyze the similarity between the fused image and the real image and yield a preliminary judgment on the fusion accuracy of each model. The correlation analysis method mainly uses four evaluation metrics: average absolute deviation (*AAD*), root mean square error (*RMSE*), correlation coefficient (*CC*), and structural similarity (*SSIM*) [26,46,47]. These indexes are used to quantitatively evaluate the similarity between the fused image and the real image.

AAD is used to measure deviation. The closer *AAD* is to 0, the smaller the deviation between the predicted value and the standard value.

$$AAD = \frac{1}{N}\sum_{i=1}^{N}|P_i - O_i| \tag{14}$$

RMSE is used to measure the difference between images, and its value ranges from 0 to 1. The smaller the *RMSE*, the higher the accuracy.

$$RMSE = \sqrt{\frac{\sum_{i=1}^{N}(P_i - O_i)^2}{N}} \tag{15}$$

CC can reflect the spectral similarity between images, and the closer *CC* is to 1, the higher the spectral similarity.

$$CC = \frac{\sum_{i=1}^{N}(P_i - \overline{P})(O_i - \overline{O})}{\sqrt{\sum_{i=1}^{N}(P_i - \overline{P})^2 \sum_{i=1}^{N}(O_i - \overline{O})^2}} \tag{16}$$

SSIM can evaluate the structural similarity between images. The closer the *SSIM* is to 1, the greater the structural similarity between images.

$$SSIM = \frac{(2\overline{PO} + C_1)\left(2_{\sigma_{po}} + C_2\right)}{(\overline{P}^2 + \overline{O}^2 + C_1)(\sigma_P^2 + \sigma_O^2 + C_2)} \tag{17}$$

In the Formulae (14)–(17), N is the total number of image pixels; P_i and O_i represent the i-th pixel of the predicted image and the observed image, respectively. $\overline{P}$ and $\overline{O}$ represent the mean of the fusion result and the observed image, respectively. $\overline{P}^2$ and $\overline{O}^2$ represent the variance between the fusion result and the observed image, respectively. σ_{po} represents the covariance between the fusion result and the observed image; C_1 and C_2 are two constants close to 0 used to stabilize the result, generally, $C_1 = (K_1L)^2$, $C_2 = (K_2L)^2$, generally $K_1 = 0.01$, $K_2 = 0.03$, $L = 255$ (dynamic range of pixel value, generally 255).

3. Results

In order to better evaluate the accuracy of the spatiotemporal fusion model under different landform types, this study mainly selected three experimental areas for algorithm comparison. The first test area belongs to the land and water boundary, and the main landform types are land and water. The second experimental region belongs to the mountainous area, and the main landform types are roads, buildings, and farmland. The third experimental region belongs to the urban area, and the landform types are mainly construction and roads. By studying the three types of terrain, we can provide more accurate support for the application of spatiotemporal fusion algorithms in different types of landforms.

3.1. The Accuracy of Land–Water Boundary

The original PS low-resolution images (Figure 2a,b) are blurry and can only roughly identify water and land—they cannot provide more detailed information. However, the results of the spatiotemporal fusion of the FSDAF, STDFA, and Fit_FC models show that FSDAF can clearly identify the water surface, land, shoal, etc., and the contours of ground objects are clear (Figure 2c). Compared with the FSDAF model, the fusion results of the STDFA model can also distinguish the water surface and the land, but there are large color spots in the results, which have a certain impact on the identification of the water–land boundary (Figure 2d). In addition, the fusion result of the Fit_FC model is very poor. Compared with the original image, it loses a large amount of detail and cannot effectively identify the land and water boundary (Figure 2e). Therefore, for the land–water boundary area, the FSDAF model has the best fusion effect, followed by the STDFA model, and the Fit_FC model has the worst effect.

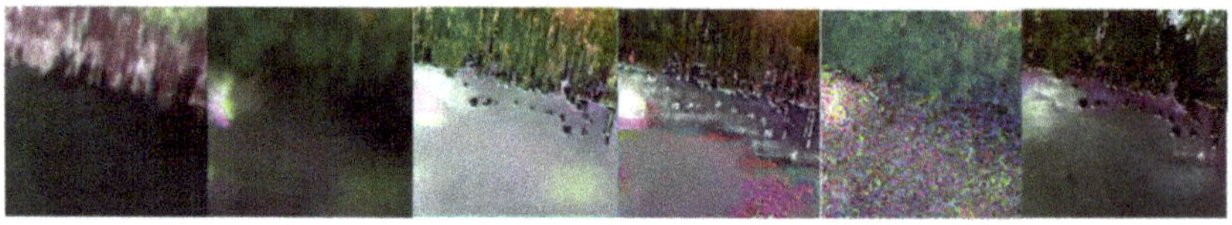

Figure 2. (**a**) PS image on 15 April; (**b**) PS image on 10 July; (**c**) fusion image by FSDAF on 10 July; (**d**) fusion image by STDFA on 10 July; (**e**) fusion image by Fit_FC on 10 July; (**f**) GF-2 verification image on 13 July.

As can be seen from Table 2, for all four bands, the images fused by the FSDAF model have a good correlation with the validation images, and the correlation coefficients are all higher than 0.6. Compared with the STDFA and Fit_FC models, the mean value of CC increased by 0.0889 and 0.3055, respectively, indicating that the fused image of FSDAF has higher spectral similarity with the validation image. At the same time, the SSIM values of the FSDAF and STDFA models are both greater than 0.7, indicating that the fusion images of the two models have good structural similarity with the predicted images. Among them, except for the near-infrared band, the FSDAF model has the highest SSIM, and its average is 0.0077 and 0.0637 higher than those of the FSDAF and STDFA models, respectively, indicating that the model has the best structural similarity. For the Fit_FC model, the RMSE of the four bands and the AAD values of the blue, green, and red bands are higher than those of the FSDAF and STDFA models, with an average of 0.1347 and 0.1028, respectively. Compared with the other two models, the average value of RMSE is increased by 0.037 and 0.036, and the average value of AAD is increased by 0.0155 and 0.0148, respectively, indicating that the fusion image of the Fit_FC model has a large deviation from the predicted image. The statistical results show that the fusion image results of the FSDAF algorithm and the STDFA algorithm in study area 3 are much better than those of the Fit_FC algorithm, which is consistent with the direct visual effect.

Table 2. The fusion accuracy evaluation of different bands for different models.

CJ5	Method	CC	SSIM	RMSE	AAD
Blue	FSDAF	0.7134	0.7449	0.0987	0.0730
	STDFA	0.6127	0.7351	0.0995	0.0733
	Fit_FC	0.2397	0.6283	0.1190	0.0941
Green	FSDAF	0.6240	0.7937	0.1123	0.0834
	STDFA	0.5258	0.7770	0.1148	0.0848
	Fit_FC	0.1247	0.6853	0.1376	0.1102
Red	FSDAF	0.6387	0.7328	0.1232	0.0970
	STDFA	0.4887	0.7135	0.1241	0.0969
	Fit_FC	0.3968	0.6609	0.1384	0.1144
NIR	FSDAF	0.6036	0.7070	0.1427	0.0939
	STDFA	0.5969	0.7221	0.1411	0.0937
	Fit_FC	0.5965	0.7185	0.1438	0.0924
Mean	FSDAF	0.6449	0.7446	0.1192	0.0868
	STDFA	0.5560	0.7369	0.1199	0.0872
	Fit_FC	0.3394	0.6733	0.1347	0.1028

3.2. The Accuracy of Mountains

Study area 2 is mainly a mountainous area. The ground objects in the basic image obtained on 15 April 2021 are mainly cultivated land, buildings, and roads. In the image obtained on 10 July 2021, the original cultivated land has undergone the process of crop coverage changes (Figure 3a,b). The original PS image can roughly identify the ground object information, but its resolution is still somewhat insufficient for the identification of more detailed information. According to the effect of the spatiotemporal fusion algorithm, the three models have better fusion effects on ground objects, and the identification of ground object information is obviously more accurate. The FSDAF algorithm and STDFA algorithm have higher fusion image accuracy, but the fusion image displays a poor response to changes in crop coverage (Figure 3c,d). Generally speaking, the fusion image and the verification image should have the similar spectral, but the color of the two models in the crop coverage area is quite different from that of the verification image. Fortunately, we were still able to distinguish vegetation cover areas by color comparisons. In terms of resolution, both models can clearly display the spatial structure information of ground objects; in particular, the land structure in the vegetation-covered area can be better observed. Compared with the original PS image, the spatial resolution of the fusion image is also improved to a certain extent, the contours of different types of objects are also clearer, and the changes in the coverage areas of crops can be better displayed (Figure 3e). However, its resolution in specific spatial details is slightly lower than that of the other two algorithms.

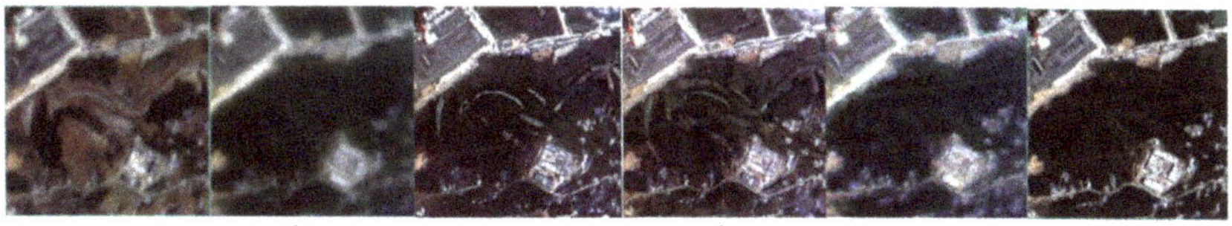

Figure 3. (**a**) PS image on 15 April; (**b**) PS image on 10 July; (**c**) fusion image by FSDAF on 10 July; (**d**) fusion image by STDFA on 10 July; (**e**) fusion image by Fit_FC on 10 July; (**f**) GF-2 verification image on 13 July.

From the statistical analysis results (Table 3), the fusion image of the Fit_FC model has a good correlation with the verification image. With the exception of the red band, the CC value of the Fit_FC model is higher than that of the FSDAF and STDFA models. The average value of CC is 0.7138, which is 0.0605 and 0.0166 higher than that of the FSDAF and STDFA models, respectively, indicating that the spectral similarity between the fusion image and the verification effect is higher. At the same time, the SSIM of the Fit_FC model in the blue, green, and red bands is also higher than that of the FSDAF and STDFA models, with an average of 0.6641, which is 0.0434 and 0.0287 higher than that of the other two models, respectively, indicating that its structural similarity is also higher than that of the other two models. In all four bands, the RMSE value of the FSDAF model is the highest, that of the Fit_FC model is the lowest, and the average values of the three models are 0.0791, 0.0052, and 0.0038 in descending order, indicating that the difference between the fusion image of the FSDAF model and the validation image is greater than others. Regarding the AAD value, the AAD of the FSDAF and STDFA models in the blue, green, and red bands is significantly higher than that of the Fit_FC model, but slightly lower than that of the Fit_FC model in the near-infrared band. The average AAD values of the FSDAF, STDFA, and Fit_FC models were 0.0382, 0.0381, and 0.0291, respectively, indicating that the Fit_FC model had less biased fusion images. The statistical results show that the Fit_FC model has the best fusion effect in mountainous areas, and the FSDAF algorithm has the worst fusion effect.

Table 3. The fusion accuracy evaluation of different bands for different models.

Band	Method	CC	SSIM	RMSE	AAD
Blue	FSDAF	0.6437	0.4863	0.2899	0.0653
	STDFA	0.7335	0.5013	0.0787	0.0652
	Fit_FC	0.7588	0.5590	0.0605	0.0445
Green	FSDAF	0.6563	0.6621	0.2817	0.0473
	STDFA	0.7107	0.6763	0.0720	0.0469
	Fit_FC	0.7648	0.7171	0.0607	0.0320
Red	FSDAF	0.6920	0.6478	0.2791	0.0288
	STDFA	0.7437	0.6680	0.0688	0.0290
	Fit_FC	0.6658	0.7041	0.0614	0.0230
NIR	FSDAF	0.6214	0.6866	0.2744	0.0115
	STDFA	0.6009	0.6961	0.0672	0.0114
	Fit_FC	0.6658	0.6760	0.0641	0.0171
Mean	FSDAF	0.6534	0.6207	0.2813	0.0382
	STDFA	0.6972	0.6354	0.0717	0.0381
	Fit_FC	0.7138	0.6641	0.0617	0.0292

3.3. *The Accuracy of Urban*

Study area 3 is mainly an urban area, and the types of ground objects in the area are mainly urban buildings, building land, roads, and vegetation greening. In the images from April and July in this area, with the exception of some areas where the land use changed (marked in yellow), the rest of the features changed little (Figure 4a,b). The original PS image can identify different types of objects, but the outlines between buildings are relatively blurred. Through direct observation of the fused image, compared to the PS image, all three models have improved spatial resolution to a certain extent, and can restore the area partially covered by shadows (red border) (Figure 4c–e). Among them, the fusion images of the STDFA and FSDAF models have a higher resolution. Fit_FC is relatively blurry on the outline of the building, and there is a more obvious block effect. For the two land use changes in the image, the fusion results of the STDFA algorithm cannot clearly reflect these. FSDAF is also extremely blurry, mainly following the original base image, making it difficult to effectively identify changes. Relatively speaking, the fusion image of the Fit_FC algorithm can better reflect the difference between the base image and the predicted image, and is more similar to the verification image. However, it does not achieve excellent result for the recognition of building outlines.

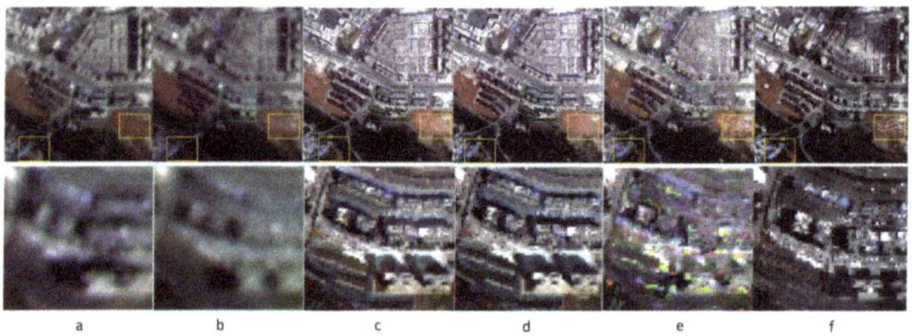

Figure 4. (**a**) PS image on 15 April; (**b**) PS image on 10 July; (**c**) fusion image by FSDAF on 10 July; (**d**) fusion image by STDFA on 10 July; (**e**) fusion image by Fit_FC on 10 July; (**f**) GF-2 verification image on 13 July.

According to the statistical analysis results (Table 4), the CC values of the three models are generally distributed in the range of 0.5–0.6, and the difference between each band is small; the FSDAF model has the highest CC values in the blue, green, and red bands, and the Fit_FC model has the highest in the near-infrared band. The mean values of CC for the FSDAF, STDFA, and Fit_FC algorithms are 0.5434, 0.5067, and 0.5362, respectively, indicating that the spectral similarity between the fused image and the validation impact is the highest for FSDAF and the lowest for the STDFA model. The SSIM gaps of the FSDAF, STDFA, and Fit_FC models are small, with average values of 0.7257, 0.7072, and 0.7323, respectively, and the Fit_FC model has the highest SSIM values in the green, red, and near-infrared bands, indicating that the structural similarity between the fusion image and the verification image is better. The RMSE value of the STDFA model is higher than that of the FSDAF and Fit_FC models in the green, red, and near-infrared bands, and the average values of the three models are 0.0040, 0.0036, and 0.0036, respectively, indicating that the STDFA model has a larger error in fused images. At the same time, the average AAD values of the three models of FSDAF, STDFA, and Fit_FC are 0.0075, 0.0058, and 0.0038, respectively, indicating that the fusion image deviation of the FSDAF model is large. In general, although the FSDAF model has the highest AAD, its fusion image still has a good prediction effect.

Table 4. The fusion accuracy evaluation of different bands for different models.

Band	Method	CC	SSIM	RMSE	AAD
Blue	FSDAF	0.5880	0.7839	0.0449	0.0091
	STDFA	0.5711	0.7835	0.0455	0.0091
	Fit_FC	0.5447	0.7775	0.0467	0.0057
Green	FSDAF	0.5092	0.7382	0.0568	0.0079
	STDFA	0.4850	0.7313	0.0581	0.0079
	Fit_FC	0.4797	0.7451	0.0578	0.0010
Red	FSDAF	0.5775	0.7238	0.0604	0.0073
	STDFA	0.5467	0.7098	0.0626	0.0005
	Fit_FC	0.5768	0.7334	0.0608	0.0030
NIR	FSDAF	0.4989	0.6570	0.0754	0.0057
	STDFA	0.4240	0.6044	0.0806	0.0057
	Fit_FC	0.5436	0.6733	0.0727	0.0053
Mean	FSDAF	0.5434	0.7257	0.0594	0.0075
	STDFA	0.5067	0.7073	0.0617	0.0058
	Fit_FC	0.5362	0.7323	0.0595	0.0038

4. Discussion

Different spatiotemporal fusion models have different fusion effects in karst areas. In order to select a more appropriate high-resolution data fusion model under different scenarios and needs, in this paper, the fusion results of the three models in different karst landforms are directly observed and statistically analyzed, and the application effects of the three models in karst areas are discussed.

4.1. FSDAF in Different Regions

The fusion image of the FSDAF model can improve the resolution of the original image and the classification accuracy of surface land use in three geomorphic types (water land border, mountain area, and urban area). Among them, the FSDAF model has a good fusion effect in the land water border area. It can not only clearly identify the water boundary, but it can also effectively identify information such as shoals. Therefore, the FSDAF model can be used for spatiotemporal fusion of target lakes, oceans, rivers, and other waters, and can accurately extract the water boundary. This advantage has important value in flood relief, remote danger monitoring of dammed lakes, and other practical applications. The FSDAF model has a large color difference between the fusion image and the verification image in mountainous areas and other areas with large seasonal changes in vegetation cover. The vegetation coverage cannot be restored well. The FSDAF model can effectively improve the resolution of urban areas. It can more effectively identify the building outline, and more effectively display the part of the original image that is blocked by shadows. However, similar to the integration performance of mountain areas, the model cannot well show the land type change in the building area. This may be because the FSDAF model mainly uses spatial prediction to retrieve pixel changes. Theoretically, spatial prediction can truly describe the surface information of the predicted date. In addition, the signals of land cover type change and local variation are retained in the fusion results [26]. However, in the actual process, the error of FSDAF mainly depends on the residual distribution under the assumption of surface uniformity. Therefore, the FSDAF model can save more detailed information through this strategy, but it limits its ability to retrieve land cover changes.

4.2. STDFA in Different Regions

The STDFA model also has a good fusion effect on the images of different landforms in karst areas, which can effectively improve the image resolution. The STDFA model has a good recognition ability in the interface area between water and land, but the accuracy of fusion results is not high due to the appearance of "patches" in the predicted image. Different from this, the fusion image resolution of the STDFA model in mountainous areas is very high, which can better identify the structural information between ground objects. However, the fusion accuracy of the STDFA model is lower than that of the Fit_FC model in areas with large changes in vegetation cover such as crops. However, based on the statistical data, the CC and SSIM of each band of the STDFA model are lower than that of the FSDAF model, while RMSE is higher than that of the FSDAF model. This indicates that in urban areas, the fusion accuracy of the STDFA model is lower than that of the FSDAF model. In addition, the STDFA model is a spatiotemporal fusion model based on the unmixing method, and its data fusion accuracy is related to two aspects: On the one hand, the STDFA model needs to classify high-resolution data in the basic period, but the classification accuracy of unsupervised classification methods (such as K-means method) will cause the fusion accuracy to decrease. On the other hand, when the resolution difference between high-resolution low-temporal data and high-temporal low-resolution data is large, the area represented by each high-resolution pixel will be more refined. For example, in the pixels of high-temporal and low-resolution data, when the richness of a certain category is very low, the fitting error will increase [48,49].

4.3. Fit_FC in Different Regions

The fusion accuracy of the Fit_FC model in different geomorphic types in karst area is quite different. The fusion effect of the Fit_FC model is poor at the interface between land and water. Compared with the original low-spatial-resolution measurement data on 10 July 2021, the spatial resolution is not significantly improved, and it is difficult to identify the boundary between water surface and land. Meanwhile, the statistical results of the Fit_FC model also show that the fusion accuracy is extremely low, and the correlation coefficient of the green band is as low as 0.12466. Therefore, the Fit_FC model is not suitable for image fusion at the interface between land and water. The Fit_FC model has a good fusion result in the mountainous area. The spatial resolution of the image is improved, and the changes of vegetation cover such as crops can be better presented. It is very suitable for spatiotemporal fusion in areas with large vegetation changes and increases the accuracy of ground class classification. Therefore, the Fit_FC model can be given priority when studying the requirements of vegetation dynamic monitoring and land use change. The fusion results of the Fit_FC model have lower resolution than the FSDAF and STDFA models in the contour of ground objects such as roads and buildings. Especially in densely built areas, the resolution gap is larger. In addition, the Fit_FC model fits the high-temporal and low-resolution data of the known and predicted periods at pixel scale, and directly applies the fitting coefficients to the high-resolution and low-temporal data. When the difference between high-resolution data and low-resolution data is large, the results of the Fit_FC model show obvious "block effect" [28,50].

4.4. Statistical Precision Analysis

For the fusion effect of the three models in different regions, it is found that the resolution is generally good in direct observation, but the statistical data of the related accuracy are obviously not high, being clearly lower than that of the related low-resolution data. This is due to the high resolution of the two sets of data that we use, which can accurately identify small changes in ground objects. Especially when the prediction data and the verification data are separated by three days, the information on the ground objects, such as vehicles, will be slightly different, and some land types will also change. At the same time, in the process of the spatiotemporal fusion of high-resolution variable images, the effect of sunlight will also have a great impact on the accuracy. Differences in shooting time, different satellite shooting angles, and changes in the incident angle of sunlight will cause the shadow areas of the basic image and the verification image to differ, which will have a certain impact on the analysis of statistical data. Under the combined effect of these factors, the fusion accuracy of the three algorithms for high-resolution data is lower than the fusion of the same model for medium- and low-resolution data. However, in general, in most cases, the resolution of the fused image becomes higher, the recognizability is greatly enhanced, and the practical application value is higher.

4.5. Classification Accuracy Verification

In order to verify the application accuracy of the fusion results in land use classification, we selected an area containing mountains, buildings, and waters for research. First, the three methods are applied to different land use types for spatiotemporal integration. Secondly, the fusion results are divided into forest land, dry land, construction land, and water through the supervised classification method. Finally, through the comparison with the data of the Third National Land Survey of China (TNLS), it can be seen from Figure 5 and Table 5 that the classification results of the fused images (Figure 5a) are highly consistent with the data of the TNLS of China (Figure 5b). However, compared with the data of TNLS, the classification results did not effectively divide the small area of water around the construction land region, which made the water area in the classification results smaller. For forest land and dry land, the classification results are scattered. At the same time, due to the time difference between the fusion image and TNLS, the original dry land is distributed with crops, which makes the classification result of the fusion image become forest land. This may result in the increase of forest land area and the decrease of dry land area. For construction land, the area may be increased due to construction activities. However, the classification results are basically consistent with the data of TNLS. In general, the classification result of the fusion image is good, and the difference of land area of each type is within 15%. Therefore, images fused by spatiotemporal fusion model can be used for land use classification.

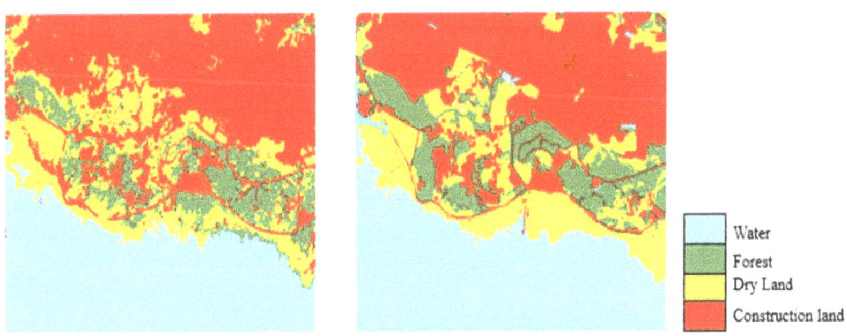

(a) (b)

Figure 5. (**a**) Supervise classification results; (**b**) Data of China's Third National Land Survey.

Table 5. Comparison between classification results and TNLS data.

Land Use	Classification (km²)	TNLS (km²)	D-Value (km²)	Ratio
Dry land	1.1301	1.3901	−0.2600	81.29%
Water	2.7325	2.6749	0.0576	102.15%
Forest land	1.0585	0.937	0.1215	112.96%
Construction land	3.1611	3.0801	0.0810	102.63%

5. Conclusions

In this study, three simple and easy-to-promote spatiotemporal fusion models, FSDAF, STDFA, and Fit_FC, are selected to fuse GF-2 and PS high-resolution satellite data. Four classical evaluation indexes, SSIM, CC, RMSE, and AAD, and visual analysis are adopted. The applicability of the three models for land use classification in karst areas is discussed comprehensively. The results show that the three models can improve the accuracy of land surface recognition, but the accuracy is different in different land use types. Among them, the fusion results of the FSDAF model can improve the recognition accuracy of land and water interface. Different from the FSDAF model, the STDFA model has the highest resolution of fusion image in mountain region, with significant improvement of fusion image resolution and rich details. The fusion effect of the Fit_FC model is poor in the boundary region of water and land. The image is blurred, and the ground feature information cannot be restored clearly, which is not conducive to the classification of land and water boundary land use. However, the Fit_FC model can clearly show land use change in vegetation covered areas. Therefore, this paper adopts a high-resolution spatiotemporal fusion algorithm to effectively improve the classification of land use in karst areas. It is of great significance to optimize the allocation of land resources and realize ecological restoration in fragile karst mountainous areas.

Author Contributions: Y.Z., C.S., S.Z., R.Y., X.L. and G.Z. contributed to the study conception and design. Material preparation, data collection, and analysis were performed by Y.Z., C.S., S.Z., R.Y., X.L. and G.Z. The first draft of the manuscript was written by Y.Z. commented on previous versions of the manuscript. All authors have read and agreed to the published version of the manuscript.

Funding: This research work was supported jointly by the Outstanding Youth of Science and Technology program of Guizhou Province of China ((2021) 5615), the Department of Science and Technology program of Guizhou Province of China ((2022) 213 and (2023) 60).

Data Availability Statement: Not applicable.

Acknowledgments: Thanks to China Resources Satellite Application Center and PlanetScope data publishing site for providing us with the data, which are important to our research.

Conflicts of Interest: The authors declare no conflict of interest.

References

1. Febles-Gonzalez, J.M.; Vega-Carreno, M.B.; Tolon-Becerra, A.; Lastra-Bravo, X. Assessment of soil erosion in karst regions of Havana, Cuba. *Land Degrad. Dev.* **2012**, *23*, 465–474. [CrossRef]
2. Zhang, S.R.; Bai, X.Y.; Zhao, C.W.; Tan, Q.; Luo, G.J.; Wang, J.F.; Li, Q.; Wu, L.H.; Chen, F.; Li, C.J.; et al. Global CO_2 consumption by silicate rock chemical weathering: Its past and future. *Earth's Futur.* **2021**, *9*, e2020EF001938. [CrossRef]
3. Li, Q.; Wang, S.; Bai, X.; Luo, G.; Song, X.; Tian, Y.; Hu, Z.; Yang, Y.; Tian, S. Change Detection of Soil Formation Rate in Space and Time Based on Multi Source Data and Geospatial Analysis Techniques. *Remote Sens.* **2020**, *12*, 121. [CrossRef]
4. Li, C.J.; Bai, X.Y.; Tan, Q.; Zhao, C.W.; Luo, G.J.; Wu, L.H.; Chen, F.; Xi, H.P.; Luo, X.L.; Ran, C.; et al. High-resolution mapping of the global silicate weathering carbon sink and its long-term changes. *Glob. Change Biol.* **2022**, *28*, 4377–4394. [CrossRef] [PubMed]
5. Zhang, S.R.; Bai, X.Y.; Zhao, C.W.; Tan, Q.; Luo, G.J.; Wu, L.H.; Xi, H.P.; Li, C.J.; Chen, F.; Ran, C.; et al. China' s carbon budget inventory from 1997 to 2017 and its challenges to achieving carbon neutral strategies. *J. Clean. Prod.* **2022**, *347*, 130966. [CrossRef]
6. Gong, S.H.; Wang, S.J.; Bai, X.Y.; Luo, G.J.; Zeng, C. Response of the weathering carbon sink in terrestrial rocks to climate variables and ecological restoration in China. *Sci. Total Environ.* **2020**, *750*, 141525. [CrossRef] [PubMed]

7. Chen, F.; Bai, X.; Liu, F.; Luo, G.; Tian, Y.; Qin, L.; Li, Y.; Xu, Y.; Wang, J.; Wu, L.; et al. Analysis Long-Term and Spatial Changes of Forest Cover in Typical Karst Areas of China. *Land* **2022**, *11*, 1349. [CrossRef]
8. Roy, D.P.; Ju, J.; Lewis, P.; Schaaf, C.; Gao, F.; Hansen, M.; Lindquist, E. Multi-temporal MODIS–Landsat data fusion for relative radiometric normalization, gap filling, and prediction of Landsat data. *Remote Sens. Environ.* **2008**, *112*, 3112–3130. [CrossRef]
9. Gevaert, C.M.; García-Haro, F.J. A comparison of STARFM and an unmixing-based algorithm for Landsat and MODIS data fusion. *Remote Sens. Environ.* **2015**, *156*, 34–44. [CrossRef]
10. Liao, C.; Wang, J.; Pritchard, I.; Liu, J.; Shang, J. A Spatio-Temporal Data Fusion Model for Generating NDVI Time Series in Heterogeneous Regions. *Remote Sens.* **2017**, *9*, 1125. [CrossRef]
11. Guan, X.; Liu, G.; Huang, C.; Liu, Q.; Wu, C.; Jin, Y.; Li, Y. An Object-Based Linear Weight Assignment Fusion Scheme to Improve Classification Accuracy Using Landsat and MODIS Data at the Decision Level. *IEEE Trans. Geosci. Remote Sens.* **2017**, *55*, 6989–7002. [CrossRef]
12. Chen, B.; Ge, Q.; Fu, D.; Yu, G.; Sun, X.; Wang, S.; Wang, H. A data-model fusion approach for upscaling gross ecosystem productivity to the landscape scale based on remote sensing and flux footprint modelling. *Biogeosciences* **2010**, *7*, 2943–2958. [CrossRef]
13. Zhang, H.K.; Huang, B.; Zhang, M.; Cao, K.; Yu, L. A generalization of spatial and temporal fusion methods for remotely sensed surface parameters. *Int. J. Remote Sens.* **2015**, *36*, 4411–4445. [CrossRef]
14. Wei, J.; Wang, L.; Liu, P.; Chen, X.; Li, W.; Zomaya, A.Y. Spatiotemporal Fusion of MODIS and Landsat-7 Reflectance Images via Compressed Sensing. *IEEE Trans. Geosci. Remote Sens.* **2017**, *55*, 7126–7139. [CrossRef]
15. Ju, J.; Roy, D.P. The availability of cloud-free Landsat ETM+ data over the conterminous United States and globally. *Remote Sens. Environ.* **2008**, *112*, 1196–1211. [CrossRef]
16. Mizuochi, H.; Hiyama, T.; Ohta, T.; Fujioka, Y.; Kambatuku, J.R.; Iijima, M.; Nasahara, K.N. Development and evaluation of a lookup-table-based approach to data fusion for seasonal wetlands monitoring: An integrated use of AMSR series, MODIS, and Landsat. *Remote Sens. Environ.* **2017**, *199*, 370–388. [CrossRef]
17. Song, H.; Huang, B. Spatiotemporal Satellite Image Fusion Through One-Pair Image Learning. *IEEE Trans. Geosience Remote Sens.* **2013**, *51*, 1883–1896. [CrossRef]
18. Liu, J.; Fan, X.; Jiang, J.; Liu, R.; Luo, Z. Learning a Deep Multi-scale Feature Ensemble and an Edge-attention Guidance for Image Fusion. *IEEE Trans. Circuits Syst. Video Technol.* **2021**, *32*, 105–119. [CrossRef]
19. Zhu, X.; Cai, F.; Tian, J.; Williams, T.K.-A. Spatiotemporal Fusion of Multisource Remote Sensing Data: Literature Survey, Taxonomy, Principles, Applications, and Future Directions. *Remote Sens.* **2018**, *10*, 527. [CrossRef]
20. Wu, B.; Huang, B.; Cao, K.; Zhuo, G. Improving spatiotemporal reflectance fusion using image inpainting and steering kernel regression techniques. *Int. J. Remote Sens.* **2017**, *38*, 706–727. [CrossRef]
21. Amorós-López, J.; Gómez-Chova, L.; Alonso, L.; Guanter, L.; Zurita-Milla, R.; Moreno, J.; Camps-Valls, G. Multitemporal fusion of Landsat/TM and ENVISAT/MERIS for crop monitoring. *Int. J. Appl. Earth Obs. Geoinf.* **2013**, *23*, 132–141. [CrossRef]
22. Zhang, W.; Li, A.; Jin, H.; Bian, J.; Zhang, Z.; Lei, G.; Qin, Z.; Huang, C. An Enhanced Spatial and Temporal Data Fusion Model for Fusing Landsat and MODIS Surface Reflectance to Generate High Temporal Landsat-Like Data. *Remote Sens.* **2013**, *5*, 5346–5368. [CrossRef]
23. Hwang, T.; Song, C.; Bolstad, P.V.; Band, L.E. Downscaling real-time vegetation dynamics by fusing multi-temporal MODIS and Landsat NDVI in topographically complex terrain. *Remote Sens. Environ.* **2011**, *115*, 2499–2512. [CrossRef]
24. Huang, B.; Song, H. Spatiotemporal reflectance fusion via sparse representation. *IEEE Trans. Geosci. Remote Sens.* **2012**, *50*, 3707–3716. [CrossRef]
25. Zhu, X.; Helmer, E.H.; Gao, F.; Liu, D.; Chen, J.; Lefsky, M.A. A flexible spatiotemporal method for fusing satellite images with different resolutions. *Remote Sens. Environ. Interdiscip. J.* **2016**, *172*, 165–177. [CrossRef]
26. Wang, Q.M.; Atkinson, P.M. Spatio-Temporal Fusion for Daily Sentinel-2 Images. *Remote Sens. Environ.* **2018**, *204*, 31–42. [CrossRef]
27. Li, Y.; Ren, Y.; Gao, W.; Jia, J.; Tao, S.; Liu, X. An enhanced spatiotemporal fusion method–Implications for DNN based time-series LAI estimation by using Sentinel-2 and MODIS. *Field Crops Res.* **2022**, *279*, 108452. [CrossRef]
28. Qiao, X.; Yang, G.; Shi, J.; Zuo, Q.; Liu, L.; Niu, M.; Wu, X.; Ben-Gal, A. Remote Sensing Data Fusion to Evaluate Patterns of Regional Evapotranspiration: A Case Study for Dynamics of Film-Mulched Drip-Irrigated Cotton in China's Manas River Basin over 20 Years. *Remote Sens.* **2022**, *14*, 3438. [CrossRef]
29. Zhang, H.; Huang, F.; Hong, X.; Wang, P. A Sensor Bias Correction Method for Reducing the Uncertainty in the Spatiotemporal Fusion of Remote Sensing Images. *Remote Sens.* **2022**, *14*, 3274. [CrossRef]
30. Chen, R.; Li, X.; Zhang, Y.; Zhou, P.; Wang, Y.; Shi, L.; Jiang, L.; Ling, F.; Du, Y. Spatiotemporal Continuous Impervious Surface Mapping by Fusion of Landsat Time Series Data and Google Earth Imagery. *Remote Sens.* **2021**, *13*, 2409. [CrossRef]
31. Morgan, B.E.; Chipman, J.W.; Bolger, D.T.; Dietrich, J.T. Spatiotemporal Analysis of Vegetation Cover Change in a Large Ephemeral River: Multi-Sensor Fusion of Unmanned Aerial Vehicle (UAV) and Landsat Imagery. *Remote Sens.* **2021**, *13*, 51. [CrossRef]
32. Luo, Y.; Guan, K.; Peng, J.; Wang, S.; Huang, Y. STAIR 2.0: A Generic and Automatic Algorithm to Fuse Modis, Landsat, and Sentinel-2 to Generate 10 m, Daily, and Cloud-/Gap-Free Surface Reflectance Product. *Remote Sens.* **2020**, *12*, 3209. [CrossRef]
33. Zhou, C.W.; Yang, R.; Yu, L.F.; Zhang, Y.; Yan, L.B. Hydrological and ecological effects of climate change in Caohai watershed based on SWAT model. *Appl. Ecol. Environ. Res.* **2019**, *17*, 161–172. [CrossRef]

34. Wu, J.; Yang, H.; Yu, W.; Yin, C.; He, Y.; Zhang, Z.; Xu, D.; Li, Q.; Chen, J. Effect of Ecosystem Degradation on the Source of Particulate Organic Matter in a Karst Lake: A Case Study of the Caohai Lake, China. *Water* **2022**, *14*, 1867. [CrossRef]
35. Yang, K.; Yu, Z.; Luo, Y.; Zhou, X.; Shang, C. Spatial-temporal variation of lake surface water temperature and its driving factors in Yunnan-Guizhou Plateau. *Water Resour. Res.* **2019**, *55*, 4688–4703. [CrossRef]
36. Xiao, L.; Chen, X.; Zhang, R.; Zhang, Z. Spatiotemporal Evolution of Droughts and Their Teleconnections with Large-Scale Climate Indices over Guizhou Province in Southwest China. *Water* **2019**, *11*, 2104. [CrossRef]
37. Lian, Y.Q.; You, G.J.Y.; Lin, K.R.; Jiang, Z.C.; Zhang, C.; Qin, X.Q. Characteristics of climate change in southwest China karst region and their potential environmental impacts. Environ. *Earth Sci.* **2014**, *74*, 937–944. [CrossRef]
38. Li, W.; Jiang, J.; Guo, T.; Zhou, M.; Tang, Y.; Wang, Y.; Zhang, Y.; Cheng, T.; Zhu, Y.; Cao, W.; et al. Generating Red-Edge Images at 3 M Spatial Resolution by Fusing Sentinel-2 and Planet Satellite Products. *Remote Sens.* **2019**, *11*, 1422. [CrossRef]
39. Kwan, C.; Zhu, X.; Gao, F.; Chou, B.; Perez, D.; Li, J.; Shen, Y.; Koperski, K.; Marchisio, G. Assessment of spatiotemporal fusion algorithms for planet and worldview images. *Sensors* **2018**, *18*, 1051. [CrossRef] [PubMed]
40. Ren, K.; Sun, W.; Meng, X.; Yang, G.; Du, Q. Fusing China GF-5 Hyperspectral Data with GF-1, GF-2 and Sentinel-2A Multispectral Data: Which Methods Should Be Used? *Remote Sens.* **2020**, *12*, 882. [CrossRef]
41. Ren, J.; Yang, W.; Yang, X.; Deng, X.; Zhao, H.; Wang, F.; Wang, L. Optimization of Fusion Method for GF-2 Satellite Remote Sensing Images based on the Classification Effect. *Earth Sci. Res. J.* **2019**, *23*, 163–169. [CrossRef]
42. Zhang, D.; Xie, F.; Zhang, L. Preprocessing and fusion analysis of GF-2 satellite Remote-sensed spatial data. In Proceedings of the 2018 International Conference on Information Systems and Computer Aided Education (ICISCAE), Changchun, China, 6–8 July 2018; IEEE: Piscataway, NJ, USA, 2018; pp. 24–29.
43. Zhu, X.; Chen, J.; Gao, F.; Chen, X.; Masek, J.G. An enhanced spatial and temporal adaptive reflectance fusion model for complex heterogeneous regions. *Remote Sens. Environ.* **2010**, *114*, 2610–2623. [CrossRef]
44. Jia, D.; Cheng, C.; Song, C.; Shen, S.; Ning, L.; Zhang, T. A Hybrid Deep Learning-Based Spatiotemporal Fusion Method for Combining Satellite Images with Different Resolutions. *Remote Sens.* **2021**, *13*, 645. [CrossRef]
45. Tang, Y.; Wang, Q.; Zhang, K.; Atkinson, P. Quantifying the Effect of Registration Error on Spatio-temporal Fusion. *IEEE J. Sel. Top. Appl. Earth Obs. Remote Sens.* **2020**, *13*, 487–503. [CrossRef]
46. Cheng, F.; Fu, Z.; Tang, B.; Huang, L.; Huang, K.; Ji, X. STF-EGFA: A Remote Sensing Spatiotemporal Fusion Network with Edge-Guided Feature Attention. *Remote Sens.* **2022**, *14*, 3057. [CrossRef]
47. Hou, S.; Sun, W.; Guo, B.; Li, C.; Li, X.; Shao, Y.; Zhang, J. Adaptive-SFSDAF for Spatiotemporal Image Fusion that Selectively Uses Class Abundance Change Information. *Remote Sens.* **2020**, *12*, 3979. [CrossRef]
48. Wu, M.; Niu, Z.; Wang, C.; Wu, C.; Wang, L. Use of MODIS and Landsat time series data to generate high-resolution temporal synthetic Landsat data using a spatial and temporal reflectance fusion model. *J. Appl. Remote Sens.* **2012**, *6*, 063507.
49. Wu, M.; Wu, C.; Huang, W.; Niu, Z.; Wang, C.; Li, W.; Hao, P. An Improved High Spatial and Temporal Data Fusion Approach for Combining Landsat and MODIS Data to Generate Daily Synthetic Landsat Imagery. *Inf. Fusion* **2016**, *31*, 14–25. [CrossRef]
50. Liu, M.; Ke, Y.; Yin, Q.; Chen, X.; Im, J. Comparison of Five Spatio-Temporal Satellite Image Fusion Models over Landscapes with Various Spatial Heterogeneity and Temporal Variation. *Remote Sens.* **2019**, *11*, 2612. [CrossRef]

land

MDPI

Article

Study on the Carbon and Nitrogen Isotope Characteristics and Sources and Their Influence on Carbon Sinks in Karst Reservoirs

Zhongfa Zhou [1,2,3,*], Jie Kong [1,2,3], Fuqiang Zhang [1,2,3], Yan Zou [1,2,3], Jiangting Xie [1,2,3] and Chaocheng Wen [1,2,3]

1 School of Karst Science, Guizhou Normal University, Guiyang 550001, China
2 School of Geography & Enviromental Science, Guizhou Normal University, Guiyang 550001, China
3 The State Key Laboratory Incubation Base for Karst Mountain Ecology Environment of Guizhou Province, Guiyang 550001, China
* Correspondence: fa6897@gznu.edu.cn; Tel.: +86-13985026897

Abstract: The hydrochemical analysis method was used to reveal the sources and spatiotemporal variations of carbon and nitrogen elements in the Pingzhai Reservoir, and the C–N coupling cycle and its influence on the karst carbon sink are discussed. The results show the following: (1) The hydrochemical type of the study area is HCO_3-Ca. (2) From the river to the reservoir and then to the reservoir outlet, the values of HCO_3^- and $\delta^{13}C_{DIC}$ showed an opposite trend. The values of NO_3^-, $\delta^{15}N$-NO_3^-, and $\delta^{18}O$-NO_3^- were different in each stage of the river. (3) HCO_3^- mainly comes from the weathering of carbonate rocks and the oxidative decomposition of organic matter. Nitrate mainly comes from chemical fertilizers, soil organic nitrogen, sewage, and livestock manure. (4) The average proportion of HCO_3^- produced by HNO_3 dissolving carbonate rock is 8.38%, but this part does not constitute a carbon sink. Compared with rivers, the proportion of HCO_3^- and (Ca^{2+} + Mg^{2+}) produced by HNO_3 dissolving carbonate rock in reservoir water is relatively large. The input of nitrate not only pollutes the water body with NO_3^- but also changes the carbon source/sink pattern of the water–rock interaction.

Keywords: karst carbon sink; carbon and nitrogen sources; dissolved inorganic carbon isotopes; nitrogen and oxygen isotopes; karst area reservoir

Citation: Zhou, Z.; Kong, J.; Zhang, F.; Zou, Y.; Xie, J.; Wen, C. Study on the Carbon and Nitrogen Isotope Characteristics and Sources and Their Influence on Carbon Sinks in Karst Reservoirs. *Land* **2023**, *12*, 429. https://doi.org/10.3390/land12020429

Academic Editor: Tilo Ziehn

Received: 1 January 2023
Revised: 31 January 2023
Accepted: 3 February 2023
Published: 6 February 2023

1. Introduction

Rivers connect terrestrial and marine ecosystems and are important channels for the transfer and transformation of nutrient elements. The transport of particulate and dissolved matter from rivers to the ocean is of great significance for the material cycle of the ecosystem, and it is estimated that the total amount of particulate sediment and dissolved matter transported into the ocean by rivers worldwide is 15.5×10^9 t/a and 4×10^9 t/a, respectively [1]. At the same time, to make full use of water energy resources, water conservation has been pursued around the world in recent decades. The construction of dams transforms a single river ecosystem into a river–reservoir ecosystem, river continuity is forced to change, and the pattern of material transport from the source to the estuary also changes. Reservoirs are usually characterized by poor flow, water temperature stratification, and large depth, and the retention and transformation of various substances in reservoirs are relatively considerable. Many studies have also been carried out on the biogeochemical cycle of river–reservoir ecosystems [2–5].

As essential elements for the growth of life, carbon, and nitrogen play an important role in ecosystems. Related studies have shown that carbon and nitrogen can reflect the construction of aquatic food chains and the division of trophic levels in aquatic ecosystems [6] and can also be used for carbon and nitrogen isotope analysis in aquatic plants and plankton [7,8]. Water bodies in karst areas are generally rich in calcium and somewhat alkaline. In terms of the impact of nutrients on water quality, even if the concentrations of nitrogen and phosphorus are low, aquatic plants (blue-green algae) and plankton can develop through the carbon fertilization effect. The input of exogenous nitrogen will contribute to the growth of living organisms such as microorganisms and plants in water, which will affect the absorption and decomposition of carbon [3,4].

However, increasing human activities such as land-use change, fossil fuel burning, and agricultural production have altered the natural state of the carbon and nitrogen cycle [9,10]. For example, the use of nitrogen fertilizer in agriculture can promote photosynthesis by increasing the net primary productivity of vegetation and microbial activities and increase the decomposition of organic matter by microorganisms, but this will produce more CO_2, part of which is re-discharged into the atmosphere as a carbon source, and the other part infiltrates into the karst surface zone and vadose zone with rainfall to form carbonic acid, increasing the dissolution of carbonate rocks [11,12]. Relevant studies also show that nitric acid produced by agricultural and urban activities interferes with the karst carbon cycle [13–15]. In contrast to carbonate rocks dissolved by carbonic acid, carbonate rocks dissolved by nitric acid do not consume CO_2 in the atmosphere/soil, leading to an increase in HCO_3^- and (Ca^{2+} + Mg^{2+}) concentrations in water, playing the opposite role in CO_2 emission reduction [16,17]. Baker et al. (2008) showed that a river flowing through the city had the highest dissolved inorganic carbon (DIC) concentration in the carbonate rock area of Britain [18]. Barnes et al. (2009) found that DIC was higher in watersheds dominated by urban land than in watersheds dominated by forestland [13]. It is estimated through laboratory simulation that fertilization of cultivated land in karst regions will lead to an additional increase in the (Ca^{2+} + Mg^{2+}) concentration in rivers by 5.7×10^{12} mmol/a and will release CO_2 to the atmosphere [19]. A study in a karst basin in southwest France found that the application of chemical fertilizers increased the concentration of nitric acid in river water and estimated that the amount of atmospheric CO_2 absorbed by the weathering of carbonate rocks in the basin decreased by 7–11%, and the karst carbon sink decreased by 5.7–13.4% in the whole region of France [20]. According to Brunet et al. (2011), the nitric acid formed by nitrification of nitrogen fertilizer can cause soil and water acidification, increase the concentration of alkaline cations, change the carbon budget, and actively participate in the weathering of carbonate rocks [21]. A study in the typical karst agricultural area of Southwest China shows that H^+ released from the nitrification of nitrogen fertilizer accelerates the weathering of carbonate rocks, which not only reduces the consumption of atmospheric CO_2 but also increases the HCO_3^- flux by approximately 20% [22].

Southwest China has the largest contiguous distribution of carbonate rocks in the world, with an exposed area of carbonate rocks of 54×10^4 km^2 [23]. Southwest China has also become a key area for studying carbon and nitrogen cycling. The Pingzhai Reservoir is located in Southwest China, and its water transfer scope involves the Yangtze River basin and the Pearl River basin [24]. Its carbon and nitrogen concentrations are of great significance for the water quality security of the Yangtze and Pearl Rivers. In terms of geological background, the Pingzhai Reservoir is located in a deep river canyon, with very thick carbonate rock strata distributed on both sides of the canyon. Affected by karstification, a multilayered karst hydrogeological structure has been formed. The karst morphology mainly includes peak-cluster depressions, dissolving gullies and troughs, falling caves, funnels, karst caves, and karst pipeline systems. In the region, in recent years, with the increase in population, frequent industrial and agricultural activities in the basin, and the large amount of agricultural fertilizers used with low utilization efficiency [24,25], nutrients have entered rivers and reservoirs through cracks in the karst and underground rivers along with the runoff and pore water generated by precipitation, leading to the accumulation of nitrogen nutrients and participating in water-rock interactions. Previous studies on carbon and nitrogen in the Pingzhai Reservoir and its inflow river were relatively isolated [25–27], but with the progress of research, it has been found that it is very important to explore the coupling of carbon and nitrogen and its environmental effects. The objectives of this study were to explore the water hydrochemical types and the spatial and temporal distribution characteristics of nitrate nitrogen and oxygen isotopes and dissolved inorganic carbon isotopes in a karst reservoir basin, determine the sources of DIC and nitrate in water, explore the carbon and nitrogen coupling cycle in different periods, and quantitatively evaluate the impact of the carbon and nitrogen coupling cycle and nitric acid from external sources on carbonate dissolution in the basin. This study can provide a reference for the study of the effect of C–N coupling and karst carbon sinks on the river–reservoir continuum in karst regions.

2. Materials and Methods

2.1. Overview of the Study Area

The Pingzhai Reservoir (105°17′3″ E–105°26′44″ E, 26°29′33″ N–26°35′38″ N) is the source reservoir of Guizhou's Central Water Control Project and undertakes the functions of irrigation, drinking water supply, and power generation in the region (Figure 1). The reservoir is formed by the convergence of five rivers (Nayong River, Shuigong River, Zhangwei River, Baishui River, and Hujia River) in the upper reaches of the reservoir, and the drainage area is 833.77 km^2. The construction of the reservoir was completed in 2015, the maximum dam height is 157.5 m, the maximum water level is 1331 m, the regulated storage capacity is 448 million m^3, and the total storage capacity is 1.089 billion m^3. The study area is located in the subtropical monsoon climate zone, summer is hot and rainy, winter is mild and slightly rainy, the annual average temperature is 14 °C, the annual average rainfall is between 1200 and 1500 mm, and the rainfall has seasonal differences under the influence of the monsoon climate. The wet season is from May to August, the dry season is from November to February, and the normal season is March, April, September, and October.

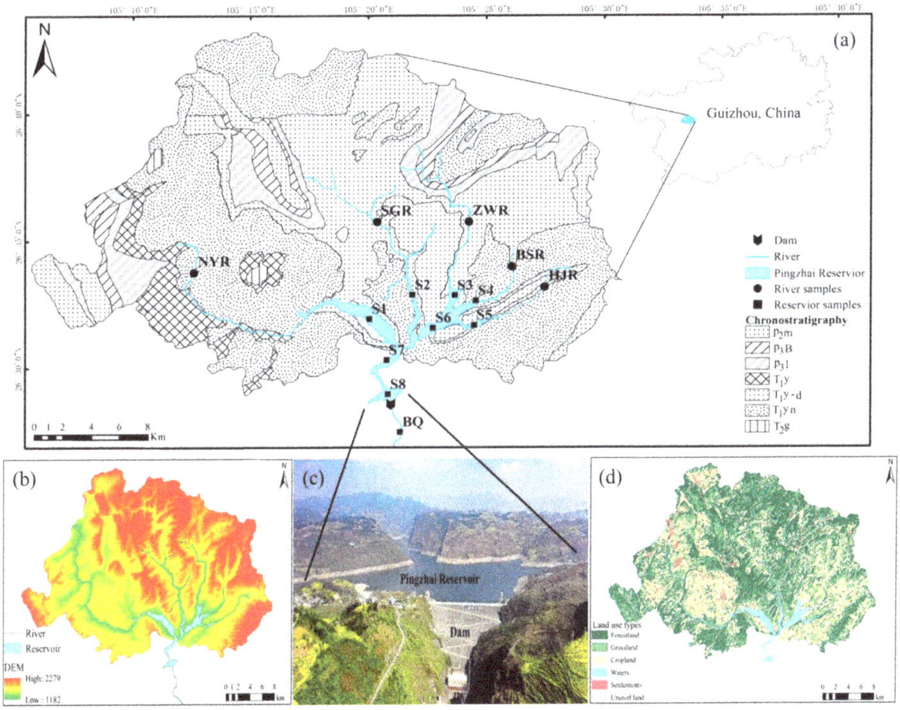

Figure 1. Overview map of the study area ((**a**) is the chronostratigraphic diagram and sampling points distribution of the study area; (**b**) is the DEM of the study area; (**c**) is the reservoir and dam; (**d**) is the land use types map).

The terrain and geomorphology of the study area are complex and belong to the middle-low mountain valley landform of tectonic dissolution and erosion. The geological structure pattern is controlled by the Yanshan movement, and anticlines, synclines, and compressive faults are formed. As shown in Figure 1, the main outcrops are the Permian Dalong Formation (P_3d), Longtan Formation (P_3l), Maokou Formation (P_2m), Triassic Yongningzhen Formation (T_1yn), Yelang Formation (T_1y), and Guanling Formation (T_2g). The stratigraphic lithology is carbonate rocks such as limestone and dolomite and clastic rocks such as sandy mudstone, shale intercalated marl, and coal. The main land-use types in the study area were cultivated land and forestland, followed by grassland, construction land, and unused land. Affected by regional lithology and crop cultivation habits, the main soil types are yellow-brown soil, lime soil, and yellow soil, in addition to a small amount of paddy soil. The main crops are rice, corn, and potato. The main fertilizers used in the region are ammonium-based N fertilizer (urea), nitrogen–phosphorus compound fertilizer, and animal manure; the applied pesticides mainly include insecticides and rust removers, which pose the risk of agricultural nonpoint source pollution.

2.2. Sample Collection and Analysis

In accordance with the sampling conditions of the study area and Technical Provisions for the Design of Water Quality Sampling Schemes (HJ495-2009), sampling points NYR, SGR, ZWR, BSR, and HJR were set up in the upper reaches of the five rivers, and sampling points S1, S2, S3, S4, and S5 were set up at the intersection of the rivers and reservoir areas. Three sampling points (S6, S7, and S8) were set up in the reservoir area, and one sampling point was set up at the outlet of the dam (BQ) (Figure 1), for a total of 14 sampling points. Water samples were collected from the study area in November 2020 and January and July 2021, representing the normal, dry, and wet periods, respectively. Water pH, water temperature (WT), electrical conductivity (EC), and dissolved oxygen (DO) were measured in the field by a WTW Multi3430 portable multiparameter water quality analyzer with an accuracy of 0.001 pH unit, 0.01 °C, 1 μs/cm, and 0.01 mg/L, respectively. The concentrations of HCO_3^- and Ca^{2+} were titrated onsite using an alkalinity kit and calcium kit (Merck, Germany) with accuracies of 0.1 mmol/L and 2 mg/L, respectively. The collected water samples were filtered through a 0.45-μm filter membrane and loaded into polyethylene sampling bottles that had been precleaned with deionized water. Concentrated nitric acid was added to the water samples to pH < 2 for determination of cation concentration, and 2 drops of $HgCl_2$ were added to the water sample to inhibit microbial activity for determination of dissolved inorganic carbon isotope ($\delta^{13}C_{DIC}$).

Anions, cations, and $\delta^{13}C_{DIC}$ were measured at the State Key Laboratory of Environmental Geochemistry, Institute of Geochemistry, Chinese Academy of Sciences, and $\delta^{15}N\text{-}NO_3^-$ and $\delta^{18}O\text{-}NO_3^-$ were measured at the Analysis and Testing Center of the Third Institute of Oceanography, Ministry of Natural Resources. The concentration of cations (K^+, Na^+, Mg^{2+}) was determined by Inductively Coupled Plasma-Emission Spectrometer (VISTA MPX, Varian, USA), and the concentration of anions (NO_3^-, Cl^-, SO_4^{2-}) was determined by ion chromatography (ICS90, Dionex, Sunnyvale, CA, USA). The limit of detection was 0.01 mmol/L. The method for the determination of water body $\delta^{13}C_{DIC}$ was to add 100% pure phosphoric acid into the injection bottle (vacuumized) and injection high purity helium gas, then inject 20 mL water sample into the injection bottle with a syringe, and heat it in a 60 °C water bath beaker. The CO_2 produced by the reaction was separated by a cold trap and then loaded with helium into a Finnigan MAT253 gas isotope mass spectrometer for determination. The bacterial denitrification method was used for the determination of $\delta^{15}N\text{-}NO_3^-$ and $\delta^{18}O\text{-}NO_3^-$. Denitrifying bacteria (ATCC 13985, DSM 6698) without nitrous oxide reductase activity were used to terminate the reaction after reducing NO_3^- to N_2O, thus obtaining nitrogen and oxygen in N_2O from the NO_3^- in the sample [28,29]. A GasBench continuous flow gas introduction instrument and MAT 253 stable isotope ratio mass spectrometer were used to determine the $\delta^{15}N$ and $\delta^{18}O$ contents in N_2O. To ensure the accuracy of the obtained measurements, reference materials USGS34

($\delta^{15}N = -1.8\%$, $\delta^{18}O = -27.9\%$), USGS32 ($\delta^{15}N = +180\%$, $\delta^{18}O = +25.7\%$), and IAEA-NO_3 were used. The test accuracy of $\delta^{13}C_{DIC}$ was 0.2%, and the result is reported as parts pel mil (%) relative to the Vienna PDB reference standard. The test accuracy of $\delta^{15}N$-NO_3^- and $\delta^{18}O$-NO_3^- was 0.3%. Atmospheric nitrogen (N_2) and Vienna standard mean seawater (V-SMOW) were used as references for the $\delta^{15}N$ and $\delta^{18}O$ results, respectively.

2.3. Flux Calculation

HCO_3^- and NO_3^- flux was calculated from the total water flow multiplied by the concentration of HCO_3^- and NO_3^- [30]. Flux is calculated using the equation:

$$flux_{C/N} = conc_{C/N} \times Q \tag{1}$$

where $flux_{C/N}$ refers to annual HCO_3^- and NO_3^- flux ($t \cdot a^{-1}$), $conc_{C/N}$ is the concentration of HCO_3^- and NO_3^- (mg/L), and Q refers to the water discharge in unit time ($m^3 \cdot a^{-1}$).

3. Results

3.1. Physicochemical Indices and Hydrochemical Characteristics of Water

The physicochemical indices of the water body of Pingzhai Reservoir and its inflow river showed seasonal changes (Table 1). The water temperature ranged from 9.15 °C to 26.65 °C, with an average temperature of 16.92 °C. The pH value of water ranged from 7.89 to 10.67, with an average value of 8.69, generally showing the characteristics of weakly alkaline water. The EC of the water body varied greatly (196–578 μs/cm) and showed the temporal pattern dry season > normal season > wet season. In terms of the water DO concentration, the annual variation ranged from 6.39 to 11.51 mg/L, with an average of 8.5 mg/L. The water body was generally in an aerobic state, which was conducive to the occurrence of nitrification, and temporally, DO was the highest in the wet season and the lowest in the dry season.

Table 1. Main hydrochemical parameters of water bodies in different seasons.

Index	Normal Season			Dry Season			Wet Season		
	River	Reservoir	Dam	River	Reservoir	Dam	River	Reservoir	Dam
WT (°C)	15.3 ± 0.9	17 ± 0.3	16.9	11.3 ± 0.9	11.4 ± 0.7	12.9	19.7 ± 0.3	25.1 ± 0.7	21.7
pH	8.3 ± 0.2	8.8 ± 0.1	8.1	8.4 ± 0.3	8.5 ± 0.1	8.6	9.6 ± 0.9	8.9 ± 0.1	7.9
EC (μs/cm)	392.6 ± 2.1	328.9 ± 1.5	390	335.2 ± 2.1	421.1 ± 1.5	434	316.6 ± 0.7	312.6 ± 0.6	421
DO (mg/L)	8.4 ± 0.2	8.7 ± 0.3	7.3	9.2 ± 0.5	6.7 ± 0.2	8.5	7.8 ± 0.2	10.4 ± 0.8	7.8
Ca^{2+} (mg/L)	48.6 ± 0.9	49.1 ± 0.6	55.2	61.1 ± 0.9	61.3 ± 0.8	69.9	52.1 ± 0.5	36.8 ± 0.3	78.3
Na^+ (mg/L)	23.9 ± 0.8	10.7 ± 0.3	12.1	34.4 ± 1.3	13.9 ± 0.3	19.6	5.6 ± 0.5	13.6 ± 0.1	3.3
Mg^{2+} (mg/L)	5.2 ± 0.9	5.7 ± 0.7	6.9	6.2 ± 0.5	6.6 ± 0.3	7.4	4.1 ± 0.4	6.9 ± 0.1	6.4
K^+ (mg/L)	2.2 ± 0.6	2.0 ± 0.2	2.4	2.1 ± 0.8	2.2 ± 0.7	2.1	3.2 ± 0.6	3.4 ± 0.4	3.0
HCO^{3-} (mg/L)	165.3 ± 3.3	133.8 ± 0.8	186.1	169.6 ± 1.7	160.1 ± 0.6	164.7	126.3 ± 1.3	91.5 ± 0.9	192.1
NO^{3-} (mg/L)	9.6 ± 0.8	11.6 ± 0.7	13.6	10.6 ± 0.4	12.6 ± 0.1	15.2	12.4 ± 0.8	9.8 ± 0.7	17.5
Cl^- (mg/L)	5.4 ± 0.7	4.5 ± 0.5	5.8	8.9 ± 0.9	6.1 ± 0.5	10.9	3.9 ± 1.1	9.1 ± 0.9	5.6
SO_4^{2-} (mg/L)	70.5 ± 0.8	49.5 ± 0.3	60.8	75.9 ± 0.7	59.1 ± 0.3	70.6	39.6 ± 1.8	57.9 ± 0.9	27.7
$\delta^{13}C_{DIC}$ (%)	−12.3 ± 0.8	−9.8 ± 0.6	−11.5	−10.9 ± 0.5	−9.9 ± 0.3	−12.7	−12.7 ± 0.6	−5.3 ± 0.4	−13.0
$\delta^{15}N$-NO_3^- (%)	2.2 ± 0.5	1.4 ± 0.7	2.3	14.4 ± 0.7	15.9 ± 0.6	14.0	5.4 ± 0.8	7.3 ± 0.6	4.3
$\delta^{18}O$-NO_3^- (%)	5.1 ± 0.6	3.0 ± 0.9	4.1	0.9 ± 0.7	2.5 ± 0.1	0.5	20.8 ± 0.3	21.8 ± 0.2	22.0

Note: mean ± standard deviation (SD).

The total cationic equivalent concentration ($TZ^+ = 2Ca^{2+} + 2Mg^{2+} + K^+ + Na^+$) in the Pingzhai Reservoir and its inflow rivers ranged from 1.92 to 14.40 meq/L, with an average value of 8.16 meq/L. The total anion equivalent concentration ($TZ^- = HCO_3^- + NO_3^- + Cl^- + 2SO_4^{2-}$) ranged from 1.97 to 10.24 meq/L, with an average value of 6.10 meq/L. Taking the river and reservoir area together, the total cationic equivalent concentration of river water ranged from 3.41 to 14.39 meq/L, with an average of 8.91 meq/L. The total anion equivalent concentration ranged from 2.37 to 10.57 meq/L, with an average of 6.47 meq/L. The total equivalent concentrations of cations and anions in the reservoir were 1.97–9.45 meq/L and 3.24–6.01 meq/L, with averages of 5.71 meq/L and 4.63 meq/L, respectively. The Piper diagram can directly reflect the composition characteristics of the main ions in water (Figure 2). The predominant cations in the Pingzhai

Reservoir were Ca^{2+} and Na^+, whose contents accounted for 68% and 21% of the total cations, respectively. The dominant anions were HCO_3^- and SO_4^{2-}, which accounted for 65% and 27% of the total anions, respectively. According to the Shukalev classification, the hydrochemical type in the study area was the HCO_3-Ca type.

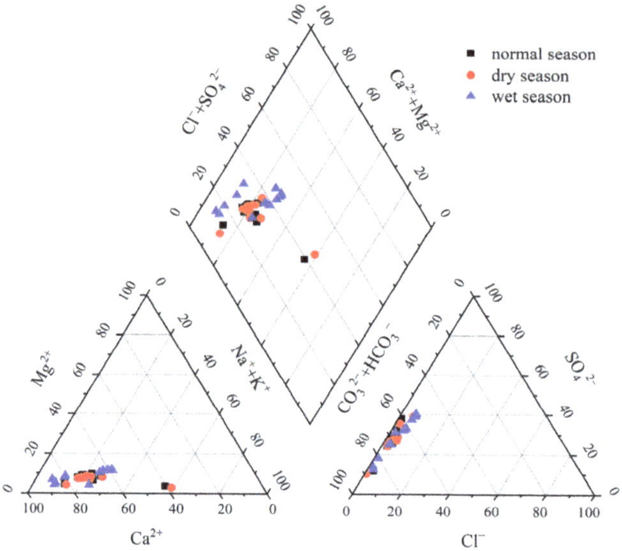

Figure 2. Piper diagram of hydrochemical types in the Pingzhai Reservoir basin.

3.2. *Characteristics of DIC and $\delta^{13}C_{DIC}$ in the Pingzhai Reservoir*

Influenced by the dissolution of carbonate rocks, the water in the study area is weakly alkaline, under which condition HCO_3^- is the main form of DIC [31]. In this study, HCO_3^- is used to represent the concentration of DIC in the water. The variation in HCO_3^- concentration in the Pingzhai Reservoir and inflow rivers ranged from 1.30 to 3.45 mmol/L, with an average of 2.32 mmol/L. The average concentrations of HCO_3^- in river, reservoir, and outlet water were 2.52 mmol/L, 2.11 mmol/L, and 2.97 mmol/L, respectively. The seasonal variation was as follows: dry season (2.69 mmol/L) > normal season (2.44 mmol/L) > wet season (1.82 mmol/L). The average annual flows of the NYR, SGR, and ZWR were 1.35×10^9 $m^3 \cdot a^{-1}$, 5.89×10^8 $m^3 \cdot a^{-1}$, and 5.22×10^8 $m^3 \cdot a^{-1}$, respectively. Flow data from BSR and HJR are lacking. According to Equation (1), combined with the HCO_3^- concentration at the monitoring section, it can be calculated that the HCO_3^- fluxes of NYR, SGR, and ZWR were 2580.25 $t \cdot a^{-1}$, 700.62 $t \cdot a^{-1}$, and 722.52 $t \cdot a^{-1}$, respectively. The $\delta^{13}C_{DIC}$ value ranged from −3.1 to −18.4%, with an average value of −9.9%, and the average $\delta^{13}C_{DIC}$ values in river, reservoir, and outlet water were −12.0%, −8.3%, and −12.4%, respectively. In terms of time, the $\delta^{13}C_{DIC}$ value was negative in the wet season and positive in the dry season. There was no correlation between HCO_3^- and $\delta^{13}C_{DIC}$ in the normal and dry periods except for a negative correlation between HCO_3^- and $\delta^{13}C_{DIC}$ in the wet season ($r = -0.692$, $p < 0.01$). This is due to the strong photosynthesis during the wet season. Phytoplankton absorb the DIC in water and fractionate $\delta^{13}C_{DIC}$, resulting in a decrease in HCO_3^- concentration and an increase in the $\delta^{13}C_{DIC}$ value. Due to the influence of the dam, stable stratification is formed in the water body due to temperature differences in summer, which blocks the material exchange between the surface water body and the deep-water body. However, with the disappearance of stable stratification of water temperature in the normal and dry periods, the high concentration of HCO_3^- at the bottom diffuses to the surface water with water turnover and internal circulation, resulting in a high concentration of HCO_3^- in the surface water.

3.3. Characteristics of NO_3^-, $\delta^{15}N$-NO_3^-, and $\delta^{18}O$-NO_3^- in the Pingzhai Reservoir

The variation in the NO_3^- concentration in the Pingzhai Reservoir ranged from 1.92 to 17.47 mg/L, with an average of 11.46 mg/L, and the time variation was wet season (12.68 mg/L) > dry season (12.09 mg/L) > normal season (11.02 mg/L). The average NO_3^- concentrations in the river, reservoir area, and outlet were 11.33 mg/L, 10.68 mg/L, and 15.4 mg/L, respectively. According to the average annual flow and concentration of NO_3^-, the NO_3^- fluxes of the NYR, SGR, and ZWR points were calculated as 210.74 $t{\cdot}a^{-1}$, 25.33 $t{\cdot}a^{-1}$, and 54.76 $t{\cdot}a^{-1}$, respectively. The $\delta^{15}N$-NO_3^- values in the dry season ranged from 13.0 to 17.1%, and the $\delta^{15}N$-NO_3^- values in the wet and normal seasons ranged from 3.4 to 8.3% and 0.4 to 2.9%, respectively. This seasonal variation indicates different nitrate sources. The $\delta^{18}O$-NO_3^- value was the highest in the wet season, intermediate in the normal season, and the lowest in the dry season.

4. Discussion

4.1. Spatial Variation and Influencing Factors of Dissolved Inorganic Carbon and Nitrate

Subject to different environmental factors, carbon and nitrogen elements and isotopes will show different spatial and temporal changes (Figure 3). Compared with the reservoir area, the $\delta^{13}C_{DIC}$ value of the river is more negative, and the DIC concentration is higher, which is due to the large proportion of farmland and woodland in the river flow area. The proportion of cultivated land area in the Baishui and Hujia River Basins is more than 50%, and the proportion of woodland area in the Shuigong River Basin is nearly half [24]. The $\delta^{13}C_{DIC}$ value of the water body inherits the $\delta^{13}C_{DIC}$ value of vegetation and soil CO_2. The amount of soil CO_2 produced by plant root respiration into the water body increased, which led to a negative $\delta^{13}C_{DIC}$ value and an increase in DIC concentration. In addition, the Narong and Zhangwei Rivers flow through cities and towns, and human activities (industrial production and agricultural cultivation) discharge sewage with negative $\delta^{13}C_{DIC}$ values into the river, which will also cause a negative $\delta^{13}C_{DIC}$ value in river water [32].

Compared with the reservoir area, the $\delta^{13}C_{DIC}$ value at the outlet of the reservoir was negative, and the DIC concentration was higher. The spatial variation in $\delta^{13}C_{DIC}$ values was negative in the river, positive in the reservoir area, and negative in the outlet. The DIC concentration first increased, then decreased, and then increased again. Yuan et al. (2021) studied DIC concentration and $\delta^{13}C_{DIC}$ in cascade reservoirs of the Yunnan section of the Lancang River and found that the DIC concentration was high in the river, low in the reservoir, and high at the outlet during the wet and dry seasons, and the $\delta^{13}C_{DIC}$ value first showed an increasing and then decreasing trend [4]. In the karst area of Southwest China, the DIC concentration of the Hongjiadu Reservoir in the Wujiang River Basin is higher at the outlet than in the river and reservoir area, and the $\delta^{13}C_{DIC}$ value is negative in the river, positive in the reservoir area, and negative at the outlet. The DIC concentration in the rainy season is lower than that in the dry season, and the $\delta^{13}C_{DIC}$ value in the rainy season is higher than that in the dry season [33]. All of these parameters showed the same change trend in the Pingzhai Reservoir basin, which was due to the following: Under the influence of the subtropical monsoon climate, the rainfall in the study area is mostly concentrated in spring and summer, the increase in rainfall leads to an increase in runoff into the reservoir, and the DIC concentration decreases due to the influence of water dilution. Generally, photosynthesis on the water surface is relatively active, and aquatic phytoplankton will produce approximately 18–20% isotope fractionation while absorbing CO_2 through photosynthesis [34]. Therefore, the DIC concentration of surface water in the reservoir area is low, and the $\delta^{13}C_{DIC}$ value is positive, while the photosynthesis of the bottom water is weak, and the degradation of organic matter at the water–sediment interface produces CO_2 with poor ^{13}C, which increases the DIC concentration and causes the $\delta^{13}C_{DIC}$ value to be negative. As a result, the DIC value of the lower water is higher than that of the upper water, and the $\delta^{13}C_{DIC}$ value of the lower water is lower than that of the upper water. The discharge mode of the Pingzhai Reservoir is bottom discharge, so bottom water with high DIC concentrations and low $\delta^{13}C_{DIC}$ values is injected downstream.

There are also relevant studies that show that the DIC concentration in the water body is lower in the rainy season than in the dry season, and the $\delta^{13}C_{DIC}$ value in the rainy season is also lower than that in the dry season, which may be related to the water nutrient level, reservoir operation mode, vegetation cover conditions, and tributary inflow [4,35,36].

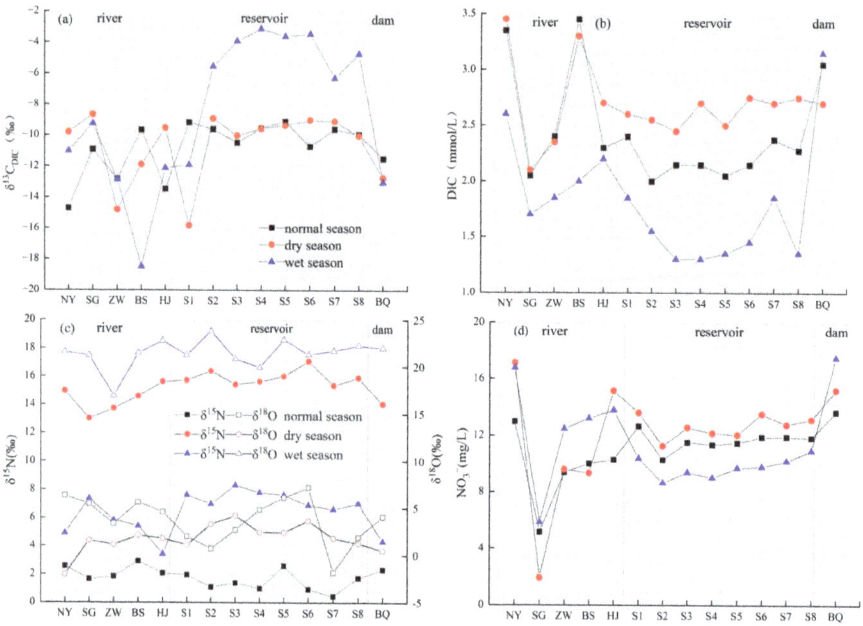

Figure 3. Variations of $\delta^{13}C_{DIC}$ and DIC, $\delta^{15}N\text{-}NO_3^-$, $\delta^{18}O\text{-}NO_3^-$, and NO_3^- along the course of water bodies in different seasons ((**a**) is the variation characteristic of $\delta^{13}C_{DIC}$; (**b**) is the variation characteristic of DIC; (**c**) is the variation characteristic of $\delta^{15}N$, $\delta^{18}O$; (**d**) is the variation characteristic of NO_3^-).

The concentration of NO_3^- has a more direct impact on the deterioration of water quality. In the river section, the concentration of nitrate in the Nayong and Hujia Rivers was higher, which is a consequence of the discharge of domestic sewage and livestock breeding wastewater increasing the concentration of nitrate in the water body. The Shuigong River sampling site, which was not surrounded by residential and cultivated land, was minimally affected by anthropogenic activities and therefore had the lowest nitrate concentration. In the reservoir area, the water was fully mixed, and the nitrate concentration differed little among the sampling points in different seasons. Similar to the distribution of the NO_3^- concentration, the $\delta^{15}N\text{-}NO_3^-$ and $\delta^{18}O\text{-}NO_3^-$ values also showed large differences in the river section and small differences in the reservoir section. It is mainly affected by crop cultivation and fertilization, domestic sewage discharge, and other factors. Fadhullah et al. (2020) used nitrate nitrogen and oxygen isotopes to study the source of nitrate in the Bukit Merah Reservoir in Southeast Asia and found that industrial and mining production and agricultural expansion in the upstream river disturbed the value of nitrate nitrogen and oxygen isotopes in the river water. The nitrogen and oxygen isotopes in the reservoir area indicated that the nitrate concentration is affected by atmospheric deposition [37]. Studies on nitrate isotopes in Chaohu Lake and its rivers showed that nitrification of soil organic nitrogen and soil erosion caused changes in nitrate isotopes [38].

4.2. Sources of Dissolved Inorganic Carbon and Nitrate

The DIC concentration and $\delta^{13}C_{DIC}$ value in water record and reflect the geochemical behavior and cycling characteristics of carbon. Because carbon is affected by many factors in the cycling process and different sources of carbon have different isotopic values, the $\delta^{13}C_{DIC}$ value can be used to trace the source of DIC. Related studies have shown that DIC in natural water mainly comes from atmospheric CO_2 (including CO_2 in atmospheric precipitation), dissolution of carbonate rock, and biogenic CO_2 (plant root respiration and organic matter decomposition) dissolution in water [38,39]. The study area is located in the acid rain region of Southwest China, the pH value of rainwater is weakly acidic, the DIC concentration is low [40], and the partial pressure of water CO_2 (pCO_2) in the region is higher than that of atmospheric CO_2 [23,35]. The influence of atmospheric precipitation and atmospheric CO_2 on DIC and $\delta^{13}C_{DIC}$ values is not considerable. Therefore, the DIC in water is mainly derived from the weathering of carbonate rocks and the oxidative decomposition of organic matter.

The sources of nitrate in aquatic ecosystems mainly include atmospheric deposition, soil organic nitrogen, chemical fertilizer, domestic sewage, and livestock manure. Atmospheric deposition of $\delta^{15}N\text{-}NO_3^-$ ranges from −8% to +15%, and of $\delta^{18}O\text{-}NO_3^-$ ranges from +60% to +95%. The nitrogen and oxygen isotope values of nitrate fertilizer range from −5% to +5% and +17% to +25%, respectively. The nitrogen and oxygen isotopes of nitrate produced by fertilizer and deposition of NH_4^+, soil organic nitrogen, sewage, and livestock manure range from −10% to +25% and −10% to +10%, respectively [41–44]. Combined with the $\delta^{15}N\text{-}NO_3^-$ and $\delta^{18}O\text{-}NO_3^-$ values of the water body in the study area, various nitrate sources could be determined (Figure 4). During the normal season, $\delta^{15}N\text{-}NO_3^-$ and $\delta^{18}O\text{-}NO_3^-$ were distributed in soil organic nitrogen and ammonium fertilizer endmembers, indicating that nitrification of soil organic nitrogen and ammonium fertilizer was the main source of nitrate in water. In the dry season, $\delta^{15}N\text{-}NO_3^-$ and $\delta^{18}O\text{-}NO_3^-$ are distributed in domestic sewage endmembers and livestock manure, and $\delta^{15}N\text{-}NO_3^-$ is higher. Ren et al. (2021) found that the $\delta^{15}N$ value of some points was higher in a study of groundwater in the Zhaotong Basin, Yunnan Province [8]. Lin et al. (2019) also found a similar phenomenon in a study of the Illinois River in Chicago, USA [45]. This is because the volatilization of NH_4^+ in sewage releases a large amount of lighter isotopes, making the remaining NO_3^- enriched in heavy $\delta^{15}N$. The $\delta^{15}N\text{-}NO_3^-$ and $\delta^{18}O\text{-}NO_3^-$ were distributed in the nitrate fertilizer endmembers during the wet season. It may be that inorganic nitrogen from the oxidation and decomposition of nitrate fertilizer enters the water with surface runoff. In general, agricultural activities (ammonium fertilizer, nitrate fertilizer, soil organic nitrogen, livestock manure, and sewage) were important sources of nitrate in the Pingzhai Reservoir. Chemical fertilizers (ammonium fertilizer, nitrate fertilizer) can improve soil fertility and increase crop yield. From 1997 to 2005, the amount of chemical fertilizer applied in China increased from 7.07 million tons to 26.21 million tons [46]. However, the irrational use of chemical fertilizers and the lack of farmland management often lead to nitrogen loss and nitrate pollution. Due to the periodic storage and discharge of the reservoir, a water-level fluctuation zone will form around the reservoir, and the water-level fluctuation zone is also the area where agricultural nonpoint source pollution and soil erosion often occur [47]. According to Han et al. (2016), the amount of soil erosion in the Yangtze River basin and Wujiang River basin reached 1.4×10^{10} t/a and 1.2×10^{9} t/a in 2014, respectively [48]. Soil erosion in the water-level fluctuation zone will transport a large amount of nutrients and soil organic nitrogen to rivers and reservoirs. This part of the lost N will be converted into nitric acid through nitrification and will participate in the geochemical cycle, which will affect the chemical composition of rivers and reservoirs, especially the concentrations of Ca^{2+} and HCO_3^- [16]. Nitric acid participates in the dissolution of carbonate rocks and directly or indirectly releases CO_2 into the atmosphere, and these anthropogenic carbon source emissions offset part of the natural carbon sink in the natural process [20]. Therefore, it is necessary to consider the impact of external acid input on the weathering of carbonate rocks in the basin.

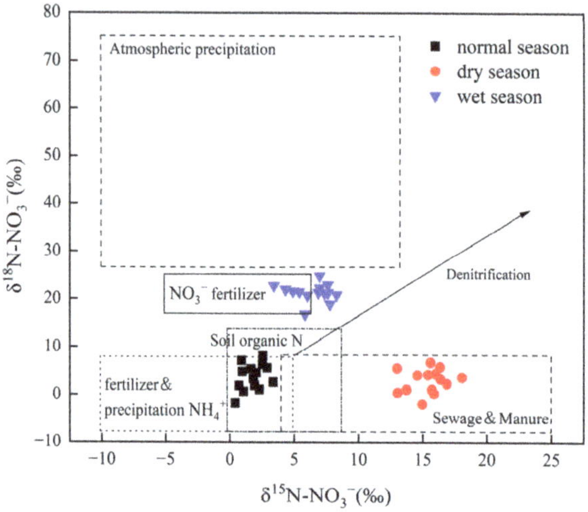

Figure 4. Distribution characteristics of $\delta^{15}N\text{-}NO_3^-$ and $\delta^{18}O\text{-}NO_3^-$ in water. The isotopic values of various nitrate sources are based on Kendall et al. (2007) [44].

4.3. Weathering of Carbonate Rocks and C–N Coupling Relationship in the Pingzhai Reservoir Basin

The average $\delta^{13}C_{DIC}$ of CO_2 produced by root respiration of C_3 vegetation is −27% because the migration and diffusion of CO_2 in the soil layer will produce approximately 4% isotope fractionation, so the $\delta^{13}C_{DIC}$ value of soil CO_2 produced by plant respiration and organic matter oxidation decomposition is approximately −23%, and the $\delta^{13}C_{DIC}$ value of marine carbonate rocks in karst areas is 0% [49–51]. According to the stoichiometric relationships of carbonate rock dissolution, with the participation of carbonic acid (H_2CO_3), 2 mol HCO_3^- will be produced in the process of carbonate rock dissolution, of which 1 mol is derived from atmospheric/soil CO_2 in the watershed and 1 mol from carbonate rocks (Equation (2)) [52].

$$Ca_{(1-x)}Mg_xCO_3 + CO_2 + H_2O = (1-x)Ca^{2+} + xMg^{2+} + 2HCO_3^- \quad (2)$$

In the process of carbonate rock dissolution by carbonic acid, carbon isotopes will produce +9% fractionation, so the water body $\delta^{13}C_{DIC}$ theoretical value is approximately −14% [53]. The variation in $\delta^{13}C_{DIC}$ in the Pingzhai Reservoir water ranges from −3.13% to −18.42%, with an average of −9.92%. Compared with the theoretical value, the $\delta^{13}C_{DIC}$ actual value of the water body is more positive, which indicates that in addition to carbonic acid, there are other sources of acid, such as nitric acid from human activities, which contribute to the dissolution of carbonate rocks in the basin (Equation (3)). In addition, the $[Ca^{2+} + Mg^{2+}]/[HCO_3^-]$ equivalent ratios in the water samples in the study area were all greater than 1 (1.37 on average), and the results indicate that Ca^{2+} and Mg^{2+} in the water sample are surplus relative to HCO_3^-, that is, Ca^{2+} and Mg^{2+} have additional sources [54]. In addition to the carbonate rocks dissolved by H_2CO_3 and HNO_3, the weathering and dissolution of silicate rocks in the watershed can also produce Ca^{2+}, Mg^{2+}, and HCO_3^- ions (Equations (4) and (5)).

$$Ca_{(1-x)}Mg_xCO_3 + HNO_3 = (1-x)Ca^{2+} + xMg^{2+} + NO_3^- + HCO_3^- \quad (3)$$

$$Ca_xMg_{(1-x)}Al_2Si_2O_8 + 2H_2CO_3 + 2H_2O = xCa^{2+} + (1-x)Mg^{2+} + 2HCO_3^- + 2SiO_2 + 2Al(OH)_3 \quad (4)$$

$$Na_xK_{(1-x)}Al_2Si_2O_8 + H_2CO_3 + H_2O = xNa^+ + (1-x)K^+ + HCO_3^- + 3SiO_2 + Al(OH)_3 \quad (5)$$

According to the dissolution equation of silicate rock, the Ca^{2+}, Mg^{2+}, and HCO_3^- produced by the dissolution of silicate rock by carbonic acid are calculated according to one-fifth and one-half of the molar concentration of SiO_4, respectively [11]. The contributions of silicate weathering to Ca^{2+}, Mg^{2+}, and HCO_3^- in the Pingzhai Reservoir basin ranged from 0.97% to 2.79% and 2.43% to 6.98%, with average values of 1.23% and 3.07%, respectively, accounting for a small proportion. This indicates that the weathering of carbonate rocks is the main process controlling the material composition and geochemical cycle of water bodies in the basin.

To show that nitric acid is indeed involved in the weathering of carbonate rocks in the basin, the ratio relationship between $[Ca^{2+} + Mg^{2+}]/[HCO_3^-]$ and $\delta^{13}C_{DIC}$ was established based on the ion and carbon isotope data in water samples. According to Figure 5, most of the sampled data fall between the endmembers of H_2CO_3 and HNO_3 dissolved carbonate rocks. This shows that H_2CO_3 and HNO_3 are jointly involved in the dissolution of carbonate rocks in the basin and have an impact on the value of carbon and nitrogen elements and ion concentration in the water body [51].

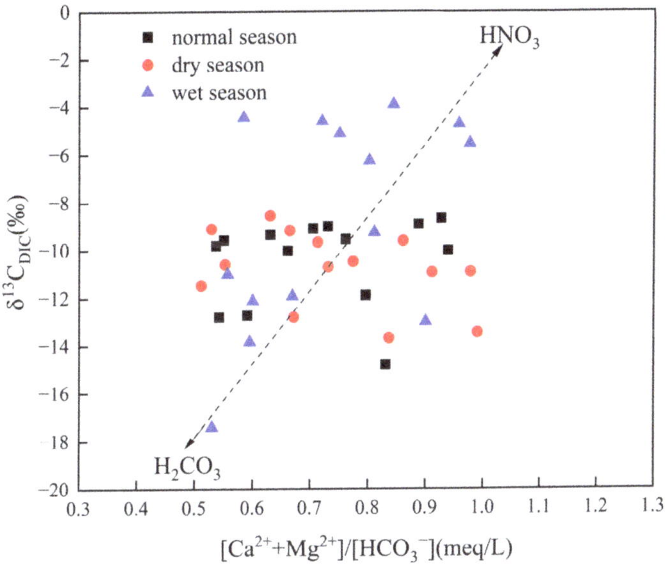

Figure 5. The relationship between $[Ca^{2+} + Mg^{2+}]/[HCO_3^-]$ and $\delta^{13}C_{DIC}$ in different seasons.

In addition to carbonate rock dissolution by carbonic acid, nitric acid produced by nitrification in agricultural activities also participates in carbonate rock dissolution. Assuming that carbonic acid and nitric acid participate in the dissolution process with equal molar ratios, the C–N coupling formula of carbonate rock dissolution can be obtained from Equation (6):

$$(a+b)Ca_xMg_{(1-x)}CO_3 + aH_2CO_3 + bHNO_3 = (a+b)xCa^{2+} + (a+b)(1-x)Mg^{2+} + bNO_3^- + (2a+b)HCO_3^- \quad (6)$$

where a and b represent the coefficients of H_2CO_3 and HNO_3 involved in the dissolution of carbonate rocks, respectively.

According to Equation (6), the molar ratio of $(Ca^{2+} + Mg^{2+})/HCO_3^-$ in water should be 2/3 (0.67), and the molar ratio of $(Ca^{2+} + Mg^{2+})/HCO_3^-$ in the water sample of the Pingzhai Reservoir basin was 0.17–0.98, with an average of 0.60. This indicates that the dissolution of carbonate rocks in the study area is controlled by C–N coupling, but H_2CO_3 and HNO_3 dissolve carbonate rocks according to the molar ratio of 1:1. Therefore, the amount of $(Ca^{2+} + Mg^{2+})$ and HCO_3^- released by HNO_3 dissolved carbonate rocks can be calculated according to the C–N coupling equation. The calculation results show that the proportion of HCO_3^- produced by HNO_3 dissolving carbonate rocks ranges from 1.50% to 13.35%, with an average of 8.38% (Table 2). The proportion of (Ca^{2+} + Mg^{2+}) ranges from 4.04% to 12.06%, with an average of 7.84%. Among the different periods, the proportion of HCO_3^- and (Ca^{2+} + Mg^{2+}) produced by HNO_3 dissolution during the wet season is the highest, with average values of 10.32% and 9.90%, respectively. The proportion in the dry season is the lowest, with averages of 7.29% and 5.21%, respectively. The average values in the normal season are 7.54% and 8.42%, respectively. This is because dam construction slows down or even stops the water flow, and the abundant water volume and appropriate temperature in the wet season make the water-rock interaction more complete. In addition, the amount of NO_3^- fertilizer from agricultural activities increased during the wet season, which combined with H^+ to generate HNO_3 to participate in the dissolution of carbonate rocks.

Table 2. The ratio of HCO_3^- and (Ca^{2+} + Mg^{2+}) produced by HNO_3 dissolution of carbonate rocks and the measured and theoretical values of $\delta^{13}C_{DIC}$.

Indexes		HCO_3^--HNO_3 (%)	(Ca^{2+} + Mg^{2+})-HNO_3 (%)	$\delta^{13}C_{DIC}$% (Measured Value)	$\delta^{13}C_{DIC}$% (Theoretical Value)
Normal season	River	5.82 ± 0.5	7.78 ± 0.4	−12.31 ± 0.4	−13.46 ± 0.2
	Reservoir	8.64 ± 0.3	8.87 ± 0.2	−9.77 ± 0.2	−13.06 ± 0.1
	Dam	7.31	8.02	−11.48	−13.25
Dry season	River	6.00 ± 0.7	5.02 ± 0.6	−10.95 ± 0.5	−13.43 ± 0.4
	Reservoir	7.85 ± 0.5	5.26 ± 0.5	−9.85 ± 0.3	−13.17 ± 0.1
	Dam	9.18	5.83	−12.74	−12.98
Wet season	River	9.66 ± 1.3	9.16 ± 0.5	−12.73 ± 0.5	−12.92 ± 0.3
	Reservoir	10.89 ± 0.6	10.88 ± 0.3	−5.32 ± 0.3	−12.75 ± 0.2
	Dam	9.03	5.74	−13.02	−13.01

Note: mean ± standard deviation (SD).

The $\delta^{13}C_{DIC}$ produced by carbonate rocks dissolved by carbonic acid is approximately −14%, and the HCO_3^- produced by carbonic acid-dissolved silicate rocks is all from soil CO_2, so the $\delta^{13}C_{DIC}$ is approximately −23%. Carbonate rocks dissolved by nitric acid do not consume soil or atmospheric CO_2, and all the HCO_3^- produced comes from carbonate rocks, so the $\delta^{13}C_{DIC}$ is approximately −0% [23,40,55]. The theoretical value of $\delta^{13}C_{DIC}$ in the Pingzhai Reservoir basin can be estimated by Equation (7):

$$\delta^{13}C_{DIC-T} = f_{cc}\delta^{13}C_{cc} + f_{cs}\delta^{13}C_{cs} + f_{nc}\delta^{13}C_{nc} \tag{7}$$

where $\delta^{13}C_{DIC\text{-}T}$ represents the theoretical value of $\delta^{13}C_{DIC}$ in water; f_{cc}, f_{cs}, and f_{nc} represent the contribution proportions of carbonate rock dissolved by carbonic acid, silicate rock dissolved by carbonic acid, and carbonate rock dissolved by nitric acid to HCO_3^- in water, respectively; $\delta^{13}C_{cc}$, $\delta^{13}C_{cs}$, and $\delta^{13}C_{nc}$ represent the values of $\delta^{13}C_{DIC}$ generated by carbonate rock dissolved by carbonic acid, silicate rock dissolved by carbonic acid, and carbonate rock dissolved by nitric acid, respectively. The calculated results show that $\delta^{13}C_{DIC\text{-}T}$ ranges from −14.06% to −12.72%, with an average value of −13.10%. The measured $\delta^{13}C_{DIC}$ ranges from −3.13% to −18.42%, with an average value of −9.32%. The measured $\delta^{13}C_{DIC}$ values in river and reservoir areas are more positive than the theoretical values in different periods, and the reservoir area is the most positive during the wet season. This is because the water flow in the reservoir area is slow and receives more light, and the photosynthesis of aquatic organisms absorbs DIC and causes isotope fractionation, making the water body $\delta^{13}C_{DIC}$ value positive [56]. Compared with the river, the proportion of HCO_3^- and (Ca^{2+} + Mg^{2+}) produced by the dissolution of carbonate rocks by nitric acid in the reservoir water is larger, which reflects that dam construction promotes the water-rock interaction and the retention of ionic substances.

As mentioned above, carbonate rock dissolved by carbonic acid in karst systems consumes atmospheric/soil CO_2 to form HCO_3^-, one part of which is used by aquatic organisms [26,57], and the other part enters the ocean with rivers for sedimentation. In addition to carbonic acid, nitric acid is also involved in the dissolution of carbonate rocks in the Pingzhai Reservoir. Nitrogen from chemical fertilizers (ammonium fertilizer and nitrate fertilizer), soil organic nitrogen, sewage, and livestock manure is lost and converted to HNO_3 (Figure 6). As a result, the concentration of HNO_3 in the water increases, and the carbonate rocks are dissolved by both carbonic and nitric acids. In this study, the average proportions of HCO_3^- and (Ca^{2+} + Mg^{2+}) produced by HNO_3-dissolved carbonate rocks were 8.38% and 7.84%, respectively. However, this part of HCO_3^- does not come from atmospheric/soil CO_2 but rather from carbonate rocks and does not constitute a carbon sink. Therefore, the environmental effect of C–N coupling not only causes water nitrate pollution but also reduces carbon sinks.

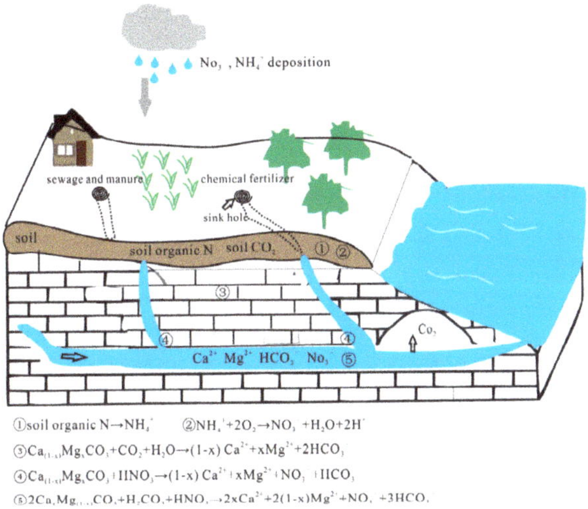

Figure 6. The C-N coupling cycle in the karst zone (adapted from Hu et al., 2017 [58]).

5. Conclusions

By monitoring the hydrochemistry and $\delta^{13}C_{DIC}$, $\delta^{15}N$-NO_3^-, and $\delta^{18}O$-NO_3^- in the Pingzhai Reservoir and its inflow rivers, we analyzed the temporal and spatial variation in water chemistry and carbon and nitrogen isotopes, explored the source of carbon and nitrogen elements in water, and analyzed the C–N coupling in water. The results show that the dominant cation in the water of the Pingzhai Reservoir is Ca^{2+}, which accounts for 68% of the total cations. The dominant anion is HCO_3^-, accounting for 65% of the total anions. The hydrochemical type was HCO_3-Ca. Dissolved inorganic carbon, nitrate, and their isotopes have different spatial and temporal variations. From the river to the reservoir area and then to the outlet, the concentration of HCO_3^- increased first, then decreased, and then increased again, while the $\delta^{13}C_{DIC}$ value was negative first, then positive, and then negative again. The values of NO_3^-, $\delta^{15}N$-NO_3^-, and $\delta^{18}O$-NO_3^- were different in each stage of the river, which were mainly affected by dam construction and water storage, surrounding land-use mode, crop cultivation and fertilization, domestic sewage discharge, and other factors. According to the characteristics of carbon and nitrogen isotopes, the HCO_3^- in the water of the study area is mainly derived from the weathering of carbonate rocks and the oxidative decomposition of organic matter. Nitrate mainly comes from agricultural activities, including chemical fertilizer (ammonium fertilizer and nitrate fertilizer), soil organic nitrogen, sewage, and livestock manure in the normal season, dry season, and wet season. The input of nitrate caused the C–N coupling cycle of hydrogeochemistry in the Pingzhai Reservoir basin and disturbed the water-rock interaction. The average proportions of HCO_3^- and (Ca^{2+} + Mg^{2+}) produced by HNO_3 dissolved carbonate rocks were 8.38% and 7.84%, respectively, but this part does not constitute a carbon sink. The proportion of HCO_3^- and (Ca^{2+} + Mg^{2+}) produced by the dissolution of carbonate rocks by HNO_3 in reservoir water was relatively large. This reflects the full water-rock interaction and retention effect due to the construction of the dam. Therefore, successful fertilizer application experience and farmland management practice should be learned, controlling nitrogen input from agricultural activities. Using new technologies can increase the capacity of domestic wastewater treatment and limit the discharge of sewage into rivers and reservoirs and prevents soil organic nitrogen loss. In addition, a limitation of the study is that it is lacking an analysis of the impact of geological conditions on the geochemical characteristics of carbon and nitrogen elements. This will be addressed in subsequent studies.

Author Contributions: Z.Z.: Resources, Data curation, Formal analysis, Project administration, Funding acquisition. J.K.: Conceptualization, Methodology, Software, Writing—original draft. F.Z.: Visualization. Y.Z.: Supervision. J.X.: Investigation. C.W.: Validation. All authors have read and agreed to the published version of the manuscript.

Funding: National Natural Science Foundation of China (42161048); National Natural Science Foundation of China (41661088); Science and Technology Plan Project of Guizhou Province (Qiankehe Jichu [2020]1Y154) financially supported this study.

Data Availability Statement: Not applicable.

Acknowledgments: This study was supported financially by the State's Key Project of Research.

Conflicts of Interest: The authors declare no conflict of interest.

References

1. Milliman, J.D.; Meade, R.H. World-Wide Delivery of River Sediment to the Oceans. *J. Geol.* **1983**, *91*, 1–21. [CrossRef]
2. Bastviken, D.; Tranvik, L.J.; Downing, J.A.; Crill, P.M.; Enrich-Prast, A. Freshwater Methane Emissions Offset the Continental Carbon Sink. *Science* **2011**, *331*, 50. [CrossRef] [PubMed]
3. Chen, S.N.; Yue, F.J.; Liu, X.L.; Zhong, J.; Yi, Y.B.; Wang, W.F.; Qi, Y.; Xiao, H.Y.; Li, S.L. Seasonal variation of nitrogen biogeochemical processes constrained by nitrate dual isotopes in cascade reservoirs, Southwestern China. *Environ. Sci. Pollut. Res.* **2021**, *28*, 26617–26627. [CrossRef] [PubMed]
4. Yuan, B.; Wu, W.; Guo, M.J.; Zhou, X.D.; Xie, S.G. Spatial-temporal dynamics and influencing factors of archaeal communities in the sediments of Lancang River cascade reservoirs (LRCR), China. *PLoS ONE* **2021**, *16*, e0253233. [CrossRef] [PubMed]
5. Wang, W.F.; Li, S.L.; Zhong, J.; Maberly, S.C.; Li, C.; Wang, F.S.; Xiao, H.Y.; Liu, C.Q. Climatic and anthropogenic regulation of carbon transport and transformation in a karst river-reservoir system. *Sci. Total Environ.* **2020**, *707*, 135628. [CrossRef]
6. Naddafi, R.; Koupayeh, N.H.; Ghorbani, R. Spatial and temporal variations in stable isotope values ($\delta^{13}C$ and $\delta^{15}N$) of the primary and secondary consumers along the southern coastline of the Caspian Sea. *Mar. Pollut. Bull.* **2021**, *164*, 112001. [CrossRef]
7. Chen, Z.D.; Huang, L.P.; Chen, L.; Liang, H.; Liu, Y.Y.; Chen, X.L.; Zhang, T.; Chen, G.J. Seasonal variation and driving factors of carbon and nitrogen stable isotope values of plankton in four lakes of Yunnan Province. *J. Lake Sci.* **2021**, *33*, 761–773. (In Chinese)
8. Ren, K.; Pan, X.D.; Liang, J.P.; Peng, C.; Zeng, J. Sources and Fate of Nitrate in Groundwater in a Typical Karst Basin: Insights from Carbon, Nitrogen, and Oxygen Isotopes. *Environ. Sci.* **2021**, *42*, 2268–2275. (In Chinese)
9. Devito, K.J.; Fitzgerald, D.; Hill, A.R.; Aravena, R. Nitrate Dynamics in Relation to Lithology and Hydrologic Flow Path in a Riparian Zone. *J. Environ. Qual.* **2000**, *29*, 1075–1084. [CrossRef]
10. Chen, Y.; Jiang, Y.J. The Effects of Agricultural Activities and Atmospheric Acid Deposition on Carbonate Weathering in a Small Karstic Agricultural catchment, Southwest China. *Acta Carsologica* **2016**, *45*, 161–172. [CrossRef]
11. Liu, Z.H. New progress and prospects in the study of rock-weathering-related carbon sinks. *Chin. Sci. Bull.* **2012**, *57*, 95–102. (In Chinese) [CrossRef]
12. Zhang, X.B.; Jiang, Y.J.; Qiu, S.L.; Cao, M.; Hu, Y.J. Agricultural Activities and Carbon Cycling in Karst Areas in Southwest China: Dissolving Carbonate Rocks and CO_2 Sink. *Adv. Earth Sci.* **2012**, *27*, 466–476. (In Chinese)
13. Barnes, R.T.; Raymond, P.A. The contribution of agricultural and urban activities to inorganic carbon fluxes within temperate watersheds. *Chem. Geol.* **2009**, *266*, 318–327. [CrossRef]
14. Aquilina, L.; Poszwa, A.; Walter, C.; Vergnaud, V.; Wickmann, A.C.P.; Ruiz, L. Long-term effects of high nitrogen loads on cation and carbon riverine export in agricultural catchments. *Environ. Sci. Technol.* **2012**, *46*, 9447–9455. [CrossRef]
15. Raymond, P.A.; Oh, N.H.; Turner, R.E.; Broussard, W. Anthropogenically enhanced fluxes of water and carbon from the Mississippi River. *Nature* **2008**, *451*, 449–452. [CrossRef]
16. Khadija, S.; Suchet, P.A.; Clauer, N.; Probst, J.L. Impact of nitrogen fertilizers on the natural weathering-erosion processes and fluvial transport in the Garonne basin. *Appl. Geochem.* **2000**, *15*, 865–878.
17. Zhao, H.J.; Xiao, Q.; Wu, X.; Liu, F.; Miao, Y.; Jiang, Y.J. Impact of Human Activities on Water-Rock Interactions in Surface Water of Lijiang River. *Environ. Sci.* **2017**, *38*, 4108–4119. (In Chinese)

18. Baker, A.; Cumberland, S.; Hudson, N. Dissolved and total organic and inorganic carbon in some British rivers. *Area* **2008**, *40*, 117–127. [CrossRef]
19. Gandois, L.; Perrin, A.S.; Probst, A. Impact of nitrogenous fertiliser-induced proton release on cultivated soils with contrasting carbonate contents: A column experiment. *Geochim. Cosmochim. Acta* **2011**, *75*, 1185–1198. [CrossRef]
20. Perrin, A.S.; Probst, A.; Probst, J.L. Impact of nitrogenous fertilizers on carbonate dissolution in small agricultural catchments: Implications for weathering CO_2 uptake at regional and global scales. *Geochim. Cosmochim. Acta* **2008**, *72*, 3105–3123. [CrossRef]
21. Brunet, F.; Potot, C.; Probst, A.; Probst, J.L. Stable carbon isotope evidence for nitrogenous fertilizer impact on carbonate weathering in a small agricultural watershed. *Rapid Commun. Mass Spectrom.* **2011**, *25*, 2682–2690. [CrossRef] [PubMed]
22. Yue, F.J.; Li, S.L.; Liu, C.Q.; Lang, Y.C.; Ding, H. Sources and transport of nitrate constrained by the isotopic technique in a karst catchment: An example from Southwest China. *Hydrol. Process.* **2015**, *29*, 1883–1893. [CrossRef]
23. Liu, C.Q.; Lang, B.C.; Li, S.L.; Piao, H.C.; Chu, C.L.; Liu, C.Z.; Zhang, W.; Zhu, S.F. Researches on biogeochemical processes and nutrient cycling in karstic ecological systems, southwest China: A review. *Front. Earth Sci.* **2009**, *16*, 1–12. (In Chinese)
24. Zhang, Y.R.; Zhou, Z.F.; Zhang, H.T.; Dan, Y.S. Quantifying the impact of human activities on water quality based on spatialization of social data: A case study of the Pingzhai Reservoir Basin. *Water Supply* **2020**, *20*, 688–699. [CrossRef]
25. Liu, X.M.; Zhou, Z.F.; Zhang, H.T.; Dan, Y.S.; Jiang, Y. Changes of Hydrochemistry and Dissolved Inorganic Carbon during Thermal Stratification in Pingzhai Reservoir. *Resour. Environ. Yangtze Basin* **2021**, *30*, 936–945. (In Chinese)
26. Ma, S.; Wei, Y.; Han, C.H.; Yan, H.; Liu, Z.H.; Sun, H.L.; Bao, Q. Hydrochemical characteristics in karst reservoirs and its implication for inorganic carbon deposition fluxes. *J. Lake Sci.* **2021**, *33*, 1701–1713. (In Chinese)
27. Dan, Y.S. Study on the Spatial Differentiation Characteristics of Nitrogen Forms and Water Quality Evaluation in Pingzhai Reservoir. Master's Thesis, Guizhou Normal University, Guiyang, China, 2020. (In Chinese).
28. Sigman, D.M.; Casciotti, K.L.; Andreani, M.; Barford, C.; Galanter, M.; Böhlke, J.K. A bacterial method for the nitrogen isotopic analysis of nitrate in seawater and freshwater. *Anal. Chem.* **2001**, *73*, 4145–4153. [CrossRef]
29. Casciotti, K.L.; Sigman, D.M.; Hastings, M.G.; Böhlke, J.K.; Hilkert, A. Measurement of the oxygen isotopic composition of nitrate in seawater and freshwater using the denitrifier method. *Anal. Chem.* **2002**, *74*, 4905–4912. [CrossRef]
30. Liu, J.; Song, X.; Wang, Z.; Yang, L.; Sun, Z.; Wang, W. Variations of carbon transport in the Yellow River, China. *Hydrol. Res.* **2014**, *46*, 746–762. [CrossRef]
31. Meng, X.; Liu, W.G. Using dual isotopes to identify sources and transformations of nitrogen in water catchments with different land uses, Loess Plateau of China. *Environ. Sci. Pollut. Res.* **2016**, *23*, 388–401.
32. Qin, C.Q.; Li, S.L.; Yue, F.J.; Xu, S.; Ding, H. Spatiotemporal variations of dissolved inorganic carbon and controlling factors in a small karstic catchment, Southwestern China. *Earth Surf. Process. Landf.* **2019**, *44*, 2423–2436. [CrossRef]
33. Yu, Y.X.; Liu, C.Q.; Wang, F.S.; Wang, B.L.; Li, J.; Li, S.L. Dissolved inorganic carbon and its isotope differentiation characteristics in cascade reservoirs in Wujiang River basin. *Chin. Sci. Bull.* **2008**, *53*, 1935–1941. (In Chinese)
34. Hélie, J.F.; Hillaire, M.C.; Rondeau, B. Seasonal changes in the sources and fluxes of dissolved inorganic carbon through the St. Lawrence River—Isotopic and chemical constraint. *Chem. Geol.* **2002**, *186*, 117–138. [CrossRef]
35. Jia, G.D.; Chen, F.J.; Deng, W.F. Seasonal Variations of Dissolved Inorganic Carbon Isotope in the Beijiang River. *J. Earth Sci.* **2012**, *37*, 365–369. (In Chinese)
36. Ding, H.; Liu, C.Q.; Lang, Y.C.; Liu, W.J. Variations of dissolved carbon and $\delta^{13}C_{DIC}$ of surface water during rainfall events in a typical karst peak cluster-depression catchment, SW China. *Front. Earth Sci.* **2011**, *18*, 182–189. (In Chinese)
37. Fadhullah, W.; Yaccob, N.S.; Syakir, M.I.; Muhammad, S.A.; Yue, F.J.; Li, S.L. Nitrate sources and processes in the surface water of a tropical reservoir by stable isotopes and mixing model. *Sci. Total Environ.* **2020**, *700*, 134517. [CrossRef]
38. Yu, Q.B.; Wang, F.; Li, X.Y.; Yan, W.J.; Li, Y.Q.; Lv, S.C. Tracking nitrate sources in the Chaohu Lake, China, using the nitrogen and oxygen isotopic approach. *Environ. Sci. Pollut. Res. Int.* **2018**, *25*, 19518–19529. [CrossRef]
39. Lee, E.S.; Krothe, N.C. A four-component mixing model for water in a karst terrain in south-central Indiana, USA. Using solute concentration and stable isotopes as tracers. *Chem. Geol.* **2001**, *179*, 129–143. [CrossRef]
40. Xiao, S.Z.; Lan, J.C.; Yuan, D.X.; Wang, Y.; Yang, L.; Ao, X.H. Hydrochemistry and Dissolved Inorganic Carbon Stable Isotope of Shibing Dolomite Karst Area in Guizhou Province. *Environ. Sci.* **2015**, *36*, 2085–2093. (In Chinese)
41. Kendall, C. Tracing nitrogen sources and cycling in catchments. In *Isotope Tracers in Catchment Hydrology*; Chapter 16; Kendall, C., Mcdonnell, J.J., Eds.; Elsevier Science: Amsterdam, The Netherlands, 1998; pp. 519–576.
42. Bu, H.M.; Zhang, Y.; Meng, W.; Song, X.F. Effects of land-use patterns on in-stream nitrogen in a highly-polluted river basin in Northeast China. *Sci. Total Environ.* **2010**, *553*, 232–242. [CrossRef]
43. Xue, D.M.; Botte, J.; Bernard, D.B.; Accoe, F.; Nestler, A.; Taylor, P.; Cleemput, O.V.; Berglund, M.; Boeckx, P. Present limitations and future prospects of stable isotope methods for nitrate source identification in surface- and groundwater. *Water Res.* **2009**, *43*, 1159–1170. [CrossRef] [PubMed]
44. Kendall, C.; Elliott, E.M.; Wankel, S.D. Tracing Anthropogenic Inputs of Nitrogen to Ecosystems. In *Stable Isotopes in Ecology and Environmental Science*, 2nd ed.; Michener, R.H., Lajtha, K., Eds.; Blackwell Publishing: Oxford, UK, 2007; pp. 375–449.
45. Lin, J.J.; Böhlke, J.K.; Huang, S.; Gonzalez-Meler, M.; Sturchio, N.C. Seasonality of nitrate sources and isotopic composition in the Upper Illinois River. *J. Hydrol.* **2019**, *568*, 849–861. [CrossRef]

46. Ju, X.T.; Xing, G.X.; Chen, X.P.; Zhang, S.L.; Zhang, L.J.; Liu, X.J.; Cui, Z.L.; Yin, B.; Christie, P.; Zhu, Z.L.; et al. Reducing environmental risk by improving N management in intensive Chinese agricultural systems. *Proc. Natl. Acad. Sci. USA* **2009**, *106*, 3041–3046. [CrossRef] [PubMed]
47. Zhao, Y.Y.; Zheng, B.H.; Jia, H.F.; Chen, Z.X. Determination sources of nitrates into the Three Gorges Reservoir using nitrogen and oxygen isotopes. *Sci. Total Environ.* **2019**, *687*, 128–136. [CrossRef] [PubMed]
48. Han, C.N.; Zheng, B.H.; Qin, Y.W.; Ma, Y.Q.; Cao, W.; Yang, C.C.; Liu, Z.C. Analysis of phosphorus import characteristics of the upstream input rivers of Three Gorges Reservoir. *Environ. Earth Sci.* **2016**, *75*, 1024. [CrossRef]
49. Cerling, T.E.; Solomon, D.K.; Quade, J.; Bowman, J.R. On the isotopic composition of carbon in soil carbon dioxide. *Geochim. Cosmochim. Acta* **1991**, *55*, 3403–3405. [CrossRef]
50. Wang, B.Y.; Liu, C.Q.; Peng, X.; Wang, F.S.; Chen, C. Stable carbon isotope as a proxy for the change of phytoplankton community structure in cascade reservoirs from Wujiang River, China. *Biogeosciences* **2011**, *8*, 831–856.
51. Kaplan, J.O.; Prentice, I.C.; Buchmann, N. The stable carbon isotope composition of the terrestrial biosphere: Modeling at scales from the leaf to the globe. *Glob. Biogeochem. Cycles* **2002**, *16*, 8-1–8-11. [CrossRef]
52. Wu, P.; Tang, C.Y.; Zhu, L.J.; Liu, C.Q.; Cha, X.F.; Tao, X.Z. Hydrogeochemical characteristics of surface water and groundwater in the karst basin, southwest China. *Hydrol. Process.* **2009**, *23*, 2012–2022. [CrossRef]
53. Jiang, Y.J. The contribution of human activities to dissolved inorganic carbon fluxes in a karst underground river system: Evidence from major elements and $\delta^{13}C_{DIC}$ in Nandong, Southwest China. *J. Contam. Hydrol.* **2013**, *152*, 1–11. [CrossRef]
54. Touhari, F.; Meddi, M.; Mehaiguene, M.; Razack, M. Hydrogeochemical assessment of the Upper Cheliff groundwater (North West Algeria). *Environ. Earth Sci.* **2015**, *73*, 3043–3061. [CrossRef]
55. Anirban, D.; Krishnaswami, S.; Bhattacharya, S.K. Carbon isotope ratio of dissolved inorganic carbon (DIC) in rivers draining the Deccan Traps, India: Sources of DIC and their magnitudes. *Earth Planet. Sci. Lett.* **2005**, *236*, 419–429.
56. Jiang, Y.J.; Hu, Y.J.; Schirmer, M. Biogeochemical controls on daily cycling of hydrochemistry and $\delta^{13}C$ of dissolved inorganic carbon in a karst spring-fed pool. *J. Hydrol.* **2013**, *478*, 157–168. [CrossRef]
57. Liu, Z.H.; Macpherson, G.L.; Groves, C.; Martin, J.B.; Yuan, D.X.; Zeng, S.B. Large and active CO_2 uptake by coupled carbonate weathering. *Earth-Sci. Rev.* **2018**, *182*, 42–49. [CrossRef]
58. Hu, L.C.; Jiang, Y.J.; Zeng, S.B.; Lei, J.Q. C-N coupling cycle and carbonate weathering in karst critical zone: A case study from Chongqing Xueyudong Observatory. *Quat. Sci.* **2017**, *37*, 1251–1265.

land

MDPI

Article

Comparative Analysis of Temporal-Spatial Variation on Mountain-Flatland Landscape Pattern in Karst Mountainous Areas of Southwest China: A Case Study of Yuxi City

Li Wu [1], Jing Zhou [1] and Binggeng Xie [2,*]

[1] Department of Geography and Land Engineering, Yuxi Normal University, Yuxi 653100, China
[2] School of Geographic Sciences, Hunan Normal University, Changsha 410081, China
* Correspondence: xbgyb1961@163.com; Tel.: +86-138-0847-3925

Abstract: Taking Yuxi City, a typical mountain-flatland in the southwestern karst mountainous area, as an example, we used six remote sensing images from 1995 to 2018 as the main data sources, and the grid scale was used to calculate the landscape pattern index in order to analyze the temporal-spatial evolution characteristics of the landscape pattern. The results are shown as follows: (1) At the class level, most landscape indices and fragmentation degrees of landscape units in the flatland area are significantly higher than those in the mountainous area. The layout of construction land and cultivated land is also more concentrated than that in the mountainous area, but the central tendency of forest and grass in the mountainous area is more eye-catching. (2) At the landscape level, although the landscape diversity index and landscape shape index of both the mountainous areas and the flatland areas decrease in the low-value area and increase in the high-value area, the proportion of high-value areas in the flatland area is noticeably greater. The proportion of the high-value areas of the largest patch index in the mountainous area is significantly greater, and in the flatland area, the low-value area continues to expand while the middle and high value areas continue to shrink. (3) The landscape shape of the flatland area is becoming more complex, and the landscape units in the mountainous area tend to be single. The natural landscape of forest and grass in the mountainous area continues to expand and tends to be contiguous, while the man-made landscape in the flatland area continually increases and shows fragmentation, reflecting the pattern characteristics formed by the coupling evolution of land use between two regions. The urban expansion and the increase in the construction land in the flatland area are mutually causal with the decrease in cultivated land and the increase in forest and grass in the mountainous area.

Keywords: mountainous areas; flatland areas; grid scale; landscape pattern evolution; Yuxi City

Citation: Wu, L.; Zhou, J.; Xie, B. Comparative Analysis of Temporal-Spatial Variation on Mountain-Flatland Landscape Pattern in Karst Mountainous Areas of Southwest China: A Case Study of Yuxi City. *Land* **2023**, *12*, 435. https://doi.org/10.3390/land12020435

Academic Editor: Xiaoyong Bai

Received: 25 December 2022
Revised: 4 February 2023
Accepted: 6 February 2023
Published: 7 February 2023

1. Introduction

Due to the imbalance between social and economic development and the natural ecological process [1], as well as the instability, sensitivity and complexity of the natural system of the mountainous area itself, the mountainous area has become the most intense region of global environmental change and ecological degradation in recent years [2], and the conflicts between human and land have become more prominent and complex. In the southwest of China, the terrain is fragmented and sensitive, and the generalized mountain landscape is an important and unique natural—human geography unit composed of "mountain" and "flatland" as the core elements [3,4], which is an important content of the scientific research of human—land systems in mountainous areas. To fully guarantee the ecological security and sustainable development of mountainous areas, it is necessary to take land resources as the main constraint condition [5], coordinate the linkage relationship between "mountain" and "flatland" in the whole region, and study the changes and correlation of "mountain" and "flatland" elements in mountainous landscapes, in order to provide more scientific and accurate strategies for rural revitalization, mountain development, spatial governance, as well as the maintenance and improvement of ecosystem services.

A landscape pattern mainly refers to the shape, proportion and spatial configuration of the elements that constitute landscape units [6]. It is not only the comprehensive expression of landscape heterogeneity in space [7,8], but also the result of various ecological processes driven by natural and social factors at different scales [9,10]. Landscape pattern change is the change in the landscape's spatial structure on the basis of the changes in various landscape elements, which are closely related to climate change, land use/land cover change and change in biodiversity [11,12]. Understanding the evolutionary characteristics of the landscape pattern is the premise and basis for landscape pattern analysis. The evolution of landscape pattern is a comprehensive reflection of the interaction and influence of natural elements and human factors in a certain region, as well as the different external characteristics and spatial combinations of various factors, which constantly affect the ecological process and marginal effect [13]. Exploring the evolution process of a landscape pattern is helpful for grasping the evolutionary characteristics and rules of the regional landscape, and to provide basic data for the assessment of the sensitivity, vulnerability and ecological risk of ecological degradation [14], which is the basis and important support for decision-makers to formulate reasonable and scientific urban planning [15]. At present, the analysis methods of landscape pattern evolution mainly include spatial statistical analysis, landscape index analysis and pattern dynamic model simulation [16,17]. From the perspective of landscape ecology, landscape index analysis regards the study area as a whole and reveals the changes in the landscape's spatial pattern characteristics in time series through various landscape indices [18]. Landscape indices include a patch level index, a patch type level index and a landscape level index. As a scale of landscape pattern characteristics, the landscape index has been widely used to analyze regional landscape patterns and dynamic evolution [19,20], including urban [21], rural [22], urban fringe areas [23] and economically developed areas [24]. However, the landscape pattern evolution of two different geographical units of "mountain" and "flatland" in the generalized mountain landscape is rarely involved. There are great differences between the physical geographical conditions and human and social activities in the mountainous area and flatland area. What are the evolutionary differences in the landscape pattern on the spatial and temporal scale? Is there a certain correlation between the evolution of the two on the space—time scale? All these need to be explained by landscape pattern analysis.

In addition, the scaling effect of a landscape pattern is also a major feature of landscape pattern analysis [25]. The results of landscape pattern evolution at different scales often vary vigorously, and the landscape index will change with the scale and have a scale effect at a finer scale [26]. Landscape pattern characteristics at different spatial scales are often different, and at the same spatial scale, a landscape pattern at different time scales will also be different [27]. The dynamic landscape characteristics of urban and rural construction land show great differences in scale [28], and the landscape pattern of cultivated land in the middle reaches of the Yangtze River has great differences in quantity, area, aggregation degree and diversity at different scales [29]. Therefore, the small-scale and refined studies on land use dynamic change and landscape pattern evolution at and below the county level have attracted ever increasing attention [30,31]. Some scholars have used the grid scale to carry out their research, and obtained more refined results compared with the scale of watershed and administrative regions [32,33].

In addition to the low level intermountain basins, valleys and depressions, "flatlands" in this study also include lacustrine plains which occupy a certain area in central Yunnan [4]. Yuxi City, with typical mountain—flatland landform features [34], is located in the mountainous area of southwest China. The flatland area is flat, suitable for farming and construction, and is an important carrier of human economic activities, while the natural conditions in the mountainous area are complex and difficult to use [35–37]. Due to the geographical connectivity of "mountains" and "flatlands", the intercrossing of human activities in different time and space has brought about the differences in the evolution of the landscape pattern on the spatial and temporal scale. Therefore, based on the grid scale and with the help of a landscape pattern index, this paper analyzes the temporal—spatial variation in mountain—flatland landscape patterns in Yuxi City, and discusses the correlation between mountains and flatlands in the evolution process of landscape patterns, so as to provide certain references for regional coordinated and sustainable development.

2. Materials and Methods

2.1. Study Area

Yuxi City is in the center of Yunnan Province on the southwest border of China, between latitude 23°19′~24°53′ north and longitude 101°16′~103°09′ east (Figure 1a). It is a prefecture-level city under the jurisdiction of Yunnan Province, bordering the provincial capital Kunming City in the north, Honghe Prefecture in the southeast, Pu'er City in the southwest and Chuxiong Prefecture in the northwest (Figure 1b). The terrain is high in the northwest and low in the southeast, with staggered distribution of mountains, canyons, plateaus, and basins. The west is mainly a deep-cut alpine valley landform. The central and eastern regions belong to the mountains of central Yunnan Province, which is dominated by a middle mountain; the terrain of most areas is undulating in the shape of waves, and there are many intermountain basins of different sizes scattered among the mountains. The eastern region is mainly dominated by plateau lake-basin landforms, with three plateau faulted lakes, Fuxian Lake, Xingyun Lake and Qilu Lake. Around three lakes, Chengjiang, Jiangchuan and Tonghai, lacustrine basins are formed, with flat and open terrain in the basins.

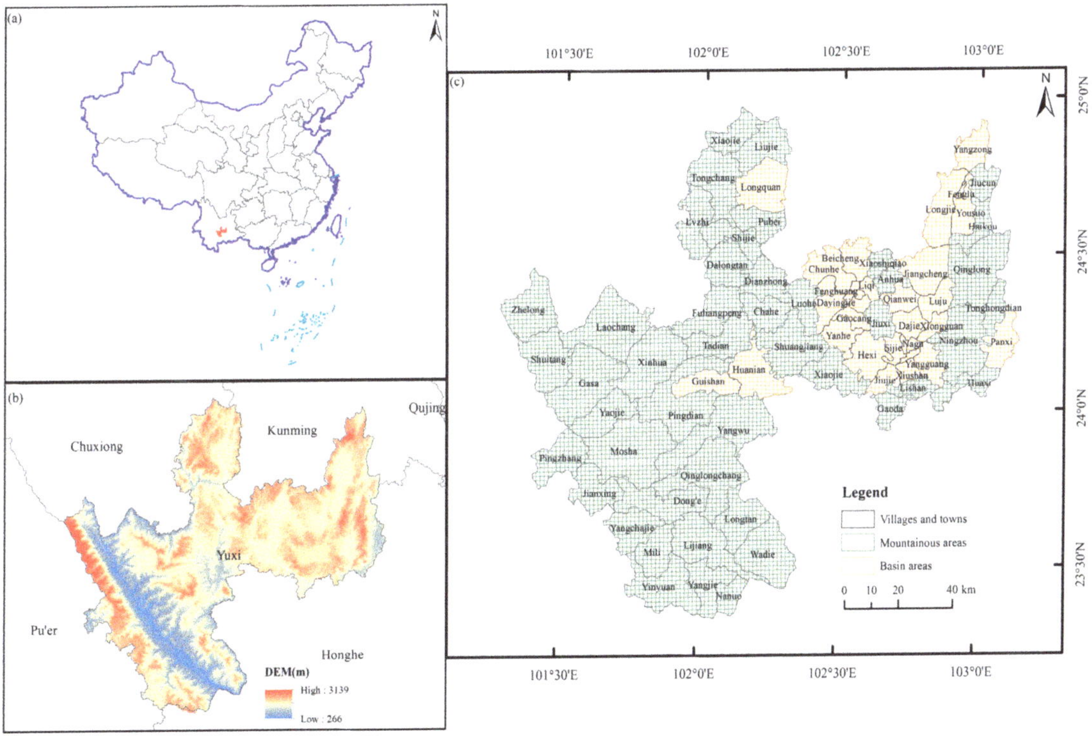

Figure 1. Overview of study area. (**a**) the location of Yuxi City in China; (**b**) the elevation of Yuxi City; (**c**) the landscape grid of mountainous areas and flatland areas.

Yuxi City is located in the low-latitude Yunnan Plateau, with long hours of sunshine and abundant heat. It belongs to the subtropical plateau monsoon climate. Due to the great difference in elevation of the terrain, the three-dimensional climate is clear. Under the comprehensive influence of the Indian Ocean and the Gulf of Tonkin, the climate in most areas is mild, with distinct dry and wet seasons, no severe cold in winter and no severe heat in summer. The annual average temperature is 15.4–24.2 °C, and the annual precipitation is 787.8–1000 mm, which mostly occurs from June to October, with heavy rain mainly from June to August. Due to the complex terrain and the great elevation difference, the rainfall is heavier, and the temperature is lower in the mountainous area than in the flatland area. From the top of the mountain to the bottom of the valley, the temperature difference is significant throughout the year and between day and night.

2.2. Data Source and Processing

In this study, six Landsat satellite remote sensing images in 1995, 2000, 2005, 2010, 2015 and 2018 were downloaded from the geospatial Data Cloud website (www.gscloud.cn, accessed on 10 June 2019), mainly including Landsat 5 TM images, Landsat 7 ETM^+ images and Landsat 8 OLI images. The selection of remote sensing images was based on the premise of little cloud and good quality. Winter and spring were selected as the image months, and the imaging times of the remote sensing images were all in January, February, and March of each year. The cloud content (CC) of all images was less than 10%, and the cloud content of most images was less than 1%. Based on the ArcGIS 10.8 software, the maximum likelihood classification method was used to classify landscape units. Referring to the standard of "Classification of Land Use Status" and combining with the research needs of landscape patterns in Yuxi City, the landscape units were divided into five types: cultivated land, forest and grass, construction land, water area and unused land (Figure 2). Kappa coefficients are all above 80%, and the accuracy meets the research requirements.

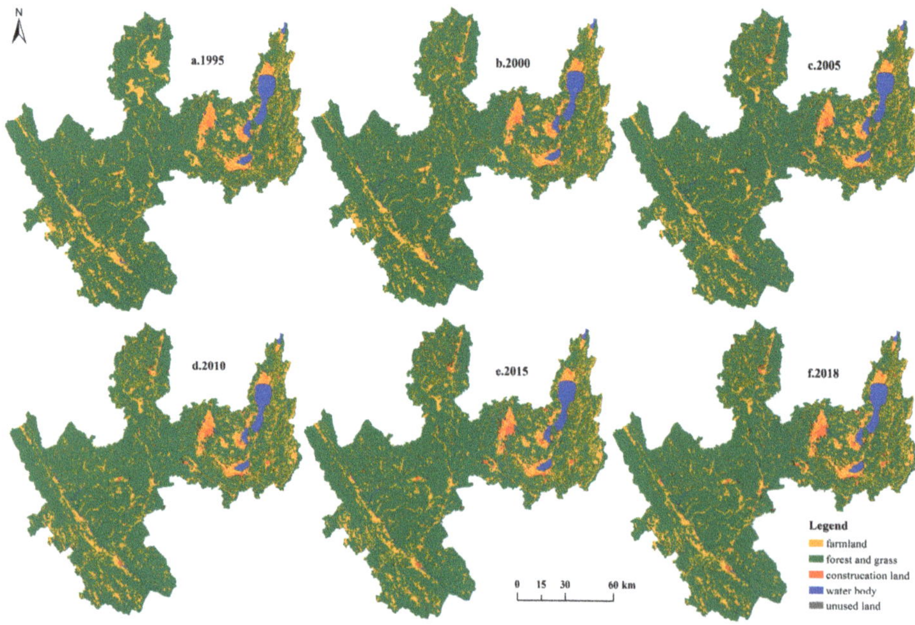

Figure 2. Interpretation of landscape types in Yuxi City from 1995 to 2018. (**a**) 1995; (**b**) 2000; (**c**) 2005; (**d**) 2010; (**e**) 2015; (**f**) 2018.

Based on the terrain characteristics, area size and elevation of Yuxi City, 500 m × 500 m, 1.5 km × 1.5 km and 3 km × 3 km grids were constructed as preselected evaluation units. With the help of analysis tools such as ArcGIS 10.8 software Create Fishnet, Dissolve, Clip and Merge, a total of 7570 grids with a size of 1.5 km × 1.5 km were finally determined. Therefore, the equal space system sampling method was used to divide the study area into 1.5 km × 1.5 km landscape plots, which were used as the basic unit for the study of landscape pattern evolution. There were 1775 landscape plots in the flatland area and 5795 landscape plots in the mountainous area (Figure 1c). The raster maps of land use in mountainous areas and flatland areas from 1995 to 2018 were cropped to obtain the raster data of landscape units over the years at the grid scale for the subsequent calculation of the landscape index. Based on the data of landscape units in each grid, the landscape index of each grid was calculated by Fragstats 4.2 software (Oregon, USA).

2.3. Index Calculation

By referring to the relevant literature [38] and combining with the unique characteristics of the mountain–flatland, landscape indices were selected respectively from the class level and landscape level to characterize and analyze the landscape pattern of Yuxi City. Plaque Density (PD), Edge Density (ED), Largest Plaque Index (LPI), and Mean Plaque Area (Area_MN) were selected at the class level. At the landscape level, Landscape Shape Index (LSI), Largest Patch Index (LPI) and Shannon Diversity Index (SHDI) were selected. The calculation formula and ecological significance of each landscape index are shown in the reference [23,39].

The Fragstats 4.2 software was used to calculate the landscape pattern index at the class level first, and excel software was used for statistics and mapping analysis. The landscape indices of each landscape plots in the mountainous area and the flatland area in six periods was then calculated by using the processed raster data of landscape units over the past years, and the spatial distribution maps of landscape indices were obtained. Finally, this study divides each landscape index into five levels according to the natural breakpoint method, which are represented by I, II, III, IV and V, respectively.

3. Measurement and Comparison of Landscape Pattern Change at Class Level

3.1. Variation Characteristics of Landscape Pattern at Class Level in Mountainous Areas

During the study period, at class level, the variation in the landscape index in the mountainous area was significantly different (Figure 3). The PD and ED of cultivated land increased first and then decreased at the turning point of 2000, and the trend of the Area_MN was in an "S" shape, while the LPI continued to decline. The PD and ED of forest and grass, water area and construction land increased first and then decreased, while the Area_MN first decreased and then increased. The patches number and density of forest and grass decreased and the Area_MN of construction land changed only slightly. The ED of cultivated land, forest and grass was higher than that of the water area, construction land and unused land. The LPI and the Area_MN of forest and grass were the largest, which had a great influence on other landscape types.

3.2. Variation Characteristics of Landscape Pattern at Class Level in Flatland Areas

During the study period, the PD of cultivated land in the flatland area increased continuously, the Area_MN decreased significantly, and the ED first increased and then continued to decrease, but it always ranked in first place among all regions, while the LPI continued to decline (Figure 4). As the first dominant type, forest and grass, the Area_MN first decreased and then increased, the PD and ED first increased and then decreased, and the LPI continued to decrease, indicating that the dominant position of large patches in the landscape gradually decreased. The PD and ED of the water area first increased and then decreased, and the degree of fragmentation gradually increased; the LPI changed little during the study period, but the Area_MN showed a downward trend. The Area_MN of construction land reached its lowest level in 2005, and then continued to rise with significant changes, while the LPI also increased significantly. However, the PD and ED first increased and then decreased with 2005 as the turning point, suggesting that the distribution of construction land was relatively scattered before 2005, but that after 2005, the construction land patches gradually concentrated with contiguous distribution.

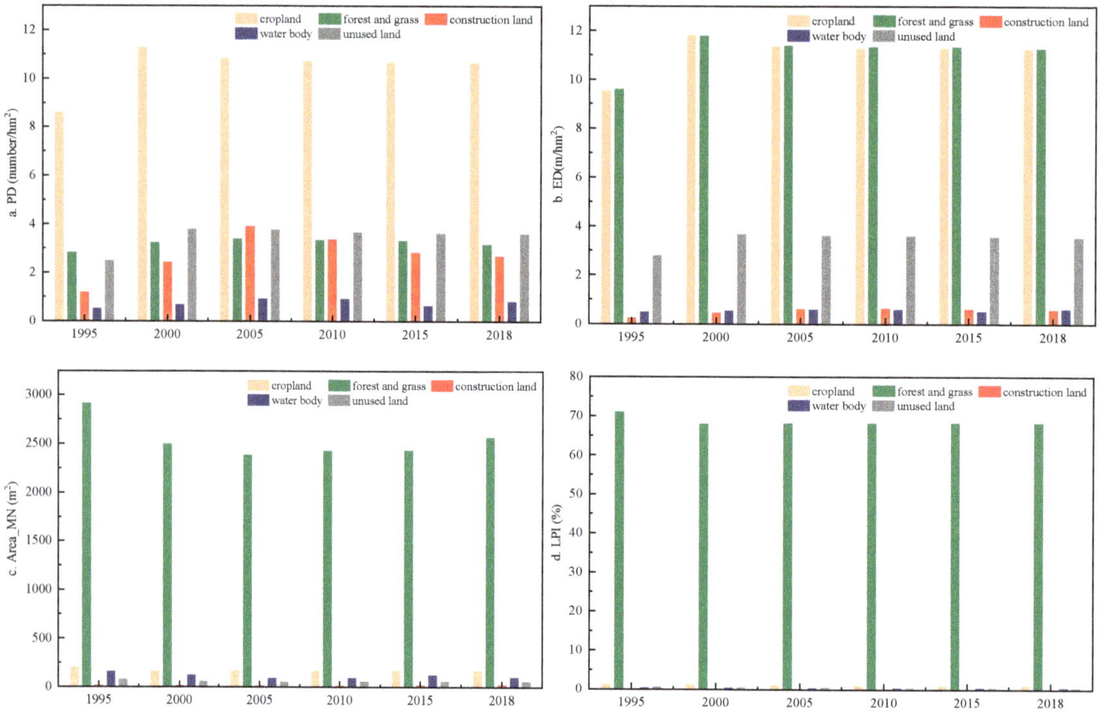

Figure 3. Landscape index at class level in mountainous areas of Yuxi City from 1995 to 2018. (**a**) Patch Density (PD); (**b**) Edge Density (ED); (**c**) Mean plaque Area (Area_MN); (**d**) Largest Patch Index (LPI).

Combined with the implementation time of regional land management policies, the changes in the landscape indices of cultivated land and forest and grass in mountainous and flatland areas were mainly affected by the policy of returning cultivated land to forest and grass during the study period. The overall fragmentation degree of construction land and forest and grass in the flatland area is clearly higher than that in the mountainous area, and the layout of the construction land is more concentrated. This is mainly because the social and economic development level of the flatland area has improved rapidly since 2005, and the urban land has expanded significantly. However, due to the migration of populations to the flatland area and for other reasons, the farmland in the mountainous area was abandoned, while the forest and grass became more concentrated and contiguous.

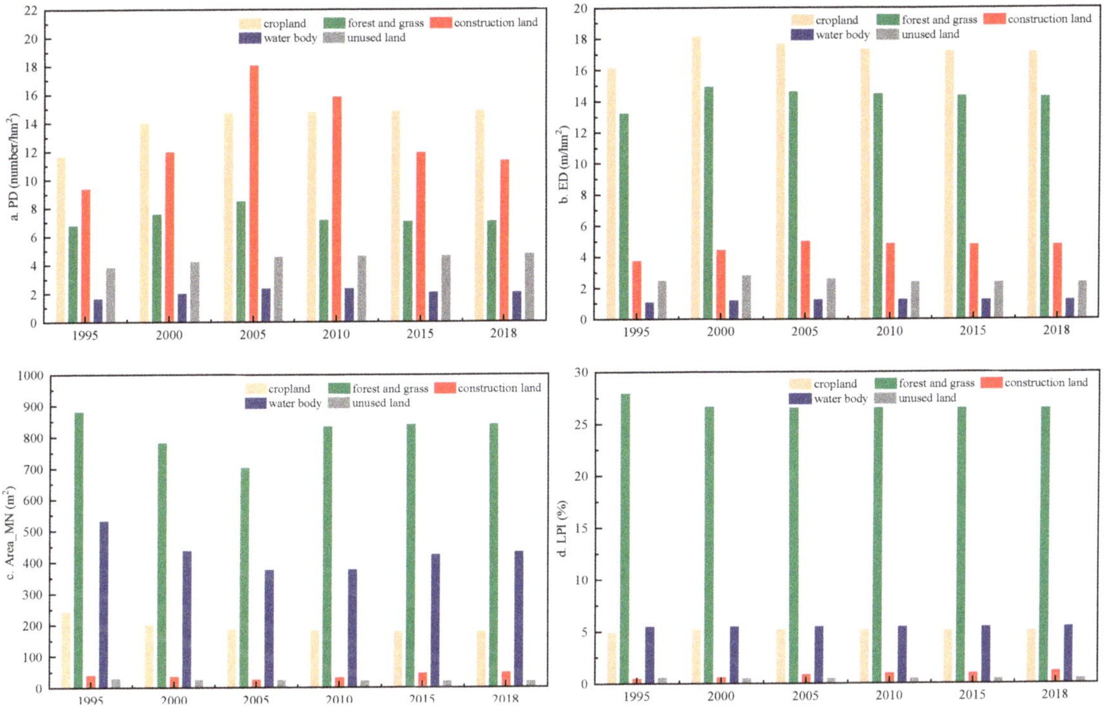

Figure 4. Landscape index at class level in flatland areas of Yuxi City from 1995 to 2018. (**a**) Patch Density (PD); (**b**) Edge Density (ED); (**c**) Mean plaque Area (Area_MN); (**d**) Largest Patch Index (LPI).

4. Measurement and Comparison of Landscape Pattern Change at Landscape Level

4.1. Variation Characteristics of Landscape Pattern at Landscape Level in Mountainous Areas

4.1.1. Landscape Diversity

During the study period, the landscape diversity in the mountainous area showed obvious phased characteristics, and the change during the period 1995–2005 was significantly higher than that during the years 2005–2018. The proportion of the SHDI was the highest in class II and III, the proportion of the low value area decreased, while the proportion of the high value area increased (Figure 5a). The area where the SHDI increased was greater than the area where the SHDI decreased in each period. From 1995 to 2018, the area where the index increased accounted for 26.74% of the total, while the index reduction areas accounted for 5.48%, indicating that the landscape diversity showed an increasing trend year by year. The high value area of the landscape diversity index mainly concentrated in the relatively low flat areas of the east and the north, where the main landscape units are cultivated land, forest and grass, while the low value area is mainly located in the central and northwest Ailao mountain area, where the main landscape unit is forest and grass (Figure 6).

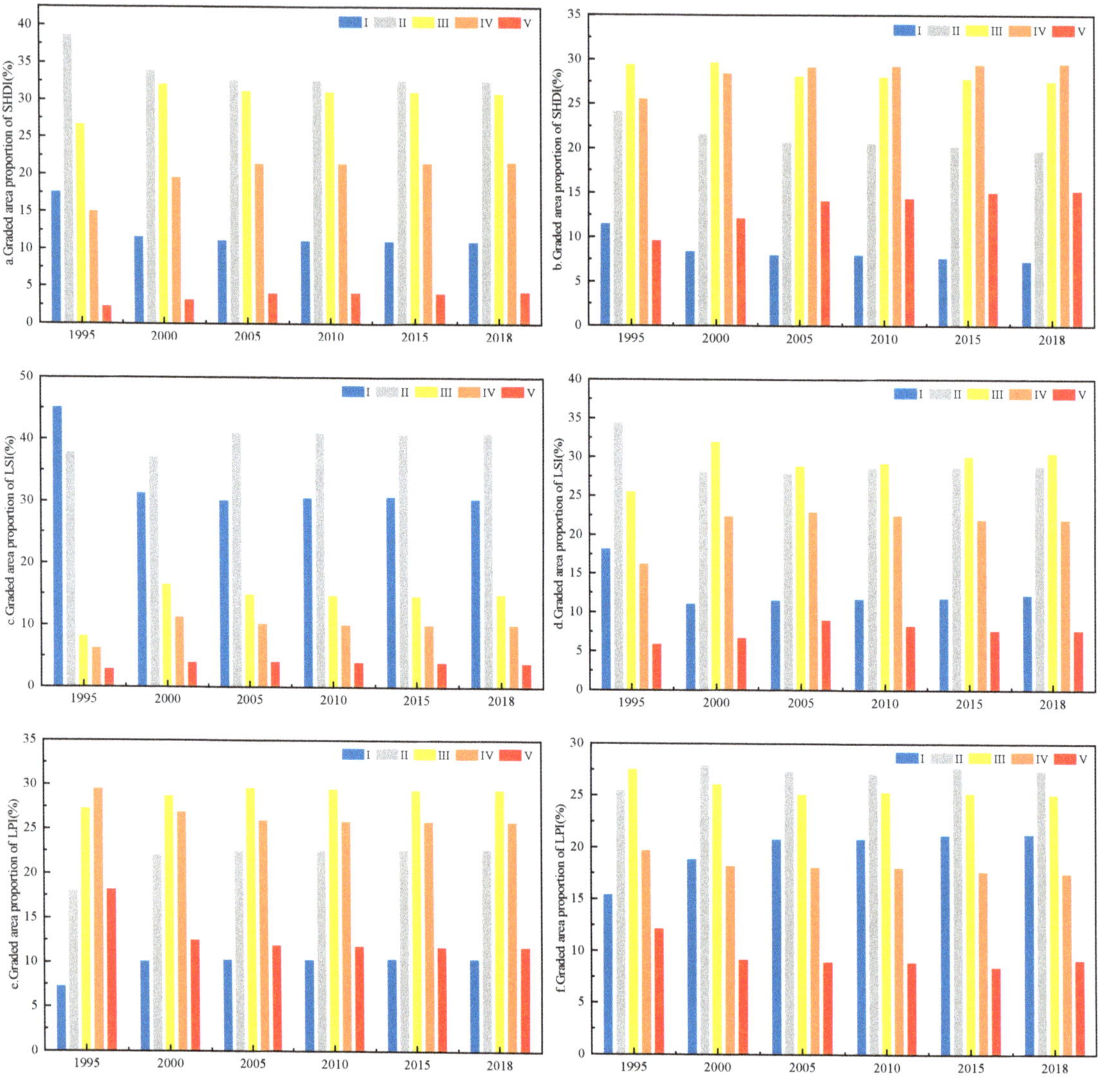

Figure 5. Graded area proportion of landscape index. (**a**) graded area proportion of SHDI in mountainous areas; (**b**) graded area proportion of SHDI in flatland areas; (**c**) graded area proportion of LSI in mountainous areas; (**d**) graded area proportion of LSI in flatland areas; (**e**) graded area proportion of LPI in mountainous areas; (**f**) graded area proportion of LPI in flatland areas.

4.1.2. Landscape Shape

Over the years, in the mountainous area, the LSI of grade I and II accounts for a relatively high proportion, and the combined proportion of the two is as high as over 70%, while the proportion of grade IV and V is about 14%, indicating that the landscape shape is relatively simple (Figure 5c). In terms of time change, the proportion of grade I showed a downward trend, while the proportion of other levels showed an upward trend. The area where the LSI increased accounted for 23.85% of the total, while the index reduction area accounted for 3.14%, indicating that from 1995 to 2018, the LSI continued to increase, and the landscape shape tended from simple to complex. However, from 2005 to 2010 and 2010 to 2015, the index increased area was smaller than the index decreased area, indicating that the landscape in these two periods was clearly contiguous, and the shape tended to be simple (Figure 7). During the study period, the areas with increased LSI were mainly distributed in the northern and southwestern regions, while the areas with decreased LSI were mainly concentrated in the eastern regions.

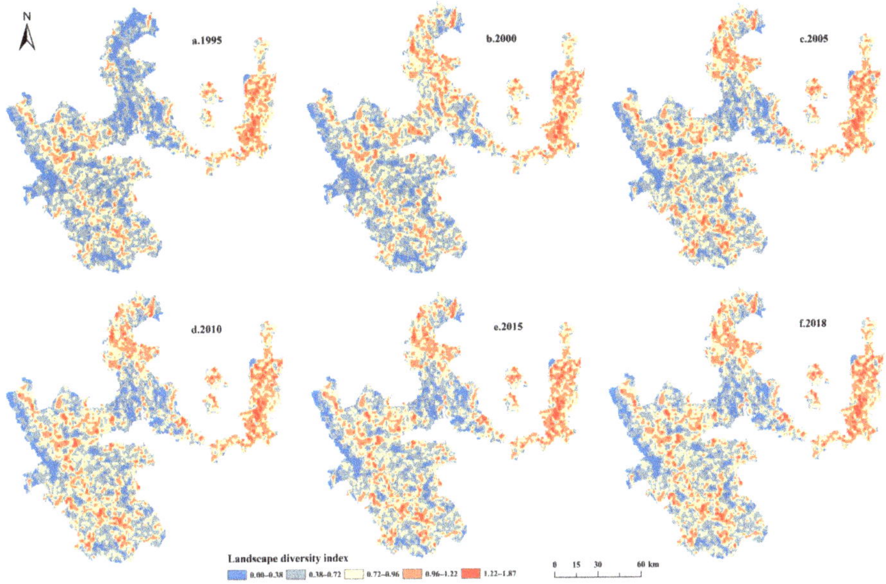

Figure 6. Spatial distribution of SHDI in mountainous areas of Yuxi City from 1995 to 2018. (**a**) 1995; (**b**) 2000; (**c**) 2005; (**d**) 2010; (**e**) 2015; (**f**) 2018.

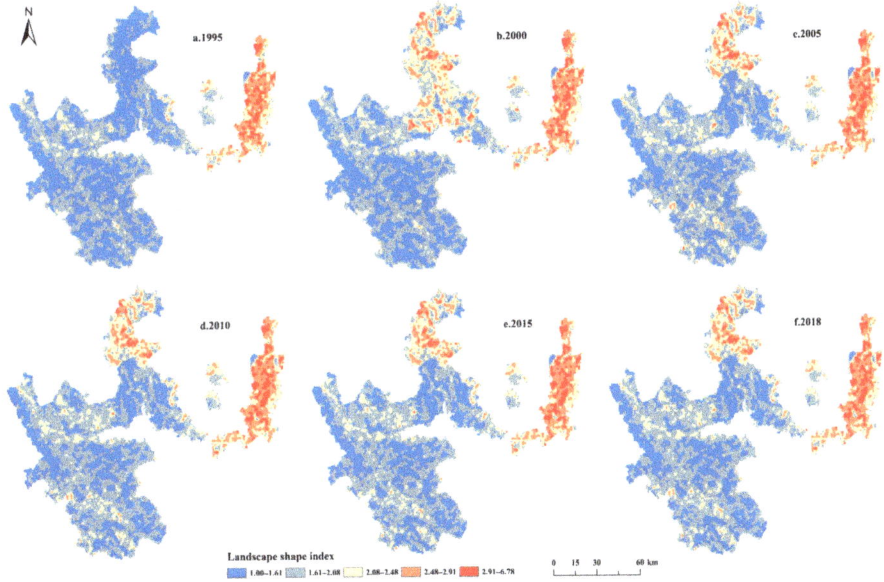

Figure 7. Spatial distribution of LSI in mountainous areas of Yuxi City from 1995 to 2018. (**a**) 1995; (**b**) 2000; (**c**) 2005; (**d**) 2010; (**e**) 2015; (**f**) 2018.

4.1.3. Largest Patch

From 1995 to 2018, the LPI levels in mountainous areas were mainly II, III and IV, but the proportion of I, II and III level increased, while the proportion of IV and V level decreased (Figure 5e). In each period, the area of the LPI that increased was much less than the decreasing area; the area of the increased LPI accounted for 7%, and the area of the index decreased accounted for 24%. The high value area of the LPI and the change in the LPI mainly occurred in the northern, central and western regions, with large topographic relief and high altitude. The low value area of the LPI with little change was mainly distributed in the eastern region, mainly in the relatively low and flat terrain area (Figure 8). From 1995 to 2018, the LPI decreased significantly in the central and western regions, especially in the northern part of the country, mainly because the forest and grass cover in the northern Yimen County was cut by a large amount of arable land, which resulted in serious fragmentation and rapid decline of the LPI.

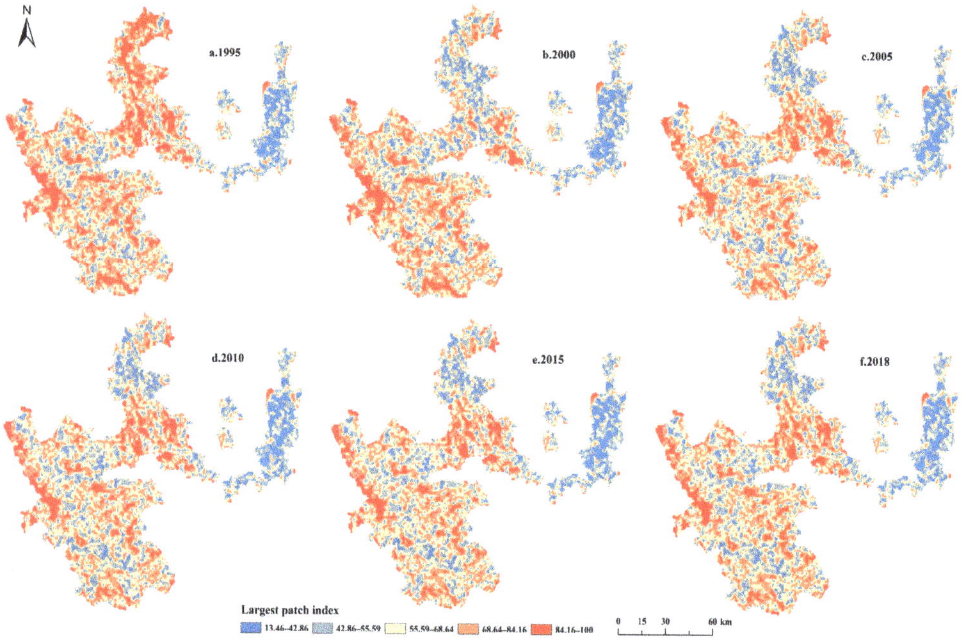

Figure 8. Spatial distribution of LPI in mountainous areas of Yuxi City from 1995 to 2018. (**a**) 1995; (**b**) 2000; (**c**) 2005; (**d**) 2010; (**e**) 2015; (**f**) 2018.

4.2. Variation Characteristics of Landscape Pattern at Landscape Level in Flatland Areas

4.2.1. Landscape Diversity

From 1995 to 2018, the SHDI in the flatland area was mainly grade II, III and IV, accounting for 78%–80% of the total, while grade I and V were relatively small, with a total proportion between 20% and 22% over the years (Figure 5b). With the construction of roads and the expansion of residential areas, the concentrated contiguity of cultivated land was divided, and the SHDI clearly increased, with the low value area decreasing and the middle and high value areas increasing. The proportion of landscape diversity level I decreased from 11.46% to 7.25%, the proportion of grade II decreased from 24.14% to 19.72%, and the proportion of grade III decreased from 29.36% to 27.46%. It can be seen that although the landscape diversity is on a downward trend, the decline rate becomes ever smaller from grade I to III, indicating that the decrease is mainly caused by the decrease in the low value area. The proportion of class IV increased from 25.52% to 29.50%, and the proportion of class V increased from 9.52% to 15.17%, with the highest increase rate of class

V. In the study period, the periodic changes were obvious, and the changes in the periods 1995–2000 and 2000–2005 were significantly higher than those in the other three periods (Figure 9). In the first four periods, the area with increased SHDI was smaller than that with decreased SHDI. However, a reversal occurred from 2015 to 2018; that is, the area with decreased SHDI was larger than that with increased SHDI, indicating that the landscape complexity began to decline at this stage and the complexity degree decreased.

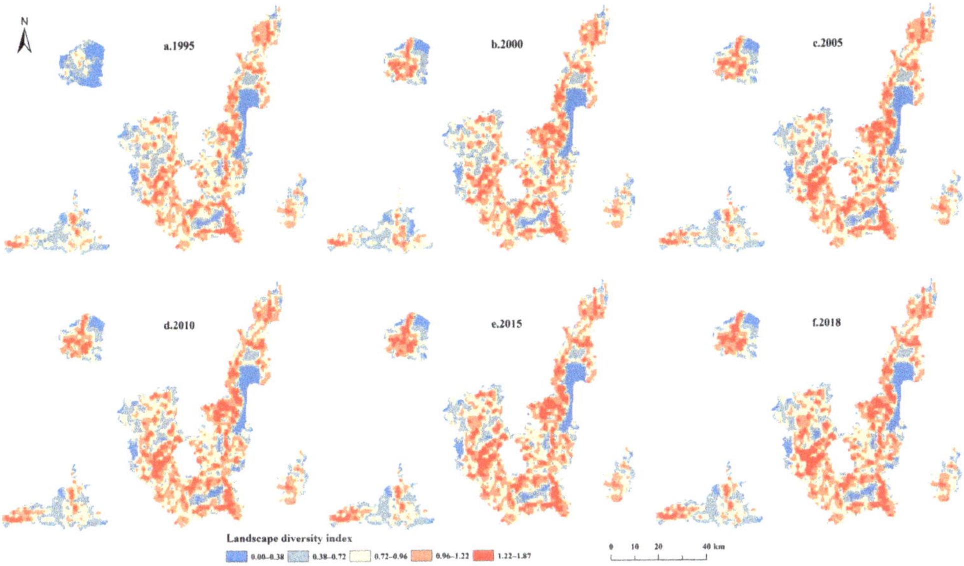

Figure 9. Spatial distribution of SHDI in flatland areas of Yuxi City from 1995 to 2018. (**a**) 1995; (**b**) 2000; (**c**) 2005; (**d**) 2010; (**e**) 2015; (**f**) 2018.

4.2.2. Landscape Shape

The LSI of the flatland area is mainly graded as II, III and IV. The proportion of grade I and II showed a downward trend, and the phased changes were first decreasing and then increasing (Figure 5d). The proportion of class III, IV and V showed an overall upward trend, and the phased changes were first increasing and then decreasing. Taking 2000 or 2005 as the boundary, the proportion of class I decreased rapidly from 1995 to 2000 and then increased and the proportion of class II decreased continuously in the period 1995–2005, then increased, while the proportion of class IV and V both peaked in 2005 and subsequently decreased. During the study period, the phased changes were clear, and the first two periods were completely the opposite to the last three periods. During the periods 1995–2000 and 2000–2005, the area of the LSI that increased was larger than that of the decreased. However, during the periods 2005–2010, 2010–2015 and 2015–2018, the area of the LSI that decreased was greater than that of the increased (Figure 10).

4.2.3. Largest Patch

From 1995 to 2018, the proportion of class I increased from 15.36% to 21.21%, and class II increased from 25.46% to 27.33%, indicating that the low value area of the LPI in the flatland area continued to expand (Figure 5f). However, the proportion of class III decreased from 27.45% to 25.01%, class IV decreased from 19.64% to 17.46%, and class V decreased from 12.09% to 9.06%. This indicated that the medium and high value area continued to shrink and that the landscape fragmentation degree clearly increased. During the study period, the overall change in the LPI decreased, mainly during the periods 1995–2000 and 2000–2005 (Figure 11). During the period 1995–2000, the area of the LPI that decreased was

45,946.49 hm^2, which was 4.01 times that of the index increased. From 2000 to 2005, the area of the LPI that decreased reached 41,649.45 hm^2, which was 1.51 times the increased area. In the first four time periods, the index decreased areas were greater than the increasing areas, but the index increased area was greater than the decreasing area in the period 2015–2018.

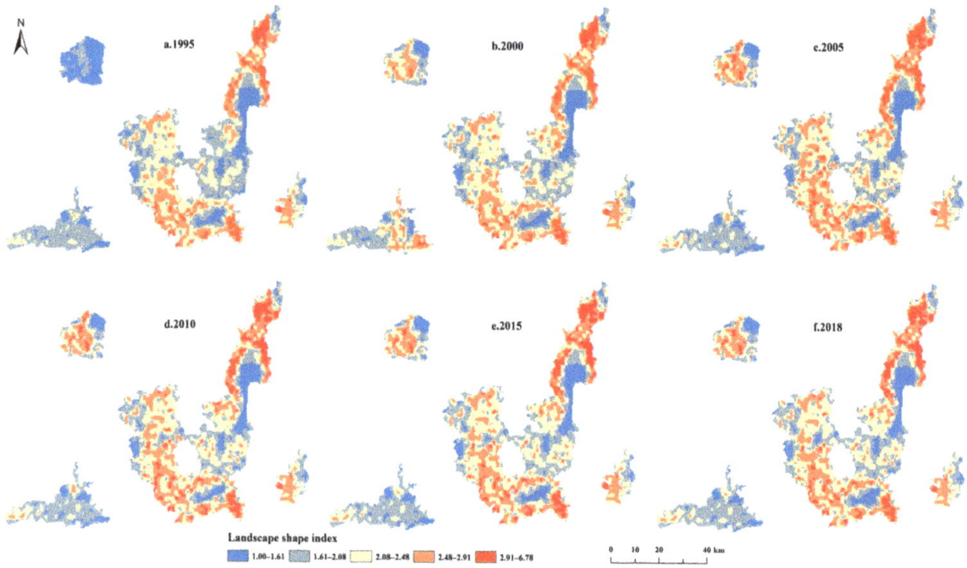

Figure 10. Spatial distribution of LSI in flatland areas of Yuxi City from 1995 to 2018. (**a**) 1995; (**b**) 2000; (**c**) 2005; (**d**) 2010; (**e**) 2015; (**f**) 2018.

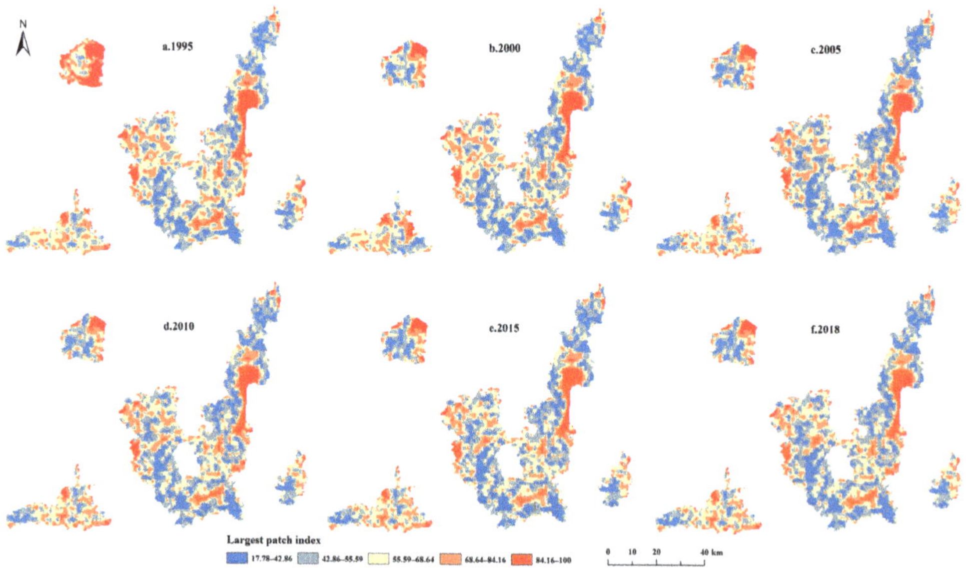

Figure 11. Spatial distribution of LPI in flatland areas of Yuxi City from 1995 to 2018. (**a**) 1995; (**b**) 2000; (**c**) 2005; (**d**) 2010; (**e**) 2015; (**f**) 2018.

5. Discussion

This study showed that the spatial and temporal trends in landscape pattern evolution in mountainous areas and flatland areas were the same: both showed an increased fragmentation degree and decreased connectivity degree, but there were also clear differences between them. At the class level, the fragmentation degree of construction land in the flatland area was significantly higher than that in the mountainous area, but its patch area was larger and the layout was more concentrated than that in the mountainous area, while the construction land in the mountainous area was more dispersed, mainly because the human activity intensity in the flatland area was significantly higher than that in the mountainous area due to the superiority of natural and socio-economic conditions [40]. Similar to the construction land, the landscape index of cultivated land in the flatland area was higher, but the Area_MN continued to decline, showing clear fragmentation but also a concentrated layout, mainly because the flatland area had flat terrain and sufficient hydrothermal conditions, which were conducive to cultivation [36]. Over time, however, human interference in the cultivated land has intensified [41]. The PD and ED of forest and grass in the flatland area were significantly higher but the Area_MN and LPI were significantly lower than those in the mountainous area, in fact only about a third of those in the mountainous area. Combined with the implementation time of regional land management policies, mainly affected by the policy of returning cultivated land to forest and grass during the study period, the degree of fragmentation of forest and grass in the flatland area was significantly higher than that in the mountainous area, while the forest and grass in the mountainous area showed an obvious trend of concentration and contiguity. At the landscape level, landscape diversity in the flatland area was significantly higher than that in the mountainous area. In the relatively low-lying area for urban construction, the urban construction and development interspersed the urban patches with the patches of cultivated land, forest and grass, with the consequence that the SHDI continued to rise, and the landscape fragmentation degree also continued to increase. The landscape shape in the mountainous area was simpler than that in the flatland area. Due to the implementation effect of the policy of returning farmland to forest and the rapid improvement of the level of social and economic development, the LPI in the flatland area continued to increase, and the landscape units in the mountainous area dominated by forest and grass tended to be more concentrated and contiguous [42].

Topography plays a key role in the formation of landscape patterns, determining the basic landscape pattern [43], and the differentiated development of the social economy will further affect the change in the landscape pattern. With the increase in topographic relief, the man-made landscape gradually gives way to the natural landscape. In terms of influencing landscape unit distribution, human factors usually dominate in flatland areas, while natural factors usually dominate in mountainous areas. The mountainous area is high in elevation and slope, and the topography is undulating, with the consequence that the accessibility is worse than in the flatland area. At the same time, it can be difficult to meet the demand for high-quality land brought by population growth and the pursuit of a prosperous life, which makes most of the population migrate to the flatland area [44], bringing about the transformation of cultivated land landscapes to forest and grass landscapes. It has been found that the landscape pattern of mountainous areas in southwest China is affected by the landforms of high mountains and river valleys, mainly forest and grass [45], while high-quality arable land mainly continues to be distributed in basins, trough valleys and low mountain valleys [46]. Before 2000, with the increase in population, in order to meet the needs of survival, the mountainous area was blindly reclaimed and the forest was destroyed [47], while the landscape diversity was higher than that of the flatland area [48]. After 2000, due to road construction and settlement expansion, concentrated and contiguous cultivated land was divided, and the landscape diversity of the flatland area increased significantly. However, due to ecological restoration or vegetation degradation, the landscape types in mountainous areas gradually became single, and the landscape diversity declined [37].

The comparison and analysis of landscape spatial patterns between mountainous areas and flatland areas based on a grid method and further microscopization and refinement from the scale [49] are of great importance for clarifying the difference in landscape patterns between mountainous areas and flatland areas on the micro scale. The difference in the spatial and temporal evolution of landscape patterns in the mountainous area and flatland area leads to the consideration of coordinated and sustainable development of mountains and flatlands. Appropriate human intervention appears to help enhance the diversity of the landscape, while inappropriate human intervention will exacerbate the problem of landscape fragmentation [50–52]. It is the differentiated characteristics of the social and economic development that form the differentiation in landscape pattern evolution between mountainous areas and flatland areas. The level of "landscape diversity" is inversely correlated with that of "biological diversity". Broken "landscape diversity" is not conducive to "biological diversity" [53], because the contact surface between the landscape system and the environment is large, and the "hinterland" is not deep, which is not conducive to the recovery of some species in the "biological chain". The decline in the landscape diversity in mountainous areas is the result of the connectivity of forest and grass. The enhancement in the connectivity of forest and grass in the mountainous area further enriches biodiversity and is conducive to the restoration of the ecosystem (animal habitat and reproduction) [54]. The landscape diversity of the flatland area is a response to the development of the society and economy. The expansion of urban construction land is conducive to population agglomeration and job creation [55]. The development of the transportation industry, although it has brought about an increase in landscape fragmentation, has facilitated the circulation of people and materials [41]. The in-depth intersection of arable land patches and urban construction land patches further expands the rural and urban interface, making it more conducive to the connection between the sales and consumption chains of agricultural products. All in all, the mountainous area provides an ecological barrier to the social and economic development of the flatland area. While the social and economic development level of the flatland area is constantly improved, it feeds the mountainous area and provides the economic foundation for the further optimization of the mountainous area and the construction of an ecological environment.

6. Conclusions

Despite the uncertainties in the interpretation accuracy of the landscape units, as well as the small extent of the study area used, some meaningful conclusions can be drawn from this research. Preliminary results show that the analysis of mountain–flatland landscape pattern evolution based on the grid scale can effectively reveal the variation difference and coupling law.

In the past 24 years, the landscape pattern of the mountainous area and the flatland area in Yuxi City has shown periodic changes, and the trend of its evolution is consistent with the laws of human social and economic development. With the further development of the social economy, the landscape fragmentation and landscape diversity in the flatland area are clearly higher than those in the mountainous area, and the degree of landscape fragmentation is further intensified, while the landscape shape in the mountainous area is simpler than that in the flatland area, and the trend in landscape concentration and contiguity is obvious. The natural landscape of forest and grass in mountainous areas continues to expand and tends to be contiguous, while the man-made landscape in flatland areas constantly increases and shows fragmentation, which reflects the pattern characteristics formed by the coupling evolution of land use between the mountain and flatland. There is a coupling linkage relationship between the landscape pattern evolution of the mountainous area and the flatland area. The urban expansion and the increase in construction land in the flatland area are mutually causal with the decrease in cultivated land and the increase in forest and grass in the mountainous area.

In future studies, we can further improve the interpretation accuracy of the landscape units and carry out further studies by taking the whole southwest mountainous area as the research area, in order to find the general coupling laws of landscape pattern evolution between mountainous areas and flatland areas. The rationality and universality of the law will be verified by statistical inspection and analysis, based on which the coordinated and sustainable development countermeasures of mountainous areas and flatland areas in the southwest karst region will be formulated.

Author Contributions: Conceptualization, L.W. and B.X.; methodology, L.W.; software, L.W.; validation, L.W. and J.Z.; formal analysis, L.W.; investigation, J.Z.; resources, J.Z.; data curation, L.W.; writing—original draft preparation, L.W.; writing—review and editing, B.X.; visualization, J.Z.; supervision, B.X.; project administration, L.W.; funding acquisition, L.W. and J.Z. All authors have read and agreed to the published version of the manuscript.

Funding: This research was funded by the National Natural Science Foundation of China, grant number 42161041, Science and Technology planning Project of Yunnan Province, grant number 202305AC160089, and the Youth Project of Science and Technology Agency in Yunnan Province, grant number 202101BA070001-275, as well as the General Project of Science and Technology Agency in Yunnan Province, grant number 202101BA070001-078.

Data Availability Statement: Not applicable.

Conflicts of Interest: The authors declare no conflict of interest.

References

1. Liu, M.; Bai, X.Y.; Tan, Q.; Luo, G.J.; Zhao, C.W.; Wu, L.H.; Luo, X.L. Climate change enhances the positive contribution of human activities to vegetation restoration in China. *Geocarto Int.* **2022**, *2022*, 2082542. [CrossRef]
2. Jing, B.; Li, T.S.; Zhu, Y.P.; Cheng, W.H. Research progress of human-earth system in the mountainous areas in China. *World Reg. Stud.* **2021**, *30*, 1230–1240.
3. Wu, L.; Zhou, J.; Xie, B.G.; Yang, S.Z.; Li, J.Z. Spatiotemporal differences of land use pattern between mountainous areas and basin areas at township scale: A case study of Yuxi City. *Front. Environ. Sci.* **2022**, *10*, 1029917. [CrossRef]
4. Wu, L.; Xie, B.G.; Xiao, X.; Xue, B.; Li, J.Z. Classification method and determination of mountainous area types at township scales: A case study of Yuxi City, Yunnan Province. *Complexity* **2020**, *2020*, 3484568. [CrossRef]
5. Song, F.J.; Wang, S.J.; Bai, X.Y.; Wu, L.H. A new indicator for global food security assessment: Harvested area rather than cropland area. *Chin. Geogr. Sci.* **2022**, *32*, 204–217. [CrossRef]
6. Liu, X.H.; An, Y.; Dong, G.H.; Jiang, M. Land use and landscape pattern changes in the Sanjiang Plain, Northeast China. *Forests* **2018**, *9*, 637. [CrossRef]
7. Fu, B.J.; Chen, L.D.; Wang, J.; Meng, Q.H.; Zhao, W.W. Land use Structure and Ecological Processes. *Quat. Sci.* **2003**, *23*, 247–255.
8. Yang, H.F.; Zhong, X.N.; Deng, S.Q.; Nie, S.N. Impact of LUCC on landscape pattern in the Yangtze River Basin during 2001-2019. *Ecol. Inform.* **2022**, *69*, 101631. [CrossRef]
9. Chen, F.; Bai, X.Y.; Liu, F.; Luo, G.J.; Tian, Y.C.; Qin, L.Y.; Li, Y. Analysis long-term and spatial changes of forest cover in typical karst areas of China. *Land* **2022**, *11*, 1349. [CrossRef]
10. Zhao, Y.Y.; Kasimu, A.; Gao, P.W.; Liang, H.W. Spatiotemporal changes in the urban landscape pattern and driving forces of LUCC characteristics in the urban agglomeration on the Northern Slope of the Tianshan Mountains from 1995 to 2018. *Land* **2022**, *11*, 1745. [CrossRef]
11. Xiao, B.Q.; Bai, X.Y.; Zhao, C.W.; Tan, Q.; Li, Y.B.; Luo, G.J. Responses of carbon and water use efficiencies to climate and land use changes in China's karst areas. *J. Hydrol.* **2023**, *2023*, 128968. [CrossRef]
12. Li, C.; Smith, P.; Bai, X.; Tan, Q.; Luo, G.; Li, Q.; Wang, J.; Wu, L.; Chen, F.; Deng, Y.; et al. Effects of carbonate minerals and exogenous acids on carbon flux from the chemical weathering of granite and basalt. *Glob. Planet. Chang.* **2023**, *22*, 104053. [CrossRef]
13. Li, J.Y.; Zheng, X.Q.; Zhang, C.X.; Deng, X.Z.; Chen, Y.M. How to evaluate the dynamic relevance between landscape pattern and thermal environment on urban agglomeration? *Ecol. Indic.* **2022**, *138*, 108795. [CrossRef]
14. Bai, X.Y.; Zhang, S.R.; Li, C.J.; Xiong, L. A carbon-neutrality-capacity index for evaluating carbon sink contributions. *Environ. Sci. Ecotechnol.* **2023**, *2023*, 100237. [CrossRef]
15. Xue, B.; Zhao, B.Y.; Li, J.Z. Urban complexity studies from the perspective of geography: A review based on the literature in the past 20 years. *Prog. Geogr.* **2022**, *41*, 157–172. [CrossRef]
16. Fu, B.J.; Chen, L.D.; Ma, K.M.; Wang, Y.L. *Principles and Applications of Landscape Ecology*; Science Press: Beijing, China, 2001.
17. Zhou, K. Wetland landscape pattern evolution and prediction in the Yellow River Delta. *Appl. Water Sci.* **2022**, *12*, 190. [CrossRef]
18. Tan, J.; Zhao, S.N.; Tan, X.L.; Dong, L.; Liu, J.R.; Ji, Q.Y. Change characteristics of land use and landscape pattern in Dongting Lake during 1996-2016. *Ecol. Sci.* **2017**, *36*, 89–97.
19. Lausch, A.; Herzog, F. Applicability of landscape metrics for the monitoring of landscape change: Issues of scale, resolution and interpretability. *Ecol. Indic.* **2002**, *2*, 3–15. [CrossRef]

20. Sun, Q.; Aliya, B.; Nurmemet, I. Effect of LUCC change on evolution of landscape pattern in Ebinur Wetland. *Soils* **2018**, *50*, 583–592.
21. Mohamed, A.; Worku, A.H. Simulating urban land use and cover dynamics using cellular automata and Markov chain approach in Addis Ababa and the surrounding. *Urban Clim.* **2020**, *31*, 100545. [CrossRef]
22. Su, K.C.; Yang, Q.Y.; Zhang, B.L.; Zhang, Z.X. The coupling mechanism between rural land use transition and small-scale peasant economy change in mountainous areas. *Geogr. Res.* **2019**, *38*, 399–413.
23. Smiraglia, D.; Ceccarelli, T.; Bajocco, S.; Perini, L.; Salvati, L. Unraveling landscape complexity: Land use/land cover changes and landscape pattern dynamics (1954-2008) in contrasting peri-urban and agro-forest regions of Northern Italy. *Environ. Manag.* **2015**, *56*, 916–932. [CrossRef] [PubMed]
24. Chen, S.L.; Yu, S.J. Analysis on land use change and its driving force in the Yangtze River Delta Region during 1985-2015. *For. Econ.* **2019**, *41*, 49–54.
25. Wu, Q.; Tan, J.X.; Guo, F.X.; Li, H.Q.; Chen, S.B. Multi-scale relationship between land surface temperature and landscape pattern based on wavelet coherence: The case of metropolitan Beijing, China. *Remote Sens.* **2019**, *11*, 3021. [CrossRef]
26. Wu, J.G. Paradigm shift in ecology: An overview. *Acta Ecol. Sin.* **1996**, *16*, 449.
27. Zhao, Y.; Tomita, M.; Harada, I.; Hara, K. Multi-scale effect on landscape pattern analysis using satelite data with a range of spatial resolutions. *J. Landsc. Ecol.* **2011**, *4*, 62–72.
28. Lei, Y.K.; He, D.; Tian, G.H.; Zhang, J.H. Spatial pattern differences between urban and rural construction land in multi-scales. *Ecol. Environ. Sci.* **2012**, *21*, 687–693.
29. Guo, S.S.; Hu, S.G.; Qu, S.J. Multi-scale analysis of farmland landscape dynamic pattern in the middle reaches of the Yangtze River. *Resour. Environ. Yangtze Basin* **2018**, *27*, 1637–1646.
30. Xie, H.L.; He, Y.F.; Xue, X. Exploring the factors influencing ecological land change for China's Beijing-Tianjin-Hebei Region using big data. *J. Clean. Prod.* **2017**, *142*, 677–687. [CrossRef]
31. Zhang, D.Y.; Su, F.L.; Wang, T.L.; Li, H.F.; Sun, D. Land use patterns and their changes of Shuangtai Estuarine Wetlands for 7 periods since 1985. *Wetl. Sci.* **2019**, *17*, 658–662.
32. Yang, G.Z.; Lv, K.; Li, F. Spatial and temporal correlation analysis of land use change and ecosystem service value in Nanchang City based on grid scale. *China Land Sci.* **2022**, *36*, 121–130.
33. Zhang, J.; Li, S.N.; Lin, N.F.; Lin, Y.; Yuan, S.F. Spatial identification and trade-off analysis of land use functions improve spatial zoning management in rapid urbanized areas, China. *Land Use Policy* **2022**, *116*, 106058. [CrossRef]
34. Wu, L.; Xie, B.G. The variation differences of cultivated land ecological security between flatland and mountainous areas based on LUCC. *PLoS ONE* **2019**, *14*, e0220747. [CrossRef] [PubMed]
35. Meybeck, M.; Green, P.; Vörösmarty, C. A new typology for mountains and other relief classes. *Mt. Res. Dev.* **2001**, *21*, 34–45. [CrossRef]
36. Liu, Y.X.; Li, Y.B. Spatial evolution of land use production function in the basin of Guizhou province under the background of rural transformation. *Geogr. Res.* **2020**, *39*, 430–446.
37. Li, Y.B.; Yao, Y.W.; Xie, J.; Wang, F.Y.; Bai, X.Y. Spatial-temporal evolution of land use and landscape pattern of the mountainbasin system in Guizhou Province. *Acta Ecol. Sin.* **2014**, *34*, 3257–3265.
38. Guo, M.; Wan, S.J.; Na, Y. Ecological land fragmentation evaluation and dynamic changes in Changbai Mountain Areas. *Resour. Dev. Mark.* **2020**, *36*, 14–22.
39. Wu, J.G. *Landscape Ecology: Patterns, Processes, Dimensions, and Grades*, 2nd ed.; Higher Education Press: Beijing, China, 2007; p. 166.
40. Xu, Y.; Tang, H.P.; Wang, B.J.; Chen, J. Effects of landscape patterns on soil erosion processes in a mountain–basin system in the North China. *Nat. Hazards* **2017**, *87*, 1567–1585. [CrossRef]
41. Wang, L.; Wu, L.; Zhang, W. Impacts of land use change on landscape patterns in mountain human settlement: The case study of Hantai District (Shanxi, China). *J. Mt. Sci.* **2021**, *18*, 749–763. [CrossRef]
42. Ma, H.Y.; Zhang, L.L.; Wei, X.Q.; Shi, T.T.; Chen, T.X. Spatial and temporal variations of land use and vegetation cover in Southwest China from 2000 to 2015. *Chin. J. Appl. Ecol.* **2021**, *32*, 618–628.
43. Zhang, J.; Zhu, W.B.; Zhao, F.; Zhu, L.Q.; Li, M.J.; Zhu, M. Spatial variations of terrain and their impacts on landscape patterns in the transition zone from mountains to plains-A case study of Qihe River Basin in the Taihang Mountains. *Sci. China-Earth Sci.* **2018**, *61*, 450–461. [CrossRef]
44. Wu, L.; Yang, Y.J.; Xie, B.G. Modeling analysis on coupling mechanisms of mountain-basin human-land systems: Take Yuxi City as an example. *Land* **2022**, *11*, 1068. [CrossRef]
45. Zhang, Y.Y.; Sun, M.Y.; Yang, R.J.; Zhang, L. Impact of land-use change on ecosystem service value in Southwest China. *J. Environ. Eng. Technol.* **2022**, *12*, 207–214.
46. Li, Y.M.; WAng, D.C.; Yuan, J.; Liu, X.Z. Temporal and spatial pattern and functional zoning of land use in Alpine Canyon Region based on terrain gradient. *Bull. Soil Water Conserv.* **2020**, *40*, 303–311.
47. Xu, Y.Q.; Luo, D.; Feng, Y.; Peng, J. A Study on land use and land cover change in karst mountainous areas in Southwestern China: A case study of the Maotiao River Watershed, Guizhou Province. *Resour. Sci.* **2010**, *32*, 1752–1760.
48. Wen, J.H.; Li, R. Temporal and spatial variation of ecosystem service value and its response to landscape pattern change in Guizhou Province, China. *Chin. J. Appl. Ecol.* **2022**, *33*, 2075–3086.

49. Xue, B.; Xiao, X.; Li, J.Z. Identification method and empirical study of urban industrial spatial relationship based on POI big data: A case of Shenyang City, China. *Geogr. Sustain.* **2020**, *1*, 152–162. [CrossRef]
50. Deng, L.Y.; Zhang, Q.; Cheng, Y.; Cao, Q.; Wang, Z.Y.; Wu, Q.; Qiao, J. Underlying the influencing factors behind the heterogeneous change of urban landscape patterns since1990. *Ecol. Indic.* **2022**, *140*, 108967. [CrossRef]
51. Chen, K.X.; Cong, P.F.; Qu, L.M.; Liang, S.X.; Sun, Z.C. Annual variation of the landscape pattern in the Liao River Delta wetland from 1976 to 2020. *Ocean. Coast. Manag.* **2022**, *224*, 106175. [CrossRef]
52. Qian, Y.; Dong, Z.; Yan, Y.; Tang, L.N. Ecological risk assessment models for simulating impacts of land use and landscape pattern on ecosystem services. *Sci. Total Environ.* **2022**, *833*, 155218. [CrossRef]
53. Wang, B.; Luo, L. Service value of a bay city ecosystem based on green buildings and landscape pattern changes. *Sustain. Comput. Inform. Syst.* **2022**, *35*, 100758. [CrossRef]
54. Huang, C.B.; Zhao, D.Y.; Deng, L. Landscape pattern simulation for ecosystem service value regulation of Three Gorges Reservoir Area, China. *Environ. Impact Assess. Rev.* **2022**, *95*, 106798. [CrossRef]
55. Wang, Q.; Wang, H.J. Spatiotemporal dynamics and evolution relationships between land-use/land cover change and landscape pattern in response to rapid urban sprawl process. *Ecol. Eng.* **2022**, *182*, 106716. [CrossRef]

land

MDPI

Article

Monitoring Roadbed Stability in Permafrost Area of Qinghai–Tibet Railway by MT-InSAR Technology

Hui Liu [1,2], Songbo Huang [1,2], Chou Xie [3,4,*], Bangsen Tian [3,4,*], Mi Chen [1,2] and Zhanqiang Chang [1,2]

1 College of Resources Environment and Tourism, Capital Normal University, Beijing 100048, China
2 Key Lab of 3D Information Acquisition and Application, Capital Normal University, Beijing 100048, China
3 Aerospace Information Research Institute, Chinese Academy of Sciences, Beijing 100094, China
4 University of Chinese Academy of Sciences, Beijing 100049, China
* Correspondence: xiechou@aircas.ac.cn (C.X.); tianbs@aircas.ac.cn (B.T.)

Abstract: Permafrost areas pose a threat to the safe operation of linear projects such as the Qinghai–Tibet railway due to the repeated alternating effects of frost heaving and thawing settlement of frozen soil in permafrost area. Time series InSAR technology can effectively obtain ground deformation information with an accuracy of up to millimeters. Therefore, it is of great significance to use time series InSAR technology to monitor the deformation of the permafrost section of the Qinghai–Tibet railway. This study uses multi-time InSAR (MT-InSAR) technology to monitor the deformation of the whole section of the Qinghai–Tibet railway, detect the uneven settlement of the railway roadbed in space, and detect the seasonal changes in the roadbed in the time domain. At the same time, the local deformation sections over the years are compared and discussed. The time series deformation monitoring results of the permafrost section Sentinel-1 data in 2020 show that the length of the railway roadbed from Tanggula station to Za'gya Zangbo station (TZ) section is approximately 620 m, the deformation of the east and west sides is uneven, and the average annual deformation difference is 60.68 mm/a. The impact of frozen soil in WangKun station to Budongquan station (WB) section on railway roadbed shows the distribution characteristics of high in the middle and low at both ends, and the maximum annual average settlement can reach −158.46 mm/a. This study shows that the deformation of permafrost varies with different ground layers. The impact of human activities on frozen soil deformation is less than that of topography and hydrothermal conditions. At the same time, the study determined that compared with other sections, the roadbed deformation of TZ and WB sections is more obvious.

Keywords: MT-InSAR; frozen soil deformation; Qinghai–Tibet corridor; roadbed deformation; climate response

Citation: Liu, H.; Huang, S.; Xie, C.; Tian, B.; Chen, M.; Chang, Z. Monitoring Roadbed Stability in Permafrost Area of Qinghai–Tibet Railway by MT-InSAR Technology. *Land* **2023**, *12*, 474. https://doi.org/10.3390/land12020474

Academic Editor: Xiaoyong Bai

Received: 5 January 2023
Revised: 10 February 2023
Accepted: 13 February 2023
Published: 14 February 2023

1. Introduction

The repeated alternating action of frost heaving and thawing settlement of frozen soil in permafrost area will cause damage to the local geological environment, and then easily lead to geological disasters such as landslide, debris flow, foundation rupture, and collapse [1,2], especially linear engineering such as railway and highway [3–5]. Uneven deformation of the roadbed is a common and serious disaster in linear engineering, which seriously affects the operation of linear engineering such as high-speed railways and highways. The Qinghai–Tibet Plateau (QTP) is the largest permafrost region except polar regions [6]. Approximately 610 km of Qinghai–Tibet railway is laid in the permafrost area, crossing national nature reserves such as Hoh Xil, Sanjiangyuan, Qiangtang. In view of the particularity and importance of the location of the Qinghai–Tibet railway, it is of great practical significance to monitor the surface deformation of the Qinghai–Tibet corridor in permafrost area [7–17].

Traditional geodetic methods, such as leveling and global positioning system (GPS) measurement, can achieve high-precision measurement of surface deformation. However, due to the special location and harsh environmental conditions of the Qinghai–Tibet plateau, both traditional leveling and GPS measurement require a lot of human and material costs, which is difficult and inefficient. Although it will not cause this problem to predict the stability of Qinghai–Tibet line foundation in combination with historical disaster data, field survey data [18], settlement index or allowable bearing capacity index [19], the accuracy is rough and low, and the prediction of influence factors cannot be better applicable to the existing situation under the external human intervention.

The wide application of synthetic aperture radar interferometry (InSAR) makes up for the shortcomings of the above method. InSAR monitors ground deformation by analyzing the phase information of two aperture radar images [20]. It has been widely used in the ground deformation caused by earthquakes, volcanoes, glaciers, landslides, and land subsidence [21–24]. With the launch of new SAR satellites such as Sentinel-1A/B, researchers can more easily obtain SAR images. Rich SAR data sources and accessibility promote the application of InSAR technology in the field of permafrost deformation monitoring. Due to its large coverage of SAR images and the relatively short return visit time, the interference coherence has been greatly improved, which better meets the accuracy requirements of deformation monitoring such as linear engineering in permafrost area, and it is used to monitor frozen soil deformation. Researchers have already demonstrated the ability to use InSAR technique to detect to this freeze/thaw-related ground motion over permafrost regions. As shown in Table 1, current researches using InSAR technology to monitor frozen soil mainly focus on the Arctic, Qinghai–Tibet Plateau and other regions of the world. It can be seen that the application of InSAR technology to permafrost monitoring has broad prospects.

Table 1. Research on monitoring frozen soil with InSAR technology.

Arctic	Qinghai–Tibet Plateau	Other Regions
Zwieback et al. [25], Bartsch et al. [26], Strozzi et al. [27], Rudy et al. [28], Liu et al. [29–33]	Zhou et al. [34], Xu et al. [35], Zou et al. [36], Xiang et al. [37], Zhang et al. [38–44], Wang et al. [45,46], Reinosch et al. [47], Lu et al. [48], Chen et al. [49], Daout et al. [50]	Chen et al. [51], Rouyet et al. [52], Antonova et al. [53], Li et al. [54], Liu et al. [29]

As can be seen from the above table, the research on permafrost in the Qinghai–Tibet plateau in recent years has been fruitful. However, over the years, most research on the frozen soil deformation along the Qinghai–Tibet railway has focused on the section from Wudaoliang to Tuotuohe, especially the Beiluhe area, and less so on the Chaerhan Salt Lake section in the north and the section from Yangbajing to Dangxiong in the south. The scope of the existing research has been very limited, and the overall deformation of the permafrost section of the Qinghai–Tibet railway has not been monitored. In addition, the existing research only involves local disaster characteristics, and there is a lack of systematic investigation and analysis of the overall disaster characteristics and laws. Most of the deformation research on the frozen soil section of the Qinghai–Tibet railway is limited to the results of more than a decade. Due to the long time, it is difficult to reflect the current situation of roadbed deformation, which is not conductive to the discovery of the existence and potential risk of deformation of railway roadbed. At the same time, the previous research results have not been compared and discussed, and the methods and effects of the measures have not been qualitatively and quantitatively evaluated. The relevant research only stays at the level of producing deformation results. Therefore, many years after the completion of the Qinghai–Tibet railway project, it is very necessary to compare the deformation monitoring results in recent years with the previous monitoring results to understand whether the deformation disaster has changed.

Most of the studies have only used permanent scatterers to obtain the deformation characteristics of ground objects along the Qinghai–Tibet railway. The scattering characteristics are directly affected by external environmental factors such as soil moisture and surface water, and almost no ground objects can maintain stable backscattering characteristics for a long time. At the same time, the vegetation coverage in the study area is low, mostly exposed sandy or mucus saline soil, and the interference coherence is high in the short term. This phenomenon makes it more difficult to obtain persistent scattering points. Therefore, it is difficult to obtain high-density stable scattering points only by obtaining traditional persistent scatterer (PS) points. Distributed scatterers (DS) refers to ground objects, mostly bare ground and sparse vegetation that resolve the backscattering coefficients of all scatterers in a unit roughly the same. The spatial density of the measurement points is increased on the region characterized by DS while retaining the high-quality information obtained using the PS technique on deterministic targets. PS and DS were properly combined to increase the density of measurement points and further improve the coherence point density and parameter estimation accuracy [55]. Considering the limitations of traditional persistent scattering points, we obtain a high-quality and high-density time series deformation point set by combining PS points and distributed scattering points (DS) with limited time baselines, which greatly improves the defects of the above methods so as to obtain the existing deformation of railway roadbed and potential risks along the route.

In this study, we used the free obtained Sentinel-1A Radar image to monitor the overall deformation of the permafrost section of the Qinghai–Tibet railway by combining PS points and DS points, compared and discussed the deformation of local areas for many years to monitor the uneven settlement of railway roadbed in space and the seasonal changes in the time domain.

2. Materials and Methods

2.1. Study Area

The repeated alternating action of frost heaving and thawing settlement of frozen soil in permafrost area will have a serious impact on linear projects such as Qinghai–Tibet railway. Approximately 610 km of Qinghai–Tibet railway is paved in permafrost area, from Nachitai to Ando station. The study area covers 32°20′–35°94′, 91°42′–94°62′, with an altitude of 3514–5544 m.

The thickness of frozen soil is approximately 60~120 m, and the thickness of active layer (ALT) is between 0.8 and 4 m, with an average of approximately 2 m It is a typical continental climate. It is cold in winter (to −20.7 °C), warm in summer (to +22.7 °C), and the annual average temperature is −0.98–1.03 °C. The ground is generally frozen for 7 months (October to April of the next year) [56]. The annual precipitation is between 191 and 485 mm, mainly concentrated in summer (May to October). The elevation fluctuates from north to south [57], and its topographic characteristics are shown in Figure 1. There are four monitoring stations close to this section, namely Golmud, Wudaoliang, Tuotuohe and Anduo meteorological stations. Land types mainly include glaciers, snow, exposed rocks and other land types [58]. Due to the melting and freezing of frozen soil, the maximum seasonal deformation of the surface can reach 10 cm. The distribution of frozen soil is shown in the upper left corner of Figure 1. The landslide within 1 km and debris flow within 5 km caused by frost heaving and thawing settlement of frozen soil will pose a threat to the railway. This study will take the permafrost section of a 5 km buffer zone as the research object.

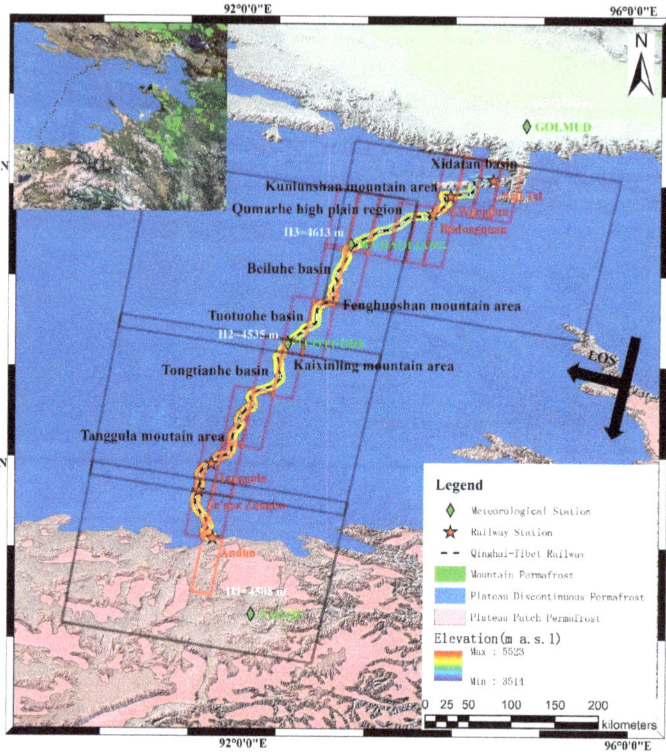

Figure 1. The green diamond is the meteorological station. The station name is marked in green font. The white mark is the altitude of each meteorological station. The red pentagram is the railway station along the way, and the name is marked in red font. The terrain distribution has been marked in black font in the Figure. The Qinghai–Tibet line passes through mountain permafrost, plateau discontinuous permafrost and island permafrost from north to south. The study area is a permafrost section passing through the blue area, which has higher risk than other sections. The black rectangular box indicates the coverage of the Sentinel-1 image, the red rectangular box is the swath of each scene used in the study, and the black arrow indicates the navigation direction and line of sight direction.

2.2. Data

In this paper, the descending image of Sentinel-1 satellite in 2020 is selected to obtain the risk situation of the section of Qinghai–Tibet line, involving 122 images in four maps, as shown in Table 2.

Table 2. Relevant parameters of four scenes of radar image.

Sequence	Start Time	End Time	Path/Frame	Image	Swath
1	12 January 2020	25 December 2020	77/475	29	7
2	5 January 2020	30 December 2020	150/475	31	7
3	5 January 2020	30 December 2020	150/480	31	6
4	5 January 2020	30 December 2020	150/485	31	3

The pixel sizes in the central azimuth and range directions of the image are 2.32 m and 13.98 m, respectively, and VV polarization mode is selected. The central incidence angle is between 33 and 34°, and the spatial range covered by each SAR image is approximately 252×190 km^2. The digital elevation model (DEM) of the study area adopts SRTM-1 with a resolution of 30 m. In addition, the Precision Orbit Data (POD) of Sentinel-1A is provided by European Space Agency (ESA). The data of glacier and frozen soil distribution originate from the "Heihe project data management center" [59].

2.3. Methods

In this study, permanent scatterers and distributed scatterers (PS, DS) were combined to select highly coherent targets and increase the coverage of coherent points. Compared with PS method, this method has a better effect and higher efficiency in the analysis of foundation deformation along the Qinghai–Tibet corridor. The processing flow chart is demonstrated in Figure 2. The deformation results largely depend on the error correction during interference processing, the selection of coherent points and phase unwrapping. Therefore, this section is expanded with the principles and methods of each part to finally obtain the desired results.

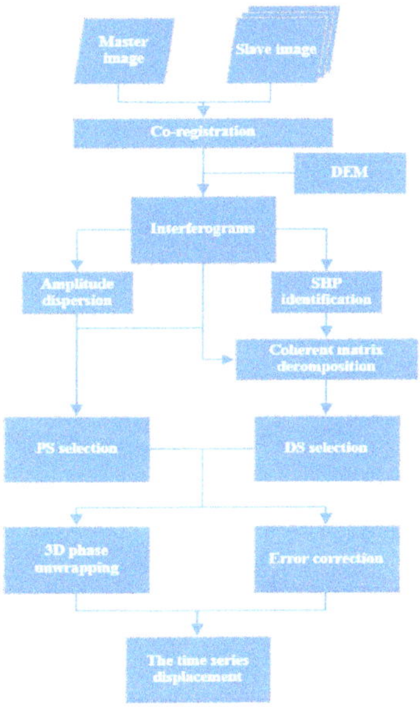

Figure 2. Process of joint point selection of PS and DS to obtain high-density point targets and roadbed time series deformation.

2.3.1. Error Correction

High coherence of target means high coherence in space domain and stable phase in time domain. For MT-InSAR, the image will be registered, and the terrain and orbit have generated phase compensation. The residual phase φ^n of unwrapped interferogram obtained after differential interference is usually composed of residual terrain, deformation, atmosphere and noise phases [60], which are wound between deformation phases.

$$\varphi^n = \varphi^n_{topo} + \varphi^n_{defo} + \varphi^n_{aps} + \varphi^n_{noise} + k \cdot 2\pi, \tag{1}$$

where φ^n_{topo} is the residual topographic phase; φ^n_{defo} is the deformation phase; φ^n_{aps} is the atmospheric phase screen, indicating the signal delay caused by weather conditions; k is the integer fuzzy number; φ^n_{noise} is the phase noise caused by time decorrelation, error registration, uncompensated spectral offset decorrelation, orbit error, soil moisture and thermal noise.

The residual topographic phase of point p is as follows:

$$\varphi^n_{p,topo} = k^z_p \cdot \epsilon^z_p, \tag{2}$$

where $k_p^z = \frac{4\pi b_n}{\lambda sin\theta R_n}$ is height-phase coefficient and ϵ_p^z is elevation error; b_n is the baseline of the nth image relative to the reference (main) image; θ is the local angle of incidence; λ is the carrier wavelength; R_n is the (zero Doppler) distance between the target and the nth orbit acquisition.

The deformation phase generated by the displacement at point p can be effectively divided into two components:

$$\varphi_{p,defo}^n = k^t v_p + \mu_{NL}, \tag{3}$$

where v is constant speed (line of sight); $k^t = \frac{4\pi t_n}{\lambda}$ is the time phase factor; t_n is the time baseline; μ_{NL} is the phase term due to possible nonlinear motion.

Atmospheric phase screen (APS) can greatly reduce the influence by considering the phase difference between two nearby points. Therefore, APS calibration can also be carried out by establishing Delaunay triangulation network.

For the two PS candidate points connected, the measured tomography signals $y_1(m)$ and $y_2(m)$ can be expressed as [61]

$$\begin{cases} y_1(m) = g_1(m, s_1)exp[j\varnothing_{APS1}(m)] \\ y_2(m) = g_2(m, s_2)exp[j\varnothing_{APS2}(m)]' \end{cases} \tag{4}$$

where $\varnothing_{APS1}(m)$ and $\varnothing_{APS2}(m)$ represent APS present in $y_1(m)$ and $y_2(m)$, respectively; s_1 and s_2 are the inclined elevations of the two PS points connected, respectively; $g_1(m, s_1)$ and $g_2(m, s_2)$ represent the math ideal measurement value of the two connection points without APS interference.

Because the spatial frequency of APS is low, the candidate points with long spatial distance usually have great differences in APS. At this time, it is difficult to deal with connection points with large APS differences. Therefore, the long connection arc is rejected by setting the distance threshold. When the arc length is short, two adjacent scatterer candidate points have similar APS:

$$\varnothing_{APS1}(m) \approx \varnothing_{APS2}(m). \tag{5}$$

Therefore, the APS can be easily calibrated by subtracting the phase of the reference point, and the relative tomographic signal Δy (m) of the connecting arc is as follows:

$$\Delta y(m) = y_2(m)exp(-j\cdot\lceil y_1(m)\rceil) = g_2(m, s_2)\cdot exp(-j\cdot\lceil g_1(m, s_1)\rceil), \tag{6}$$

where $\lceil\ \rceil$ is the operation of phase retention.

2.3.2. Selection of Coherent Points

Permanent and distributed scatterers are combined to increase the coverage of coherent points. Ferretti et al. proposed the Dispersion of Amplitude (DA) based on the definition of PS [62]. When the main scatterer exists in the pixel, its phase is mainly determined by the phase of the main scatterer, which is less affected by noise, and the phase standard deviation has the following relationship with the amplitude:

$$\sigma_\varphi \approx \frac{\sigma_A}{m_A} \equiv D_A, \tag{7}$$

where σ_φ is the standard deviation of phase; σ_A is the standard deviation of amplitude A; m_A is the mean amplitude of N SAR images in time dimension; D_A is the dispersion index.

When extracting DS, considering its statistical distribution characteristics [63], the extracted DS is transformed into extracting several pixels subject to the same statistical distribution, which can be called homogeneous pixels. Then, the recognition of planar targets is transformed into identifying homogeneous pixels first, and then estimating their phase. The Kolmogorov–Smirnov algorithm is used to identify homogeneous pixels. By determining whether their cumulative distribution function (CDF) is the same, we can judge whether they are homogeneous pixels. K–S algorithm defines D_N as the maximum absolute value of the difference between the two cumulative distribution functions:

$$D_N = \max_{-\infty<x<\infty} |S_N(x) - P_N(x)|, \tag{8}$$

where S_N and P_N are the cumulative distribution functions of two different SAR image pixels, respectively. By setting a certain threshold (D_{thr}), when $D_N \leq D_{thr}$, it is considered that the two samples obey the same statistical distribution, that is, they are judged to be homogeneous pixels so as to ensure that the identified homogeneous pixel is directly or indirectly connected with the central pixel P, and there is no independent region. In addition, the algorithm takes the identified homogeneous pixel as DS, which makes the subsequent phase estimation more accurate, so the results are more reliable.

2.3.3. Phase Unwrapping

The accuracy of surface deformation information acquisition mainly depends on phase unwrapping. The main idea of phase unwrapping algorithm based on network flow is to minimize the difference between the derivative of unwrapping phase and the derivative of winding phase. This method can not only greatly reduce the time and space complexity of phase unwrapping algorithm and improve the calculation speed, but also limit the whole error to a small range and prevent the retransmission of error, so as to improve the accuracy of unwrapping results. In this study, the minimum cost flow algorithm is based on irregular networks: first, the phase with high coherence coefficient is extracted as a high-quality phase data set. Next, a Delaunay triangulation is established according to the position of these phase points. Then, the residual points in the triangulation are identified, and the minimum cost flow algorithm is used to connect the positive and negative residual point pairs to establish the branch tangent. Finally, the unwrapping phase value is obtained by the method of adding and subtracting $2n\pi$ around or through the branch tangent.

The phase difference between two adjacent phase points (such as targets p and q) can be written as [64]

$$\Delta\varphi_{pq}^{n} = W\left\{k_p^z \Delta\epsilon_{pq}^z + k^t \Delta v_{pq} + \Delta w_{pq}^n\right\}, \tag{9}$$

where W is the winding operator, $\Delta\epsilon_{pq}^z$ is the relative elevation error, Δv_{pq} is the relatively constant velocity, and $\Delta w_{pq}^n = \mu_{pq,NL}^n + \varphi_{pq,aps}^n + \varphi_{pq,noise}^n$ is the phase difference between the model and measured values between p and q points in the nth interferogram.

Therefore, the Delaunay triangulation connection network is constructed in the spatial domain as follows. Assuming that the largest connected network is searched after K iterations, and there are $P^{(n)}$ arcs and $Q^{(n)}$ PS points (n = 1, $\cdots$, K) in the nth search result of the connected network, then the maximum connected network contains $P^{(k)}$ connected arcs and $Q^{(k)}$ points. The integration of the nth search connection network can be modeled as

$$\Delta S^{(n)} = G^{(n)} \cdot S^{(n)}, \tag{10}$$

where $\Delta S^{(n)}$ is the relative height of the $P^{(n)}$ point arc of the nth search connection network; $G^{(n)}$ is the transformation matrix from point arc to point, which is composed of $0, -1, 1$; $S^{(n)}$ is the absolute height of the $Q^{(n)}$ point in the nth search connection network:

$$\Delta S^{(n)} = \begin{bmatrix} \Delta \hat{S}_1^{(n)} \\ \Delta \hat{S}_2^{(n)} \\ \vdots \\ \Delta \hat{S}_p^{(n)} \\ p \end{bmatrix}_p^{(n)} \times 1, G^{(n)} = \begin{bmatrix} 1 & \cdots & -1 & \cdots & 0 & \cdots \\ \vdots & & & & & \\ \cdots & 1 & 0 & \cdots & -1 & \cdots \\ \vdots & & & & & \\ \cdots & 0 & \cdots & 1 & -1 & \cdots \\ \vdots & & & & & \end{bmatrix}_{p^{(n)} \times Q^{(n)}}, S^{(n)} = \begin{bmatrix} \Delta S_1^{(n)} \\ \Delta S_2^{(n)} \\ \vdots \\ \Delta S_{Q(n)}^{(n)} \end{bmatrix}_{Q^{(n)} \times 1}. \quad (11)$$

3. Results

In order to improve the phase solution accuracy, it is necessary to check whether the spatio-temporal baseline meets the requirements, and reduce the impact on the deformation phase. In the study area, 21 swaths were analyzed by MT-InSAR, and a total of 118 pairs of interference pairs were generated. The temporal and spatial baseline distribution and combination mode are shown in Figure 3. The re-entry period of SAR satellite is 12 days, the vertical baseline is less than 130 m, and the coherence is relatively high. The finally obtained surface time series deformation information along the Qinghai–Tibet line is shown in Figure 4. The areas with serious deformation are mainly distributed from WangKun station to Budongquan station (WB) in the north section of permafrost area and from Tanggula station to Za'gya Zangbo station (TZ) in the south section.

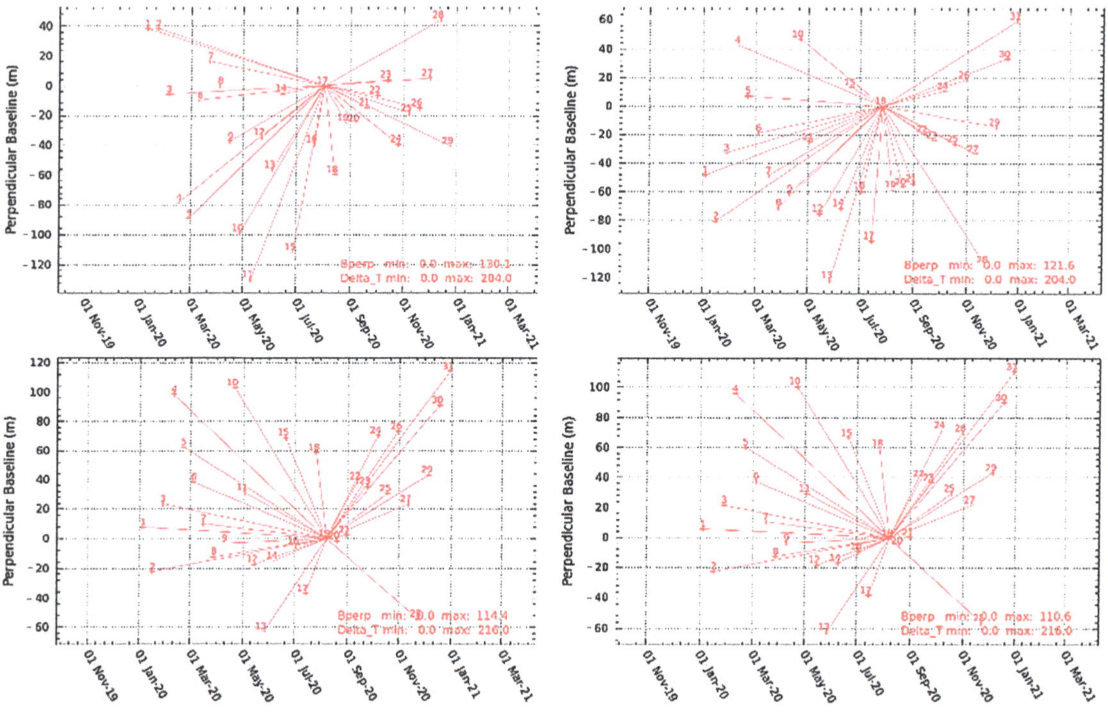

Figure 3. The center point is the reference main image, each line represents an interference pair, and the X and Y axes represent its spatio-temporal baseline distribution.

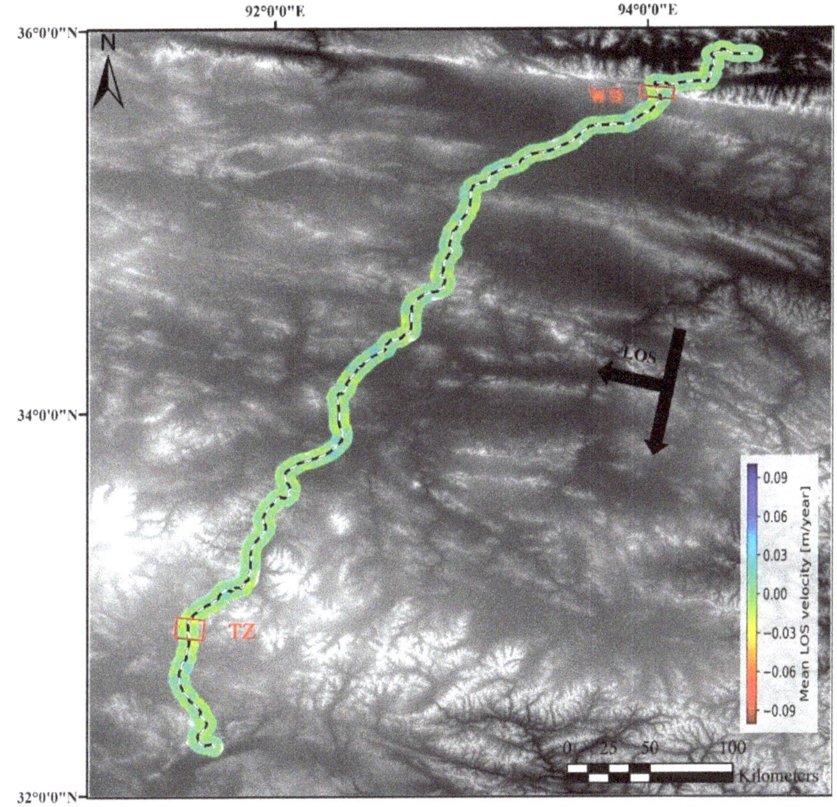

Figure 4. Overall deformation results obtained by time series InSAR processing. The surface deformation of frozen soil section of Qinghai–Tibet corridor is mainly distributed in WB in the north and TZ in the south, which is represented by red rectangular box in the Figure. The black arrow indicates the Los direction.

4. Discussion

4.1. Risk Section and Deformation Law

Affected by the thermal thawing and frost heaving of the active layer of permafrost, there are roadbed deformation and frozen soil collapse in WB and TZ of the two sections, as shown in Figures 5a and 5a′, respectively, representing the original amplitude map and annual average deformation of WB section in 2020. The deformation is mostly concentrated in the valley area in the north and appears to be lesser in the flat area in the south. Figure 5b,b′ is the original amplitude map and annual average deformation map of TZ section. As it passes through the inland river, it is greatly affected by runoff and thermal melting, resulting in melting collapse in many places in the southern section of the area.

4.1.1. Deformation from WangKun Station to Budongquan Station

The deformation section between WangKun station and Budongquan station is located in the valley area, and its terrain, deformation and geological structure, is shown in Figure 6.

Affected by global warming and human activities, the phenomenon of frost heaving and thawing settlement of permafrost active layer in this area is obvious. Because the terrain of this section is high on both sides and low in the middle, when the shear force reaches a certain threshold due to the thawing settlement of frozen soil, a shallow landslide will occur at the front of the mountains on both sides. At the same time, because this section is located in the valley, there is a large amount of glacier melt water and precipitation, and the surface and underground runoff is more abundant. The hydrothermal effect will continuously thicken the frozen soil active layer, aggravate the frost heave and thaw settlement, and further promote the collapse on both sides of this section. It can also be clearly seen in Figure 6d that melting collapse occurs in R1 and R2 areas along both sides of the railway section. The center latitude and longitude of R1 is 35.66984° N, 94.05412° E; the average annual deformation rate is −62.91 mm/a; and the collapse area reaches 0.36 km^2. The deformation on both sides of the railway at R2 is uneven, with roadbed lifting on the left and thawing collapse of frozen soil on the right. The annual deformation is 48.13 mm/a and −158.46 mm/a, respectively. The existence of large-scale melt collapse body and the difference in roadbed deformation affect the normal operation of railway. In addition, since the railway section in flat areas is basically free of deformation, it can be inferred that the influence of human activities of railway operation on frozen soil deformation is smaller than that of topography and hydrothermal. In other words, topography and hydrothermal action are the main causes of railway deformation in this section.

Considering that geological factors may also affect it, we compared the geological distribution Figure 6c with the deformation distribution Figure 6d. It was determined that the surface deformation of this section is mainly concentrated between the two faults, and the deformation is weak in the north of the North fault. At the same time, there are Qp2gl, Qp3gl, TB3, P1QQ3 and other strata between the two fault layers. The severely deformed strata are mainly middle Pleistocene ice deposit Qp2gl and late Pleistocene ice deposit Qp3gl, which are mainly ice water accumulation. The moraine is composed of boulders, gravel, sand and clay. The soil is soft, sensitive to hydrothermal changes, and prone to hot melting and frost heaving. Therefore, the roadbed deformation in frozen soil section may be affected by thawing settlement of frozen soil, fault, and lithology.

Due to the complex physical movement along the slope of this section, more information (movement rate, direction, and trend) about terrain (gradient and direction), geology (mantle composition and surface coverage), hydrology (surface and underground runoff and ice melting of permafrost) is needed to monitor the evolution of active layer of permafrost so as to better grasp the deformation inducement and law of this area.

4.1.2. Deformation from Tanggula Station to Za'gya Zangbo Station

The section with serious roadbed deformation from Tanggula station to Zajiazangbu station is selected. The whole section shows a downward trend, with uneven deformation on both east and west sides. The total length of the section is approximately 620 m, and the central latitude and longitude are 32.90665° N, 91.52807° E. This section passes through Za'gya Zangbo, the longest inland river in Tibet. The runoff is supplemented by the water melted by ice and snow. As many sections of runoff pass through this section, the river has a great impact on the thermal thawing of permafrost, resulting in changes in the hydrothermal status of the active layer above the permafrost [65]. Coupled with the joint impact of railway operation activities, the phenomenon of frost heaving and thawing settlement of the frozen soil layer is obvious.

The red arrow shows the connection between the north and south sections. It can be observed that the deformation of the south section is larger than that of the north section as a whole. In Figure 7a, the annual average deformation phase at P3 is smaller than that of P1 and P2. The water activity and temperature distribution of railway roadbed in frozen soil area are the key factors affecting roadbed frost heaving and thawing settlement. In unsaturated state, the higher the water content, the greater the frost heaving amount of soil with the same density, the greater the corresponding frost heaving and thawing settlement rate; that is, water supply is the fundamental reason for frost heaving and thawing settlement of frozen soil [6,66]. There are two branches of runoff in the south, which have a connecting trend, and the runoff at the arrow is expanding, which has a great impact on the frozen soil. The frost heaving and thawing settlement process of frozen soil active layer is more intense, resulting in uneven roadbed deformation on the north and south sides.

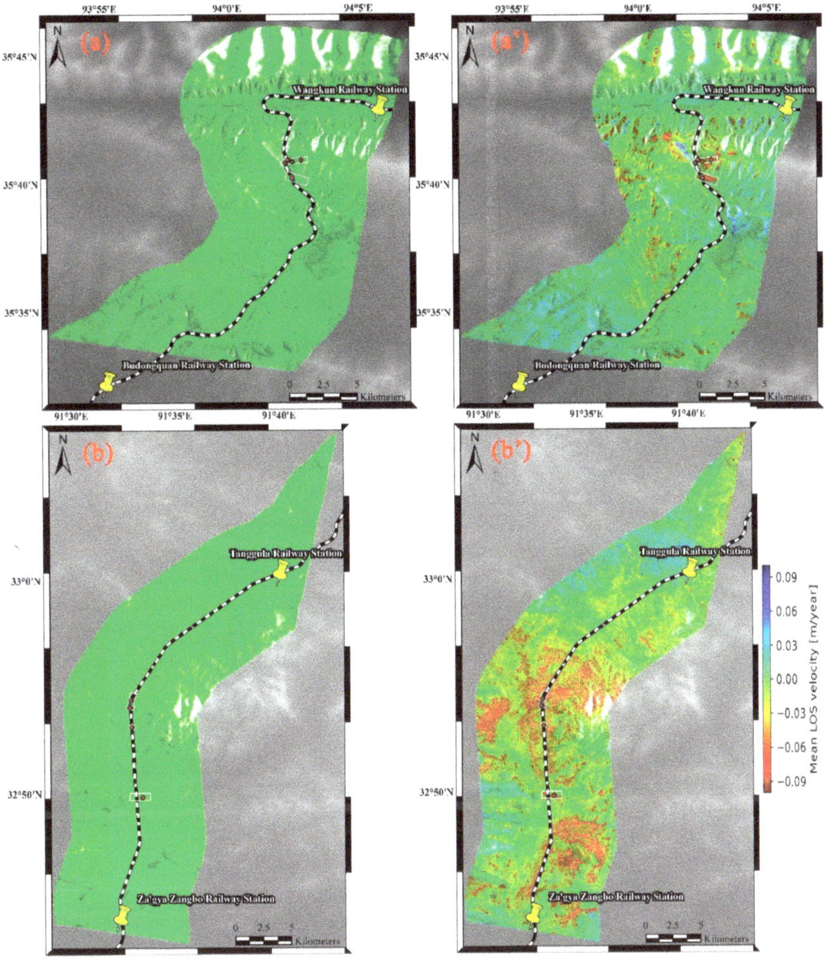

Figure 5. (**a**,**a'**) are the original amplitude map and annual average deformation map of WB deformation section, respectively (the annual average deformation of the surface after being affected by the thermal thawing and frost heaving of permafrost active layer), (**b**,**b'**) are the original amplitude map and annual average deformation map of TZ deformation section, respectively, In the Figure, the red solid dot indicates the roadbed deformation or serious deformation area threatening the roadbed, and the white box indicates the deformation risk section.

In the south section, the deformation on the left and right sides is uneven and the difference is obvious, which poses a threat to the railway operation. As shown in P1 and P2 in Figure 7c, the cumulative deformation is 113.54 mm and 54.76 mm, respectively, and the annual average deformation is −112.768 mm/a and −52.084 mm/a, respectively. The uneven deformation may be affected by the thickness difference in the active layer of permafrost. The thermal melting of frozen soil in summer leads to different collapse degrees on both sides of the railway, and it cannot be completely frozen in winter, resulting in greater and greater differences on both sides of the roadbed and affecting the stability of the roadbed.

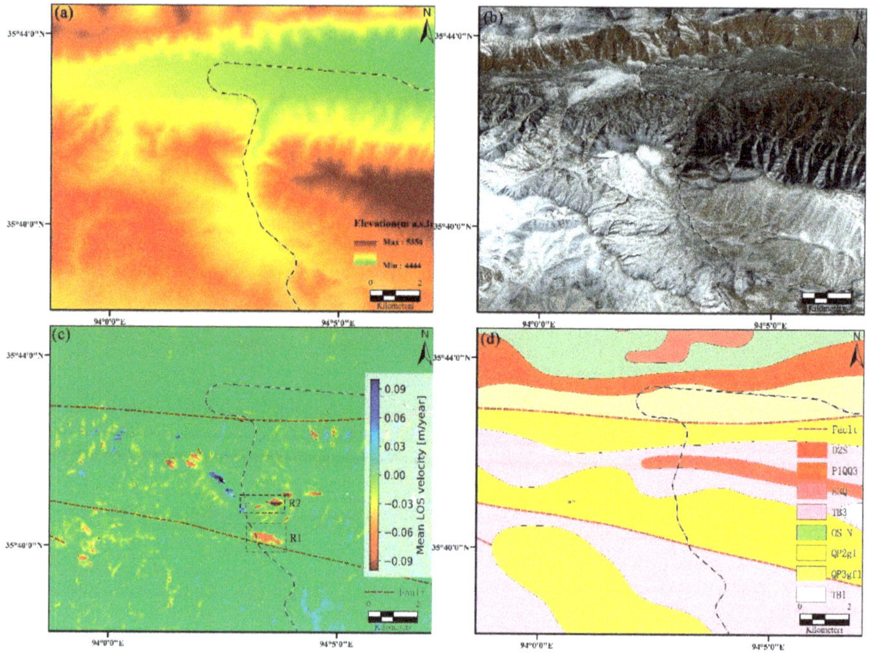

Figure 6. (**a**) shows the DEM of the railway section; (**b**) is the optical image of Google Earth; (**c**) is the distribution of surface deformation at unfrozen spring station; (**d**) is a geological distribution map, and the fault zone is shown in red dotted line.

4.2. Comparison and Discussion on Deformation Law of Road Section

4.2.1. Comparison of Different Deformation of Roadbed Frozen Soil

In order to compare the different time series deformation laws of WB and TZ deformation sections and understand their differences in climate response, we obtain the surface time series deformation information from the deformation risk area, as shown in Figure 8. At the same time, in order to more accurately judge the response of time series deformation and climate change, we take the actual distance from each risk area to the meteorological station as the weight factor to obtain the weighted average temperature and precipitation.

In addition to the thawing and collapse of frozen soil in WB section, at the west side of the railway passing through Kunlun Mountain, the central latitude and longitude are 35.67944° N and 94.04924° E. The roadbed has been raised due to frost heaving of some frozen soil, with cumulative deformation of 50.507 mm and annual average deformation of 20.09 mm/a. From January to early March, the ground surface is constantly lifted due to the frost heaving of frozen soil. In summer, the frost heaving slows down due to the increase in temperature and precipitation, which is basically in a constant trend. In autumn and winter, when the temperature decreases, the surface temperature also decreases. The water in the active layer of frozen soil under the surface solidifies, and the frozen soil heaves. The maximum value of frozen soil heave is rising, as shown in Figure 8c. This shows that the active layer thickens, and because it is located on the shady slope, the frozen soil layer has good development conditions, which belongs to the developing permafrost.

The settlement area of TZ section is rapidly affected by the climate. From the end of April, the precipitation and temperature began to rise, the frozen soil active layer gradually began to melt, and the surface settlement is serious. Its trend line is shown in Figure 8d. The annual average time series deformation rate is 38.47 mm/a, the cumulative deformation variable is 86.39 mm, and the latitude and longitude of the deformation center are 32.8847° N and 91.5283° E. The seasonality of the climate is seriously affected by the southeast monsoon. Most of the precipitation occurs from June to August in the form of rainstorm, which leads to flash floods and extensive surface erosion. At the same time, it also leads to the intensification of the melting of frozen soil active layer. The remaining precipitation occurs in the form of snow or hail. Sometimes snow will be generated during the ice period (usually 7–8 months, from September to April of the next year), which is consistent with the deformation. From September to April of the next year, the deformation will obviously slow down, while from June to August, the frozen soil will continue to melt under the influence of temperature and precipitation, resulting in intensified surface settlement. As the settlement area is located in the roadbed section of Qinghai–Tibet railway, it poses a threat to railway operation.

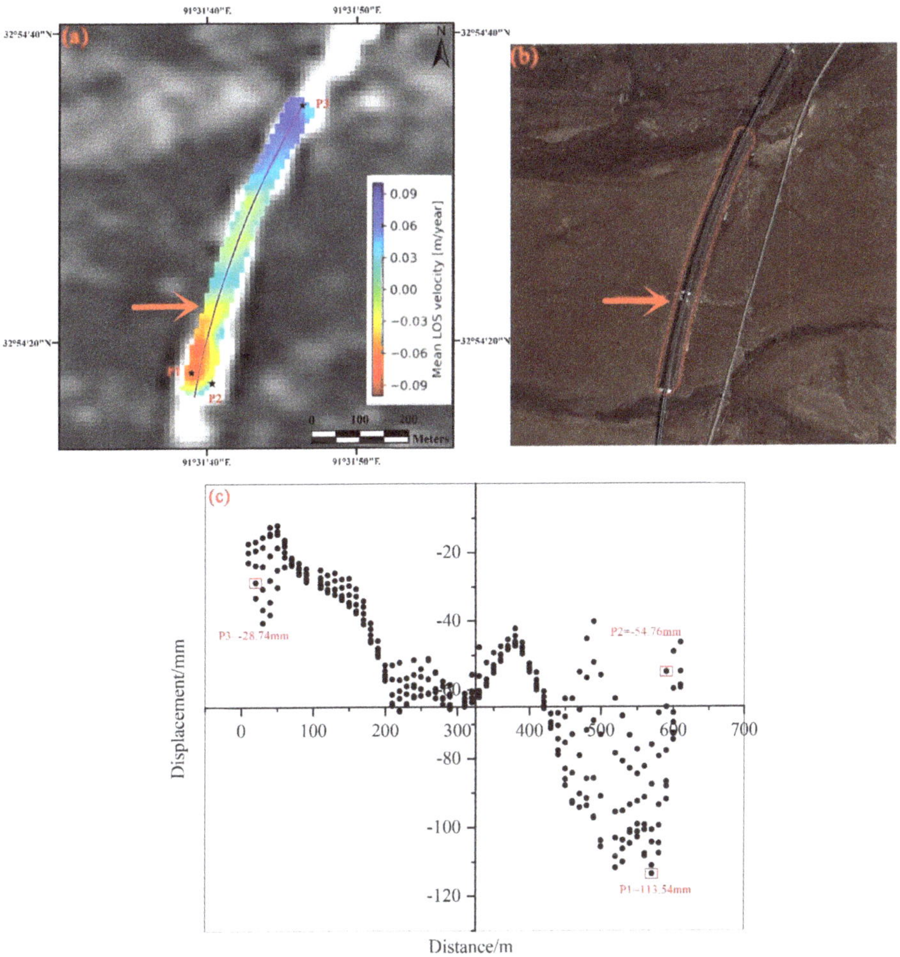

Figure 7. (**a–c**), respectively show the annual average deformation, topographic and temporal deformation of railway roadbed. The red arrows in (**a**,**b**) show the runoff formed by the thermal melting of frozen soil and the boundary of different deformation mechanisms at the north and south ends. The red box indicates the roadbed range of the railway section. The abscissa in (**c**) represents the distance of p3–p1 point from north to south, and the ordinate is the deformation.

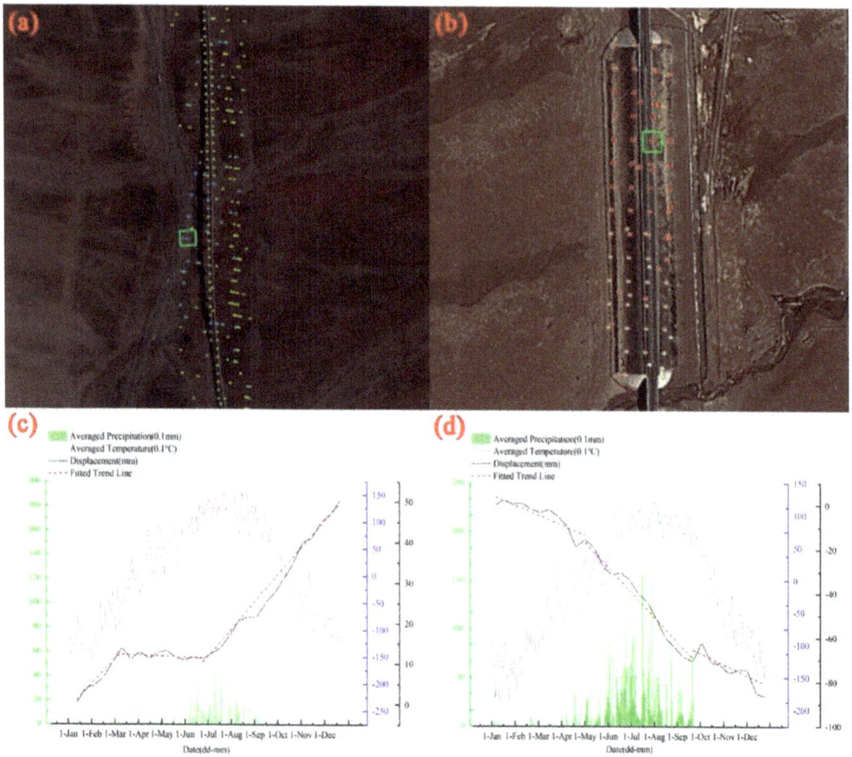

Figure 8. (**a**,**b**) depict the distribution of roadbed deformation in WB and TZ sections, respectively, (**c**,**d**) depict the time series deformation of railway roadbed in WB and TZ sections under the influence of climate factors, respectively. The green box in the optical diagram is the selection point of time series deformation, and the red dotted line in the broken line diagram is the deformation subsection fitting line.

4.2.2. Comparison of the Latest Available Results with Previous Results

Because Beilu River and Tuotuo River are located in the basin area, the river system has a great impact on the frozen soil, and the frozen soil section often becomes a research hotspot. This paper compares the deformation of the reach from Wudaoliang to Tuotuo with the study of Zhang et al. from 2009 to 2018, as shown in Figure 9. The three places C1, C2 and C3 pass through Beiluhe basin, Fenghuoshan area and Tuotuohe basin, respectively. C1 and C3 are located in the valley basin, rich in water resources and geographical environment, and there is no high mountain shelter. The frozen soil is greatly affected by thermal melting and forms a large number of thermal melting lakes, as shown in Figure 10a,b. Therefore, these two areas are scattered point deformation areas in InSAR detection results. There is no obvious frozen soil collapse at the railway roadbed. However, under the global warming environment and the expansion of inland rivers and thermal melting lakes, the frozen soil under the railway roadbed will also be affected, and there is still a certain risk of thawing collapse. The deformation at C2 is small, and it has been significantly improved compared with the results in 2018. This area not only takes heat dissipation measures for frozen soil, as shown in Figure 10c, but is also located in high mountain and valley area (Figure 10d), with an altitude of 5000–5200 m and a height drop of 400–600 m with the railway section. The risk of frozen soil collapse is further reduced. In short, although there is no serious deformation of railway roadbed at C1, C2 and C3, considering that there are still many thermal melting lakes near C1 and C3, it is still necessary to monitor the frozen soil section.

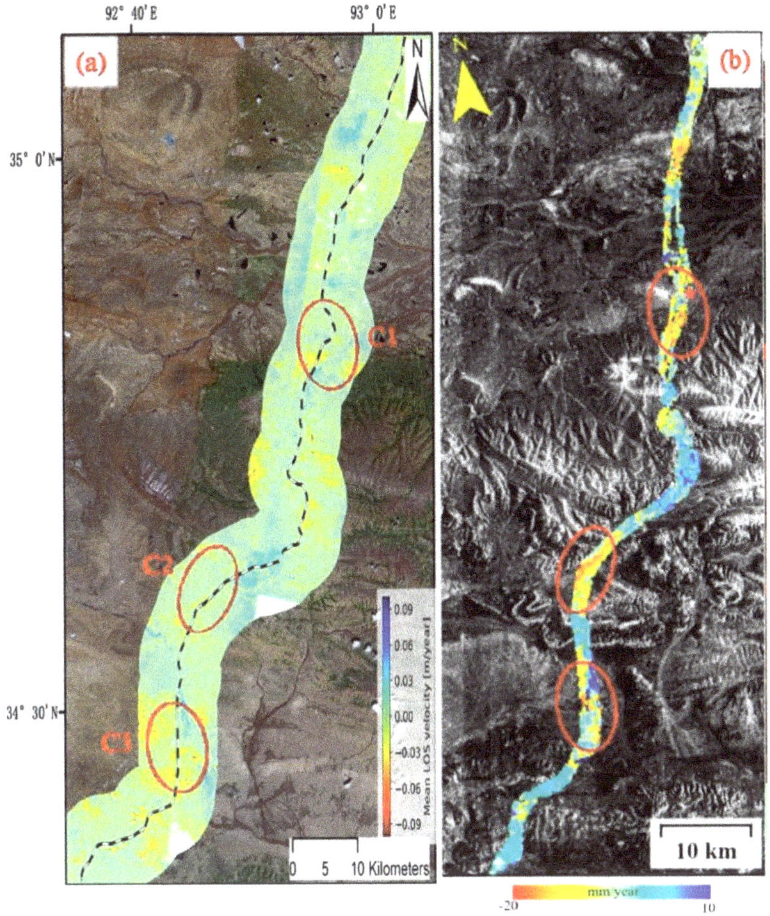

Figure 9. (**a**) shows the annual average time series deformation of the railway corridor from Beilu River to Tuotuo River in 2020, and (**b**) shows the deformation detection results of Zhang et al. in 2018. The three deformations detected by Zhang are marked with circles in the Figure, namely C1, C2 and C3. The deformation area obviously decreases, especially at C2. No obvious deformation is detected at C1 and C3 railway roadbed.

4.3. Uncertainty Analysis of Results

Due to the destructive effects of atmospheric precipitation (especially snowfall), the formal application of InSAR technique to monitor structures generating scattering may provide incorrect results. The corrupting impact of atmospheric precipitation on the phase of reference targets has been mentioned in many studies. In winter, the main source of alteration for the propagating signal properties is the growth in snow depth between the SAR observations [67].

As shown in Figure 10, part of our study area is covered by snow because it is located in the Qinghai–Tibet plateau. Snow can affect radar interference and skew results. Therefore, in addition to removing atmospheric phase, terrain phase and noise, we also need to remove the influence of snow removal. As the influence of snow in the main deformation area is small, this study does not deal with the influence of snow temporarily. However, the impact of snow removal will be the focus of our next study.

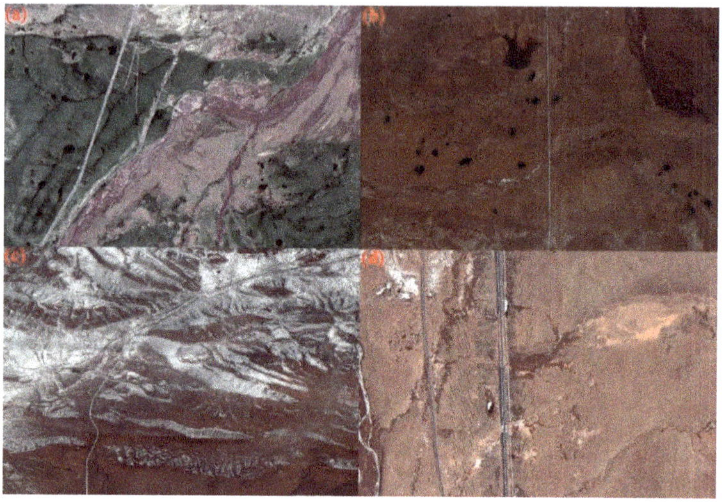

Figure 10. In (**a**,**b**), there are widely distributed hot melt lakes formed by ground collapse after frozen soil melting. (**c**,**d**) are the topographic distribution and solutions of Fenghuo Mountain section, respectively. It can be seen that hot rods are arranged on the left and right sides of the road to dissipate heat and reduce the impact of frozen soil hot melt and frost heaving on the roadbed.

5. Conclusions

In this study, four Sentinel-1 satellite maps are selected to fully cover the study area, with a total of 122 images. The MT-InSAR technology is carried out for the 610 km-long Qinghai–Tibet railway section (from Naij Tal railway station to Anduo railway station) in the permafrost area so as to obtain the time series deformation information of the surface along the Qinghai–Tibet railway and compare its deformation with climate factors. The main conclusions are as follows:

(1) The areas with serious deformation of the Qinghai–Tibet corridor are mainly distributed in the railway section from WangKun station to Budongquan station and the section from Tanggula station to Za'gya Zangbo station, and there are many areas of railway roadbed subsidence and mountain collapse.

(2) The influence of the frozen soil section from WangKun station to Budongquan station on the railway roadbed is high in the middle and low at both ends. The influence of human activities of railway operation on the frozen soil deformation is smaller than that of topography and hydrothermal. At the same time, the geological strata and fault zone of this section also have a certain impact on the roadbed deformation. D2S, OS, and S3Q frozen soil layers are more stable than Qp2gl and Qp3gl strata.

(3) Between the Tanggula and Za'gya Zangbo station, there was a 620 m-long railway roadbed with uneven deformation on the east and west sides, with an average annual difference of 60.68 mm/a. At the same time, uneven deformation also exists in the railway roadbed on the north and south sides.

(4) Through comparison, it is determined that the roadbed deformation does not exist in the area from Beilu River to Tuotuo River, and the permafrost has no great impact on the railway roadbed.

(5) Under the situation of global warming, the frozen soil will continue to undergo thermal thawing and frost heaving. At the same time, the permafrost will continue to decrease and the active layer will continue to thicken. In order to ensure the stable operation of the Qinghai–Tibet railway in the permafrost section, it is necessary to regularly monitor the deformation of the permafrost area up to 610 km by MT-InSAR.

At the same time, our research has some shortcomings. The impact of snow cover on InSAR interference in the study area needs further treatment. This is the focus of our next research work.

Author Contributions: H.L. and S.H. performed experiments, analyzed the data and prepared the manuscript. C.X. and B.T. provided crucial guidance and support through the research. M.C. significantly contributed to the validation work and data interpretation. Z.C. provided valuable suggestions for this study. All authors have read and agreed to the published version of the manuscript.

Funding: This research was funded by Fujian Provincial Science and Technology Project (Science and Technology Service Network Initiative, CAS) (2020T3011), jointly funded by the Outstanding Youth Science and Technology Program of Guizhou Province of China ([2021]5615).

Data Availability Statement: The data supporting the findings of this study are available from the first author (H.L.) upon reasonable request.

Acknowledgments: The authors would like to thank ESA (European Space Agency) and Alaska satellite facility (https://asf.alaska.edu/, accessed on 30 July 2021) for providing the Sentinel-1 datasets of the Copernicus mission.

Conflicts of Interest: The authors declare no conflict of interest.

References

1. Wang, C.; Zhang, Z.; Zhang, H.; Wu, Q.; Zhang, B.; Tang, Y. Seasonal deformation features on Qinghai-Tibet railway observed using time-series InSAR technique with high-resolution TerraSAR-X images. *Remote Sens. Lett.* **2017**, *8*, 1–10. [CrossRef]
2. Wang, C.; Zhang, Z.; Zhang, H.; Zhang, B.; Tang, Y.; Wu, Q. Active Layer Thickness Retrieval of Qinghai–Tibet Permafrost Using the TerraSAR-X InSAR Technique. *IEEE J. Sel. Top. Appl. Earth Obs. Remote Sens.* **2018**, *11*, 4403–4413. [CrossRef]
3. Luo, Q.; Li, J.; Zhang, Y. Monitoring Subsidence over the Planned Jakarta–Bandung (Indonesia) High-Speed Railway Using Sentinel-1 Multi-Temporal InSAR Data. *Remote Sens.* **2022**, *14*, 4138. [CrossRef]
4. Zhang, J.; Ke, C.; Shen, X.; Lin, J.; Wang, R. Monitoring Land Subsidence along the Subways in Shanghai on the Basis of Time-Series InSAR. *Remote Sens.* **2023**, *15*, 908. [CrossRef]
5. Shao, Z. The characteristics of high-speed railway subgrade's temperature, moisture and frost heave deformation in high altitude and seasonal frozen region: Taking the Minle section of Lanzhou-Xinjiang passenger railway line as an example. *J. Glaciol. Geocryol.* **2018**, *40*, 588–597.
6. Wu, L.; Qi, W.; Niu, F.; Niu, Y. A review of studies on roadbed frozen damage and countermeasures in seasonal frozen ground regions in China. *J. Glaciol. Geocryol.* **2015**, *37*, 1283–1293. [CrossRef]
7. Wang, Y.; Lv, W.; Xue, K.; Wang, S.; Zhang, L.; Hu, R.; Zeng, H.; Xu, X.; Li, Y.; Jiang, L.; et al. Grassland changes and adaptive management on the Qinghai–Tibetan Plateau. *Nat. Rev. Earth Environ.* **2022**, *3*, 668–683. [CrossRef]
8. Ni, J.; Wu, T.; Zhu, X.; Chen, J.; Wu, X.; Hu, G.; Zou, D.; Li, R.; Du, Y. Quantifying the Relationship Between Human Activities Intensity and Thawing Hazards of the Frozen Ground on the Qinghai–Tibet Plateau. *Ecol. Impacts Degrad. Permafr.* **2022**, *10*, 845873. [CrossRef]
9. Li, C.; Bai, X.; Tan, Q.; Luo, G.; Wu, L.; Chen, F.; Xi, H.; Luo, X.; Ran, C.; Chen, H.; et al. High-resolution mapping of the global silicate weathering carbon sink and its long-term changes. *Glob. Chang. Biol.* **2022**, *28*, 4377–4394. [CrossRef]
10. Xiong, L.; Bai, X.; Zhao, C.; Li, Y.; Tan, Q.; Luo, G.; Wu, L.; Chen, F.; Li, C.; Ran, C.; et al. High-Resolution Data Sets for Global Carbonate and Silicate Rock Weathering Carbon Sinks and Their Change Trends. *Earth's Futur.* **2022**, *10*, e2022EF002746. [CrossRef]
11. Song, F.; Wang, S.; Bai, X.; Wu, L.; Wang, J.; Li, C.; Chen, H.; Luo, X.; Xi, H.; Zhang, S.; et al. A New Indicator for Global Food Security Assessment: Harvested Area Rather Than Cropland Area. *Chin. Geogr. Sci.* **2022**, *32*, 204–217. [CrossRef]
12. Zhang, S.; Bai, X.; Zhao, C.; Tan, Q.; Luo, G.; Wu, L.; Xi, H.; Li, C.; Chen, F.; Ran, C.; et al. China's carbon budget inventory from 1997 to 2017 and its challenges to achieving carbon neutral strategies. *J. Clean. Prod.* **2022**, *347*, 130966. [CrossRef]
13. Liu, M.; Bai, X.; Tan, Q.; Luo, G.; Zhao, C.; Wu, L.; Luo, X.; Ran, C.; Zhang, S. Climate change enhances the positive contribution of human activities to vegetation restoration in China. *Geocarto Int.* **2022**, 1–21. [CrossRef]
14. Du, C.; Bai, X.; Li, Y.; Tan, Q.; Zhao, C.; Luo, G.; Wu, L.; Chen, F.; Li, C.; Ran, C.; et al. Inventory of China's Net Biome Productivity since the 21st Century. *Land* **2022**, *11*, 1244. [CrossRef]
15. Chen, F.; Bai, X.; Liu, F.; Luo, G.; Tian, Y.; Qin, L.; Li, Y.; Xu, Y.; Wang, J.; Wu, L.; et al. Analysis Long-Term and Spatial Changes of Forest Cover in Typical Karst Areas of China. *Land* **2022**, *11*, 1349. [CrossRef]
16. Li, Z.; Zhao, L.; Wang, L.; Zou, D.; Liu, G.; Hu, G.; Du, E.; Xiao, Y.; Liu, S.; Zhou, H.; et al. Retrieving Soil Moisture in the Permafrost Environment by Sentinel-1/2 Temporal Data on the Qinghai–Tibet Plateau. *Remote Sens.* **2022**, *14*, 5966. [CrossRef]
17. Wu, L.; Wang, S.; Bai, X.; Chen, F.; Li, C.; Ran, C.; Zhang, S. Identifying the Multi-Scale Influences of Climate Factors on Runoff Changes in a Typical Karst Watershed Using Wavelet Analysis. *Land* **2022**, *11*, 1284. [CrossRef]

18. Fei, D.-Q.; Liu, F.-G.; Zhou, Q. Risk analysis of landslide and debris flow disasters along the Qinghai-Tibet Railway. *Arid. Land Geogr.* **2016**, *39*, 345–352.
19. Ni, J.; Wu, T.; Zhu, X.; Wu, X.; Pang, Q.; Zou, D.; Chen, J.; Li, R.; Hu, G.; Du, Y.; et al. Risk assessment of potential thaw settlement hazard in the permafrost regions of Qinghai-Tibet Plateau. *Sci. Total. Environ.* **2021**, *776*, 145855. [CrossRef]
20. Olsen, K.M.; Calef, M.T.; Agram, P.S. Contextual uncertainty assessments for InSAR-based deformation retrieval using an ensemble approach. *Remote Sens. Environ.* **2023**, *287*, 113456. [CrossRef]
21. Pezzo, G.; Palano, M.; Beccaro, L.; Tolomei, C.; Albano, M.; Atzori, S.; Chiarabba, C. Coupling Flank Collapse and Magma Dynamics on Stratovolcanoes: The Mt. Etna Example from InSAR and GNSS Observations. *Remote Sens.* **2023**, *15*, 847. [CrossRef]
22. Ma, S.; Qiu, H.; Zhu, Y.; Yang, D.; Tang, B.; Wang, D.; Wang, L.; Cao, M. Topographic Changes, Surface Deformation and Movement Process before, during and after a Rotational Landslide. *Remote Sens.* **2023**, *15*, 662. [CrossRef]
23. Feng, X.; Chen, Z.; Li, G.; Ju, Q.; Yang, Z.; Cheng, X. Improving the capability of D-InSAR combined with offset-tracking for monitoring glacier velocity. *Remote Sens. Environ.* **2023**, *285*, 113394. [CrossRef]
24. Ma, D.; Zhao, R.; Li, Y.; Li, Z. Land Subsidence Assessment of an Archipelago Based on the InSAR Time Series Analysis Method. *Water* **2023**, *15*, 465. [CrossRef]
25. Zwieback, S.; Meyer, F.J. Top-of-permafrost ground ice indicated by remotely sensed late-season subsidence. *Cryosphere* **2021**, *15*, 2041–2055. [CrossRef]
26. Bartsch, A.; Leibman, M.; Strozzi, T.; Khomutov, A.; Widhalm, B.; Babkina, E.; Mullanurov, D.; Ermokhina, K.; Kroisleitner, C.; Bergstedt, H. Seasonal Progression of Ground Displacement Identified with Satellite Radar Interferometry and the Impact of Unusually Warm Conditions on Permafrost at the Yamal Peninsula in 2016. *Remote Sens.* **2019**, *11*, 1865. [CrossRef]
27. Strozzi, T.; Antonova, S.; Günther, F.; Mätzler, E.; Vieira, G.; Wegmüller, U.; Westermann, S.; Bartsch, A. Sentinel-1 SAR Interferometry for Surface Deformation Monitoring in Low-Land Permafrost Areas. *Remote Sens.* **2018**, *10*, 1360. [CrossRef]
28. Rudy, A.C.; Lamoureux, S.F.; Treitz, P.; Short, N.; Brisco, B. Seasonal and multi-year surface displacements measured by DInSAR in a High Arctic permafrost environment. *Int. J. Appl. Earth Obs. Geoinformation* **2018**, *64*, 51–61. [CrossRef]
29. Liu, L.; Zhang, T.; Wahr, J. InSAR measurements of surface deformation over permafrost on the North Slope of Alaska. *J. Geophys. Res. Earth Surf.* **2010**, *115*, F3. [CrossRef]
30. Liu, L.; Schaefer, K.; Zhang, T.; Wahr, J. Estimating 1992–2000 average active layer thickness on the Alaskan North Slope from remotely sensed surface subsidence. *J. Geophys. Res. Atmos.* **2010**, *117*. [CrossRef]
31. Liu, L.; Jafarov, E.E.; Schaefer, K.M.; Jones, B.M.; Zebker, H.A.; Williams, C.A.; Rogan, J.; Zhang, T. InSAR detects increase in surface subsidence caused by an Arctic tundra fire. *Geophys. Res. Lett.* **2014**, *41*, 3906–3913. [CrossRef]
32. Liu, L.; Schaefer, K.; Gusmeroli, A.; Grosse, G.; Jones, B.M.; Zhang, T.; Parsekian, A.D.; Zebker, H.A. Seasonal thaw settlement at drained thermokarst lake basins, Arctic Alaska. *Cryosphere* **2014**, *8*, 815–826. [CrossRef]
33. Liu, L.; Schaefer, K.M.; Chen, A.C.; Gusmeroli, A.; Zebker, H.A.; Zhang, T. Remote sensing measurements of thermokarst subsidence using InSAR. *J. Geophys. Res. Earth Surf.* **2015**, *120*, 1935–1948. [CrossRef]
34. Zhou, P.; Liu, W.; Zhang, X.; Wang, J. Evaluating Permafrost Degradation in the Tuotuo River Basin by MT-InSAR and LSTM Methods. *Sensors* **2023**, *23*, 1215. [CrossRef]
35. Xu, Z.; Jiang, L.; Niu, F.; Guo, R.; Huang, R.; Zhou, Z.; Jiao, Z. Monitoring Regional-Scale Surface Deformation of the Continuous Permafrost in the Qinghai–Tibet Plateau with Time-Series InSAR Analysis. *Remote Sens.* **2022**, *14*, 2987. [CrossRef]
36. Zou, L.; Wang, C.; Tang, Y.; Zhang, B.; Zhang, H.; Dong, L. Interferometric SAR Observation of Permafrost Status in the Northern Qinghai-Tibet Plateau by ALOS, ALOS-2 and Sentinel-1 between 2007 and 2021. *Remote Sens.* **2022**, *14*, 1870. [CrossRef]
37. Xiang, W.; Zhang, R.; Liu, G.; Wang, X.; Mao, W.; Zhang, B.; Cai, J.; Bao, J.; Fu, Y. Extraction and analysis of saline soil deformation in the Qarhan Salt Lake region (in Qinghai, China) by the sentinel SBAS-InSAR technique. *Geodesy Geodyn.* **2021**, *13*, 127–137. [CrossRef]
38. Zhang, Z.; Wang, M.; Liu, X.; Wang, C.; Zhang, H. Map and Quantify the Ground Deformation Around Salt Lake in Hoh Xil, Qinghai-Tibet Plateau Using Time-Series InSAR From 2006 to 2018. *IEEE J. Sel. Top. Appl. Earth Obs. Remote Sens.* **2021**, *14*, 858–869. [CrossRef]
39. Zhang, Z.; Wang, M.; Liu, X.; Wang, C.; Zhang, H.; Tang, Y.; Zhang, B. Deformation Feature Analysis of Qinghai–Tibet Railway Using TerraSAR-X and Sentinel-1A Time-Series Interferometry. *IEEE J. Sel. Top. Appl. Earth Obs. Remote Sens.* **2019**, *12*, 5199–5212. [CrossRef]
40. Zhang, X.; Zhang, H.; Wang, C.; Tang, Y.; Zhang, B.; Wu, F.; Wang, J.; Zhang, Z. Active Layer Thickness Retrieval Over the Qinghai-Tibet Plateau Using Sentinel-1 Multitemporal InSAR Monitored Permafrost Subsidence and Temporal-Spatial Multilayer Soil Moisture Data. *IEEE Access* **2020**, *8*, 84336–84351. [CrossRef]
41. Zhang, Z.; Wang, M.; Liu, X. Deformation Monitoring of Qinghai-Tibet Railway from 1997–2018 using SAR Interferometry with Multi-sensors SAR Datasets. In Proceedings of the 2019 SAR in Big Data Era (BIGSARDATA), Beijing, China, 5–6 August 2019; pp. 1–4. [CrossRef]
42. Zhang, Z.; Lin, H.; Wang, M.; Liu, X.; Chen, Q.; Wang, C.; Zhang, H. A Review of Satellite Synthetic Aperture Radar Interferometry Applications in Permafrost Regions: Current status, challenges, and trends. *IEEE Geosci. Remote Sens. Mag.* **2022**, *10*, 93–114. [CrossRef]
43. Zhang, Z.; Wang, M.; Wu, Z.; Liu, X. Permafrost Deformation Monitoring Along the Qinghai-Tibet Plateau Engineering Corridor Using InSAR Observations with Multi-Sensor SAR Datasets from 1997–2018. *Sensors* **2019**, *19*, 5306. [CrossRef] [PubMed]

44. Zhang, X.; Zhang, H.; Wang, C.; Tang, Y.; Zhang, B.; Wu, F.; Wang, J.; Zhang, Z. Time-Series InSAR Monitoring of Permafrost Freeze-Thaw Seasonal Displacement over Qinghai–Tibetan Plateau Using Sentinel-1 Data. *Remote Sens.* **2019**, *11*, 1000. [CrossRef]
45. Wang, L.; Marzahn, P.; Bernier, M.; Ludwig, R. Sentinel-1 InSAR measurements of deformation over discontinuous permafrost terrain, Northern Quebec, Canada. *Remote Sens. Environ.* **2020**, *248*, 111965. [CrossRef]
46. Wang, J.; Wang, C.; Zhang, H.; Tang, Y.; Zhang, X.; Zhang, Z. Small-Baseline Approach for Monitoring the Freezing and Thawing Deformation of Permafrost on the Beiluhe Basin, Tibetan Plateau Using TerraSAR-X and Sentinel-1 Data. *Sensors* **2020**, *20*, 4464. [CrossRef]
47. Reinosch, E.; Buckel, J.; Dong, J.; Gerke, M.; Baade, J.; Riedel, B. InSAR time series analysis of seasonal surface displacement dynamics on the Tibetan Plateau. *Cryosphere* **2020**, *14*, 1633–1650. [CrossRef]
48. Lu, P.; Han, J.; Li, Z.; Xu, R.; Li, R.; Hao, T.; Qiao, G. Lake outburst accelerated permafrost degradation on Qinghai-Tibet Plateau. *Remote Sens. Environ.* **2020**, *249*, 112011. [CrossRef]
49. Chen, J.; Liu, L.; Zhang, T.; Cao, B.; Lin, H. Using Persistent Scatterer Interferometry to Map and Quantify Permafrost Thaw Subsidence: A Case Study of Eboling Mountain on the Qinghai-Tibet Plateau. *J. Geophys. Res. Earth Surf.* **2018**, *123*, 2663–2676. [CrossRef]
50. Daout, S.; Dini, B.; Haetberli, W.; Doin, M.-P.; Palrsons, B. Ice loss in the Northeastern Tibetan Plateau permafrost as seen by 16 yr of ESA SAR missions. Earth Planet. *Sci. Lett.* **2020**, *545*, 116404. [CrossRef]
51. Chen, Y.; Wang, L.; Bernier, M.; Ludwig, R. Retrieving Freeze/Thaw Cycles Using Sentinel-1 Data in Eastern Nunavik (Québec, Canada). *Remote Sens.* **2022**, *14*, 802. [CrossRef]
52. Rouyet, L.; Lauknes, T.R.; Christiansen, H.H.; Strand, S.M.; Larsen, Y. Seasonal dynamics of a permafrost landscape, Adventdalen, Svalbard, investigated by InSAR. *Remote Sens. Environ.* **2019**, *231*, 111236. [CrossRef]
53. Antonova, S.; Sudhaus, H.; Strozzi, T.; Zwieback, S.; Kääb, A.; Heim, B.; Langer, M.; Bornemann, N.; Boike, J. Thaw Subsidence of a Yedoma Landscape in Northern Siberia, Measured In Situ and Estimated from TerraSAR-X Interferometry. *Remote Sens.* **2018**, *10*, 494. [CrossRef]
54. Li, Z.-W.; Zhao, R.; Hu, J.; Wetn, L.; Feng, G.; Zhalng, Z.; Wang, Q. InSAR analysis of surface deformation over permafrost to estimate active layer thickness based on one-dimensional heat transfer model of soils. *Sci. Rep.* **2015**, *5*, 1–9. [CrossRef]
55. Li, Y.; Song, W.; Jin, B.; Zuo, X.; Li, Y.; Chen, K. A SqueeSAR Spatially Adaptive Filtering Algorithm Based on Hadoop Distributed Cluster Environment. *Appl. Sci.* **2023**, *13*, 1869. [CrossRef]
56. Chai, M.; Mu, Y.; Zhang, J.; Ma, W.; Liu, G.; Chen, J. Characteristics of Asphalt Pavement Damage in Degrading Permafrost Regions: Case Study of the Qinghai–Tibet Highway, China. *J. Cold Reg. Eng.* **2018**, *32*, 5018003. [CrossRef]
57. Zhang, S.; Niu, F.; Wang, J.; Dong, T. Evaluation of damage probability of railway embankments in permafrost regions in Qinghai–Tibet Plateau. *Eng. Geol.* **2021**, *284*, 106027. [CrossRef]
58. Zhang, Q.; Li, Y.; Zhang, J.; Luo, Y. InSAR technique applied to the monitoring of the Qinghai–Tibet Railway. *Nat. Hazards Earth Syst. Sci.* **2019**, *19*, 2229–2240. [CrossRef]
59. Wang, T. (Ed.) *1:4000000 Map of the Glaciers, Frozen Ground and Deserts in China*; Beijing Science Press: Beijing, China, 2006.
60. Hanssen, R.F. *Radar Interferometry Data Interpretation and Error Analysis*; Springer Science & Business Media: Berlin/Heidelberg, Germany, 2001.
61. Zhu, X.; Dong, Z.; Yu, A.; Wu, M.; Li, D.; Zhang, Y. New Approaches for Robust and Efficient Detection of Persistent Scatterers in SAR Tomography. *Remote Sens.* **2019**, *11*, 356. [CrossRef]
62. Ferretti, A.; Prati, C.; Rocca, F. Analysis of Permanent Scatterers in SAR interferometry. In Proceedings of the IGARSS 2000. IEEE 2000 International Geoscience and Remote Sensing Symposium. Taking the Pulse of the Planet: The Role of Remote Sensing in Managing the Environment. Proceedings (Cat. No.00CH37120), Honolulu, HI, USA, 24–28 July 2000; Volume 2, pp. 761–763. [CrossRef]
63. Ferretti, A.; Fumagalli, A.; Novali, F. A New Algorithm for Processing Interferometric Data-Stacks: SqueeSAR. *IEEE Trans. Geosci. Remote Sens.* **2011**, *49*, 3460–3470. [CrossRef]
64. Minh, D.H.T.; Hanssen, R.; Rocca, F. Radar Interferometry: 20 Years of Development in Time Series Techniques and Future Perspectives. *Remote Sens.* **2020**, *12*, 1364. [CrossRef]
65. Chang, J. *Research the Impacts of Surface Coverage Change on the Hydrological Process in Permafrost Watershed*; Lanzhou University: Lanzhou, China, 2012.
66. Wu, L.-Z.; Xu, Q.; Huang, R.-Q. Analysis of freezing-thawing test process of unsaturated clay. *Rock Soil Mech.* **2011**, *32*, 1025–1028.
67. Zakharov, A.; Zakharova, L. An Influence of Snow Covers on the Radar Interferometry Observations of Industrial Infrastructure: Norilsk Thermal Power Plant Case. *Remote Sens.* **2023**, *15*, 654. [CrossRef]

land

MDPI

Review

Horst-Controlled Karstification in the Bakony Region (Hungary)

Márton Veress

Department of Geography, Savaria University Centre, Eötvös Lóránd University, 9700 Szombathely, Hungary; veress.marton@sek.elte.hu

Abstract: The karst of the horsts of the Bakony Region belonging to horst types of different development is described. Horsts elevated to the summit position are characterised by the most widespread and diverse karstification (with covered karst, cave-ins and shafts). Cryptopeneplains affect the karstification of horsts elevated to the summit position, but karst features also occur in their areas. On threshold surfaces of marginal position, the hypogene branches of regional flow systems influenced the development of spring caves. The karstification of basaltic mountains has been mainly governed by the properties of the basalt caprock (the major features being ponors with blind valleys and caprock dolines).

Keywords: horst type; epigene karst system; hypogene branch; cave-in; subsidence doline; ponor with blind valley; spring cave

Citation: Veress, M. Horst-Controlled Karstification in the Bakony Region (Hungary). *Land* **2023**, *12*, 682. https://doi.org/10.3390/land12030682

Academic Editor: Xiaoyong Bai

Received: 17 February 2023
Revised: 7 March 2023
Accepted: 10 March 2023
Published: 14 March 2023

1. Introduction

In this study, the karst of the horsts of the Bakony Region belonging to horst types of different development is described. Horsts belonging to different types have different elevations, expansions and coveredness (the expansion and quality of the cover), degrees and methods of its exposure, environments and positions as compared to karstwater level, bedrock morphologies and the presence of the aquifuge that is intercalated into the bedrock. Therefore, the reason for the different karstification of horst types is found in one (or even several) of the above factors and thus, the karstification of the horsts is traced back to such properties. The karstifications of the horst types differ from each other to a larger and smaller degree, which can also be interpreted by the consideration of the above-mentioned factors.

Modern karst research focuses on the detailed study (hydrology, geomorphology) of different karst areas [1–6]. This will be followed in the case of the description of the karst of the Bakony Region. The flow systems of the karst of the Bakony Region are classified based on the work of Klimchouk [7] and Goldscheider et al. [8], its karst features and their characteristics are described considering the work of Sweeting [9], White [10], Waltham et al. [11], Ford and Williams [12], Gutierrez [13] and Veress [14].

Basic features of karsts are features that can be found nearly on all karst types, but in different densities. The karst features are the following: karren, dolines and uvalas (they may have different varieties on different karst types), ponors with blind valleys, poljes, concretions, karstic gorges, inselberg karst and subsurface karst features [11]. Their common characteristic feature is that they develop by dissolution (caves are partly formed by erosion) and, with the exception of gorges and some karren features, surface features are closed.

Karren may develop on bare surfaces (rillenkarren, rinnenkarren, grikes) or below the soil (kamenitza, grike, rootkarren). Some of them may develop on both bare surfaces and soil-covered surfaces, or on karst covered with sediment too. Among dolines, solution dolines (mainly drawdown dolines) are distinguished, which are widespread on soil-covered karst. Caprock dolines are formed on the caprock, on consolidated rock, while the several varieties of subsidence dolines develop on permeable, unconsolidated rock. Collapse dolines develop by collapse processes. Poljes are large features with independent hydrography (with springs, surface streams and ponors). Concretions are accumulation features that develop at karst springs and in the channels of streams. Ponors are formed along rock boundaries where the water of the streams from non-karstic areas flows into the karst. Inselberg karst is a complex landform assemblage and is constituted by residuals (inselbergs) of various shapes that developed by dissolution.

There are caves that develop by dissolution (spring caves, shafts), while others are of both dissolution and erosion origin (inflow cave, through cave, spring cave).

Among karst types, evaporate karsts and carbonate karsts are distinguished. Carbonate karsts include temperate karsts, tropical karsts, glaciokarsts, mixed allogenic-autogenic karsts and covered karsts. Their features are partly different. On temperate karsts, solution dolines are specific, while on mixed autogenic–allogenic karsts, there are ponors with blind valleys, on glaciokarsts, karren and shafts, on covered karsts, caprock dolines and subsidence dolines, on tropical karsts, inselberg karst, and on Mediterranean karst and tropical karsts, poljes are common.

A variety of temperate karsts have a horst structure, such as the Bakony Region. Since this karst is constituted by horsts of different structures and development, there occur features that are characteristic of covered karst, partly of soil-covered karst and of mixed allogenic–autogenic karst. However, it has to be noted that some features as epigenetic-antecedent gorges and cave-ins related to them are particularly specific to the Bakony Region.

2. Study Area

The description below is based on Veress's work [15]. The parts of the Bakony Region and the names of the features that are mentioned in this study are described in Figure 1. The Bakony Region is the southwestern part of the Transdanubian Mountains. It is surrounded by the Little Hungarian Plain in the NW, by Lake Balaton in the SE and the micro-regions of the Balaton Basin (Balaton Riviera, Tapolca Basin), the Great Hungarian Plain (Mezőföld) and the Vértes Mountains in the NE (and Mór Graben with graben structure). Its elevation is 150–700 m and its area is 4300 km^2. Its largest area is the Bakony Mountains (2200 km^2). Parts of the Bakony Region are the Keszthely Mountains, Northern Bakony, Southern Bakony, Balaton Uplands and Bakonyalja (Figure 1).

The mountains are of horst structure, where more elevated horsts are mountains, while less elevated horsts are basins. They are mainly built up of Triassic dolomite which is significantly widespread on the surface too [16]. However, Triassic, Dachstein and Jurassic limestones frequently occur in great extension on the horsts of higher elevation. Cretaceous and Eocene limestones of smaller thickness can mainly be found in larger and smaller areas on horsts with medium elevation, mostly covered with loess. Horsts of lower elevation (basins) are mostly covered by non-karstic rocks (Middle Oligocene–Lower Miocene gravels, Figure 2).

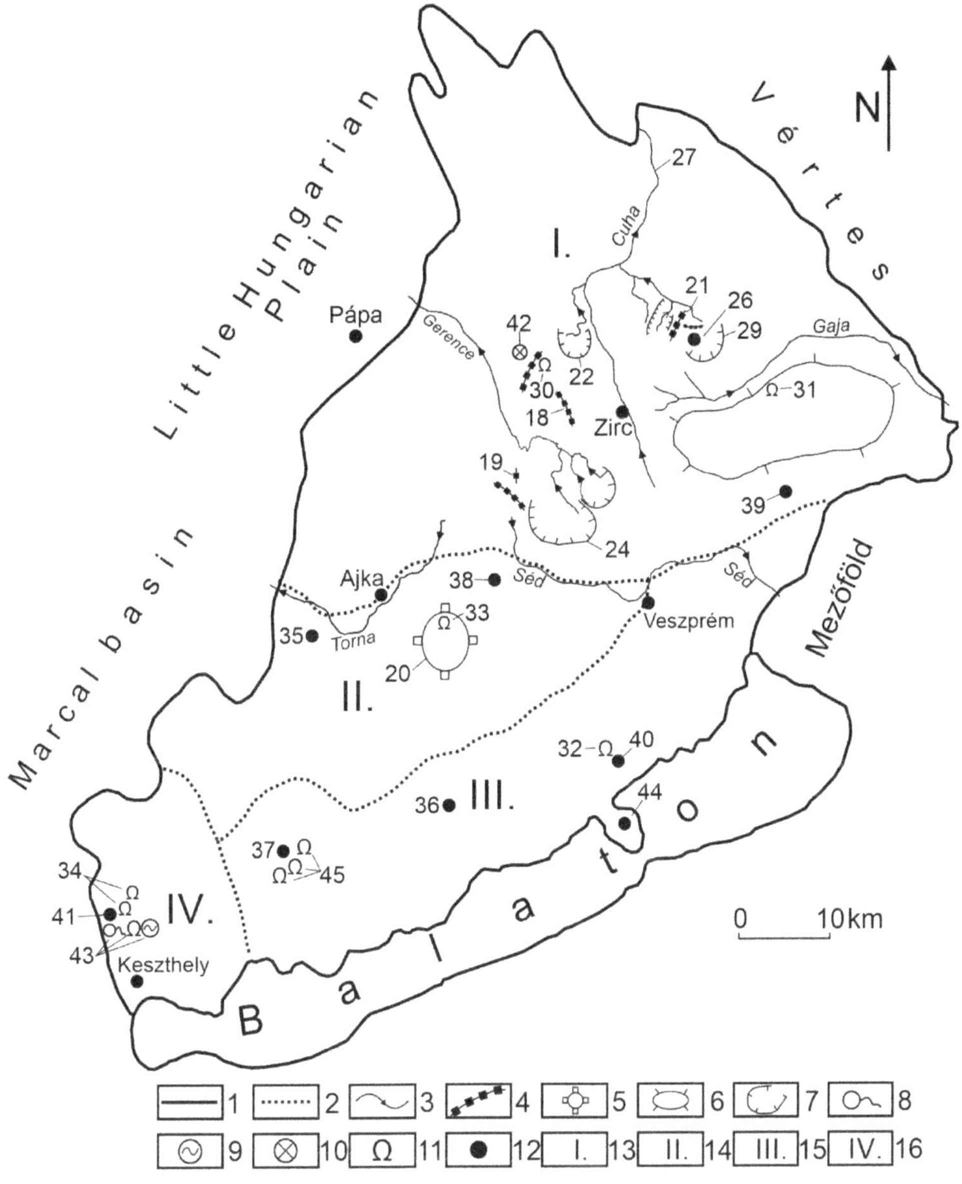

Figure 1. Parts of the Bakony Region. Legend: 1. boundary of mountains, 2. boundary of micro-region group, 3. stream, 4. horst, 5. basaltic horst, 6. plateau, 7. basin, 8. karst spring, 9. karst lake, 10. doline group, 11. cave, 12. settlement, 13. Northern Bakony, 14. Southern Bakony, 15. Balaton Uplands, 16. Keszthely Mountains, 17. Kőris Mountain, 18. Som Mountain, 19. Mester-Hajag, 20. Kab Mountain, 21. Sűrű mountain group, 22. Porva basin, 23. Lókút basin, 24. Hárskút basin, 25. Tapolca basin, 26. Dudar basin, 27. Cuha valley, 28. Gerence valley, 29. Ördög valley, 30. Ördög-lik of Kőris Mountain, 31. Alba-Regia cave, 32. Lóczy cave, 33. Öreg-köves inflow cave, 34. Cserszegtomaj caves, 35. Devecser, 36. Szentbékálla, 37. Tapolca, 38. Márkó, 39. Hajmáskér, 40. Balatonfüred, 41. Hévíz, 42. Eleven-Förtés doline group, 43. Lake Hévíz and its spring cave, 44. Tihany Peninsula, 45. Tapolca cave system.

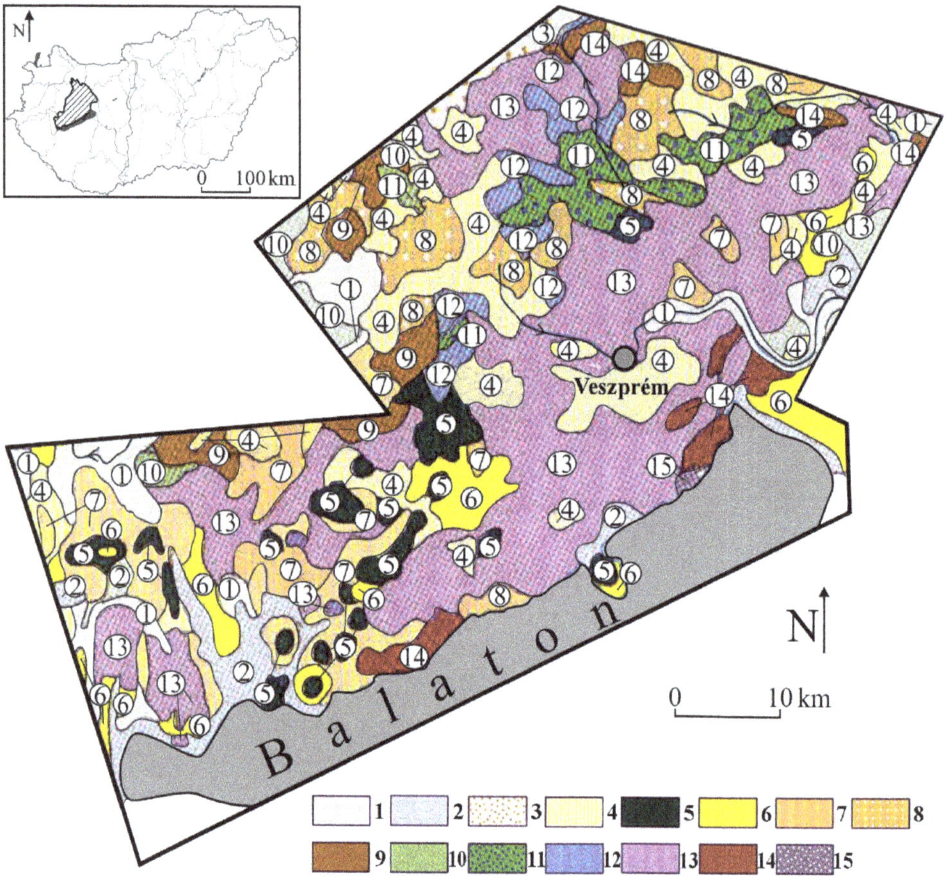

Figure 2. Geological map of the Bakony Region [17]. Legend: 1. fluvial sediment, 2. Holocene peat, 3. Holocene wind-blown sand, 4. Pleistocene loess, 5. Pliocene basalt, 6. Upper-Miocene freshwater limestone, 7. Middle Miocene limestone, 8. Oligocene gravel, 9. Eocene limestone, 10. Upper Cretaceous limestone, 11. Lower Cretaceous limestone, 12. Jurassic limestone, 13. Triassic limestone, dolomite, 14. Permian sandstone, 15. old Paleozoic phyllite.

The uniformly developed karstwater reserves of the mountains (regional groundwater, Figure 3), mostly stored in the main dolomite and preceding karstwater; pumping it resurged not only in karst springs, but also fed karstic swamps and lakes (Lake Balaton) [18,19]. However, it is also transmitted into the basin sediments of the environs [20,21] (Figure 3).

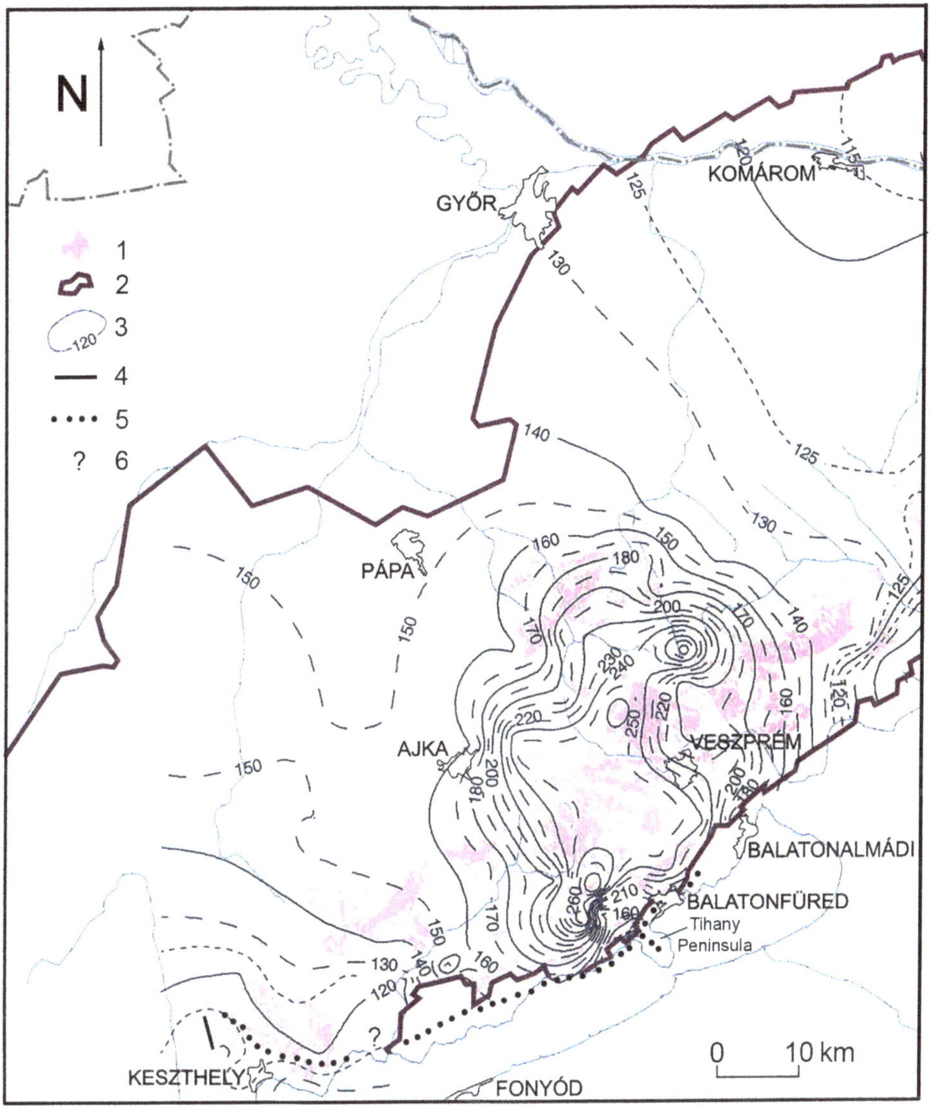

Figure 3. Karstwater map of the Bakony Mountains [21]. Regional groundwater of the mountains according to the state of January 2006. Legend: 1. infiltration area, 2. boundary of groundwater storage, 3. altitude of the isoline of groundwater level, 4. hypogene branch, 5. former hypogene branch, 6. termination boundary is not known.

In the Bakony Region, where impermeable intercalations are mainly in the Middle-Eocene limestone (Szőc Limestone Formation) and in Cretaceous limestones, perched water tables developed in some horsts and horst groups [22,23]. Particularly, the percolating water of the gorges of the Bakony Mountains [22] may significantly feed the perched water tables, however, at sites where the aquifuge is absent or already removed by erosion, it feeds the regional groundwater [24].

As the horsts of the mountains performed oscillating movements since the Late Cretaceous period, Pécsi [25] put them into different genetic types (Figure 4). He distinguished cryptopeneplain (up to 300 m elevation, it is covered with non-karstic impermeable rocks, morphologically basin), low-threshold surface (up to 300 m, its surface is built up of Triassic carbonates), horst elevated to summit position (surface elevation is 400–550 m, constituting rocks are Cretaceous and Eocene limestones, on which mainly permeable loess accumulations are found, morphologically mountain), horst in summit position (surface elevation is 600–700 m, Triassic and Jurassic carbonates are widespread at its surface, cover only occurs in patches) and horst type with basalt cover (basalt caprock and widespread loess cover).

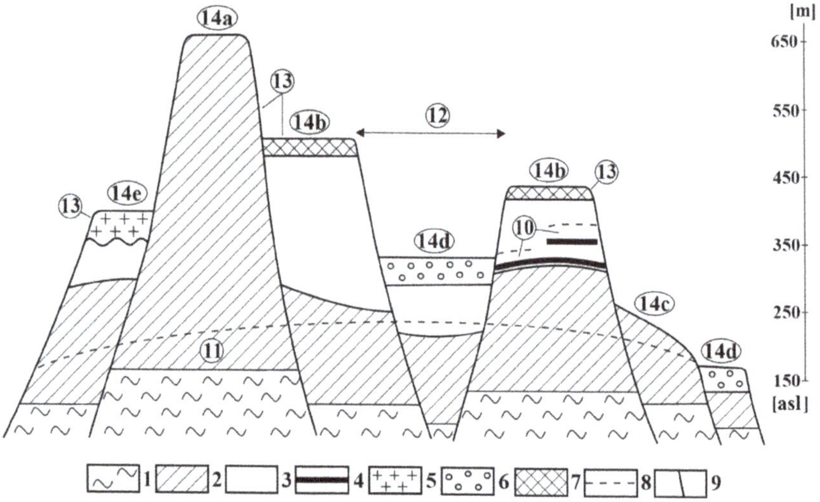

Figure 4. Horst types of the mountains [26]. 1. non-karstic rock, 2. Triassic carbonates, 3. Jurassic, Cretaceous and Eocene limestones, 4. impermeable intercalation, clay, marly limestone, silica, etc., 5. basalt, 6. gravel, 7. loess, 8. karst water table, 9. fault, 10. karst water storey, 11. main karst water, 12. basin, graben, 13. mount, 14a. horst in summit position, 14b. horst elevated to summit position, 14c. threshold surface, 14d. cryptopeneplain, 14e. horst covered with basalt.

3. Factors Influencing Its Karstification

In the Bakony Region, two flow systems are present, a regional and a local. At the ascending branches of regional flows, hypogene karst systems are present, while at the descending branches of regional flows and at local flows, epigene karst systems are present. The main constituent of regional flow is the regional groundwater, which is not exclusively fed by descending waters, but also by the water of the perched water table, which percolates along fractures and faults. The regional flow with a NW–SE direction partly resurges in the karst springs of mountain margins, and, to a larger extent, it arrives at the sediment and at carbonate rocks on the floor of surrounding basins. In the SW, the former hypogene branch extended from the Keszthely Mountains nearly to the NE termination of Lake Balaton (Figure 3). Members of this branch are the Cserszegtomaj Caves [27,28] and the Lóczy Cave [29,30]. The hypogene branch still exists at Lake Hévíz, which is represented by the water of the spring cave of Lake Hévíz with a temperature of 40 °C [31].

Local flow occurs at sites where the limestones constituting the horsts are interbedded with impermeable or partly impermeable intercalations (marl, chert, clay). Such local flow is particularly characteristic of the Cretaceous limestone horsts of Northern Bakony (for example on Mester-Hajag Figure 5).

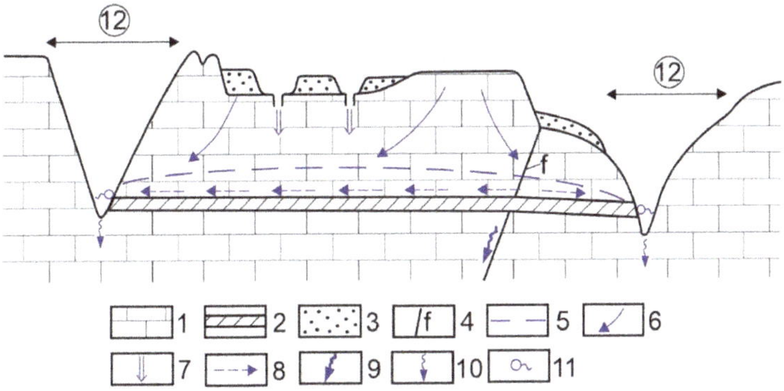

Figure 5. Water flow system of Mester-Hajag: Legend: 1. limestone, 2. impermeable intercalation, 3. permeable cover, 4. fault, 5. karstwater level, 6. infiltration, 7. local water inflow, 8. water motion above aquifuge, 9. water drainage along fault, 10. linear seepage, 11. spring, 12. epigenetic valley.

As has already been mentioned, the mountains are separated into horsts of different elevations, expansions and evolutions [21]. The distribution of horst types in the mountains is described in Figure 6.

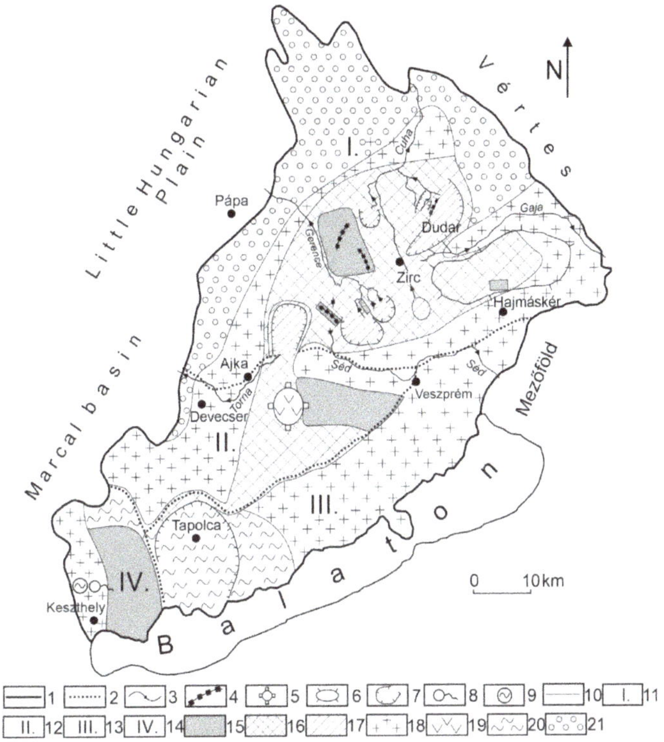

Figure 6. Distribution of horst types in the Bakony Region. Legend: 1. boundary of the mountains, 2. boundary of micro-region group, 3. stream, 4. horst, 5. basaltic horst, 6. plateau, 7. basin, 8. karst spring, 9. karst lake, 10. boundary of horst type, 11. Northern Bakony, 12. Southern Bakony, 13. Balaton Uplands, 14. Keszthely Mountains, 15. horst in summit position, 16. horst elevated to summit position, 17. cryptopeneplain, 18. threshold surface, 19. horst with basalt cap, 20. buttes with basalt cap in karstic and non-karstic environments, 21. hilly country covered with superficial deposit (fluvial sediment, dune sand).

The majority of the mountains are covered by impermeable or permeable non-karstic sediment (mainly loess). The caprock thins out in some places (at the mounds of the bedrock or at creeks) as compared to its environment. On horsts covered with impermeable beds, epigenetic valleys of different stages of maturity emerged. These valleys may also have permanent streams at sites where a horst is surrounded by terrains covered with impermeable beds. As a result of the elevated position of the horst relative to its environment of covered terrain, some parts of these valleys are of epigenetic-antecedent origin and constitute gorges. On terrains covered with impermeable beds (for example on Kab Hill which is covered with basalt), mixed allogenic–autogenic karst developed.

The bedrock, therefore, became karstified and dissected into mounds and depressions. The mounds of Cretaceous limestones are partly exhumed. On the terrains between them, the caprock thinned out (Mester-Hajag). However, some parts of threshold surfaces that are uncovered and have a low elevation (below 300 m) are also dissected by mounds.

4. Materials and Methods

Based on geological maps, topographic maps and literary data, the areas of the mountains were classified as horst development types. Field studies and karstmorphological mappings were made in different areas of the mountains and types were determined with the help of their results, features and feature assemblages belonging to different karst types. Thus, there was an opportunity to describe the karst of certain horst types as well as to compare the different karstifications of horst types.

5. Discussion

The karst features are subsoil karren (these are grike and rootkarren on sandstone pseudokarren), subsidence dolines or sinkholes (mainly suffosion dolines), drawdown dolines, collapse dolines, caprock dolines, ponors with blind valleys, cave-ins, spring caves reflecting a thermal effect of various degrees, shafts, caves with wreathing, pseudokarst caves, freshwater limestones (in brook channels) and spring cones.

Among karst features, subsidence dolines are the most widespread. They occur in patches (altogether, in 22 patches, out of which, 19 patches are in the Northern Bakony) and in small numbers (there are about 700 subsidence dolines). Their size and density is small and a lot of them are inactive filled dolines. They occur at sites where the horst is covered with permeable sediment and where the caprock thinned out or was thin in the first place. These sites are the floors of creeks, terrains between exposed mounds, the covered mounds of the bedrock and limestone sites veneered with loess at which a larger quantity of water arrives from the surrounding impermeable rock or from the impermeable and wedging caprock [26] (Figure 7). On its soil-covered karst, some drawdown dolines and collapse dolines also occur (Tapolca Karst and the environs of Devecser). On the mixed allogenic–autogenic karst, ponors with blind valleys occur at the termination of the basalt cap (Figure 8). Caprock dolines also developed on the basalt by the collapse of the rock [32].

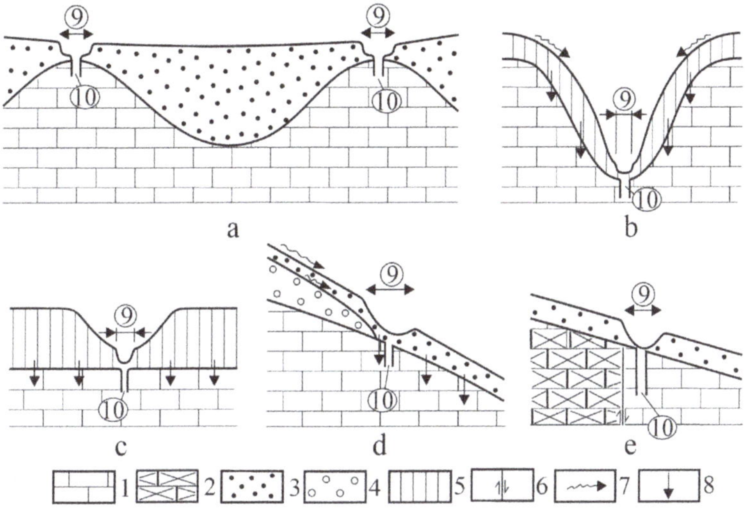

Figure 7. Development of subsidence dolines [33]. Legend: Development of subsidence dolines at various development environments: (**a**) above the mound of the bedrock at low accumulation cover thickness, (**b**,**c**) at cover that thinned out in an erosional way, (**d**) at the termination of covered impermeable cover, (**e**) at the margin of covered, intercalated non-karstic rock, 1. limestone, 2. non-karstic rock, 3. permeable cover, 4. impermeable cover, 5. reworked, partly impermeable cover, 6. fault, 7. water flow, 8. infiltration, 9. subsidence doline, 10. shaft.

Figure 8. Ponor of Macskalik cave (Kab Mountain).

Shaft caves are common and of relatively great dimension in the mountains (102 shaft caves) (Figure 9). Among them, there occur caves with significant length, for example, the Alba-Regia Cave which is on the Tés Plateau is 3.6 km long. Shafts open out of subsidence dolines and are of dissolution origin.

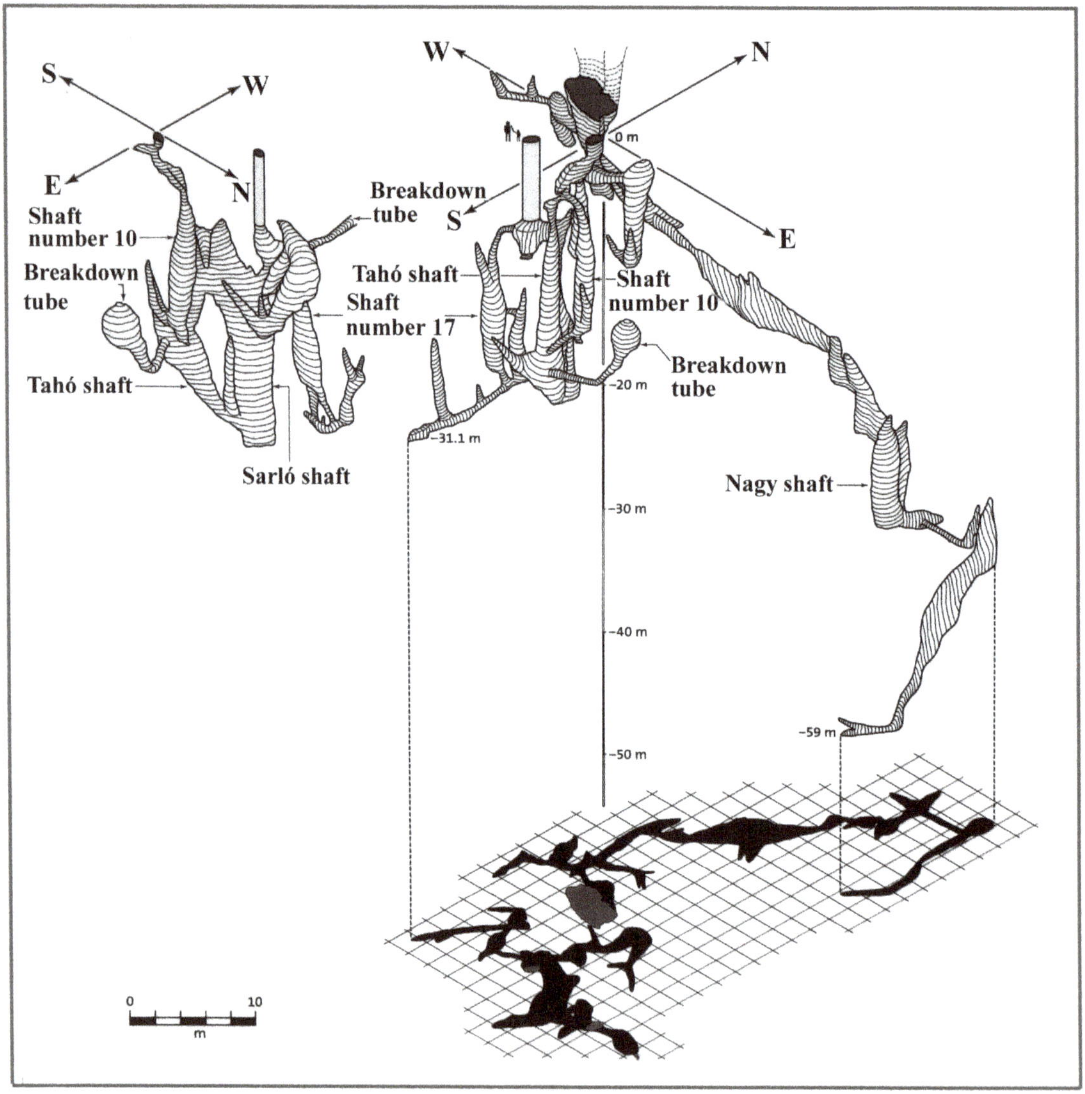

Figure 9. Öregköves ponor shaft cave [34].

Among their caves, cave-ins are widespread (about 225 caves), which mostly occur in epigenetic-antecedent valley sides (about 212 caves); there are few spring caves (about 10 caves), and they were affected by thermal effects of various degrees. They are situated at the hypogene branch of the regional flow. The pseudokarren of the sandstone are also related to a former hypogene branch (Szentbékálla, Figure 10). Here, the silica that precipitated from warm water solution cemented Pannonian sand [35]. Amorphous silica that developed by precipitation is dissolved much more intensively than crystallised silica [9,36,37]. Karren (mainly kamenitzas) resulted from the dissolution of amorphous silica [32].

Figure 10. Karren near Szentbékkálla. Legend: 1. sandstone polygon, 2. kamenitza, 3. rinnenkarren.

Pseudokarst caves are most widespread on basaltic terrains. From the ascending waters of the early hypogene branch, spring cones were built on the Tihany Peninsula (Figure 10) during the precipitation of dissolved limestone and from the diatom skeletons of lakes [38]. Materials of spring cones grew round and round space and thus, cavities developed (Figure 10).

The karst and karstification of the individual horst types are different, which can be traced back to the variations in geology, elevation, expansion, coveredness and hydrology of the horsts.

At the surface of horsts in the summit position (for example Kőris Mountain, Som Mountain), there are no local flows due to the distribution of Triassic Main Dolomite without impermeable intercalations, but descending branches of regional flows are present. Surface karstification is of low intensity; disregarding karren formation (on Dachstein limestones), it is only present at sites where sediment patches occur on the surface of such horsts.

The complete lack of surface karst features (drawdown dolines) can be traced back to the fact that in their area, the presence of low-inclined terrains is insignificant or they are completely absent. On steep terrains, no drawdown dolines develop. Based on literary data, according to Zámbó [39], at a surface with an inclination larger than 20°, no solution dolines occur. According to other data, doline development is the most intensive on surfaces with an inclination of 2–7° and there are no dolines on surfaces steeper than 13° in the Mecsek Mountains [40]. Telbisz et al. [41] state that in the Serbian Miroč Mountains, only 23% of surfaces with an inclination of 12° have dolines. All this can be traced back to the fact that on steep surfaces, the ratio of surface runoff increases and infiltration decreases. To our knowledge, there is only one subsidence doline group on this horst type (the Eleven-Förtés doline group on Kőris Mountain) where dolines developed on a terrain of low inclination, on a superficial deposit patch [42]. The infiltration and the survival of the superficial deposit was promoted by the fact that it is situated in a depression of the bedrock. Here, the Keszthely Mountains can also be mentioned, which are a lower elevated variety of these horst types (some mounds are at an altitude of 400–500 m), where 20 subsidence dolines occur [26]. There are also some inactive shafts on this horst type (for example, the Ördög-lik of Kőris Mountain). These were formed below subsidence dolines, but they became truncated after the denudation of the superficial deposit.

The epigenetic valleys of horsts deepened into the rock below the former regional groundwater level and exposed inactive, presently dry cavities (cave-ins, the medium section of Cuha Valley and the northern section of Gerence Valley).

In the mountains, the number and expansion of their horsts, which are horsts elevated to the summit position, is large. On these horsts, perched water tables overlie impermeable intercalations. Local flows developed at them, their descending branches are fed by the water of brooks percolating on valley floors, the water flowing into subsidence dolines that occur on the horst surface and by the infiltrating meteoric water. Waters of the outlet branch emerge in karst springs situated high above the regional groundwater level or they get into the regional groundwater along the fractures or faults that dissect the aquifuge (Figure 5). The subsidence dolines and cave-ins of the horst belong to the karst systems of local flows. Permeable superficial deposits are widespread on them and locally thin at several places for the already mentioned reasons.

The majority of subsidence dolines occur on horsts belonging to this type. As regards the horsts of Northern Bakony, subsidence dolines occur on 30% of horsts elevated to the summit position, while they can be found only on 9.1% of the horsts in the summit position and on 12.5% of cryptopeneplains [26]. These features developed at places where the cover is thin or thinned out (Figure 7). Since the shafts of the mountains occur below subsidence dolines, 90–95% of shafts can also be found on this horst type. Tés Plateau is a good example of the distribution of subsidence dolines and shafts, with 137 dolines and 46 shafts. However, the majority of cave-ins (estimated 80–90%) can also be found on the horsts of this type.

Cave-ins and cavities also occur subordinately in the sides of some epigenetic-regression valleys, but their majority and those longer than 1–2 m are exclusively in the walls of epigenetic-antecedent valley sections without exception. These valleys are formed on uplifting horsts which are surrounded by cryptopeneplain. The streams of the cryptopeneplain carve an antecedent valley section in the uplifting area of the horst. With their percolating water, streams generate a perched water table and trigger its cavity formation, then they deepen and open up the cavities at the top (inactive part) of the perched water table (cave-in). However, the perched water table can continue to accumulate if there is an aquifuge below the antecedent valley section. The Dudar Basin (cryptopeneplain) and the Sűrű hill group (horst elevated to summit position) constitute such a system where, for example, in the Ördög-árok there are more than 40 cave-ins of various sizes.

With their streams, cryptopeneplains have an important role in the hydrological and karstic development of the horsts in their environs. However, in the area of some cryptopeneplains (Hárskút Basin, Lókút Basin, Porva Basin), the material of the impermeable beds (Csatka Gravel Formation) has been partially eroded and the limestone outcrops became partially covered with loess. At these sites, for example, on the floor of epigenetic valleys, subsidence dolines, shafts and inactive shafts are common.

On the basaltic horst such as the Kab Hill, mixed allogenic–autogenic karst developed with ponors with blind valleys (Figure 8) and with inflow cave-like shafts (Figure 9). With the lack of erosion, they are of dissolution origin [32,43]. These features may have formed at the margin of the basalt cap or at the limestone outcrops within the basalt cap. On the basalt cover caprock, dolines are also common, which may develop into ponors with blind valleys [26], but on terrains with loess, subsidence dolines also occur.

On threshold surfaces, drawdown doline groups (Tapolca Karst, the environs of Devecser) are typical. The dolomite terrains of this horst type are dissected into mounds, but some dolines of small depth also occur on these surfaces (between Márkó and Hajmáskér).

Some spring caves can also be mentioned from this horst type. These are connected to a former (Lóczy Cave) or present (the spring cave of Lake Hévíz) hypogene branch, but cavities opened up due to anthropogenic activity can also be found (Cserszegtomaj Caves). In case of the latter case, as well as the spring cave of Hévíz and the Lóczy Cave in Balatonfüred, a hot water effect is dominant, but in the development of the Tapolca cave system, lukewarm waters and cold waters also played a role. The landscape of threshold surfaces also includes the spring cones (Tihany Peninsula, Figure 10) and pseudokarren (Szentbékkálla, Figure 11).

Figure 11. (**A**) spring cone, (**B**) exposed cavity (Tihany Peninsula).

In Hungarian karst literature, both the Transdanubian Mountains with the Bakony Region (which it is a part of) together are regarded as an independent karst type [44,45]. The Transdanubian Mountains are called Transdanubian type, while the other karst areas are called Aggtelek type. A more appropriate term is character instead of type. (Characteristics of karsts of Aggtelek character coincide with those of temperate soil-covered karst.) The difference may be traced back to the fact that the Transdanubian Mountains are poor in karst features, but hypogene caves are characteristic [46] and erosion caves and solution dolines are subordinate (they are absent). In contrast to the karsts of Transdanubian character, erosion caves, rootkarren and solution dolines are typical of the karsts of Aggtelek character [44]. However, morphological characteristics have become more accurate as compared to an earlier classification (for example rootkarren can also be found in the Transdanubian Mountains); basically, this differentiation is substantial.

The karst features of the Bakony Region are not widespread uniformly, but in patches of variable size. Although it is the most intensively karstified part of the Transdanubian Mountains, the frequency and size of their karst features is limited. The above characteristic feature can be traced back to its separation into horsts and to recent uplift. As a result of the former, karstification sites are localised (and thus, for example no ponors with significant catchment area developed), while, as a result of the latter karstification being young, there was little time for the development of karst features.

Karstic patches are constituted by subsidence dolines because this doline type is related to permeable superficial deposit patches, and the water infiltrated into the caprock ensures abundant (the water does not flow down low-inclined surfaces), permanent and uniform water supply to the epikarst (the already mentioned low resistances are evidence for this). This favours cavity formation and thus, shaft development in the epikarst. Abundant infiltration ensures favourable conditions for the transportation of superficial deposits into the karst. At sites where impermeable beds occur, neither subsidence dolines, nor other karst features occur. However, the small size of the horsts also results in a patchy appearance. On the other hand, the occurrence of solution dolines, a dominant variety of temperate soil-covered karst, is subordinate. All this results from the fact that, as has already been mentioned, the extension of low-inclined terrains is restricted in its uncovered terrain. In spite of the position of threshold surfaces at low elevation, this doline type occurs on them since their surface is low-inclined and hardly dissected.

Impermeable caprock is widespread in the mountain region. If it is the material of the Csatka Gravel Formation, a stream network develops on it because it is of clayey composition (cryptopeneplain). Their stream and valleys stretch across the surrounding horsts as epigenetic-regression valleys, on the floor of which, significant percolation may take place. However, percolation sites are not ponors. It may occur that the gravel cover already terminates in the area of the horst and its flowing water arrives at the margin of the cover. However, since the cover is of small extension, not even valleys develop on its surface. Waters arriving at the termination of the gravel cover form subsidence dolines on the loess surface that interacts with the cover. Ponors with blind valleys only developed at the termination of the basalt cover of the Kab Hill. Therefore, ponors with blind valleys are subordinate and inflow caves that constitute the continuation of such features are absent in the mountains. Concretions are subordinate either in the caves or on the surface, while they are missing from the poljes.

As mentioned above, the most widespread cave type in the mountain region is the cave-in. These are widespread in the sides of epigenetic-antecedent gorges. Their evolution contributed to the erosion of the impermeable beds (along which phreatic cavity formation took place) and of the stream since the area of the surrounding cryptopeneplain ensured abundant water supply to this.

Shafts are also characteristic; their frequency is comparable to that on other temperate karsts. The reason for this is the relatively high number of subsidence dolines. Subsidence doline development and shaft development interact. Shafts enable the local transportation of superficial deposits into the karst and thus, the development of indentations, while the developing subsidence dolines collect surface waters and transmit them into the epikarst. This favours the transformation of epikarst cavities into shafts.

There are no erosional inflow caves in the mountain region. The development of such caves can only be expected on Kab Hill, but the low inclination of the basaltic terrain and the lack of gravel do not favour erosion. Spring caves are also absent except for some which were transformed by a warm water effect. The reason for this is that there are only some karst springs in the mountains since their karst water, as mentioned above, flows in the surrounding basins [20], and the water emerged gallery-like and not point-like before karstwater extractions [18]

The stability of the surface karst of the mountains is also promoted by the fact that there has been a land use change and return to forestry and grazing in the past decades. These cultivation methods decrease the transport of superficial deposits into karst features. Thus, these features become filled to a lesser degree, or, if they do become filled, it takes place by natural processes.

Intensive mining activity took place in the mountains which could only be practiced safely by artificially sinking the karstwater level. In the past decades, the karstwater level has been rising and gradually reaching its former elevation. Thus, the operation of springs and spas (including medicinal waters), which are fed by the karstwater of the mountains, is ensured in the long term.

6. Conclusions

The karst of the mountains can primarily be related to various local flows, at which epikarst systems developed.

The karstification of horst types and thus their karst features are different. The karstification of horsts in the summit position is not significant since the water leaves the area of these horsts, and there is no permeable cover, or it hardly ever occurs in their area. The karstification of horsts elevated to the summit position is diverse where subsidence dolines are predominant because the permeable cover is widespread. The development of cave-ins was enabled by the impermeable intercalations of the horst and intensive valley deepening.

Cryptopeneplains only karstify at sites where their impermeable cover was denuded. Streams starting from these terrains cause gorge development on horsts elevated to the summit position and their cavity formation. Low-inclined surfaces of mountain marginal threshold surfaces favoured the development of some solution dolines and spring caves. Features reflecting a hot water effect (spring caves, sandstone karren, spring cones) also occur on this horst type. On the basaltic Kab Hill, ponor with blind valleys also developed.

Predominant karst features of the mountains are subsidence dolines, cave-ins and shafts. One of the important directions of future research may be the study of the epikarst, which helps obtain a better understanding of the development or lack of some karst features on different horst types.

Funding: This research received no external funding.

Data Availability Statement: No new data were created or analyzed in this study. Data sharing is not applicable to this article.

Conflicts of Interest: The author declare no conflict of interest.

References

1. Denizman, C. Caves, sinkholes, and fractures in the eogenetic karst of Florida, a GIS-based spatial analysis. *Acta Carsologica* **2022**, *51*, 33–46. [CrossRef]
2. Fernandes Ferreira, C.; Hussain, Y.; Uagoda, R. A semi-automatic approach for doline mapping in Brazilian covered karst: The way forward to vulnerability assessment. *Acta Carsologica* **2022**, *51*, 19–31. [CrossRef]
3. Jones, B. Karst on the Cayman Islands. In *Geology of the Cayman Islands*; Springer: Cham, Switzerland, 2022. [CrossRef]
4. Durán-Valsero, J.J.; Pardo-Igúzquiza, E.; Morales-García, R.; Luque-Espinar, J.A.; Durán-Laforet, S.R.; Balard, D.; Quiroga, E.; Borrás, J.; Ferreres, J. The Importance of Snow in the Hydrogeology of a High Relief Karst System: Sierra De Tendeñera, in the Pyrenees Mountain Range (Huesca, Northern Spain). In *EuroKarst 2022, Málaga*; Advances in Karst Science; Andreo, B., Barberá, J.A., Durán-Valsero, J.J., Gil-Márquez, J.M., Mudarra, M., Eds.; Springer: Cham, Swizterland, 2023. [CrossRef]
5. Pardo-Igúzquiza, E.; Gil-Márquez, J.M.; Mudarra, M.; Andreo, B.; Durán-Valsero, J.J. Morphometric Comparison of Dolines in Three Karst Landscapes Developed on Different Lithologies. In *EuroKarst 2022, Málaga*; Advances in Karst, Science; Andreo, B., Barberá, J.A., Durán-Valsero, J.J., Gil-Márquez, J.M., Mudarra, M., Eds.; Springer: Cham, Switzerland, 2023. [CrossRef]
6. Romano, G.; De Girolamo, M.; Parise, M. Geophysical Researches to Detect Karst Caves in the Main Polje of Apulia. In *EuroKarst 2022, Málaga*; Advances in Karst Science; Andreo, B., Barberá, J.A., Durán-Valsero, J.J., Gil-Márquez, J.M., Mudarra, M., Eds.; Springer: Cham, Switzerland, 2023. [CrossRef]
7. Sweeting, M.M. *Karst Landforms*; Columbia University Press: New York, NY, USA, 1973; 362p.
8. White, W.B. *Geomorphology and Hydrology of Karst Terrains*; Oxford University Press: New York, NY, USA; Oxford, UK, 1988; 463p.
9. Waltham, T.; Bell, F.; Culshaw, M. *Sinkholes and Subsidence*; Springer: Berlin/Heidelberg, Germany, 2005; 382p.
10. Klimchouk, A.B. *Hypogene Speleogenesis: Hydrogeological and Morphogenetic Parspective*; Special Paper No 1; National Cave and Karst Research Institute: Carlsbad, NM, USA, 2007; 106p.
11. Ford, D.C.; Williams, P.W. *Karst Hydrogeology and Geomorphology*; John Wiley & Sons: Chichester, UK, 2007; 561p.
12. Goldscheider, N.; Mádl-Szőnyi, J.; Erőss, A.; Schill, E. Review: Thermal water resources in carbonate rock aquifers. *Hidrogeol. J.* **2010**, *18*, 1303–1318. [CrossRef]
13. Gutierrez, F.; Parise, M.; De Waele, J.; Jourde, H. A review on natural and human-induced geohazards and impacts in karst. *Earth Sci. Rev.* **2014**, *138*, 61–88. [CrossRef]
14. Veress, M. *Covered Karst*; Springer: Dordrecht, The Netherlands, 2016; 536p. [CrossRef]
15. Veress, M. The Effect of Mining Activities on the Paleokarstic Features, Recent Karstic Features, and Karst Water of the Bakony Region (Hungary). *Mining* **2022**, *2*, 757–768. [CrossRef]
16. Kercsmár, Z.; Selmeczi, I.; Budai, T.; Less, G.; Konrád, G. Geology of the karst terrain sin Hungary. In *Cave and Karst Systems of Hungary*; Veress, M., Leél-Őssy, S., Eds.; Springer: Berlin/Heidelberg, Germany, 2022.
17. Haas, J. Földtan (Geology). In *Magyarország Térképekben*; Kocsis, K., Schweitzer, F., Eds.; MTA, Földrajztudományi Kutatóintézet: Budapest, Hungary, 2011; pp. 35–38. (In Hungarian)
18. Szádeczky-Kardos, E. A Dunántúli középhegység karsztvízének néhány problémájáról (On some problems of the karst water of the Transdanubian Mountains). *Hidrológiai Közlöny* **1941**, *21*, 67–92. (In Hungarian)
19. Papp, F. Dunántúl karsztvizei és a feltárás lehetősége Budapesten (The karstwaters of Transdanubia and the opportunity of exploration in Budapest). *Hidrológiai Közlöny* **1942**, *21*, 146–196. (In Hungarian)
20. Mádl-Szőnyi, J.; Erőss, A.; Csondor, K.; Iván, V.; Tóth, Á. Hydrogeology of the Karst regions in Hungary. In *Cave and Karst Systems of Hungary*; Veress, M., Leél-Őssy, S., Eds.; Springer: Cham, Switzerland, 2022; pp. 137–160.
21. Csepregi, A. *A karsztvíz kiemelés hatása a Dunántúli-Középhegység vízháztartására. (The effect of karst water extraction on the water balance of the Transdanubian Mountains) In Bányászati Karsztvízszint Süllyesztés a Dunántúli-Középhegységben*; Alföldi, L., Kapolyi, L., Eds.; MTA Földrajztudományi Kutatóintézet: Budapest, Hungary, 2007; pp. 77–112. (In Hungarian)

22. Jaskó, J. A balatonfelvidéki és északbakonyi patakok vízhozamának kapcsolata a földtani felépítéssel (The relation of discharge of the streams of Balaton Highlands and Northern Bakony with geological structure). *Hidrológiai Közlöny* **1961**, *41*, 75–84. (In Hungarian)
23. Veress, M. *Covered Karst Evolution Northern Bakony Mountains, W-Hungary*; A Bakony Természettud. Kut. Eredményei 23; Bakonyi Természettudományi Múzeum: Zirc, Hungary, 2000; 167p.
24. Bratán, M.; Mohos, P.; Zsuffa, I. A Gerence-patak hidrológiai tanulmánya (The hydrological study of Gerence stream). *Hidrológiai Közlöny* **1967**, *47*, 284–296. (In Hungarian)
25. Bulla, B. *Magyarország Természeti Földrajza*; Physical Geography of Hungary; Tankönyvkiadó: Budapest, Hungary, 1964; 424p. (In Hungarian)
26. Pécsi, M. A Pannóniai-medence morfogenetikája. (The Morphogenesis of the Pannonian basin). *Földrajzi Értesítő* **1980**, *29*, 105–127. (In Hungarian)
27. Veress, M. The surface morphology of karsts in Hungary. In *Cave and Karst Systems of Hungary*; Veress, M., Leél-Őssy, S., Eds.; Springer: Cham, Switzerland, 2022; pp. 179–248.
28. Leél-Őssy, S. Karszt és barlangkutatás a Szalonnai-karszton. (Karst and cave research on Szalonna karst). *Hidrológiai Közlöny* **1953**, *33*, 67–70. (In Hungarian)
29. Kárpát, J. A Cserszegtomaji-kútbarlang (The well cave of Cserszegtomaj). *Karszt és Barlang* **1982**, *1*, 35–40. (In Hungarian)
30. Leél-Őssy, S. Karsztmorfológiai vizsgálatok a balatonfüredi Lóczy-barlangban és környékén (Karstmorphological investigations in the Lóczy cave in Balatonfüred and in its environs). *Földrajzi Értesítő* **1958**, *7*, 379–381. (In Hungarian)
31. Láng, S. A Bakony geomorfológiai képe (The geomorphology of the Bakony Mountains). *Földrajzi Közlemények* (In Hungarian). **1958**, *82*, 324–346.
32. Plózer, I. A Hévízi-tó forrásbarlangjának feltárása (Exploration of the spring caves of Lake Hévíz). *Karszt és Barlang* **1977**, *1–2*, 65–66. (In Hungarian)
33. Veress, M.; Unger, Z. Kab Mountain: Karst Under a Basalt Cap. In *Landscapes and Landforms of Hungary*; Lóczy, D., Ed.; Springer: Cham, Switzerland, 2015; pp. 55–62. [CrossRef]
34. Veress, M. Development Environments and Factors of Subsidence Dolines. *Geosciences* **2021**, *11*, 513. [CrossRef]
35. Győrffy, D. Geomorfológiai tanulmányok a Káli-medencében. (Geomorphological studies in the Káli basin). *Földrajzi Értesítő* **1957**, *6*, 265–299. (In Hungarian)
36. Siffert, B. *Quelques Reactions de la Silice en Solution: La Formation des Argiles*; Memories du Service de la Carte: Paris, France, 1962.
37. Siever, R. Silica solubility 0–200 °C and diagenesis of siliceous sediments. *J. Geol.* **1969**, *70*, 35–46. [CrossRef]
38. Kovács-Pálffy, P.; Kalmár, J.; Veress, M. A Tihanyi-félsziget algás eredetű forrásüledékeinek ásvány-kőzettani vizsgálata. (Mineralogical-petrographic analyses of spring deposits of algal origin on the Tihany Peninsula). *Karsztfejlődés* **2007**, *7*, 111–135. (In Hungarian)
39. Zámbó, L. Az Aggteleki-karszt felszínalaktani jellemzése. (Geomorphological description of Aggtelek Karst). *Földrajzi Értesítő* **1998**, *47*, 359–378. (In Hungarian)
40. Lippmann, L.; Kiss, K.; Móga, J. Az Abaliget-Orfűi karszt karsztos felszínformáinak vizsgálata térinformatikai módszerekkel (Investigation of the karstic phenomenon near Orfű and Abaliget by GIS methods). *Karsztfejlődés* **2008**, *13*, 151–166. (In Hungarian)
41. Telbisz, T.; Mari, L.; Kohán, B.; Čalić, J. A szerbiai Miroč-hegység töbreinek térinformatikai GPS-es terepi vizsgálata (A spatial informatics GPS study of the dolines of the Serbian Miroč Mountains). *Karsztfejlődés* **2007**, *12*, 71–90. (In Hungarian)
42. Veress, M.; Puskás, J. Adalékok az Eleven-Förtési töbörcsoport (Bakony-hegység) karsztosodásához (Contributions to the karstification of the eleven-Förtés doline group, Bakony Mountains). *Karsztfejlődés* **2007**, *12*, 171–193. (In Hungarian)
43. Móga, J.; Németh, R. The morphological research of the basalt and loess covered plateaus in the Bakony Mts. (Transdanubian middle mts.-Hungary). *Acta Carsologica* **2005**, *34*, 397–414.
44. Jakucs, L. A magyarországi karsztok fejlődéstörténeti típusai (Evolutionary types of Hungarian karsts). *Karszt és Barlang* **1977**, *1–2*, 1–16. (In Hungarian)
45. Hevesi, A. Magyarország karsztvidékeinek kialakulása és formakincse, I. rész (Development and landforms of karst regions in Hungary I). *Földrajzi Közlemények* **1991**, *15*, 25–35. (In Hungarian)
46. Leél-Őssy, S.; Virág, M. The Caves of the Northeastern Part of the Transdanubian Mountains. In *Cave and Karst Systems of Hungary*; Cave and Karst Systems of the World; Veress, M., Leél-Őssy, S., Eds.; Springer: Cham, Switzerland, 2022. [CrossRef]

land

Article

Factors Influencing Four Decades of Forest Change in Guizhou Province, China

Xiaona Guo [1], Ruishan Chen [1,*], Michael E. Meadows [2,3], Qiang Li [1,4], Zilong Xia [3] and Zhenzhen Pan [5]

1 School of Design, Shanghai Jiao Tong University, Shanghai 200240, China
2 Department of Environmental and Geographical Science, University of Cape Town, Cape Town 7701, South Africa
3 School of Geography and Ocean Sciences, Nanjing University, Nanjing 210023, China
4 School of Agriculture and Biology, Shanghai Jiao Tong University, Shanghai 200240, China
5 School of Life Science and Technology, Central South University of Forestry and Technology, Changsha 410004, China
* Correspondence: chenrsh04@gmail.com

Abstract: Globally, the loss of forest vegetation is a significant concern due to the crucial roles that forests play in the Earth's system, including the provision of ecosystem services, participation in biogeochemical cycles, and support for human well-being. Forests are especially critical in mountains environments, where deforestation can lead to accelerated biodiversity loss, soil erosion, flooding, and reduced agricultural productivity, as well as increased poverty rates. In response to these problems, China has implemented a series of ecological restoration programs aimed at restoring forests. However, there is a lack of knowledge as to whether the forest cover is increasing or decreasing, as well as the relative roles played by natural and human factors in forest change. Here, we aim to address these issues by analyzing the pattern and process of the forest changes in Guizhou province, a typical mountainous karst area with a fragile environment in southwestern China, between 1980 and 2018, and evaluating the extent to which these forest changes were influenced by natural and anthropogenic driving forces. Using a temporal sequence of satellite images and a Markov model, we found that the forest cover increased by 468 km^2, and that over 33% of the cropland in Guizhou province was converted into forest between 1980 and 2018, with the most significant increases in the forest cover occurring in Qiandongnan. Through correlation analyses and generalized linear model (GLM) regression, we demonstrate that management factors exerted a more significant positive impact on the forest cover than climate change. While the mean annual precipitation and temperature were mostly stable during the period studied, the effects of population and gross domestic product (GDP) on the forest changes weakened, and the influence of land-use change markedly increased. These findings provide valuable information for resource managers engaging in forest protection, deforestation prevention, and ecological restoration in similar regions.

Keywords: factors; forest change; Guizhou

Citation: Guo, X.; Chen, R.; Meadows, M.E.; Li, Q.; Xia, Z.; Pan, Z. Factors Influencing Four Decades of Forest Change in Guizhou Province, China. *Land* **2023**, *12*, 1004. https://doi.org/10.3390/land12051004

Academic Editors: Xiaoyong Bai and Adrianos Retalis

Received: 29 March 2023
Revised: 15 April 2023
Accepted: 27 April 2023
Published: 3 May 2023

1. Introduction

Globally, forest loss due to plantation forestry, agriculture, mining-related wildfires, and urbanization has enormous implications, particularly for climate change and biodiversity. As a result, governments, conservationists, and even private corporations are engaged in efforts to curb these losses and promote forest recovery [1–4]. In China, a series of ecological restoration programs have been implemented at the national, regional, and local scales over the past several decades, including the Grain for Green Program (1999–2020) and the Rocky Desertification Treatment Program (2008–2020) [5]. These interventions have greatly improved the sustainability of China's land systems, with the rate of forest cover increasing from 8.6% in 1949 to 23.04% in 2020 [6]. Substantial forest recovery has been detected through remote-sensing imagery, revealing an overall increase in greening since

2000, most notably in China and India [7]. This trend is particularly prominent in certain provinces, including Guizhou province, where the forest cover increased from 11.98% in 1949 to 61.5% in 2020 [8]. Ecological restoration interventions have significantly increased the vegetation growth and carbon stock in China more generally [9,10]. It is clear that continuous and long-term ecological restoration projects can, among other benefits, help forests accumulate nutrients [11], and that embracing the implications of restoration interventions can contribute to the United Nation's Sustainable Development Goals. It is the interaction between the natural environmental and socio-economic factors that determines forest dynamics, including recovery. Natural factors include those related to soil [12] and climate, especially the mean annual temperature and rainfall [13–15]. However, the spatial and temporal aspects of forest change in remote and environmentally fragile regions are not fully understood, and the trajectory of the forest changes in China as a whole is still subject to debate. While some locations have undergone 'greening,' others remain subject to forest clearance [16–18]. Therefore, it is important to establish the details of recent trends in forest cover, and their driving forces, especially in environmentally vulnerable regions, such as the karst area of southwestern China, which has historically endured significant levels of rocky desertification [19].

Guizhou province has a total area of 176,167 km^2, of which 92.5% is hilly and 61.9% is karst [20], and it is considered to be among the most environmentally vulnerable regions in China. Karst topsoil is typically shallow, so if the forest vegetation is cleared, it is highly susceptible to erosion [21] and produces a particular form of land degradation, known as rocky desertification, which, in turn affects regional socioeconomic development. This has led Guizhou to become the least developed province in China [22]. China has responded to this land-degradation crisis in the karst region of its southwestern area, including Guizhou province, through an integrated portfolio of ecosystem-restoration programs since the 1980s. A number of previous studies described rocky desertification and associated spatio-temporal variations in land-use change, the mechanisms underlying these processes, and restoration responses [23–25]. Accelerated soil erosion and its underlying causes have been a particular focus [26–28]. However, relatively little attention has been paid to forest loss, which is an important element in land degradation and rocky desertification, particularly in Guizhou province. The forests of Guizhou Province, lying in the central part of southwestern China's karst area, play a crucial role in the ecological security and ecosystem services of a region that forms a part of an ecological safety barrier between the Pearl River and the Yangtze River catchments, which makes it ecologically critical, but highly vulnerable [29]. Understanding forest change is central to achieving sustainability and providing support for decisions regarding land-use management in the region.

Land-use and land-cover change (LULCC) is the alteration of natural or semi-natural landscapes due to human activities, such as urbanization, agricultural expansion, and deforestation [30]. In previous research, developed numerous models were developed to explore LULCC, in order to detect the changes in land use at specific locations and analyze its drivers [31,32]. From the perspective of landscape ecology, these models can be classified into three types: whole-landscape models, distributional models, and spatial-landscape models [33]. However, these focus mainly on ecological processes while tending to underplay or even ignore the role of human decision-making [34]. By the end of the 1990s, a considerable amount of tropical-deforestation-modeling work, represented by Lambin [35] and Kaimowitz and Angels [36] emerged that considered the role of human decision-making. Models of LULCC, including empirical–statistical models, stochastic models, optimization models, dynamic simulation models, and integrated models [37], can be categorized according to different criteria. Agarwal et al. [38] listed 19 models, including Markov models, spatial-simulation models, and regression models, based on their space, time, and decision-making characteristics. Among these, the Markov chain (MC) model is widely used in the spatiotemporal evaluation of LULC changes [39]. The choice of the model depends, to a large extent, on the particular scientific questions to be answered, along with data availability. Although LULCC modeling has made significant progress in understanding the dynamics and effects of land-use change [40], there remains a pressing need for more interdisciplinary research that integrates multiple drivers and factors affecting land-use decisions. This includes the development of more advanced modeling techniques that combine multiple methods [41]. The understanding of the interaction between scales and across scales is likely to remain the research frontier of the modeling of land-use/cover changes in the future.

The aim of this paper is to evaluate the change in forest cover and the relative importance of selected contributing factors in Guizhou province over four decades (1980–2018), with a view to determining the relative influence of human and natural factors. Using a remote-sensing monitoring dataset of multi-period land use and land cover from Landsat, we employed a Markov model to analyze the forest change in the study area. Additionally, we considered a range of environmental (e.g., soil erosion, karstification intensity, drought index) and socio-economic (e.g., population, gross domestic product (GDP), and accessibility) data to investigate the factors that influence forest change through a correlation analysis and a generalized linear model (GLM) regression. The systematic understanding of the forest change in Guizhou province in this paper has the potential to be used more widely to develop ecological restoration strategies and promote more sustainable land-use management in the future.

2. Materials and Methods

2.1. Study Area

Guizhou (24°37′–29°13′ N, 103°36′–109°35′ E) is representative of China's southwestern karst region, with over 60% of its land area consisting of the karst landform [42] (Figure 1). The region encompasses a variety of landforms, including mountains, hilly areas, plateaus, basins, and river valleys. Unlike other karst provinces, there are no extensive plain areas, and the mean elevation is approximately 1100 m. The climate is classified as subtropical humid monsoon, with an average annual temperature of around 15 °C and an annual precipitation of approximately 1200 mm [43]. The environment is highly susceptible to degradation, and it is particularly prone to accelerated soil erosion, resulting in rocky desertification [44]. By 2016, karst-rock desertification in Guizhou was reported to extend across almost 250,000 km^2 of the province, making it was the most degraded karst province in the country [45]. The region's economy has experienced significant growth in recent years. Agriculture, particularly cash crops such as oranges, peaches, and dragon fruit, contributes greatly to rural livelihoods, although, due to the area's ecological vulnerability, this focus on agriculture has led to environmental problems, such as ecosystem fragmentation, a decline in biodiversity, soil erosion, and reduced surface runoff [46]. With a population of 360 million in 2018, including a rural population of 189 million, the pressure on the land has become unbearable, exacerbating land degradation in the study area [47]. Furthermore, national policies have led to the improvement and proliferation of highways and high-speed railways, which have contributed to economic development, but they have also led to the removal of vegetation and loss of ecosystem services [44].

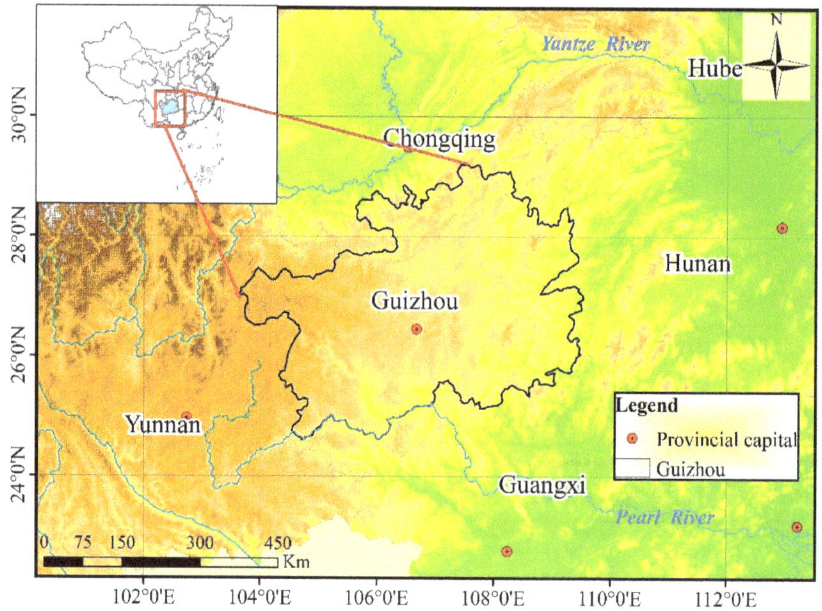

Figure 1. The geographical location of Guizhou province.

2.2. Data

2.2.1. Land-Use Data

Land use (1-km-resolution raster) data, based on visual interpretation of Landsat TM/ETM imagery, were obtained for the years 1980, 1990, 2000, 2010, and 2018 from the Resource and Environment Data Cloud Platform (China's multi-period land-use–land-cover remote-sensing data-monitoring set (CNLUCC); Resource and Environment Data Registration and Publishing System) [48]. The dataset is the most freely available dataset in China and has been widely used for detecting land-use change and analyzing ecosystem services from local to national scale. Its accuracy in identifying cropland and built-up areas is over 85%; its average accuracy for other land-use types exceeds 75%. Primary land-use categories identified were cropland, forest, grassland, water, built-up, and 'others' (Table 1); secondary categories included 25 sub-types of land use.

Table 1. Land-use–land-cover classification in Guizhou, China.

Class 1	Class 2/25 Land Use Sub-Types
Cropland	Paddy field, dryland
Forest	Forest land, shrubland, sparse woods, other forest areas
Grassland	Highly covered grassland, middle-covered grassland, low-covered grass land
Water	Canals, lakes, reservoirs and ponds, permanent ice and snow, intertidal zone, shoals
Built-up area	Urban land, rural residential land, other built-up areas
Others	Sand, Gobi, saline–alkali land, marshland, bare land, bare rocky land, others

2.2.2. Forest-Change Drivers

In addition to the mapping of land-use changes, several other drivers were considered in developing the model. Given the vulnerability of the region to rocky desertification, karstification intensity (KI) was included as a potentially important factor. The KI was obtained from the Guizhou Institute of Mountain Resources. Slopes were also considered important, as these influence the spread of forests that affect forest growth [49]. We derived the slope factor from a digital elevation model (DEM) (2009) at a 30-m resolution from

Geospatial Data Cloud (https://www.gscloud.cn/ (accessed on 30 January 2020)). Climate characteristics, particularly drought frequency and intensity, also have a significant impact on vegetation cover [50], and Guizhou is frequently affected by drought, which restricts forest growth. As a result, both the drought index (the ratio of annual evaporation capacity to annual precipitation) and mean annual precipitation were included as potential drivers. The drought index was provided by the Guizhou Institute of Mountainous Climate and Environment. Other factors relating to human activities, including urbanization, are known to play significant roles in forest change [51]. Land-use change and ecological restoration projects are considered direct human-activity factors [52,53] and, given that Guizhou has been at the forefront of China's economic growth since 2000, with the highest growth rate in the country for the last three consecutive years, balancing economic development with environmental protection is highly challenging [54]. Accordingly, we also included factors associated the anthropogenic influence: GDP, population, and accessibility for analyzing forest dynamics. The mean annual temperature/precipitation, accessibility, population (people/km^2), and GDP of nine municipalities in Guizhou were obtained from the Resource and Environment Data Cloud Platform (REDCP) [48].

2.3. Methods

2.3.1. Data Preprocessing

Primary data were obtained and processed according to the methods presented in Table 2, while a flow chart illustrating the methodology employed in this study is presented as Figure 2:

Table 2. Data and methods of potential factors.

Drivers	Original Data	Source	Processing Method	Period
LUC	Landsat TM/ETM	Resource and Environment Data Cloud Platform	Markov model and R	1980, 1990, 2000, 2010, 2018
P	Land use, night light, settlement density	Resource and Environment Data Cloud Platform	Spatial analysis	1995, 2000, 2010, 2015
GDP	GDP, land use, night light, settlement density	Resource and Environment Data Cloud Platform	Spatial analysis	1995, 2000, 2010, 2015
A	State, county, and township roads	Guizhou Institute of Mountainous Resources	Spatial analysis	2010
KI	Lithological data	Guizhou Institute of Mountainous Resources	Spatial analysis	2010
DI, MAP, MAT	Precipitation, evaporation	Guizhou Institute of Mountainous Climate and Environment, Resource and Environment Data Cloud Platform	Spatial analysis	1980–2015
S	DEM	Geospatial Data Cloud	Spatial analysis	2009

Notes that LUC, P, GDP, A, KI, DI, MAP, MAT, and S represent land-use change, population, gross domestic product, accessibility, karstification intensity, drought index, mean annual precipitation, mean annual temperature, and slope, respectively.

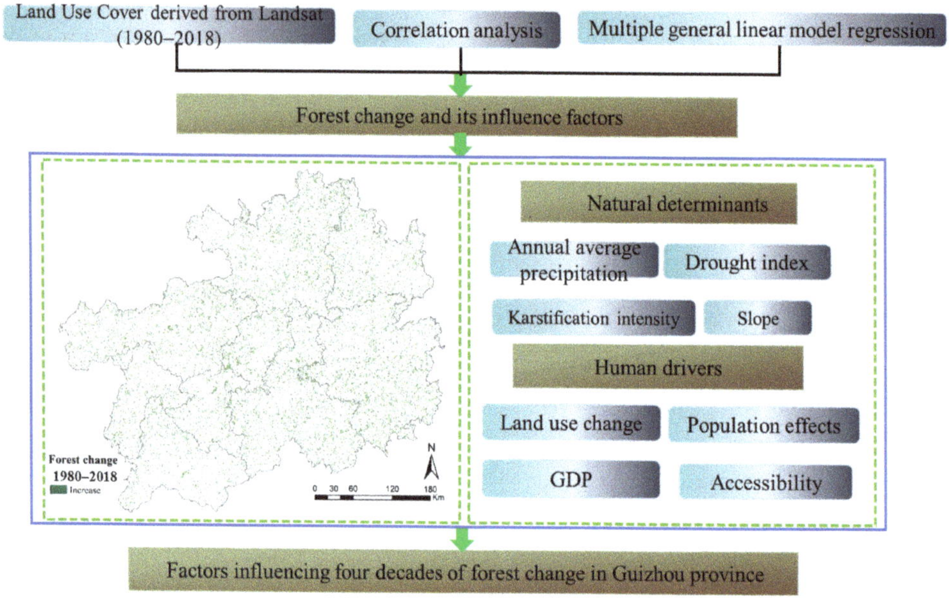

Figure 2. Flow chart of methodology employed.

Forest change: Based on a Markov transition matrix [55] and spatial analysis function of ArcGIS, a transfer matrix of different land types was obtained, and on this basis, land-use changes, including forest cover, were estimated. Chord diagrams of land transformation were constructed for Guizhou province and each of its nine major municipalities using R.

Land-use transition involves the changes in regional land-use patterns, and they are significant components of land-use-transition studies [56]. Markov modeling is commonly used to consider the processes and mechanisms of landscape-dynamics changes over the longer term [55,57]. We adopted a transition matrix as the core part of the Markov model (see Formula (1)), which is generally applied in estimations of land-cover changes [58,59]. While different types of conversion may occur, more attention was paid to those that account for most of the forest change. In addition, we calculated the conversion ratio of the main converted types through Formula (2). All analyses were conducted by ArcGIS and R.

$$\mathrm{T} = \begin{bmatrix} D_{11} & D_{12} & \dots & D_{1n} \\ D_{21} & D_{22} & \dots & D_{2n} \\ \dots & \dots & \dots & \dots \\ D_{n1} & D_{n2} & \dots & D_{nn} \end{bmatrix} \tag{1}$$

where T refers to the conversion matrix of different types of land-use change from 1980 to 2018, D_{nn} refers to the change in the area (unit: km^2) from one land-use type to another during the study period, and n refers to the area of a certain type of land that was involved in the computation.

$$R_{ij} = A_{ij}/B_i \tag{2}$$

where R_{ij} refers to the rate of the i type of land use converted to j type, A_{ij} represents the area of i type of land use converted to j grade, and B_i is the total area of i type of land in 1980 (unit: km^2).

Population and Gross Domestic Product (GDP): These data were analyzed spatially in ArcGIS 10.3. According to REDCP, population data were processed as follows:

$$\mathrm{POP}_{ij} = \mathrm{POP} \times \left(\mathrm{Q}_{ij}/Q\right) \tag{3}$$

where POP_{ij} is the population-spatial-distribution data in a 1 km × 1 km grid, Q_{ij} is the total weighting of land-use type, night light, and settlement density in a grid, POP is the population of the county-level administrative unit in which the grid is located, and Q is the total weighting of land-use type, night light, and settlement density for the county-level administrative unit in which the grid is located.

According to REDCP, GDP data were processed as follows:

$$GDP_{ij} = GDP \times \left(Q_{ij}/Q\right) \quad (4)$$

where GDP_{ij} is the GDP-spatial-distribution data in a 1 km × 1 km grid, Q_{ij} is the total weighting of land-use type, night light, and settlement density in a grid, GDP is the GDP of the county-level administrative unit in which the grid is located, and Q is the total weighting of land-use type, night light, and settlement density for the county-level administrative unit in which the grid is located.

Accessibility (A): This parameter refers to the accessibility of a location in terms of transportation, including national, provincial, county, and township roads. The Euclidean distance was calculated for each of the different levels of road, weighted according to ranking of their importance (national > provincial > county > township), and then analyzed spatially.

Karstification intensity (KI): The degree of karstification was classified as follows, according to the purity of carbonate: detrital carbonate-reservoir rock (DCRR), carbonate-reservoir rock with detrital reservoir rock (CRR-DRR), and non-carbonate rock (NCR). Weights were assigned according to karstification intensity, whereby DCRR > CRR-DRR > NCR.

Drought index and mean annual precipitation/temperature: A drought index [60] was obtained from Formulae (5)–(7). Next, based on Kriging interpolation in ArcGIS10.3, the map of drought index was made.

$$K = E'/P' \quad (5)$$

$$P' = P/P_P \quad (6)$$

$$E' = E/E_P \quad (7)$$

where K denotes the drought index of any period, P' describes the relative rate of change in precipitation during the period (1980–2015), P represents annual total precipitation for 2015, P_P is the mean annual precipitation during the period (1980–2015), E' describes the relative rate of change in evaporation during the period (1980–2015), E represents the evaporation in 2015, and E_P is the mean annual evaporation during the period (1980–2015).

Slope: Following image cutting and splicing, DEM was used to classify slopes, as follows: 0–6°, 6–15°, 15–25°, 25–35°, and >35°.

2.3.2. Analysis of Drivers

Firstly, we explored the relationship between forest area in 2018 and its drivers. Data for the most recent available year were used to consider potential drivers as indicated: land use change (2018), population (2015), gross domestic product (GDP, 2015), accessibility (2010), karstification intensity (2010), drought index (2015), mean annual precipitation (2015), and slope (2009). Using them as baseline values, we then determined the influence of these factors on forest change over time.

To assess the influence of these factors on forest change over time, we conducted a correlation analysis between forest changes and the various drivers and then applied generalized linear model (GLM) regression to quantify the relative contribution of each variable. The GLM regression extends linear model regressions by expanding the distribution range of dependent variables and introducing a continuous function, and it is generally applicable to non-normally distributed data [61]. As Formulae (8)–(10) show, the model is a function of mean μ with a linear combination x_β formed from regressor x and coefficient vector β.

$$\mu_i = E(Y_I|X_1, X_2, \ldots, X_k),\ i = 1, \ldots, n \quad (8)$$

$$\eta_i = g(\mu_i) \quad (9)$$

$$g(\mu_i) = \eta_i = \beta_0 + \beta_1 X_{i1} + \beta_2 X_{i2} + \beta_3 X_{i3} + \ldots + \beta_k X_{ik} \quad (10)$$

where X is explanatory variables (factors driving forest change), Y_I is dependent variables (the area of forest), μ_i is n independent samples subject to exponential distribution; η_i represents k linear combinations of explanatory variables; $g(\mu_i)$ refers to a function linking μ_i and η_i, and k is the number of explanatory variables.

3. Results

3.1. Spatio-Temporal Patterns of Forest Change in Guizhou Province

3.1.1. Forest Transition

Forest transition describes the range of forest change, from shrinking to expansion [62,63]. According to Table 3, forests were the largest land-use type (53%) from 1980 to 2018 and, while their distribution remained relatively stable, some increases over time were evident (Figure 3a: Forest change in Guizhou Province). Notably, the forest cover reached its lowest value in 2000. In terms of the forest subtypes, Table 4 shows that the greatest increase was in the category 'forest land' (i.e., forests with greater biomass and substantial tree-canopy cover).

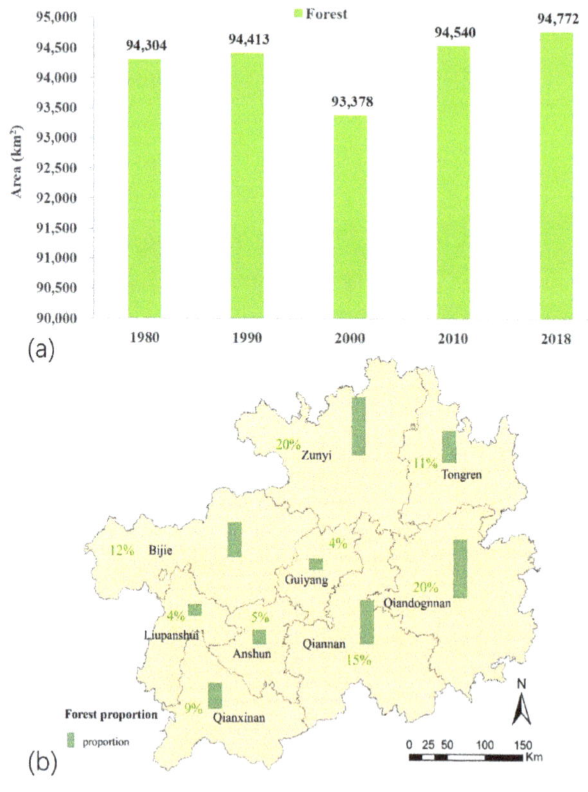

Figure 3. (**a**) Forest changes in Guizhou Province. (**b**) Average forest cover in nine municipalities.

Table 3. Guizhou land-use changes from 1980 to 2018.

Year	Cropland		Forest		Grassland		Water		Built-Up Area		Others		Total Area
	Area (km^2)	(%)	Area (km^2)	(%)	Area (km^2)	(%)	Area (km^2)	(%)	Area (km^2)	(%)	Area (km^2)	(%)	Area (km^2)
1980	49,037	27.92	94,304	53.69	31,423	17.89	363	0.21	484	0.28	36	0.02	175,647
1990	48,926	27.85	94,413	53.75	31,366	17.86	380	0.22	517	0.29	44	0.03	175,646
2000	49,318	28.08	93,378	53.16	31,951	18.19	395	0.22	561	0.32	44	0.03	175,647
2010	49,184	28.01	94,540	53.84	30,718	17.49	467	0.27	641	0.37	37	0.02	175,587
2018	48,552	27.64	94,772	53.95	29,382	16.72	721	0.41	2225	1.27	30	0.02	175,682

Table 4. Forest subtypes in Guizhou from 1980 to 2018 (km^2).

Forest Subtypes	1980	1990	2000	2010	2018	Changes
Forest land	24,038	24,048	23,673	23,818	24,392	354
Shrubland	43,617	43,675	43,163	43,339	43,469	−148
Sparse woods	26,364	26,403	26,238	27,073	26,615	251
Other forest areas	285	287	304	310	296	11

3.1.2. Spatial Changes

The distribution of the forest area in Guizhou is uneven, decreasing from east to west. Figure 3b shows the mean annual proportion of total forest cover in the province's nine major municipalities from 1980 to 2018. Qiandongnan, in the southeast, and Zunyi, in the north, have the greatest forest cover, accounting for 40% of the overall total, followed by Qiannan (south), Bijie (northwest), Tongren (northeast), and Qianxinan (southwest) with 15%, 12%, 11%, and 9%, respectively. Liupanshui (west), Anshun (next to Liupanshui), and Guiyang (central) have the lowest forest cover.

Based on temporal changes over the last 40 years, Figure 4 and Table 5 suggest that the forest cover was either maintained or increased, with the exception of Tongren, Qiannan, Qianxinan, and Anshun, all of which experienced some degree of forest loss. Qiandongnan experienced the greatest degree of forest increase, with 478 km^2 (2.6%), followed by Bijie and Guiyang, with increases of 202 km^2 (2.5%) and 138 km^2 (3.6%), respectively. Zunyi and Liupanshui both experienced only very limited increases in forest area (34 km^2 and 14 km^2, respectively).

Table 5. Forest changes from 1980 to 2018 in nine municipalities (km^2).

Year	1980	1990	2000	2010	2018	Changes
Anshun	4442	4447	4417	4457	4432	−10
Bijie	11,388	11,397	11,380	11,548	11,590	202
Guiyang	3872	3874	3869	3883	4010	138
Liupanshui	3947	3959	3970	4021	3961	14
Qiandongnan	18,563	18,587	18,458	18,773	19,041	478
Qiannan	14,640	14,651	14,279	14,422	14,515	−125
Qianxinan	8010	8017	7948	8067	7980	−30
Tongren	10,155	10,156	9846	9972	9922	−233
Zunyi	19,287	19,325	19,211	19,397	19,321	34
Total	94,304	94,413	93,378	94,540	94,772	468

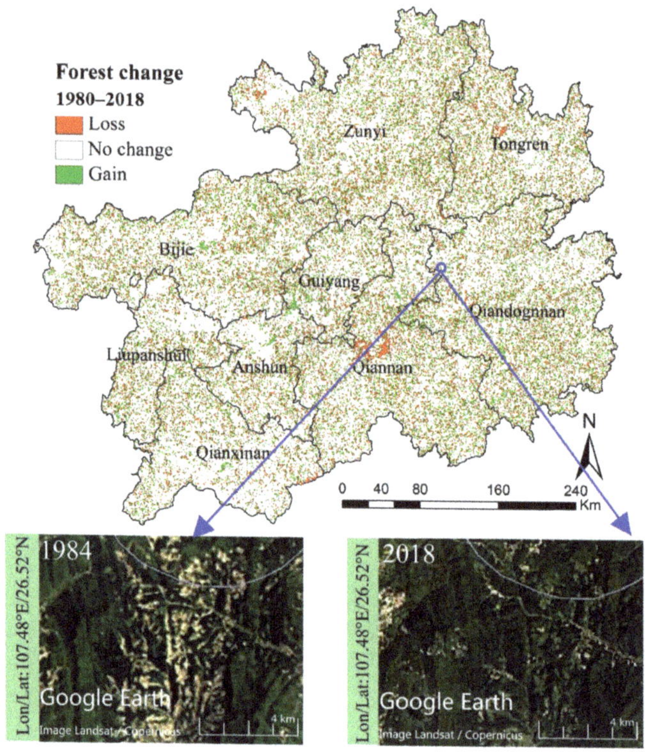

Figure 4. Spatial pattern of forest changes in nine municipalities.

3.2. Possible Drivers of Forest Change

3.2.1. Land-Use Change

Although the overall forest increase during the study period was relatively small across the province as a whole, in very substantial areas, forests replaced agriculture. Figure 5 demonstrates that forests replacing cropland happened in all nine municipalities in Guizhou Province, which is attributable largely to the implementation of the Grain for Green (GFG) project. Indeed, 36% of the cropland was converted into forests, which is significantly higher than the equivalent values for the grassland and construction land. With respect to individual municipalities, 47% of the farmland was converted into forests in Qiandongnan, followed by Zunyi, Qiannan, and Tongren, where 40%, 39%, and 37% of the land, respectively, was converted from agricultural land. Bijie had the smallest portion of cropland converted into forest land (29%).

The implementation of the Grain for Green program accounts for a considerable, and indeed increasing, proportion of the total forest changes in Guizhou (Figure 6). Prior to the implementation of this policy in Guizhou in 2000, the forest cover was largely unchanged. For instance, the forest area increased by only 109 km^2 from 1980 to 1990, which was much less than the change during the periods of 1990–2000, 2000–2010, and 2010–2018 (see Figure 3a) and, indeed, the forest cover actually decreased across the province in 2000. In Figure 6, it can be seen that, in 2018, 17% of the cropland was converted into forests in Guizhou. Given that Guizhou is located in the upper and middle reaches of the Yangtze and Pearl Rivers, the forest-cover change has also been brought about by the implementation of two of China's eight major shelterbelt projects (the Shelterbelt Program for Upper and Middle Reaches of the Yangtze River, 1989 and the Shelterbelt Program for the Pearl River, 1996).

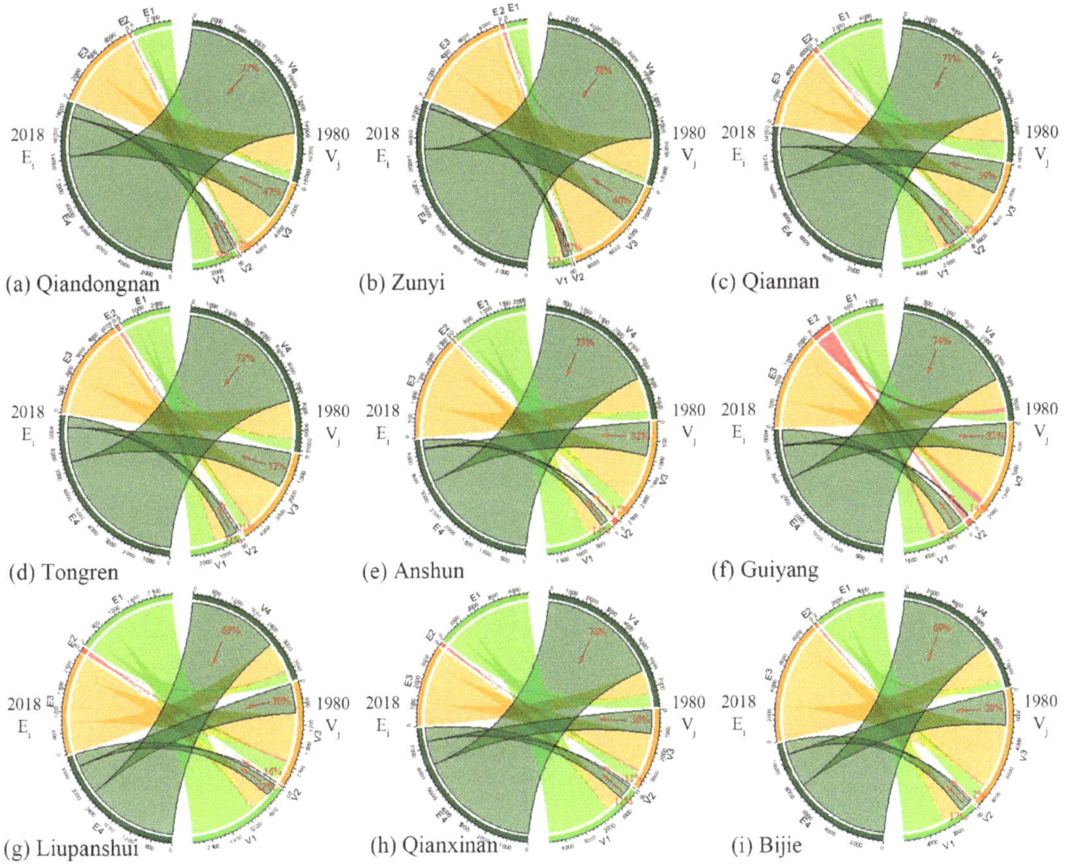

Figure 5. Percentages of other land-use types converted into forests in Guizhou and its nine major municipalities 1980–2018. Ei (i = 1, 2, 3, 4) denotes the area of the i category of land in 2018, which transformed from the j category of land in 1980; Vj (j = 1, 2, 3, 4) represents the area of the j class of land in 1980, which converted into the i class of land in 2018; 1 = grassland, 2 = construction land, 3 = cropland, 4 = forest. Red arrows indicate percentages of i-type land in 1980 converted into forests by 2018.

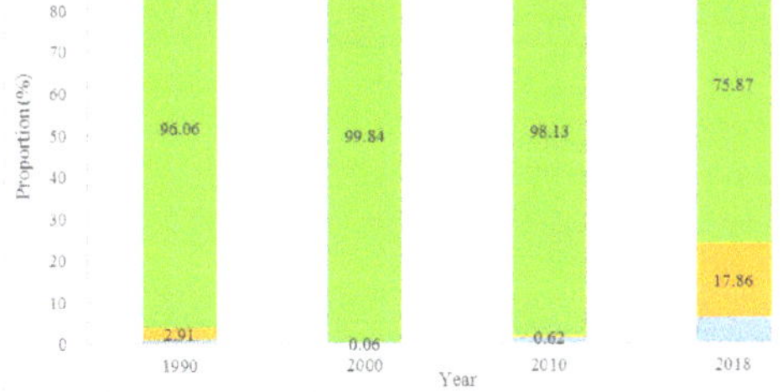

Figure 6. Percentages of four major land use types converting into forests in Guizhou: 1980 to 2018.

3.2.2. Population Effects

Figure 7a indicates that the population densities are generally higher in the western part of the province and this, in effect, reflects the forest distribution. At a very basic level, therefore, population density influences forest disturbance or clearance, a relationship that is further illustrated through the correlation analysis in Table 6.

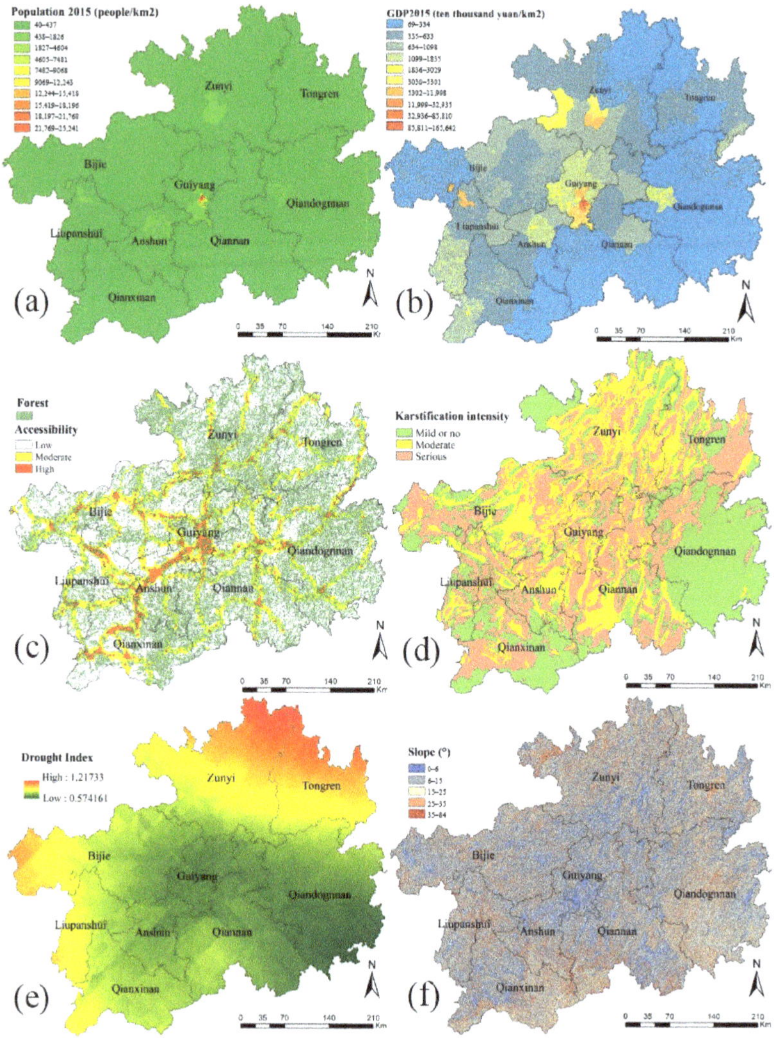

Figure 7. Drivers of forest change. (**a**) Population density in 2015 (people per km^2). (**b**) Distribution of GDP in 2015 (RMB $10{,}000/km^2$). (**c**) Accessibility and forest area. (**d**) Karstification intensity. (**e**) Spatial distribution of drought index. (**f**) Slope in 2009.

Table 6. Correlations between and multiple GLM analyses of the relationships between forest areas and factors.

Method	Correlation Analysis		Multiple-GLM Regression
Variable	r	p	SS, %
Drought index	0.084	0.460	0.64
Karstification intensity	−0.097	0.394	0.85
Mean annual precipitation	−0.296 **	0.008	0.97
GDP	−0.255 *	0.024	1.88
Population	−0.281 *	0.012	2.23
Land-use change (LUC)	0.580 **	0.000	2.27
Accessibility	0.388 **	0.000	2.29
Slope of 15–25°	0.882 **	0.000	88.87

* $p < 0.05$, ** $p < 0.01$; SS, proportion of variances explained by the variable.

3.2.3. GDP

Figure 7b illustrates the GDP per km^2, which also takes land use, night light, and settlement density into account (for more details, see sections on methods and data). In parallel with the population density, the western parts of Guizhou province also have high GDP values and lower forest cover, so there is an inverse relationship between GDP and forest area (Table 5). The areas with higher GDP exhibit forest loss as a consequence of economic development and urban expansion.

3.2.4. Accessibility

Figure 7c indicates that access to local transport routes negatively influences forest cover. The road and railway density are greatest in the western and southwestern parts of the province, and the effect of this on forest change is clearly evident. Qiandongnan is characterized by lower accessibility levels and is richer in forest resources, while Guiyang, Anshun, and Liupanshui have greater densities of both road and railway routes, which negatively affect forest recovery.

3.2.5. Karstification Intensity

The substantial karst area in Guizhou province is an important factor because the lack of surface water and relatively thin soils constrain forest development. Indeed, the distribution of karst (Figure 7d) has a negative influence on the forest cover (Table 6).

3.2.6. Drought Index (DI)

The balance between moisture inputs, in the form of precipitation, and outputs, in the form of evaporation, is a key determinant of the vegetation type, and the drought index is used here to account for this balance. Figure 7e illustrates marked spatial patterns in the occurrence of drought in Guizhou province, and suggests that the lower values in Qiandongnan are associated with greater forest cover. Statistically, however, the effect of moisture stress is less marked than may have been expected (Table 6).

3.2.7. Slope

In this study, the slope angle was classified into five categories: 0–6°, 6–15°,15–25°, 25–35°, and >35° (Figure 7f). Specifically, the forests are distributed preferentially on slopes of 15–25° and occur less frequently on lands with lower slope angles, presumably because these lands are more suited to agriculture and urban development. Table 6 illustrates a very strongly positive correlation between slopes of 15–25° and forest area.

From the correlation analysis and multiple general linear models, it can be seen (Table 6) that 15–25° slopes play a dominant role in the forest areas, explaining 88.87% of the variation, while accessibility and land-use change account for 2.29% and 2.27%, respectively, followed by population effects (2.23%), GDP (1.88%), mean annual precipitation (0.97%), karstification intensity (0.85%), and drought index (0.64%).

Table 6 further reveals that the mean annual precipitation, GDP, population effects, land-use changes, accessibility, and slopes of 15–25° all play a role in forest change. Due to data-availability constraints, we ultimately selected mean annual precipitation, GDP, population, and land-use change (LUC) for further analysis as drivers of forest-cover changes.

3.3. Relative Importance of Drivers Changes over Time

To determine the relative importance of the drivers for forest change over time, we conducted a correlation analysis and multiple-GLM regression for the different periods (Table 7), taking the mean annual precipitation (MAP), annual mean temperature (MAT), population, GDP, and cropland conversion into forest as key drivers of forest changes over time, although, given the data limitations, the analysis was conducted only from 1990 to 2018, and the data for GDP and population in 1995 were used for 1990.

Table 7. Changes in drivers of forest variation over time.

Year	Variable		MAP	MAT	Population	GDP	LUC
1990	Correlation analysis	r	0.095	0.064	−0.379 **	−0.150	–
		sig	0.391	0.565	0.000	0.174	–
	Multiple-GLM regression	SS, %	3.34%	3.59%	74.44%	18.62%	–
2000	Correlation analysis	r	−0.013	0.095	−0.283 **	−0.261 *	–
		sig	0.908	0.391	0.009	0.016	–
	Multiple-GLM regression	SS, %	25.63%	11.12%	63.08%	0.18%	–
2010	Correlation analysis	r	0.111	0.006	−0.317 **	−0.310 **	0.236 *
		sig	0.345	0.960	0.006	0.007	0.042
	Multiple-GLM regression	SS, %	9.77%	13.70%	9.28%	0.07%	67.18%
2018	Correlation analysis	r	−0.064	0.012	−0.312 **	−0.282 *	0.784 **
		sig	0.583	0.921	0.006	0.014	0.000
	Multiple-GLM regression	SS, %	0.34%	2.78%	3.04%	2.04%	91.81%

Notes: Annual precipitation (AP, mm), annual mean temperature (AMT, °C), population (per person/km^2), GDP (RMB 10,000/km^2), LCU (km^2); * $p < 0.05$, ** $p < 0.01$; SS, proportion of variances explained by the variable.

In Table 8, it can be seen that, prior to 2000, the population exerted the most significant impact on the forests but, after 2000, its influence was reduced to 3.04%. This may be attributed to the type of economic development in Guizhou before 2000 [64], whereby the inhabitants exploited forests for firewood [65] or settled on unused land [46]. Currently, the substitution of gas and hydropower for firewood helps to reduce the pressures of the population on the forests [65]. The negative effect of GDP on forest change diminished over the years, probably due to the transformations associated with economic development, which has reduced dependence on the direct consumption of natural resources, including forests [66]. The impact of the land-use change, mainly the conversion of cropland into forests, increased to 91.81% by 2018, probably as a consequence of the implementation of GFG [67]. While some fluctuations were observed during dry periods, the influence of MAP and MAT on the forest cover did not vary substantially in recent years.

Table 8. Comparison of results of this study with other products (areas in km^2).

Data	Resolution	1980	1990	2000	2010	2018	Changes 2000–2018	Changes 1980–2018
CNLUCC (this study)	1 km	94,304	94,413	93,378	994,540	94,772	1394	468
GlobeLand30	30 m	–	–	83,079	84,472	83,329	250	–
GLASS-GLC	5 km	4868	6110	6270	6969	7018	748	2150
MODIS/006/MCD12Q1	500 m	–	–	10,696	12,707	22,762	12,066	–

4. Discussion

4.1. Validation of Forest Change in Guizhou through Comparison with Other Data Sources

We compared the results of our study with those of other land-use products to verify the trends in the forest changes (Tables 8 and 9). We used ArcGIS spatial analysis to determine the forest change according to MODIS/006/MCD12Q1, GlobeLand30, and GLASS-GLC. It can be seen in Table 6 that all the products show an increasing trend in forest cover during recent decades, albeit with some interannual differences. Table 8 reveals that the data sources, definition/classification criteria, classification technique, and spatial resolution of the land-use data underlie the differences in the estimation of different annual forest areas and forest changes [68,69]. For example, the differences in spatial resolution between GLASS-GLC and CNLUCC may affect the land-cover classification and explain the minor differences in forest-change estimation. Moreover, shrubland is a single class in Global Land 30, and it is not classified as forest land, which may explain why the forest-change increase in Global Land 30 is lower than that in this study. Some grasslands are misclassified as shrublands in MODIS/006/MCD12Q1 and, since shrublands are components of 'forest' [70], the forest change from 2000 to 2018 in that product is much greater than in the results presented by CNLUCC, in which grassland is a single category, independent of forest cover. Moreover, CNLUCC is obtained through detailed field data [71], and it has been widely applied in many major projects, such as the Western Development of China and the second national soil-erosion survey of China, among others (http://www.resdc.cn/data.aspx?DATAID=95 (accessed on 30 January 2020)), which indicates its reliability. It is noteworthy that in CNLUCC, the forest area decreased in 2000, while in GLASS-GLC, it marginally increased. This difference has two possible causes, *viz.* the two products differ in terms of spatial resolution and classification technique. Notably, the spring and summer droughts in 1989 and the serious drought in southwestern China in 2000 may have interrupted the otherwise consistent increase in forest cover over time [72].

Table 9. Parameters of data products relating to forest-cover estimation.

Data	Spatial Resolution	Data Source	Classification Technique	Accuracy	Subclass or Description
CNLUCC	1 km	HJ-1A/B, Landsat TM/ETM+/OLI	Visual interpretation	Above 75%	Forest land, shrubland, sparse woods, other forest areas
GlobeLand30	30 m	Landsat TM/ETM+	POK-based method	2000/2010: 80.33 ± 0.2% 2020: 85.72%	Over 30% of land covered with trees and vegetation, including deciduous broad-leaved forests, evergreen broad-leaved forests, deciduous coniferous forests, evergreen coniferous forests, mixed forests, and sparse forests with crown coverage of 10–30%
GLASS-GLC	5 km	Landsat TM/ETM+	Conventional maximum-likelihood classifier, J4.8 decision-tree classifier, Random Forest classifier, and support-vector-machine classifier	82.81%	Broad leaf, leaf on; broad leaf, leaf-off; needle leaf, leaf on; needle leaf, leaf off; mixed leaf type, leaf on; mixed leaf type, leaf off.
MODIS/006/MCD12Q1	500 m	MODIS	Decision-tree classification algorithm	66.42%	Evergreen needleleaf forests; evergreen broadleaf forests; deciduous needleleaf forests; deciduous broadleaf forests; mixed forests; closed shrublands; open shrublands

Notes: sources of the table are from [67,68].

4.2. The Effects of Ecological Restoration Policy on Forest Change

Guizhou has the largest karst area in the world [73]. Since the release of the "Decision on Basic Greening of Guizhou in Ten Years" in 1990, China has implemented a batch of key projects to protect and restore the fragile ecological environment. These include the Shelterbelt Program for Upper and Middle Reaches of the Yangtze River (1989), the Natural Forest Protection Project (1998), and the Grain for Green program (1999), all of which were applied in Guizhou and contribute to greening [74,75]. In 2008, China imposed the "Outline of Comprehensive Control Plan for Rocky Desertification in Karst Areas," which included 55 counties in Guizhou Province among China's 100 pilot counties for the comprehensive control of rocky desertification, with the aim of further restoring the ecological environment. In subsequent years, Guizhou vigorously supported the development and protection of forest resources through various initiatives, such as the Afforestation Planning across County and Township and Village in Guizhou Province (2014–2017), the Three Year Action Plan for Green Guizhou Construction (2015–2017), the Guizhou Forestry Industry Three Year Multiplication Plan (2015–2017), the Implementation Plan on Promoting the Development of Forestry Industry in Guizhou Province, and the Ten Forestry Industry Bases Construction Plan of Guizhou Province (2018–2020). Additionally, to strengthen the protection of forest resources, Guizhou strictly implemented a forest-cutting quota system to ensure that the total growth of trees was far greater than the total consumption. It also carried out forest-ecological-benefit compensation (2004) and enforced a special law enforcement campaign to protect forest (2014). Guizhou has also reviewed and approved the use of forest land and defined the forestry ecological red line to strictly protect and rationally use forest land resources [76].

In short, China as a whole, and Guizhou in particular, have implemented various targeted policies to hasten vegetation restoration and protect their forests, all of which have contributed to the increase in forest cover (Figure 8). With the implementation of ecological-restoration and protection programs, this trend seems set to continue [16]. According to the Guizhou Statistical Yearbook, forest cover is defined as the ratio of the forest area to the total land area, expressed as a percentage. According to national regulations, this also includes shrublands and farmland–forest mosaic areas. Therefore, the level of forest resources and greening is even greater than that recorded in our study.

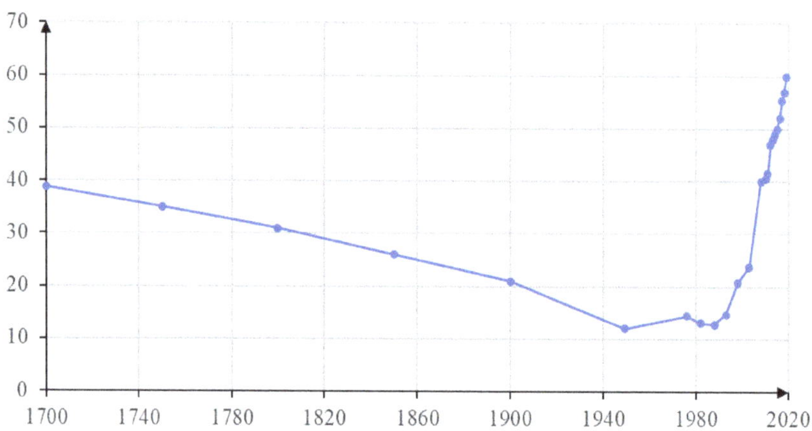

Figure 8. Changes in forest cover (%) in Guizhou province (source: Guizhou Statistical Yearbook).

4.3. Limitations and Prospects

Although forest changes may be driven by multiple factors, not all of which are addressed in this study, our analysis, based on the conditions of the study area, considers the factors that are most likely to be significant. The findings offer important support for the government in identifying key areas for forest conservation and restoration.

Nevertheless, there are some limitations. For example, the data relating to some of the explanatory variables were not available for the entire period, meaning that in some cases, we used data from the closest suitable year in the analysis. Other relevant driving factors should be the subject of future research and analysis, such as the choice of afforestation species, the method of production of seedlings, and specific planting-environment conditions. Species selection, in particular, is a key challenge in afforestation [77,78], and the selection of the correct species mixtures can markedly increase the success of the restoration. In addition, the choice of appropriate species for specific environments, which can adapt to current and future environmental conditions, is crucial [79]. In degraded ecosystems, planting species that can withstand particular environmental constraints should be used. Other factors, such as the occurrence of vegetation fires, also need to be considered [80], as these may affect the rate of tree recruitment, forest-age structure, and species composition [81,82]. The effects of soil humidity on forest recovery are complex and may be important in seed germination [83], while soil moisture is a further constraint on successful regeneration [84].

5. Conclusions

In Guizhou province, forests are a prominent land type due to the favorable hydrothermal conditions, and the results of this study show that the forest cover has increased over the last few decades. In terms of area, Qiandongnan holds the largest share of forest, and experienced the most substantial increase of all the nine municipalities during the study period. On the other hand, Liupanshui, in the west of Guizhou, has the lowest forest cover and exhibited very little change overall. While forest changes are the result of both natural and artificial factors, the relative influence of these factors shifted over time. Prior to 2000, the population exerted a much stronger influence on the forests but, since then, the function of other factors has increased, particularly land-use changes. The nine major municipalities in Guizhou experienced different outcomes as a result, with Qiandongnan exhibiting the highest percentage of farmland converted into forest, at 47%, followed by Zunyi, with 40%, Qiannan, with 39%, and Tongren, with 37%. Bijie has the smallest portion of cropland converted into forest (29%). These results emphasize the dynamic nature of driving forces in determining forest cover and demonstrate the value of geospatial analysis in understanding their emerging influence. The methodology and modeling approach adopted here are used to illustrate the relative roles of natural and management factors and may be applied in other similar regions to reduce forest degradation and increase forest restoration.

Author Contributions: Conceptualization: R.C.; methodology: X.G., Q.L. and Z.X.; writing—original draft: X.G. and R.C.; writing—review and editing: X.G., R.C., M.E.M., Q.L., Z.X. and Z.P. All authors have read and agreed to the published version of the manuscript.

Funding: This study was conducted with the support of the National Social Science Fund of China (Grant No.20ZDA085), the National Key Research and Development Program of China (Grant No. 2017YFC1503001), and the China Postdoctoral Science Foundation (Grant No. 2022M722055 & 2022TQ0205).

Data Availability Statement: The source of relevant data acquisition has been described in the text.

Conflicts of Interest: The authors declare no conflict of interest.

References

1. Curtis, P.G.; Slay, C.M.; Harris, N.L.; Tyukavina, A.; Hansen, M.C. Classifying drivers of global forest loss. *Science* **2018**, *361*, 1108. [CrossRef] [PubMed]
2. McAlpine, C.; Johnson, A.; Salazar, A.; Syktus, J.; Wilson, K.; Meijaard, E.; Seabrook, L.; Dargusch, P.; Nordin, H.; Sheil, D. Forest loss and Borneo's climate. *Environ. Res. Lett.* **2018**, *13*, 044009. [CrossRef]
3. Viña, A.; McConnell, W.J.; Yang, H.; Xu, Z.; Liu, J. Effects of conservation policy on China's forest recovery. *Sci. Adv.* **2016**, *2*, e1500965. [CrossRef] [PubMed]
4. Siqueira-Gay, J.; Sonter, L.J.; Sánchez, L.E. Exploring potential impacts of mining on forest loss and fragmentation within a biodiverse region of Brazil's northeastern Amazon. *Resour. Policy* **2020**, *67*, 101662. [CrossRef]
5. Bryan, B.A.; Gao, L.; Ye, Y.; Sun, X.; Connor, J.D.; Crossman, N.D.; Stafford-Smith, M.; Wu, J.; He, C.; Yu, D.; et al. China's response to a national land-system sustainability emergency. *Nature* **2018**, *559*, 193–204. [CrossRef]
6. Hong, Y.; Du, M. Greening Efforts Push China's Forest Coverage Rate to over 23 Pct. Available online: http://en.people.cn/n3/2021/0629/c90000-9866358.html (accessed on 30 December 2021).
7. Chen, C.; Park, T.; Wang, X.; Piao, S.; Xu, B.; Chaturvedi, R.K.; Fuchs, R.; Brovkin, V.; Ciais, P.; Fensholt, R.; et al. China and India lead in greening of the world through land-use management. *Nat. Sustain.* **2019**, *2*, 122–129. [CrossRef]
8. Guizhou Statistics Bureau. *Guizhou Statistical Yearbook*; China Statistics Press: Beijing, China, 2021.

9. Tong, X.; Brandt, M.; Yue, Y.; Ciais, P.; Rudbeck Jepsen, M.; Penuelas, J.; Wigneron, J.-P.; Xiao, X.; Song, X.-P.; Horion, S.; et al. Forest management in southern China generates short term extensive carbon sequestration. *Nat. Commun.* **2020**, *11*, 129. [CrossRef]
10. Tong, X.; Brandt, M.; Yue, Y.; Horion, S.; Wang, K.; Keersmaecker, W.D.; Tian, F.; Schurgers, G.; Xiao, X.; Luo, Y.; et al. Increased vegetation growth and carbon stock in China karst via ecological engineering. *Nat. Sustain.* **2018**, *1*, 44–50. [CrossRef]
11. Huang, L.; Wang, B.; Niu, X.; Gao, P.; Song, Q. Changes in ecosystem services and an analysis of driving factors for China's Natural Forest Conservation Program. *Ecol. Evol.* **2019**, *9*, 3700–3716. [CrossRef]
12. Xiong, Q.; Li, L.; Luo, X.; He, X.; Zhang, L.; Pan, K.; Liu, C.; Sun, H. Driving forces for recovery of forest vegetation after harvesting a subalpine oak forest in eastern Tibetan Plateau. *Environ. Sci. Pollut. Res.* **2021**, *28*, 67748–67763. [CrossRef]
13. Zhao, Y.; Ren, H.; Li, X. Forest Transition and Its Driving Forces in the Qian-Gui Karst Mountainous Areas. *J. Resour. Ecol.* **2020**, *11*, 59–68.
14. Sitters, J.; Holmgren, M.; Stoorvogel, J.J.; López, B.C. Rainfall-Tuned Management Facilitates Dry Forest Recovery. *Restor. Ecol.* **2012**, *20*, 33–42. [CrossRef]
15. Bronson, D.R.; Gower, S.T.; Tanner, M.; Van Herk, I. Effect of ecosystem warming on boreal black spruce bud burst and shoot growth. *Glob. Change Biol.* **2009**, *15*, 1534–1543. [CrossRef]
16. Wang, H.; Lv, Z.; Gu, L.; Wen, C. Observations of China's forest change (2000–2013) based on Global Forest Watch dataset. *Biodivers. Sci.* **2015**, *23*, 575–582. [CrossRef]
17. Li, H.; Ma, Y.; Liu, W.; Wenjun, L. Clearance and fragmentation of tropical rain forest in Xishuangbanna, SW, China. *Biodivers. Conserv.* **2009**, *18*, 3421–3440. [CrossRef]
18. Zhai, D.; Xu, J.; Dai, Z.; Cannon, C.H.; Grumbine, R.E. Increasing tree cover while losing diverse natural forests in tropical Hainan, China. *Reg. Environ. Change* **2013**, *14*, 611–621. [CrossRef]
19. Zhang, J.Y.; Dai, M.H.; Wang, L.C.; Zeng, C.F.; Su, W.C. The challenge and future of rocky desertification control in karst areas in southwest China. *Solid Earth* **2016**, *7*, 83–91. [CrossRef]
20. Xiong, K. *Remote-Sensing and GIS-Based Typical Study of Rock Desertification in Karst Areas*; Chinese Geological Press: Beijing, China, 2002.
21. Wang, S.; Ji, H.; Ouyang, Z.; Zhou, D.; Zhen, L.; Li, T. Preliminary study on carbonate rock weathering pedogenesis. *Sci. China* **1999**, *42*, 572–581. [CrossRef]
22. Chen, R.; Ye, C.; Cai, Y.; Xing, X. Integrated Restoration of Small Watershed in Karst Regions of Southwest China. *AMBIO* **2012**, *41*, 907–912. [CrossRef]
23. Yan, X.; Cai, Y. Multi-scale anthropogenic driving forces of karst rocky desertification in southwest China. *Land Degrad. Dev.* **2015**, *26*, 193–200. [CrossRef]
24. Li, D.; Liu, J.; Chen, H.; Zheng, L.; Wen, L.; Wang, K. Forage grass cultivation increases soil organic carbon and nitrogen pools in a karst region, southwest China. *Land Degrad. Dev.* **2018**, *29*, 4397–4404. [CrossRef]
25. Liu, X.; Zhang, W.; Wu, M.; Ye, Y.; Wang, K.; Li, D. Changes in soil nitrogen stocks following vegetation restoration in a typical karst catchment. *Land Degrad. Dev.* **2019**, *30*, 60–72. [CrossRef]
26. Dai, Q.; Peng, X.; Wang, P.; Li, C.; Shao, H. Surface erosion and underground leakage of yellow soil on slopes in karst regionS of southwest China. *Land Degrad. Dev.* **2018**, *29*, 2438–2448. [CrossRef]
27. Bai, X.Y.; Zhang, X.B.; Chen, H.; He, Y.B. Using Cs-137 fingerprinting technique to estimate sediment deposition and erosion rates from Yongkang depression in the karst region of Southwest China. *Land Degrad. Dev.* **2010**, *21*, 474–479. [CrossRef]
28. Dai, Q.; Peng, X.; Zhao, L.; Shao, H.; Yang, Z. Effects of underground pore fissures on soil erosion and sediment yield on karst slopes: Soil erosion and sediment on karst slopes. *Land Degrad. Dev.* **2017**, *28*, 1922–1932. [CrossRef]
29. Guo, F.; Jiang, G.; Yuan, D.; Polk, J.S. Evolution of major environmental geological problems in karst areas of Southwestern China. *Environ. Earth Sci.* **2013**, *69*, 2427–2435. [CrossRef]
30. Liu, B.; Pan, L.; Qi, Y.; Guan, X.; Li, J. Land Use and Land Cover Change in the Yellow River Basin from 1980 to 2015 and Its Impact on the Ecosystem Services. *Land* **2021**, *10*, 1080. [CrossRef]
31. Heistermann, M.; Müller, C.; Ronneberger, K. Land in sight? Achievements, deficits and potentials of continental to global scale land-use modeling. *Agric. Ecosyst. Environ.* **2006**, *114*, 141–158. [CrossRef]
32. Verburg, P.H.; Kok, K.; Pontius, R.G.; Veldkamp, A. Modeling Land-Use and Land-Cover Change. In *Land-Use and Land-Cover Change: Local Processes and Global Impacts*; Lambin, E.F., Geist, H., Eds.; Springer: Berlin/Heidelberg, Germany, 2006; pp. 117–135. [CrossRef]
33. Baker, W.L. A review of models of landscape change. *Landsc. Ecol.* **1989**, *2*, 111–133. [CrossRef]
34. Brown, D.G.; Walker, R.; Manson, S.; Seto, K. Modeling Land Use and Land Cover Change. In *Land Change Science: Observing, Monitoring and Understanding Trajectories of Change on the Earth's Surface*; Gutman, G., Janetos, A.C., Justice, C.O., Moran, E.F., Mustard, J.F., Rindfuss, R.R., Skole, D., Turner, B.L., Cochrane, M.A., Eds.; Springer: Dordrecht, The Netherlands, 2012; pp. 395–409. [CrossRef]
35. Lambin, E.F. Modelling and monitoring land-cover change processes in tropical regions. *Prog. Phys. Geogr. Earth Environ.* **1997**, *21*, 375–393. [CrossRef]
36. Kaimowitz, D.; Angelsen, A. *Economic Models of Tropical Deforestation: A Review*; Center for International Forestry Research (CIFOR): Bogor, Indonesia, 1989.

37. Lambin, E.F.; Rounsevell, M.D.A.; Geist, H.J. Are agricultural land-use models able to predict changes in land-use intensity? *Agric. Ecosyst. Environ.* **2000**, *82*, 321–331. [CrossRef]
38. Agarwal, C.; Green, G.M.; Grove, J.M.; Evans, T.P.; Schweik, C.M. *A Review and Assessment of Land-Use Change Models: Dynamics of Space, Time, and Human Choice*; U.S. Department of Agriculture, Forest Service, Northeastern Research Station: Burlington, NJ, USA, 2002.
39. Verburg, P.H.; Alexander, P.; Evans, T.; Magliocca, N.R.; Malek, Z.; Rounsevell, M.D.A.; van Vliet, J. Beyond land cover change: Towards a new generation of land use models. *Curr. Opin. Environ. Sustain.* **2019**, *38*, 77–85. [CrossRef]
40. Regasa, M.S.; Nones, M.; Adeba, D. A Review on Land Use and Land Cover Change in Ethiopian Basins. *Land* **2021**, *10*, 585. [CrossRef]
41. Wang, J.; Bretz, M.; Dewan, M.A.A.; Delavar, M.A. Machine learning in modelling land-use and land cover-change (LULCC): Current status, challenges and prospects. *Sci. Total Environ.* **2022**, *822*, 153559. [CrossRef]
42. Bai, X.; Zhang, S.; Li, C.; Xiong, L.; Song, F.; Du, C.; Li, M.; Luo, Q.; Xue, Y.; Wang, S. A carbon neutrality capacity index for evaluating carbon sink contributions. *Environ. Sci. Ecotechnol.* **2023**, *15*, 100237. [CrossRef]
43. Wu, J.; Zhang, F.; Cao, G.; Li, W.; Zhao, X. Temporal and spatial analysis of precipitation in Guizhou based on TRMM 3B42 satellite data. *IOP Conf. Ser. Earth Environ. Sci.* **2017**, *81*, 012076. [CrossRef]
44. Tian, Y.; Wang, S.; Bai, X.; Luo, G.; Xu, Y. Trade-offs among ecosystem services in a typical Karst watershed, SW China. *Sci. Total Environ.* **2016**, *566–567*, 1297–1308. [CrossRef]
45. China's State Forestry Administration. *The Bulletin of Rocky Desertification in China*; China's State Forestry Administration: Beijing, China, 2012.
46. Zhao, L.; Hou, R. Human causes of soil loss in rural karst environments: A case study of Guizhou, China. *Sci. Rep.* **2019**, *9*, 3225. [CrossRef]
47. Guizhou Statistics Bureau. *Guizhou Statistical Yearbook*; China Statistics Press: Beijing, China, 2019.
48. Xu, X.; Liu, J.; Zhang, S.; Li, R.; Yan, C.; Wu, S. *China's Multi-Period Land Use Land Cover Remote Sensing Monitoring Data Set (CNLUCC)*; Resource and Environment Data Registration and Publishing System: Beijing, China, 2018. [CrossRef]
49. Baltaci, U.; Yildirim, F. Effect of Slope on the Analysis of Forest Fire Risk. *Hacet. J. Biol. Chem.* **2020**, *48*, 373–379. [CrossRef]
50. Ding, Y.; Xu, J.; Wang, X.; Peng, X.; Cai, H. Spatial and temporal effects of drought on Chinese vegetation under different coverage levels. *Sci. Total Environ.* **2020**, *716*, 137166. [CrossRef]
51. Yin, L.; Dai, E.; Zheng, D.; Wang, Y.; Ma, L.; Tong, M. What drives the vegetation dynamics in the Hengduan Mountain region, southwest China: Climate change or human activity? *Ecol. Indic.* **2020**, *112*, 106013. [CrossRef]
52. Padilla, F.M.; Vidal, B.; Sanchez, J.; Pugnaire, F.I. Land-use changes and carbon sequestration through the twentieth century in a Mediterranean mountain ecosystem: Implications for land management. *J. Environ. Manag.* **2010**, *91*, 2688–2695. [CrossRef] [PubMed]
53. Yang, H.; Mu, S.; Li, J. Effects of ecological restoration projects on land use and land cover change and its influences on territorial NPP in Xinjiang, China. *Catena* **2014**, *115*, 85–95. [CrossRef]
54. Rao, C.; Yan, B. Study on the interactive influence between economic growth and environmental pollution. *Environ. Sci. Pollut. Res.* **2020**, *27*, 39442–39465. [CrossRef]
55. Muller, M.R.; Middleton, J. A Markov model of land-use change dynamics in the Niagara Region, Ontario, Canada. *Landsc. Ecol.* **1994**, *9*, 151–157.
56. Long, H.; Qu, Y. Land use transitions and land management: A mutual feedback perspective. *Land Use Policy* **2018**, *74*, 111–120. [CrossRef]
57. Urban, D.L.; Wallin, D.O. Introduction to Markov models. In *Learning Landscape Ecology: A Practical Guide to Concepts and Techniques*; Gergel, S.E., Turner, M.G., Eds.; Springer: New York, NY, USA, 2017; pp. 129–142. [CrossRef]
58. Zhu, E.; Deng, J.; Zhou, M.; Gan, M.; Jiang, R.; Wang, K.; Shahtahmassebi, A. Carbon emissions induced by land-use and land-cover change from 1970 to 2010 in Zhejiang, China. *Sci. Total Environ.* **2018**, *646*, 930–939. [CrossRef]
59. Bai, X.Y.; Wang, S.J.; Xiong, K.N. Assessing spatial-temporal evolution processes of karst rocky desertification land: Indications for restoration strategies. *Land Degrad. Dev.* **2013**, *24*, 47–56. [CrossRef]
60. Wang, J.; Guo, J.; Qing, J. Application of a kind of k drought index in the spring drought analysis in northwest China. *J. Nat. Resour.* **2011**, *22*, 709–717. [CrossRef]
61. Araromi, D.O.; Majekodunmi, O.T.; Adeniran, J.A.; Salawudeen, T.O. Modeling of an activated sludge process for effluent prediction—A comparative study using ANFIS and GLM regression. *Environ. Monit. Assess.* **2018**, *190*, 495. [CrossRef]
62. Mather, A.S. The Forest Transition. *Area* **1992**, *24*, 367–379.
63. Oliveira, T.M.; Guiomar, N.; Baptista, F.O.; Pereira, J.M.C.; Claro, J. Is Portugal's forest transition going up in smoke? *Land Use Policy* **2017**, *66*, 214–226. [CrossRef]
64. Yu, X. Environmental development and governance in western China since 2000: A case from Guizhou Province. *Int. J. Environ. Stud.* **2016**, *73*, 791–805. [CrossRef]
65. Zhang, X.; Zha, T.; Zhao, Y.; Qin, J.; Lyv, Z.; Ma, Z.; Yu, H.; Zhu, Y.; Wang, G.; Tettenborn, F.; et al. Sustainable effects of small hydropower substituting firewood program in Majiang County, Guizhou Province, China. *Sustainability* **2017**, *9*, 988. [CrossRef]
66. Dong, S.; Zhao, Y.; Li, X. Spatial Differentiation Characteristics and Driving Forces of Forest Transition: A Case Study of Zunyi City, Guizhou. *J. Resour. Ecol.* **2018**, *9*, 341–351.

67. Wang, B.; Gao, P.; Niu, X.; Sun, J. Policy-driven China's Grain to Green Program: Implications for ecosystem services. *Ecosyst. Serv.* **2017**, *27*, 38–47. [CrossRef]
68. Zeng, Z.; Estes, L.; Ziegler, A.; Chen, A.; Searchinger, T.; Hua, F.; Guan, K.; Jintrawet, A.; Wood, E. Highland cropland expansion and forest loss in Southeast Asia in the twenty-first century. *Nat. Geosci.* **2018**, *11*, 556–562. [CrossRef]
69. Lei, G.; Li, A.; Bian, J.; Zhang, Z. The roles of criteria, data and classification methods in designing land cover classification systems: Evidence from existing land cover data sets. *Int. J. Remote Sens.* **2020**, *41*, 5062–5082. [CrossRef]
70. Zeng, T.; Zhang, Z.; Zhao, X.; Wang, X.; Zuo, L. Evaluation of the 2010 MODIS Collection 5.1 Land Cover Type Product over China. *Remote Sens.* **2015**, *7*, 1981–2006. [CrossRef]
71. Wu, B.; Yuan, Q.; Yan, C.; Wang, Z.; Xinfang, Y.; Li, A.; Ma, R.; Huang, J.; Chen, J.; Chang, C.; et al. Land Cover Changes of China Form 2000 to 2010. *Quat. Sci.* **2014**, *34*, 723–731.
72. Yan, X.; Li, Y.; Xia, Y.; Hu, Y.; Yang, C. Analysis of characteristics and causes of persistent drought in Guizhou during 1961—2016. *Mid-Low Latit. Mt. Meteorol.* **2019**, *43*, 1–7.
73. Han, G.; Liu, C. Water geochemistry controlled by carbonate dissolution: A study of the river waters draining karst-dominated terrain, Guizhou Province, China. *Chem. Geol.* **2004**, *204*, 1–21. [CrossRef]
74. An, H.; Liu, M. Analysis on the benefits of Grain for Green Project and sustainable development of Guizhou Province. *Subtrop. Soil Water Conserv.* **2008**, *20*, 1–4.
75. Zhang, B.; Nie, C.; Zhu, J.; Yao, Y.; Mo, S.-g.; Luo, Y.; Cen, G. Dynamic change of forest resources in Guizhou province. *Geogr. Res.* **2003**, *22*, 725–732.
76. Han, D.; Yang, T.; Pan, T.; Chen, T.; Zhang, X. Analysis on measures of forest resources growth and change in guizhou province. *Agric. Technol.* **2020**, *40*, 73–75.
77. Sabir, M.; Ali, Y.; Khan, I.; Salman, A. Plants Species Selection for Afforestation: A Case Study of the Billion Tree Tsunami Project of Pakistan. *J. Sustain. For.* **2022**, *41*, 537–549. [CrossRef]
78. Dey, D.C.; Gardiner, E.S.; Kabrick, J.M.; Stanturf, J.A.; Jacobs, D.F. Innovations in afforestation of agricultural bottomlands to restore native forests in the eastern USA. *Scand. J. For. Res.* **2010**, *25*, 31–42. [CrossRef]
79. Cunningham, S.C.; Mac Nally, R.; Baker, P.J.; Cavagnaro, T.R.; Beringer, J.; Thomson, J.R.; Thompson, R.M. Balancing the environmental benefits of reforestation in agricultural regions. *Perspect. Plant Ecol. Evol. Syst.* **2015**, *17*, 301–317. [CrossRef]
80. Babintseva, R.M.; Titova, Y.V. Effects of Fire on the Regeneration of Larch Forests in the Lake Baikal Basin. In *Fire in Ecosystems of Boreal Eurasia*; Goldammer, J.G., Furyaev, V.V., Eds.; Springer: Dordrecht, The Netherlands, 1996; pp. 358–365. [CrossRef]
81. Goldammer, J.G.; Furyaev, V.V. Fire in Ecosystems of Boreal Eurasia: Ecological Impacts and Links to the Global System. In *Fire in Ecosystems of Boreal Eurasia*; Goldammer, J.G., Furyaev, V.V., Eds.; Springer: Dordrecht, The Netherlands, 1996; pp. 1–20. [CrossRef]
82. Chu, T.; Guo, X. Remote Sensing Techniques in Monitoring Post-Fire Effects and Patterns of Forest Recovery in Boreal Forest Regions: A Review. *Remote Sens.* **2014**, *6*, 470–520. [CrossRef]
83. Xiao, S.-C.; Xiao, H.-L.; Peng, X.-M.; Tian, Q.-Y. Daily and seasonal stem radial activity of Populus euphratica and its association with hydroclimatic factors in the lower reaches of China's Heihe River basin. *Environ. Earth Sci.* **2014**, *72*, 609–621. [CrossRef]
84. Curtin, D.; Beare, M.H.; Hernandez-Ramirez, G. Temperature and Moisture Effects on Microbial Biomass and Soil Organic Matter Mineralization. *Soil Sci. Soc. Am. J.* **2012**, *76*, 2055–2067. [CrossRef]

land

MDPI

Article

NDVI-Based Vegetation Dynamics and Their Responses to Climate Change and Human Activities from 2000 to 2020 in Miaoling Karst Mountain Area, SW China

Yangyang Wu [1,2], Jinli Yang [3], Siliang Li [2], Chunzi Guo [2,4], Xiaodong Yang [3,5], Yue Xu [6], Fujun Yue [2], Haijun Peng [6], Yinchuan Chen [7], Lei Gu [8], Zhenghua Shi [3] and Guangjie Luo [1,9,*]

1 School of Geography and Resources, Guizhou Education University, Guiyang 550018, China; wuyangyang@gznc.edu.cn
2 School of Earth System Science, Tianjin University, Tianjin 300072, China
3 College of Ecology and Environment, Xinjiang University, Urumqi 830017, China
4 Administration of Ecology and Environment of Haihe River Basin and Beihai Sea Area, Ministry of Ecology and Environment of People's Republic of China, Tianjin 300061, China
5 Department of Geography and Spatial Information Techniques, Center for Land and Marine Spatial Utilization and Governance Research, Ningbo University, Ningbo 315211, China
6 State Key Laboratory of Environmental Geochemistry, Institute of Geochemistry, Chinese Academy of Sciences, Guiyang 550081, China
7 Shanghai Ecology and Environment Scientific Research Center, Yangtze River Basin Ecological Environment Supervision and Administration Bureau, Ministry of Ecology and Environment of People's Republic of China, Shanghai 200120, China
8 College of Geography and Remote Sensing Sciences, Xinjiang University, Urumqi 830017, China
9 Guizhou Provincial Key Laboratory of Geographic State Monitoring of Watershed, Guizhou Education University, Guiyang 550018, China
* Correspondence: luoguangjie@gznc.edu.cn

Citation: Wu, Y.; Yang, J.; Li, S.; Guo, C.; Yang, X.; Xu, Y.; Yue, F.; Peng, H.; Chen, Y.; Gu, L.; et al. NDVI-Based Vegetation Dynamics and Their Responses to Climate Change and Human Activities from 2000 to 2020 in Miaoling Karst Mountain Area, SW China. *Land* **2023**, *12*, 1267. https://doi.org/10.3390/land12071267

Academic Editor: Eusebio Cano Carmona

Received: 29 March 2023
Revised: 26 May 2023
Accepted: 16 June 2023
Published: 21 June 2023

Abstract: Understanding spatiotemporal shifts in vegetation and their climatic and anthropogenic regulatory factors can offer a crucial theoretical basis for environmental conservation and restoration. In this article, the normalized difference vegetation index (NDVI) of the Miaoling area from 2000 to 2020 is studied using a trend analysis and the Mann–Kendall mutation test (MK test) to review the vegetation's dynamic changes. Our study uses the Hurst index, a partial correlation analysis, and a geographic detector to investigate the contributions of climate change and human activities to regional vegetation changes and their drivers. We found that Miaoling's annual average NDVI was between 0.66 and 0.83 in 2000–2020, with a mean of 0.766. The overall trend was slow upward (0.0009/year), and 53.82% of the region continued to grow and gradually increased from west to east in the spatial domain, among which the karst regional NDVI distribution area and its growth rate were higher than those of non-karst sites. Based on correlations between climatic factors and NDVI, precipitation seasonality (coefficient of variation, CV) had the strongest correlation (positive correlation) with NDVI, while vapor pressure deficit (VPD) had a negative correlation with NDVI. In the interaction, human activities played a dominant role in the influence of NDVI on the vegetation of Miaoling. The night light index had the most explanatory power on the NDVI ($q = 0.422$), and the interaction between anthropogenic factors and other factors dominated its explanatory power. This study has academic and practical importance for the management, protection, and sustainable development of karst basins.

Keywords: vegetation dynamics; Miaoling; karst plateau; trend analysis; geographic detector

1. Introduction

Since the Anthropocene, human-induced global climate change has had a significant impact on terrestrial ecosystems [1]. As carbon, water, and energy exchangers between land and air, plants can provide people with oxygen, food, fiber, fuel, carbon sinks, and other valuable ecosystem services [2–4]. Accordingly, significant changes in the global climate and their impacts on vegetation growth over recent decades have received increasing attention [5]. These include changes in climate and environmental conditions as well as human activities, such as land use change. Environmental factors include a wider range of chemical, physical, and biological elements that can affect ecosystems, including disturbances such as droughts, fires, and floods [6]. The normalized difference vegetation index (NDVI) is becoming a significant indicator in studying the spatial vegetation dynamics of regional ecological changes [7–9]. Thus, the role of climate and human factors in vegetation dynamics is one of the hottest topics in global change science [10,11].

The impact of climate change and human activity on vegetation change has been studied by many scientists. Research shows that climate change mainly impacts vegetation by changing climate, temperature, precipitation, soil moisture, and seasonal variations [7,10,12,13]. Amongst these, temperature controls the growth and distribution of vegetation by preventing the onset, termination, and distribution of the photosynthetic process [14,15], which is considered the main cause of the greening trend in northern latitudinal and elevated regions (Qinghai–Tibetan Plateau) [16–18]. However, in the middle and lower latitudes, temperature does not have an important impact on plant growth. However, precipitation has a clear role in boosting vegetation growth in arid and semi-arid regions [19,20]. For example, in southwest Chinese karsts [21,22] and Xinjiang [23], precipitation seasonality and variability significantly affect vegetation change. Precipitation, rather than temperature, has become the key factor controlling vegetation growth [20,24]. Soil moisture, which is strongly related to temperature and precipitation, has become an important limiting factor in arid and semi-arid areas. Furthermore, with high temperatures and heatwaves, the atmospheric vapor pressure deficit (VPD) is becoming an increasingly important driver of plant community function. High VPD can induce plant stomatal closure to prevent high water loss [25,26], which has been identified as a major factor in extreme drought-induced plant death [25]. Clearly, the impacts of climate change on vegetation cover are diverse and complex because of regional variations in climate change and ecological environmental conditions.

At present, vegetation changes and their spatial patterns are strongly affected by human activities. With the rapid development of the world's population and economy, the impact of anthropogenic activities on changes in surface vegetation cover is increasing, affecting the balance of terrestrial ecosystems on a large scale [8]. Among these, land use change is a significant and strongly spatially changing factor in vegetation change. On the one hand, unreasonable agricultural activities, excessive reclamation, grazing, and urban expansion significantly reduce vegetation cover [27,28]. On the other hand, human activities can increase vegetation cover by planting trees, closing hills to reforestation, and improving agricultural technology [29,30]. Besides land use change, many studies have used the night light index (NLI) to characterize regional economic development or urbanization and the intensity of surface human activities [29–31]. Therefore, the NLI was included as one of the factors influencing vegetation in the NDVI.

There are few studies on the connection between vegetation changes and their drivers in important watersheds in the karst region, although there have been many studies on the impact of climate variation and human behaviors on vegetation change [3,7,10,15,19,20]. Miaoling Mountain is an important watershed ecological security barrier in the karst region of southwest China. It is located in the center of the core area of southwest China's karst plateau. The contradiction between man and land is pronounced due to the high ecological sensitivity and vulnerability of karst areas, plus climatic factors. Biomass above ground [32,33] and biomass below ground [34,35] are significantly lower in karstic forests than in non-karstic forests. Therefore, vegetation plays an important role in managing karst desertification and restoring ecosystems [36]. However, to our knowledge, the driving mechanism of human activities on NDVI changes in vegetation in the Miaoling's karst watershed is still unclear; there is no research to quantify how climatic and human factors affect vegetation in the area. Additionally, the impact of each driver on vegetation change has not been quantified.

Based on this deficiency, this article aims to study the vegetation dynamics of important watersheds (Miaoling) in karst areas from the following four perspectives: climate, human activity, topography, and soil. The aims of this article are as follows: (1) to analyze the variation characteristics and trends of the NDVI's spatial distribution in Miaoling from 2000 to 2020; (2) to explore the correlation between climate factors and NDVI; and (3) to explore the key driving factors affecting the NDVI of Miaoling vegetation. The results of this study have far-reaching significance for the sustainable ecological development of karst areas in southwest China and the realization of China's strategic goal of "carbon peaking and carbon neutrality". At the same time, it provides a reference value for the vegetation's dynamic driving factors in the important karst watershed and theoretical guidance for ecological environment management and sustainable development of the watershed.

2. Materials and Methods

2.1. Study Area

Mt. Miaoling, the research site, is located in the center of the core region of the Karst Plateau in southwest China. It is an important watershed for the ecological security barrier of the watershed in the karst region of southwest China. Its main range (25.73°–27.16° N, 103.82°–109.48° E) takes the watershed of the Pearl River system and the Yangtze River system as the main axis and then extracts the Miaoling study area from the small watershed in the north and south of the main axis and divides it into east, central, and west sections (Figure 1). From east to west, Miaoling is the watershed of the Yuanjiang River Basin, the Liujiang River Basin, the Wujiang River Basin, the Hongshui River Basin, the Niulan-Jiang-Heng He River Basin, and the Beipanjiang River Basin. The altitude in the area is 146–2877 m, and the peaks are often above 1500–2000 m. The main peak of the eastern section is Leigong Mountain, at up to 2179 m. The middle section of Doupeng Mountain is 1961 m high, and the western section with Laowang Mountain is 2127 m above sea level. The eastern, central, and western parts reach an altitude of 2179 m. Miaoling is a humid, mountainous area with a sub-tropical monsoon climate. The main vegetation in the Miaoling Mountain area consists of mixed evergreen and mixed broadleaf forests as the primary forest type and shrubs, grass, and forbs as the vegetation degradation type, which differs significantly from the zonal non-karst vegetation of evergreen broadleaved forests in subtropical China (ECVC 1980) [37,38]. It is one of the important forest areas in Guizhou Province. Miaoling has no tectonic veins, and the geology and topography of the sections are very different. Additionally, it is formed by a number of north–south anticlines composed of hard rock formations and a combination of uplifted and high regions. The layered landforms composed of planes and large karst basins are the most prominent regional landforms, thus forming unique landscapes such as planes with concentrated arable, basin areas, and terraced fields located high on the hillsides.

2.2. Data

The NDVI is based on the Google Earth Engine cloud computing platform and selects US remote sensing imagery from the Landsat 5, Landsat 7, and Landsat 8 satellites with 16-day temporal resolution and 30 m spatial resolution. At the same time, the dataset was referenced by the Land Use and Global Change Remote Sensing Team of the Institute of Geographical Sciences and Natural Resources Research, Chinese Academy of Sciences, with a spatial resolution of 30 m and a temporal resolution of each year [39]. Through methods such as series data preprocessing and data smoothing, the maximum NDVI value of each pixel in a year from 2000 to 2020 was obtained. The annual mean NDVI values from 2000 to 2020 were generated using the maximum synthesis method (MSV). When generating long-time series NDVIs, the maximum value synthesis method can decrease the effect of cloud cover, shadows, suspended particles in the atmosphere, etc., so that the error is reduced and the accuracy is improved [39].

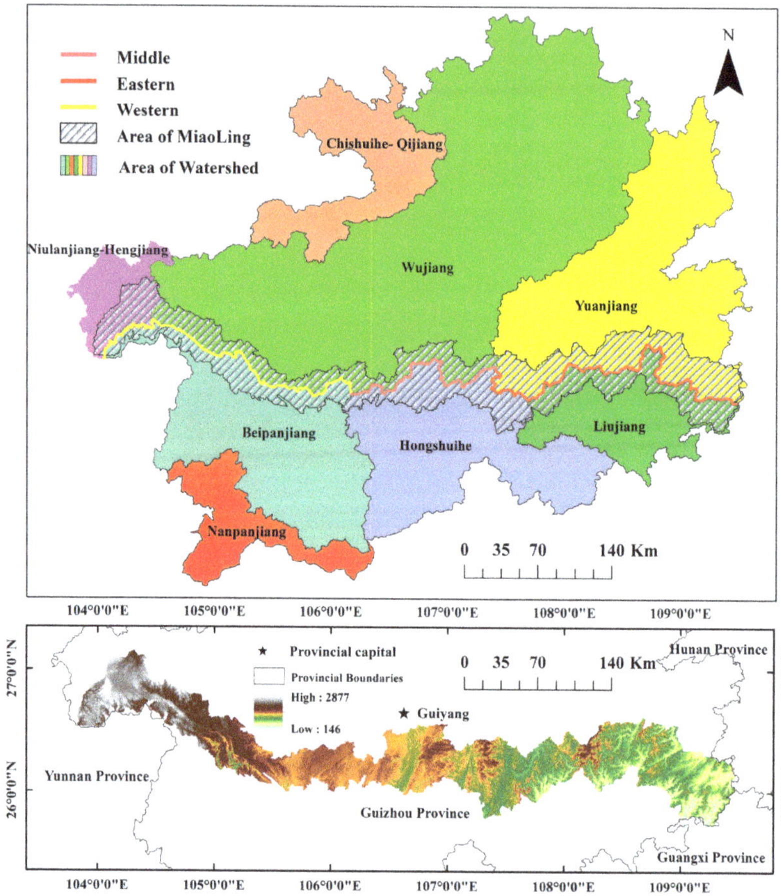

Figure 1. Location of the Miaoling Mountain area.

The DEM data involved in this study are derived from the geospatial data cloud, which is downloaded using a rectangular frame clipping area and then transformed into raster data through an ArcGIS overlay, which is the basis for the calculation of the total height, slope, and aspect of the research region. The land use change data are sourced from Zenodo (https://zenodo.org/, accessed on 15 June 2020) and published by Huang Xin et al. from Wuhan University in China [40]. The biggest advantage of this dataset is its continuous 30 m land use classification results. The NLI can be used to describe the intensity of human activity and is closely linked to economic development. This study selected it as a factor in human activity and conducted a related driving analysis.

The meteorological data use raster data for precipitation and air temperature with a resolution of 1 km. This dataset introduces the influence of terrain on the climate in the temperature and precipitation data generated by the ANUSPLIN interpolation tool. The interpolation error is the smallest, and the accuracy is greatly improved compared with other interpolation methods [41]. The analysis of meteorological elements is more suitable. The meteorological interpolation software ANUSPLIN is used to interpolate the data into monthly synthetic precipitation multi-band data with a spatial resolution of 1 km. It is highly accurate, has high resolution, is a long-time series, and has better scientific research and application potential [42]. The above-mentioned monthly synthetic temperature and precipitation data are extracted by ArcGIS and then resampled to 30 m.

Soil moisture is an important link between the atmosphere and the terrestrial ecosystem. The soil moisture data included a set of neural nets using a data fusion of up to 11 microwave remote sensing-based soil moisture products. These data were obtained by satellite for the global surface soil moisture for the period 2003–2018 with a spatial resolution of 0.1° [43]. The bioclimatic variable data are available at a spatial resolution of 1 km, with historical monthly weather data for 1960–2018 [44]. The range of data downloaded for this study was 2000–2018. The VPD dataset mixes the high spatial resolution climate normal of the Worldclim dataset with the less accurate but time-varying data from CRU Ts4.0 and Japan's 55-year reanalysis (JRA55) with climate-assisted interpolation. Conceptually, the interpolation time variation from CRU Ts4.0/JRA55 is often applied to Worldclim high spatial resolution climatology to create a high spatial resolution dataset covering a wider range of time records. The above data and the data sources involved in the study are as follows (Table 1):

Table 1. Data and sources.

Data Type	Factors	Time	URL
Vegetation	NDVI	2000–2020	https://code.earthengine.google.com
Climate	Precipitation seasonality(Coefficient of variation)	2000–2018	https://www.worldclim.org/data/
	Precipitation of wettest quarter	2000–2018	
	Mean annual precipitation	2000–2020	http://www.resdc.cn
	Mean annual temperature	2000–2020	
	Vapor pressure deficit	2000–2020	https://climate.northwestknowledge.net/
Soil	Soil moisture	2003–2018	https://doi.org/10.1594/PANGAEA.912597
Landform	Elevation	2020	http://www.gscloud.cn/
Topography	Karst	2010	http://www.csdata.org/
	Non-karst	2010	
Human activity	Land use change	2000–2020	https://zenodo.org/
	Night light index	2000–2020	https://ngdc.noaa.gov/

2.3. Methods

In this study, in order to explore the characteristics, trends, and driving factors of vegetation change in Miaoling, we used the following methods to analyze the relevant data in sequence and with specific applications: We first conducted trend analysis and a Mann–Kendall (MK) trend test on NDVI changes in Miaoling's vegetation. Additionally, we used the Hurst index to perform a continuous analysis of the change trends of the NDVI to explore the continuity of the Miaoling NDVI trends in the spatial distribution and change trend. The correlation study of the regional distribution and variation trends of NDVI in Miaoling with climatic conditions was completed by using a partial correlation analysis. The association characteristics of the climatic elements, human activities, terrain, and vegetation changes in Miaoling were also studied using a geographic detector.

2.3.1. Theil–Sen Median Trend Analysis

We conducted a trend analysis on the NDVI of Miaoling's vegetation using the Theil–Sen median method, which is referenced in this literature [45] and is expressed in Equation (1) as follows:

$$\beta = \text{median}\left(\frac{x_j - x_i}{j - i}\right), \forall j > i \quad (1)$$

where j and i are the time series data. A value greater than 0 means that the time series shows an upward trend. A value less than 0 means that the time series shows a downward trend, and a value closer to 0 means that the time series changes are not significant. The Theil–Sen median trend analysis and MK test significance test results were superimposed and analyzed. As Table 2 shows, the results are grouped into six categories. Based on the above results, we also conducted breakpoint detection on the dynamic trend of vegetation in Miaoling.

Table 2. Type of change trend of DNVI.

Slope	p	Trend	Slope	p	Trend
Slope < 0	$p < 0.01$	significant decrease	Slope > 0	$p < 0.01$	significant increase
Slope < 0	$0.01 < p < 0.05$	minimal decrease	Slope > 0	$0.01 < p < 0.05$	moderate increase
Slope < 0	$p > 0.05$	non-significant decrease	Slope > 0	$p > 0.05$	non-significant increase

2.3.2. Mann–Kendall Test

The Mann–Kendall trend test is a nonparametric statistical test used to test significant trends of change. The samples do not need to have a specific distribution, nor are they affected by a few outliers [46]. Equation (3) is as follows:

$$\begin{gathered} S_k = \sum_{i=1}^{k}\sum_{j=1}^{i}\text{sgn}(X_i - X_j) \\ UF_k = [S_k - E(S_k)]/\sqrt{\text{Var}(S_k)} \end{gathered} \tag{2}$$

where X_i and X_j represent the NDVI values of time i and j, respectively. S_k is the cumulative count of $X_i > X_j$. $E(S_k)$ and $\text{Var}(S_k)$ are the mean and variance of S_k, respectively. $UF_k > 0$ indicates an upward trend of the NDVI sequence, while $UF_k < 0$ indicates a downward trend of the NDVI sequence. Combining the NDVI trend classification results (Table 2), the non-significant decrease and non-significant increase are classified into one category (i.e., no change). The results are divided into the following five levels [47]: significant decrease, moderate decrease, no change, moderate decrease, and significant decrease.

2.3.3. Hurst Index

The Hurst index is an effective method to describe the information dependence of long time series [48]. For the NDVI time series, NDVI (τ), = 1, 2, 3, 4, . . . , n. For any positive integer ≥ 1, the mean series of the time series is defined as follows:

$$\overline{\text{NDVI}_{(\tau)}} = \frac{1}{\tau}\sum_{t=1}^{\tau}\text{NDVI}_{(\tau)} \qquad \tau = 1, 2, \cdots, n \tag{3}$$

The Hurst index is calculated as follows:

$$\frac{R_{(\tau)}}{S_{(\tau)}} = (c\tau)^H \tag{4}$$

The relevant values involved in the calculation of the Hurst index are calculated as follows:

(1) The cumulative deviation is as follows:

$$X_{(t,\tau)} = \sum_{t=1}^{t}(\text{NDVI}_{(t)} - \overline{\text{NDVI}_{(\tau)}}), 1 \leqslant t \leqslant \tau \tag{5}$$

(2) The range sequence is as follows:

$$R_{(\tau)} = \max_{1\leqslant t\leqslant \tau} X_{(t,\tau)} - \min_{1\leqslant t\leqslant \tau} X_{(t,\tau)}, \tau = 1, 2, \cdots, n \tag{6}$$

(3) The standard difference order is as follows:

$$S_{(\tau)} = [\frac{1}{\tau}\sum_{t=1}^{\tau}(\text{NDVI}_{(t)} - \text{NDVI}_{\tau})^2]^{\frac{1}{2}}, \tau = 1, 2, \cdots, n \quad (7)$$

The Hurst index (value) may reflect the persistent nature of the NDVI time series. In the Hurst exponent (value), when 0 < H < 0.5, this indicates that the change is continuing to decline, meaning that the future change trend is opposite to the past change trend. When = 0.5, this indicates that the NDVI time series is a random series and there is no long-term correlation. When 0.5 < H < 1, the time series are characterized by long-term dependence and persistence, meaning that the future change is consistent with the past trend. In other words, areas that have tended to increase in past years are likely to increase in years to come, and vice versa. The closer it is to 1, the stronger the persistence.

2.3.4. Analysis of Correlation

To explore the response of vegetation dynamic changes in Miaoling to climate change factors, we conducted a partial correlation analysis between NDVI and climate factors (Table 3) to reveal the main driving forces controlling the interannual changes of NDVI from 2000 to 2020.

$$R_{12,3} = \frac{r_{12} - r_{13}r_{23}}{\sqrt{(1 - r_{13}^2)(1 - r_{23}^2)}} \quad (8)$$

where $R_{12,3}$, $R_{13,2}$, $R_{23,1}$ are the correlation coefficients among the variables; $R_{12,3}$ is the partial correlation coefficient between r_1 and r_2 after fixing the variable r_3. The value range of the partial correlation coefficient ranges from −1 to 1. When $R_{12,3} > 0$, the correlation is positive, meaning that both factors correlate in the same direction. When $R_{12,3} < 0$, the correlation is negative. The higher the partial correlation coefficient, the stronger the correlation between the two elements at the pixel.

Table 3. Statistics of the partial correlation coefficient and significance reveal the intricate relationship between climate factors and NDVI.

Factors	Mean	Correlation Coefficient Range	Extremely Significant Positive Correlation (%)	Significant Positive Correlation (%)	Non-Significant Correlation (%)	Significant Negative Correlation (%)	Extremely Significant Negative Correlation (%)
X1	0.32	−0.63~0.82	64.5	13	10.1	8.7	3.7
X2	−0.26	−0.68~0.76	11.4	7.5	9.2	17.2	54.7
X3	0.23	−0.57~0.66	23.6	34.6	16.9	18.7	6.2
X4	0.21	−0.67~0.79	49.2	5	17.7	19.1	9.1
X5	0.17	−0.58~0.66	40.2	13.5	15.1	14.2	17
X6	0.15	−0.59~0.65	15.3	32.2	25.4	12.5	15.3

Note: X1, precipitation seasonality (CV); X2, VPD; X3, precipitation of wettest quarter; X4, MAP; X5, soil moisture; X6, MAT.

2.3.5. Geographic Detector

Geographic detection is a new spatial statistical method to detect spatial differentiation and reveal driving factors [49]. It contains the following four detectors: factor detection, interaction detection, risk detection, and ecological detection. The first two parts apply here. The model is shown below:

(1) Factor detector

A factor detector could determine the effect of detecting the spatial heterogeneity of vegetation change. The spatial heterogeneity of X to Y could be expressed as q × 100%, and the greater the number, the greater the influence of the detection factors on vegetation change [50], which is as follows:

$$q = 1 - \frac{\sum_{h=1}^{L} N_h \sigma_h^2}{N\sigma^2} \quad (9)$$

where h is the vegetation change or detection factor hierarchy; N is the number of class h or total region units; and Y is the change in class h or total region Y value.

(2) Interaction detector

The interaction detector is appropriate to identify the impact of the combination of the detection drivers X_a and X_b on the heterogeneity of the spatial variation of vegetation. The five interaction results are as follows: [50].

Miaoling's vegetation change trend from 2000 to 2020 was regarded as the dependent variable Y. The selected driving factors mainly include the key factors selected by partial correlation analysis as the detection factor X, as shown in Table 4.

Table 4. The q value of each driving factors.

Factors	X1	X2	X3	X4	X5	X6	X7	X8	X9	X10	X11	X12	X13
q value	0.355	0.322	0.243	0.21	0.17	0.155	0.272	0.173	0.125	0.331	0.401	0.397	0.422
p value	0.000	0.000	0.000	0.000	0.000	0.000	0.000	0.000	0.000	0.000	0.000	0.000	0.000

Note: X1, precipitation seasonality (CV); X2, VPD; X3, precipitation of wettest quarter; X4, MAP; X5, soil moisture; X6, MAT; X7, elevation; X8, slope; X9, aspect; X10, karst; X11, non-karst; X12, land use change; X13, NLI.

3. Results

3.1. Spatial Distribution and Trend Change of NDVI

This article uses a combination of the Theil–Sen test and the MK method, as well as breakpoint detection, to count pixels of vegetation coverage from 2000 to 2020. Research results show that Miaoling has a mean NDVI of 0.766 (Figure 2c). In addition, the average NDVI variation range of Miaoling over the years is 0.659–0.827, with a rising trend with an increase of 0.9×10^{-3}/year ($R^2 = 0.053$, $p > 0.05$) (Figure 2a). In addition, the breakpoint detection results indicate that there is a breakpoint in the Miaoling study time series (Figure 2b). As shown in Figure 2c, the change trend before and after 2011 was opposite. Before 2011, the NDVI in Miaoling showed an increasing trend, while it tended to decline after 2011. For the period 2000–2020, the overall NDVI in the Miaoling region slowly increased. However, there are significant fluctuations, mainly reflected in the sharp decline after 2010 and 2017 and the sharp growth in 2013. Especially after 2011 and 2018, the NDVI values tended to decrease, with a decrease rate of over 10% compared to the previous year's NDVI value in 2019.

The characteristics of the spatial distribution of the NDVI in Miaoling are shown in Figure 3. The eastern section is mainly characterized by a significant growth distribution of NDVI, accounting for 40.6% of the area. The NDVI in the middle segment mainly shows moderate growth, with an area ratio of 30.5%. The northern and western sections of the middle section are mainly characterized by a decrease and no change, with area proportions of 12.9% and 16%, respectively.

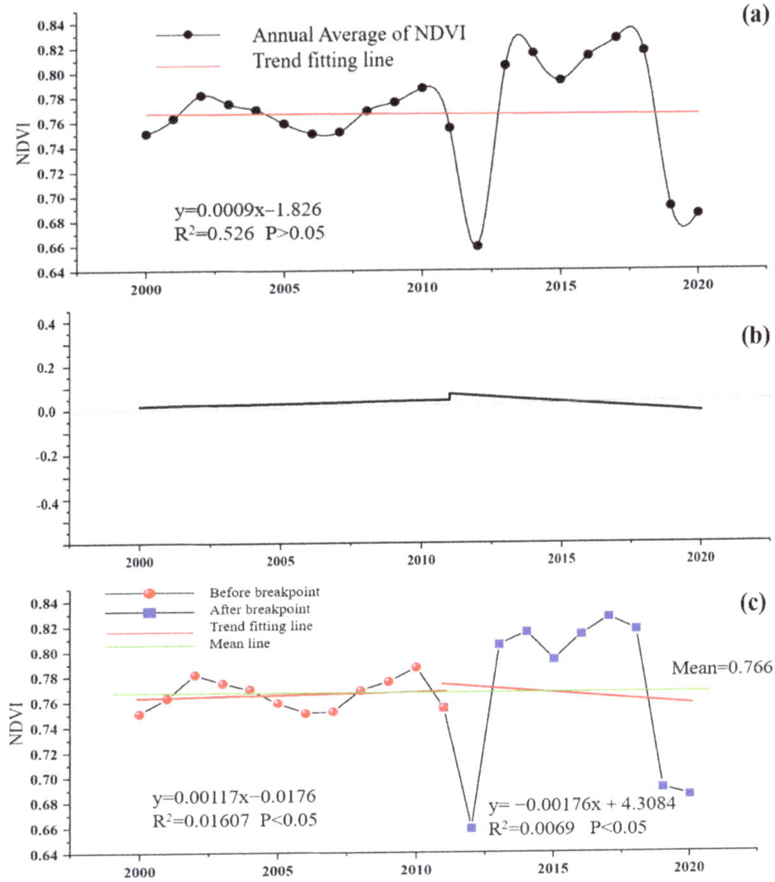

Figure 2. The variation characteristics of NDVI values in Miaoling from 2000 to 2020. (**a**) The interannual variation and trend fitting of NDVI values in Miaoling; (**b**) Breakpoint detection results; (**c**) Trend fitting of NDVI before and after breakpoints.

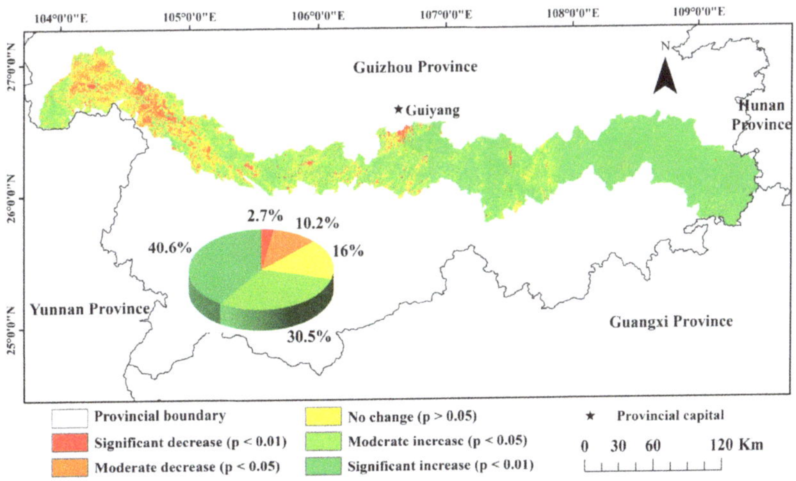

Figure 3. Distribution of the inter-annual NDVI in the Miaoling region from 2000 to 2020.

The annual average NDVI change can reflect the dynamic characteristics of vegetation and serve as an important indicator of vegetation health and ecosystem stability. Thus, the Theil–Sen median trend analysis was used to analyze the NDVI trend over the last 21 years at the pixel scale. We found that the spatial distribution of NDVI is highly heterogeneous (Figure 4a). The NDVI in Miaoling increased in most areas and decreased in some local areas. The change trend of NDVI in most areas is an increasing trend (accounting for 68.73%). Among them, the highest growth is concentrated in the east, accounting for about half of the total growth. In addition, the slightly degraded and significantly degraded areas account for 32.61%. The significantly degraded areas are mainly distributed in the western part of the study area. The remaining 8.66% did not show a significant change.

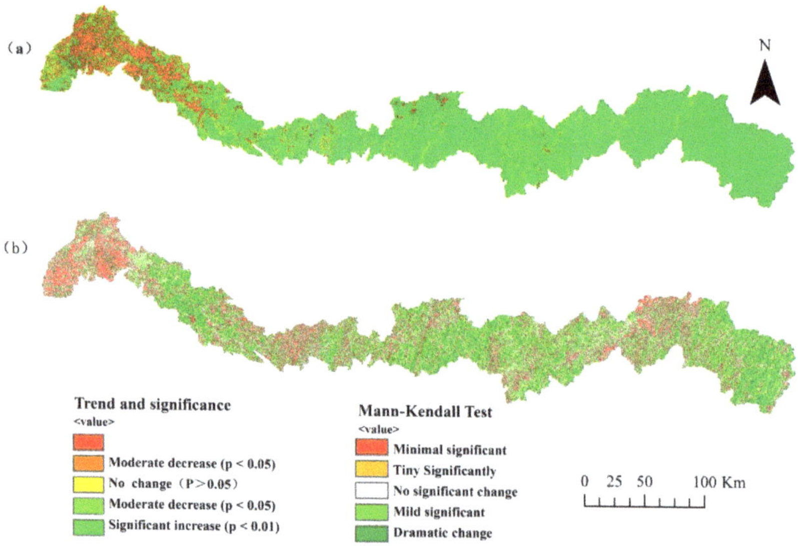

Figure 4. The trend of NDVI distribution (**a**) and MK test results (**b**) in the Miaoling area from 2000 to 2020. The results divide the change trend into five grades: significant degradation, $p < 0.01$; mild degradation, $p < 0.05$; no significant change, $p > 0.05$; mild improvement, $p < 0.01$; significant improvement, $p < 0.05$.

The MK test results indicate that from 2000–2020 (Figure 4b), the change in the NDVI in most areas of Miaoling is relatively strong, with a mainly strong distribution in the eastern, central–southern, and a small part of the western areas, while the area with an insignificant change is mainly distributed across the majority of the western Miaoling area.

We also tested the persistence of NDVI changes in Miaoling through the Hurst index, and the mean Hurst index of NDVI in Miaoling was found to be 0.56 (Figure 5). Additionally, the region with continuous growth ($0.5 < H < 1$) has the widest distribution (53.82%), mainly in the eastern and central parts of the study area. However, most of the west area is still trending down. As can be seen, the Hurst index of the NDVI has similar spatial heterogeneity to the distribution of the NDVI in the study area.

3.2. *Impact Factor of Land Use Change on the NDVI*

The karst area is mainly concentrated in the eastern and middle sections of the Miaoling area (Figure 6a), dominated by woodland and followed by grass. The non-karst area is mainly distributed in the western section and is mostly arable land, followed by grass (Figure 6b). The NDVI of the karst area is higher than that of the non-karst area in the Miaoling area.

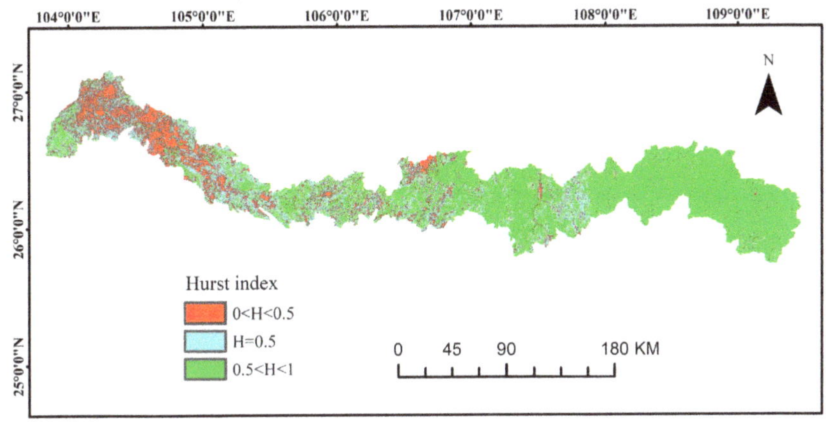

Figure 5. The Hurst index distribution of NDVI in Miaoling. 0 < H < 0.5; continuous decline; H = 0.5, no significant change; 0.5 < H < 1, continuous growth.

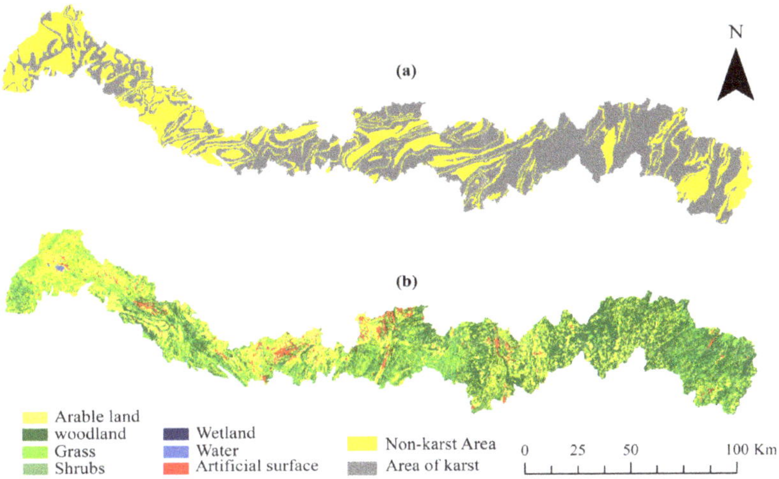

Figure 6. Karst distribution in Miaoling area of Guizhou Province (**a**) and spatial distribution characteristics of land use types (**b**) in 2000 to 2020.

The overall average proportion of land use types in the Miaoling area (Figure 7) is as follows: wood land (50%) > arable land (31.60%) > grass (14%) > artificial surface (1.93%) > shrub (1.45%) > waters (0.93%) > wetlands (0.09%). The total proportion of forest and arable land fluctuates, and the proportion of artificial surface continues to increase; grass, water, wetlands, and shrubs fluctuate slightly. Comparing the spatial distribution of NDVI in Miaoling (Figure 3) with land use types, it can be roughly divided into the following characteristics: The land use types corresponding to NDVI reduction areas are mainly concentrated in areas such as artificial surfaces, cultivated land, and wetlands. The NDVI of forested areas mainly manifests as growth.

The most immediate effect of land use change on NDVI is change. The most notable changes in NDVI are specifically in arable land and woods. Figure 8 shows that the Miaoling region's degraded regions of wood land, arable land, and grass are primarily located in the center region. As a result, it is clear from Figure 6b that the primary land use categories (wood land, grass, and arable land) in the Miaoling region are spread in regions with high concentrations of human activity (artificial surface areas).

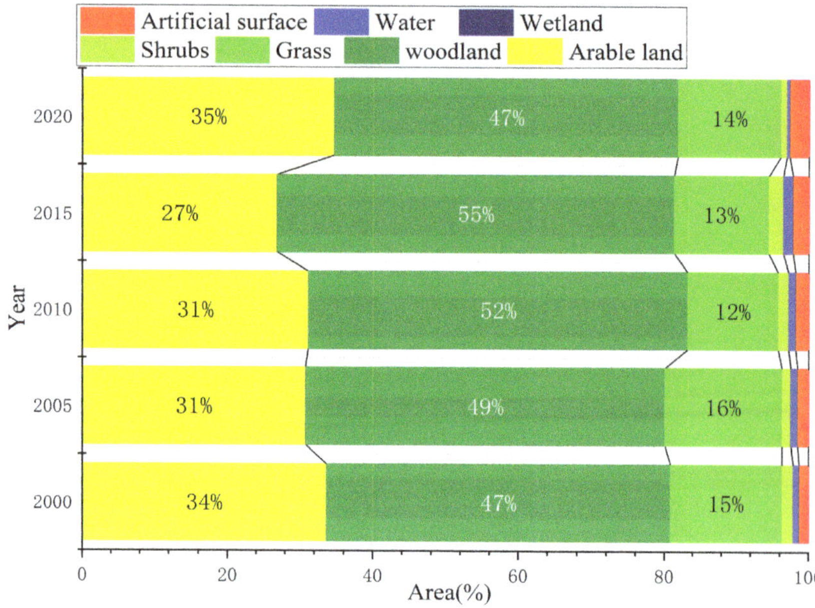

Figure 7. Total area ratio of each land use type in Miaoling, Guizhou, from 2000 to 2020.

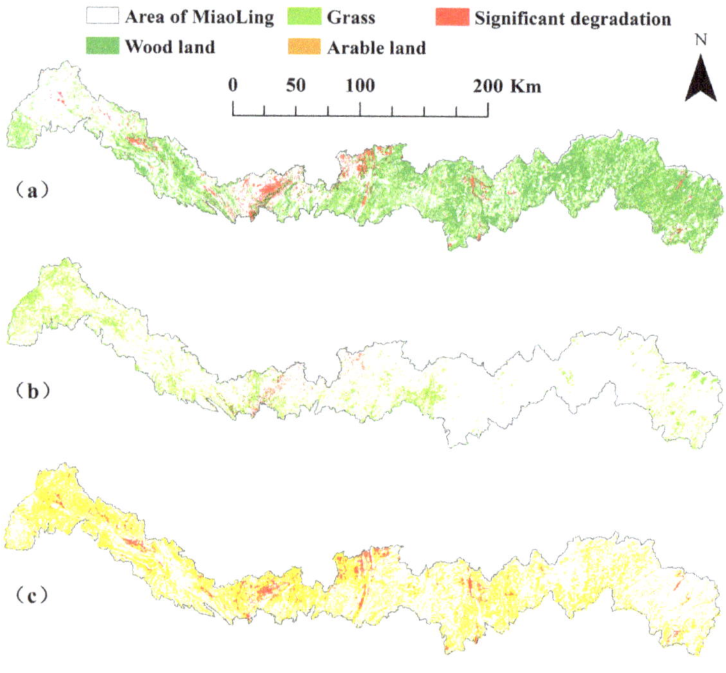

Figure 8. Woodland (**a**), grass (**b**), and arable land (**c**) change distribution characteristics in Miaoling.

3.3. Impact Climate Factors of the NDVI

Climatic factors are as important as human activities (land use change) in influencing vegetation dynamics. Our analysis of the partial correlation between NDVI and climate factors was calculated at the image metric scale for the studied region from the year 2000 to 2020. The findings of the partial correlation study make it clear that the NDVI and precipitation seasonality (CV) have the strongest association, whose mean correlation coefficient was 0.32. As shown in Table 3, their correlation strength is in order of precipitation seasonality (CV) (0.32) > VPD (−0.26) > precipitation in the wettest quarter (0.23) > MAP (0.21) > soil moisture (0.17) > MAT (0.15). The other five factors, with the exception of VPD, have an overall positive correlation with the NDVI.

The results showed significant spatial heterogeneity in the NDVI vegetation–climate correlation in the Miaoling region. In combination with Figure 9 and Table 3, 77.8% of the precipitation seasonality (CV) area was positively correlated with the NDVI (Figure 9a), which was widely distributed in the study area. A total of 71.9% of the VPD was significantly negatively correlated with the NDVI, and only 18.9% of the area was positively correlated with the NDVI, which was mainly distributed in the western and central parts of the study area (Figure 9b). A total of 58.2% of Precipitation of wettest quarter was significantly positively correlated with NDVI over the total area, and the areas with a significant negative correlation were mainly concentrated in the western part of the study area (Figure 9c). Precipitation was significantly positively correlated with NDVI in 54.2% of the spatial area and distributed in the westernmost, easternmost, and northernmost parts of the study area (Figure 9d). A total of 53.7% of the soil moisture area was significantly positively correlated with NDVI, while the easternmost area was significantly negatively correlated (Figure 9e). The least correlated climatic factor with NDVI in Miaoling was temperature, which was significantly positively correlated with NDVI in 47.5% of the study area, while MAT was significantly negatively correlated with NDVI in the far west and north (Figure 9f).

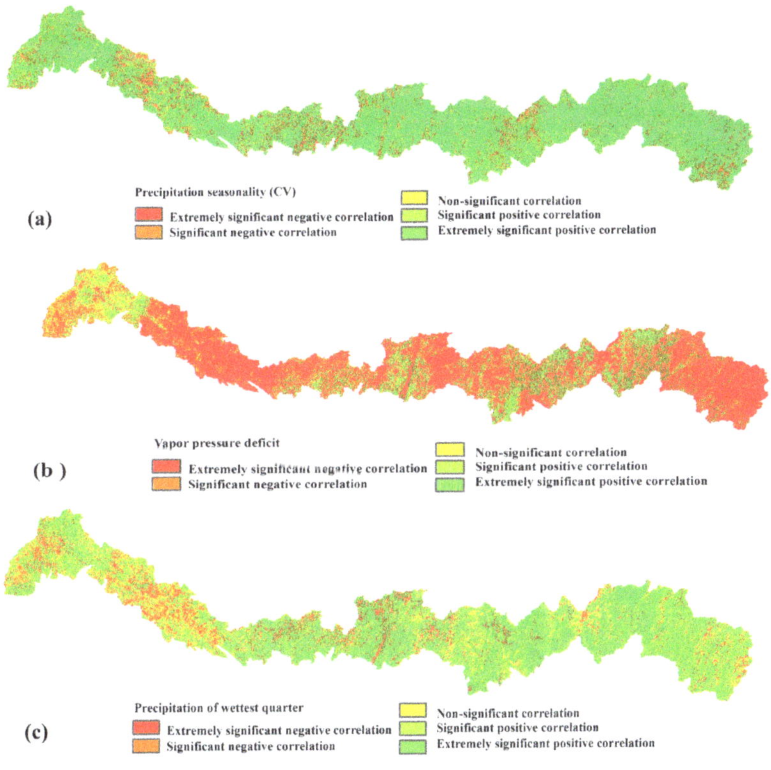

Figure 9. *Cont.*

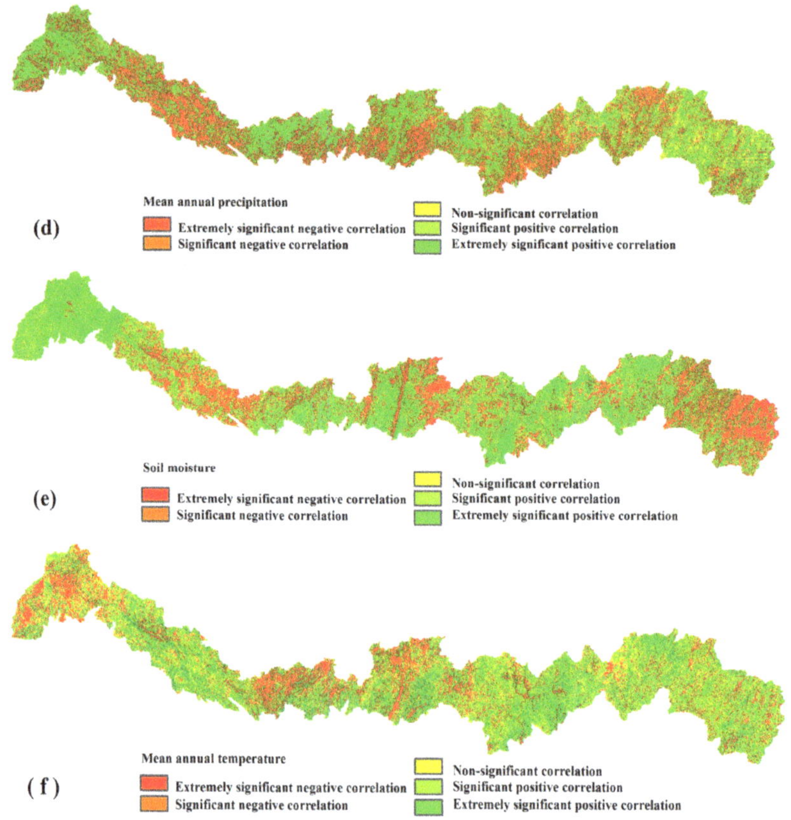

Figure 9. Spatial distribution characteristics of partial correlation between NDVI in Miaoling and precipitation seasonality (CV) (**a**), VPD (**b**), precipitation of wettest quarter (**c**), MAP (**d**), soil moisture (**e**), and MAT (**f**).

In summary, precipitation seasonality (CV) and VPD in climate factors showed positive and negative related relationships with NDVI distribution and were all widely distributed across the study area, but the spatial heterogeneity of other climatic factors with NDVI was significantly complicated with them.

On the time scale, the average annual value of NDVI shows significant fluctuations in the study time series, but overall, it shows a slow growth trend (Figure 2a). The linear fitting trend of precipitation seasonality (CV) and the NDVI is upward (Figure 10a). Figure 10b shows that the NDVI value and precipitation seasonality (CV) have basically similar dynamic changes, and the NDVI increases and decreases with the increase and decrease in precipitation seasonality (CV). This result shows that precipitation seasonality (CV) has a positive effect on the increase and decrease in NDVI. On the contrary, the fitting trend of VPD and NDVI is decreasing (Figure 10c) because the general feature of VPD change is a significant upward trend, while the change of NDVI is contrary to the fluctuation of VPD. Therefore, it can also be concluded that there is a negative correlation between VPD and NDVI.

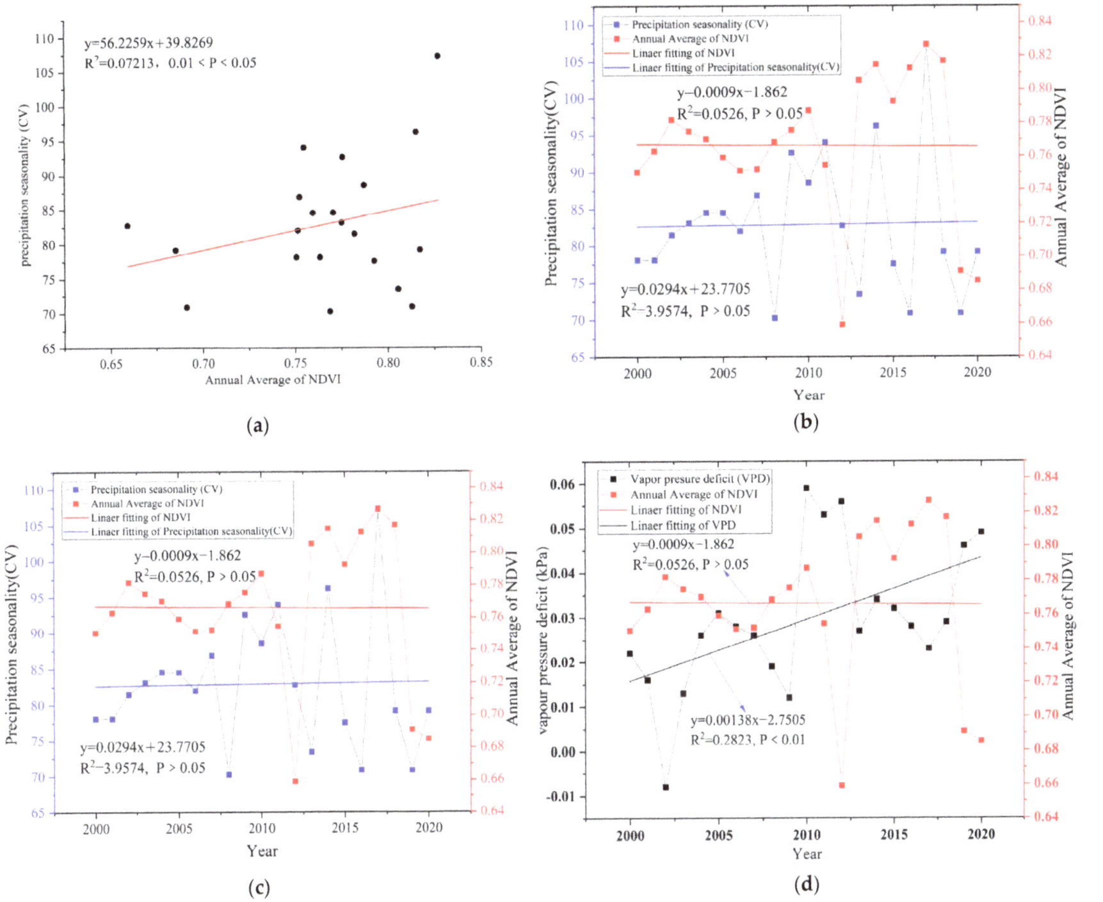

Figure 10. Variation trends and correlations of precipitation seasonality (CV), VPD, and NDVI. (**a**) Linear fitting of NDVI and precipitation seasonality (CV) trend; (**b**) Annual average change of NDVI and precipitation seasonality (CV); (**c**) Linear fitting of NDVI and VPD trend; (**d**) Annual average change of NDVI and VPD.

3.4. Detection of the Impact of Key Factors on NDVI

3.4.1. Detection Factor Influence

In order to further explore the driving characteristics of various influencing factors on the dynamic changes of Miaoling's vegetation, this section used 13 factors (Table 2) covering climate, soil, terrain, geomorphology, and human activities to conduct factor driving force and factor interaction detection on the NDVI of Miaoling's vegetation. To explore the contribution rate and interaction of climate factors and human activities to vegetation change in the Miao Mountains and further verify the differences in the main impact factors of NDVI in the Miao Mountains. The q value of the factor detection results reflects the influence of each factor on the NDVI of Miaoling's vegetation (explanatory power). The factor detection results show that the explanatory power of each factor on the NDVI is in the following order: NLI > non-karst > land use change > precipitation seasonality (CV) > karst > VPD > elevation > precipitation of the wettest quarter > MAP > slope > soil moisture > MAT > aspect. Specifically, NLI (X13) has the strongest explanatory power for NDVI (q = 0.422), and aspect (X9) has the smallest explanatory power for NDVI (q = 0.125).

The explanatory power of the above factors passed the 0.05 test with a confidence level of 95%. Overall, the explanatory power of human activities on vegetation NDVI is greater than the climactic factors in Miaoling. Additionally, the order of partial correlation and explanatory power of each factor for NDVI is consistent.

3.4.2. Detection Factor Interaction Analysis

The change in vegetation and the existence of spatial heterogeneity are driven by many factors. The interaction detection results of various influencing factors in Miaoling (Figure 11) show that their interaction is manifested as dual factor enhancement and non-linear enhancement, and all interaction factors have obvious enhancement characteristics on the driving force of NDVI compared to a single influencing factor. Among them, the interaction between land use change and NLI [q (X12∩X13) = 0.459] has the strongest explanatory power for the spatial distribution of NDVI, showing a driving feature of dual factor enhancement. On the contrary, the interaction between soil moisture and aspect orientation [q (X5∩X9) = 0.112] has the least explanatory power on NDVI. In addition, the research results also indicate that the interaction between human activity factors and other factors is significantly greater than that between other factors. It can be seen that the explanatory power of the interaction between human activities and other factors in the Miaoling area is dominant and that it is a key driving factor affecting the vegetation change in the Miaoling area.

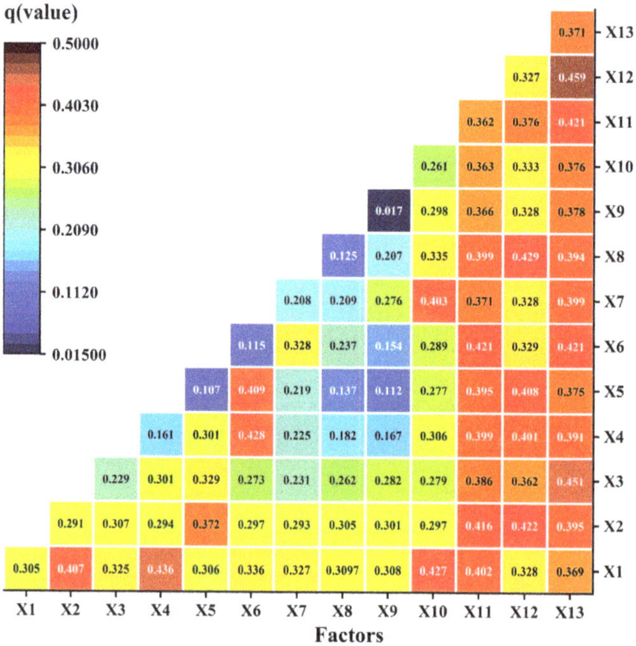

Figure 11. Explanatory power of interaction between key factors. X1, precipitation seasonality (CV); X2, VPD; X3, precipitation of wettest quarter; X4, MAP; X5, soil moisture; X6, MAT; X7, elevation; X8, slope; X9, aspect; X10, karst; X11, non-karst; X12, land use change; X13, NLI.

4. Discussion

4.1. Analysis of the Spatial Distribution Trend of NDVI

Previous research has confirmed that the ecological environment of karst regions has greatly improved [32–34,38,51–53]. Our research findings indicate that the Miaoling region has exhibited sluggish growth in NDVI over time compared to non-karst areas, indicating an improvement in their ecological environment [53,54]. This positive ecological change is attributed to the residents' strong awareness of environmental protection as well as the preservation of subtropical evergreen broad-leaved forests, evergreen and deciduous broad-leaved mixed forests, and evergreen shrubs in the area [34,38]. Furthermore, the vegetation coverage exhibited a distribution pattern of "gradually increasing from west to east", as confirmed by an average Hurst index of 0.56. The slow and incremental increase in NDVI observed in our study area is consistent with the continuous improvement in vegetation coverage seen in other study areas across China [20–22,55]. The spatial heterogeneity of the impact of all climate factors on vegetation cover change is evident. Nevertheless, as demonstrated by Figures 6 and 7, the utilization of arable land and the increase in artificial surfaces have resulted in a consistent decline in NDVI values in the related areas.

As mentioned above, the NDVI variation in Miaoling not only shows a slow upward trend over the whole period, but there are also strong downward processes. As shown in Figure 2, NDVI declined considerably after 2010 and 2018. Notably, a considerable decrease occurred in 2012, and this phenomenon can be primarily attributed to the backdrop of global climate change, paralleling the NDVI changes seen in other karst regions of China. Investigations of this phenomenon reveal that the most severe drought and extreme weather in 50 years occurred in the karst region of southwest China in 2010 [56], leading to a sharp decline in Miaoling's NDVI that continued through 2012 and culminated in the lowest NDVI value of the entire research period. Furthermore, it is likely that the natural disasters that occurred in 2017 have had a negative impact on vegetation dynamics, thereby comprehensively influencing the overall NDVI trend in the Miaoling Mountains during the past 21 years.

4.2. NDVI Variation and Land Use Change

Human activity is widely recognized as an important driver of vegetation cover. Land use change, representing human activity, has been shown to be highly correlated with vegetation change [24,26,49,57]. Previous research studies have demonstrated the importance of land use change as a key factor in the spatial distribution of the NDVI, which significantly influences vegetation change, regional ecological security [58], and ecosystem services [59]. Our study, depicted in Figure 6b, reveals that forests in the Miaoling area have the highest NDVI value, followed by grass, shrub areas, arable land, wetland, water areas, and artificial surfaces. The changes in NDVI's distribution across different spatial and temporal scales are driven by both climatic factors and human activities [1,12,15]. The explanatory power of human activities on NDVI changes is much higher than that of climate factors, as shown in Table 4, indicating that NLI, land use change, and human activities play significant roles in the ecosystem. Land use change represent ongoing challenges for vegetation variation and the effects of anthropogenic activity [19].

The land use change characteristics of human activities in Miaoling reveal that vegetation coverage has been significantly impacted by land use change over the past 21 years. Human activities such as returning farmland to forests and grass, traditional farming practices, and the combination of human activities and climate change have played a significant role in improving or degrading the vegetation [53]. Changes in forest, arable land, and artificial land surfaces have had a considerable impact on regional vegetation coverage, primarily due to the tradeoff between forests and arable land. The distribution of land use change and changes significantly influence the NDVI's spatial distribution in the region.

4.3. Impact Climatic Factors of NDVI Variation

Understanding the relationship between regional NDVI changes and climatic factors is critical for predicting regional vegetation changes and for effective ecological restoration management [56]. Climate change affects vegetation growth and change through dynamic changes and interactions between different weather factors. Miaoling's spatial heterogeneity is evident from its correlation distributions between NDVIs and six climate factors. This is due, in part, to the spatial variability of climate change. The strongest correlation among the climatic factors is precipitation seasonality (CV), as shown in Table 3. Compared with results from other regions, vegetation cover (0.0009/year) in the Miaoling's karst region is more sensitive to changes in climatic factors, including precipitation seasonality (CV) and VPD.

Seasonal changes in precipitation have a significant effect on NDVI, as shown by research [60]. Precipitation is the main driver of vegetation change [61]. Seasonal changes in precipitation can also influence vegetative phenology and cover [20]. In areas of high humidity, the risk of drought is lower, making the growing season of vegetation more sensitive to precipitation seasonality (CV) in order to maximize water benefits [53,55]. The results of this study show that the regularity of the temporal change of NDVI in Miaoling is similar to that of precipitation seasonality (CV) (Figure 10b). Obviously, in the period of precipitation seasonality (CV) and precipitation of the wettest quarter (i.e., plant growth season) (Figure 12a), when precipitation is abundant, with an increase in temperature to a certain extent, the photosynthesis, respiration, and transpiration processes of plants can be increased, and plant growth can be promoted. This is based on the strong precipitation seasonality in Miaoling, in the subtropical monsoon region. The rate of change of precipitation and the precipitation in the wettest season strongly promote the growing season of plants in Miaoling, and the trend of change is consistent, so it has a significant regulating effect on NDVI. NDVI is also highly positively correlated with precipitation seasonality (CV) and precipitation in the wettest season in terms of spatial heterogeneity (Figure 9a,c).

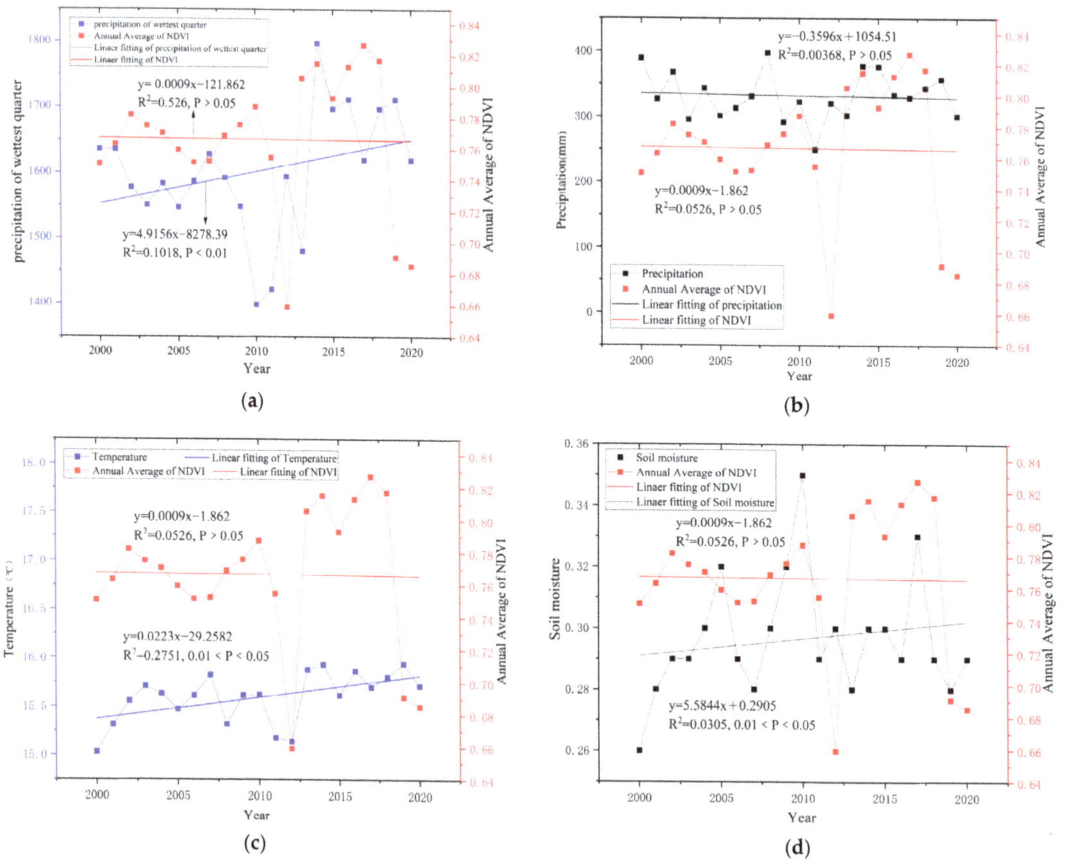

Figure 12. Time variation and trend of NDVI and precipitation of wettest quarter (**a**); MAP (**b**); MAT (**c**); soil moisture (**d**).

Furthermore, there is a significant negative correlation between VPD and NDVI, which is second only to precipitation seasonality (CV) in explaining NDVI changes. VPD also plays an important role in the interannual variability of NDVI, as an increase in VPD causes a decrease in NDVI (Figure 10c). However, current Earth system models underestimate VPD's interannual variability and its effect on GPP and NEP [50] by ignoring VPD's indirect influence on NDVI. Due to global warming, VPD is increasing, and vegetation is severely affected [55,62]. In addition, previous research has found that the vegetative landscape is "browning", i.e., plant growth is decreasing [7,63]. Above a certain threshold, plant photosynthesis and growth in most species are limited, which leads to a higher risk of hydraulic failure and a decrease in NDVI.

Temperature and precipitation are the most significant climatic drivers of vegetation growth, as per previous studies [9,11,16,20]. The change in vegetation phenology, structure, and coverage has been observed to be significant from tropical to northern regions and from coastal to inland areas [7,12,19]. However, the karst landscape, which is complex and heterogeneous, has also contributed to vegetation change, including drought and land degradation, due to rocky desertification [64,65]. The results of this study show that NDVI increases with the increase in temperature in the study time sequence (Figure 12c). The increase in temperature can increase the photosynthetic efficiency of vegetation and prolong its growth period, thereby improving the status of NDVI [6]. However, the partial correlation is slightly smaller than that of precipitation (Table 3). Contrary to the results of other studies, "temperature has a stronger impact on NDVI than precipitation" [55,64].

A study conducted on the interaction between NDVI and the climate of karst vegetation in Guizhou revealed that the effect of MAT on NDVI is stronger than that of MAP [53,60]. The study concluded that the explanatory power of MAT on NDVI in the Miaoling area is lower than that of MAP. The binary hydrological structure commonly found in karst regions leads to a substantial loss of precipitation [23], suggesting that water resources are not efficiently utilized for the thriving of vegetation. The study's findings are consistent with the fact that precipitation has a lower positive correlation with NDVI changes. This is particularly relevant in karst areas that are prone to drought, where precipitation changes have a significant impact on vegetation [34,62,66]. It is confirmed that NDVI is more sensitive to precipitation than temperature in Miaoling.

Soil moisture mainly comes from precipitation and affects vegetation growth [8,17]. Therefore, it is also one of the important factors limiting vegetation growth in karst areas. Due to the relatively thin nature of the karst soil layer, soil water is easy to lose, which limits plant growth and ecological recovery in karst areas [50,67]. In this study, the explanatory power of soil moisture is relatively weak, which also proves that the characteristics of soil moisture loss under the dual structure in karst areas are easy and the positive effect on NDVI is not significant. As shown in Figures 9e and 12d, the soil moisture of Miaoling's arable areas (mainly in non-karst areas) is highly positively correlated with NDVI, which accounts for 43.7% of Miaoling.

4.4. *Influence of Factor Interaction on NDVI*

Changes in vegetation growth are inextricably linked to climate variation and human activities. Based on the detection of geographical detectors, we further analyzed the interaction of 13 factors on NDVI. We found that NLI between 2000 and 2020 is the strongest driving factor for explaining the changes in vegetation in Miaoling. Additionally, the results showed that the interaction between various factors has significantly higher explanatory power for the changes in the NDVI than itself (Figure 11), which is consistent with other research results [13,16,66].

In the study, NLI has the greatest explanatory power for the trend of vegetation cover change ($q = 0.422$, Table 4). There are a few studies that combine NLI with NDVI, indicating that urbanization has a negative impact on vegetation coverage or the ecological environment by increasing NLI [30,68,69]. There is a significant negative correlation between vegetation coverage and NLI values. From Figures 5 and 13, it can be seen that there is a significant overlap between the areas where NDVI continues to degrade and the areas where NLI significantly increases. The areas where NDVI improves mostly correspond to areas with a low or no nighttime light index.

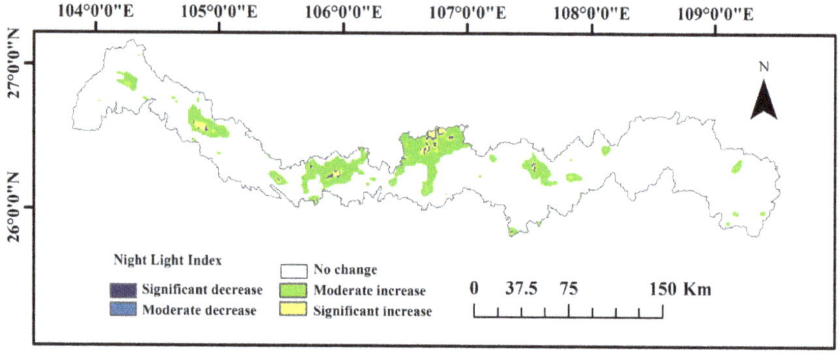

Figure 13. Characteristics and spatial distribution of NLI changes from 2000 to 2020.

Studies have shown that the interaction between precipitation seasonality (CV) (q = 0.355), land use change (q = 0.0.397), non-karst areas (q = 0.401), and NLI (q = 0.422) has the most obvious effect on the NDVI of vegetation (Table 4). Unlike other studies [5,7,43], we found that two factors of human activity and their interaction with other factors have a high explanatory power for the changes in NDVI, and it mainly shows a bifactor enhancement; the contribution rate is very large. This also shows that human activity can significantly influence the NDVI in the karst region [37,39,55]. Among them, the interaction between NLI and land use change (q (X13∩X12) = 0.459) has the largest impact on NDVI, and it mainly shows a bifactor enhancement. Therefore, human activities can be identified as the dominant factor in vegetation dynamics, while other factors only serve as constraints in karst basins. Relevant studies on the Loess Plateau have also shown this [15,20,58].

Temperature and precipitation are considered the foremost drivers of vegetation growth with regards to climatic factors, as cited by previous research [8,10,13,19]. Although the influence of mean annual temperature (MAT) (q = 0.155) and mean annual precipitation (MAP) (q = 0.21) on vegetation in Miaoling does not differ significantly, their interactions (q (X4∩X6) = 0.428) exhibit a significantly higher impact on NDVI compared to their individual effects. Furthermore, soil moisture and VPD are strongly linked to temperature and precipitation [69,70]. A few researchers have reported a direct or indirect dependence of VPD's impact on NDVI on the prevailing temperature and soil moisture conditions [25,71,72]. Hence, when the response of plants in an ecosystem decreases the evaporation capacity due to atmospheric drying, the conservation of soil moisture improves, along with some evidence of NDVI growth. This suggests that hydrothermal conditions in subtropical regions significantly influence vegetation growth and change. Moreover, the impact of factors such as altitude, slope, and aspect interaction are relatively minor in driving vegetation trends in Miaoling, but this impact increases significantly under the influence of human activities, reinforcing the crucial role of human activities in vegetation change in the Miaoling (karst area) region. In other words, the varying responses of vegetation NDVI to climate factors and human activities may be explained by the interaction of various factors in terms of temporal and spatial scale differences.

4.5. Limitations of This Study

In this study, we explored the dynamic change characteristics and trends of the regional vegetation's NDVI in Miaoling and conducted a driving analysis using climate factors, human activities, and topography. However, there are still some limitations. Firstly, we did not analyze the seasonal characteristics of vegetation growth in Miaoling, such as growing season and non-growing season. In future research, we should pay attention to the differences between these vegetation changes and their comprehensive relationship. Secondly, due to the limitation of spatial resolution differences in NDVI and climate factors, the NDVI variation trend of some pixels may be overestimated or underestimated, as may also be found in correlation analyses. Despite these shortcomings, this work is helpful to comprehensively understand the spatiotemporal characteristics of Miaoling's vegetation and the driving factors of vegetation dynamics. At the same time, it provides a reference value for the dynamic driving factors of vegetation in important karst basins.

5. Conclusions

Based on NDVI data, the present study examines the alterations in vegetation coverage within a karst basin watershed from 2000 to 2020. Furthermore, the research also investigates the impact exerted by climate factors, topography, and human activities on Miaoling's vegetation and how they interact with each other. The key results of this study are as follows:

(1) Under the pixel scale and spatial distribution in Miaoling, the vegetation coverage gradually increases from west to east. During the study period, the NDVI of Miaoling's vegetation showed an overall upward trend (0.0009/year), with an average value of 0.766, 53.82% of the region continuing to grow, and a distribution pattern of "gradually increasing from west to east". The vegetation in NDVI showing an upward trend is much larger than the area showing a downward trend, and the site with a downward trend is mainly in the western and central parts of Miaoling.

(2) The correlation between the vegetation's NDVI and meteorological factors presents significant spatial heterogeneity. Climate change has a two-sided impact on NDVI changes in vegetation in the study area because there is a positive promoting effect and a relatively inhibitory effect for the NDVI. The NDVI and VPD of the vegetation in the study area show a negative correlation and a positive correlation with the other five climate factors as a whole, with the greatest correlation being with precipitation seasonality (CV).

(3) Compared with climate change and landform factors, human activity factors have a greater driving force on the NDVI of Miaoling's vegetation, and their interaction with other factors is also significantly higher, which also shows that the dynamic change and development trend of the NDVI of Miaoling's vegetation are strongly affected by human activities. Therefore, human activities can be considered the dominant factor driving NDVI changes in the Miaoling area.

Author Contributions: Conceptualization, Y.W. and S.L.; methodology, G.L.; software, L.G. and J.Y.; formal analysis, C.G. and F.Y.; investigation, J.Y., L.G., Z.S. and X.Y.; data curation, Y.C., H.P. and X.Y.; writing—original draft preparation, Y.W.; writing—review and editing, C.G. and S.L.; visualization, J.Y. and L.G.; supervision, Y.X.; project administration, G.L.; funding acquisition, Y.W. All authors have read and agreed to the published version of the manuscript.

Funding: This research was funded by Guizhou Provincial Science and Technology Projects (QKHJC-ZK [2022] YB334); Guizhou Provincial Science and Technology Projects (QKHZC [2023] YB228) and Doctoral program of Guizhou Education University (X2023024).

Data Availability Statement: Not applicable.

Acknowledgments: We thank the anonymous reviewers for their valuable comments. We gratefully acknowledge the design of S.L. and the contributions of the co-authors.

Conflicts of Interest: The authors declare no conflict of interest.

References

1. Karl, T.R.; Trenberth, K.E. Modern Global Climate Change. *Science* **2003**, *302*, 1719–1723. [CrossRef] [PubMed]
2. Bonan, G.B.; Pollard, D.; Thompson, S.L. Effects of Boreal Forest Vegetation on Global Climate. *Nature* **1992**, *359*, 716–718. [CrossRef]
3. Haberl, H.; Erb, K.H.; Krausmann, F.; Gaube, V.; Bondeau, A.; Plutzar, C.; Gingrich, S.; Lucht, W.; Fischer-Kowalski, M. Quantifying and Mapping the Human Appropriation of Net Primary Production in Earth's Terrestrial Ecosystems. *Proc. Natl. Acad. Sci. USA* **2007**, *104*, 12942–12947. [CrossRef]

4. Zhang, S.R.; Bai, X.Y.; Zhao, C.; Tan, Q.; Luo, G.; Wu, L.; Xi, H.; Li, C.; Chen, F.; Ran, C.; et al. China's Carbon Budget Inventory from 1997 to 2017 and Its Challenges to Achieving Carbon Neutral Strategies. *J. Clean. Prod.* **2022**, *347*, 130966. [CrossRef]
5. Kong, D.; Miao, C.; Wu, J.; Zheng, H.; Wu, S. Time Lag of Vegetation Growth on the Loess Plateau in Response to Climate Factors: Estimation, Distribution, and Influence. *Sci. Total Environ.* **2020**, *744*, 140726. [CrossRef]
6. Piao, S.L.; Wang, X.H.; Park, T.; Chen, C.; Myneni, R.B. Characteristics, Drivers and Feedbacks of Global Greening. *Nat. Rev. Earth Environ.* **2019**, *1*, 14–27. [CrossRef]
7. Sun, Z.H.; Mao, Z.G.; Yang, L.Y.; Liu, Z.; Han, J.C.; Wanag, H.Y.; He, W. Impacts of Climate Change and Afforestation on Vegetation Dynamic in the Mu Us Desert, China. *Ecol. Indic.* **2021**, *129*, 108020. [CrossRef]
8. Liu, Y.; Li, Y.; Li, S.C.; Motesharrei, S. Spatial and Temporal Patterns of Global NDVI Trends: Correlations with Climate and Human Factors. *Remote Sens.* **2015**, *7*, 13233–13250. [CrossRef]
9. Tian, F.; Fensholt, R.; Verbesselt, J.; Grogan, K.; Horion, S.; Wang, Y.J. Evaluating Temporal Consistency of Long-Term Global NDVI Datasets for Trend Analysis. *Remote Sens. Environ.* **2015**, *163*, 326–340. [CrossRef]
10. Zheng, K.Y.; Tan, L.H.; Sun, Y.W.; Wu, Y.J.; Duan, Z.; Xu, Y.; Gao, C. Impacts of Climate Change and Anthropogenic Activities on Vegetation Change: Evidence from Typical Areas in China. *Ecol. Indic.* **2021**, *126*, 107648. [CrossRef]
11. Zhu, Z.C.; Piao, S.L.; Myneni, R.B.; Huang, M.; Zeng, Z.Z.; Canadell, J.G.; Ciais, P.; Sitch, S.; Friedlingstein, P.; Arneth, A.; et al. Greening of the Earth and its drivers. *Nat. Clim. Chang.* **2016**, *6*, 791. [CrossRef]
12. Piao, S.L.; Friedlingstein, P.; Ciais, P.; Zhou, L.M.; Chen, A.B. Effect of Climate and CO_2 Changes on the Greening of the Northern Hemisphere over the Past Two Decades. *Geophys. Res. Lett.* **2006**, *33*, 432–436. [CrossRef]
13. Rogier, D.J.; Jan, V.; Achim, Z.; Michael, S. Shifts in Global Vegetation Activity Trends. *Remote Sens.* **2013**, *5*, 1117–1133. [CrossRef]
14. Braswell, B.H.; Schimel, D.S.; Under, E.; Iii, B.M. The Response of Global Terrestrial Ecosystems to Interannual Temperature Variability. *Science* **1997**, *278*, 870–872. [CrossRef]
15. Liu, Z.J.; Wang, J.Y.; Wang, X.Y.; Wang, Y.S. Understanding the impacts of 'Grain for Green' land management practice on land greening dynamics over the Loess Plateau of China. *Land Use Policy* **2020**, *99*, 105084. [CrossRef]
16. Bogaert, J. Evidence for a Persistent and Extensive Greening Trend in Eurasia Inferred from Satellite Vegetation Index Data. *J. Geophys. Res. Atmos.* **2002**, *107*, ACL 4-1–ACL 4-14. [CrossRef]
17. Shen, M.G.; Piao, S.L.; Cong, N.; Zhang, G.G.; Jassens, I.A. Precipitation impacts on vegetation spring phenology on the Tibetan Plateau. *Glob. Chang. Biol.* **2015**, *21*, 3647–3656. [CrossRef]
18. Sun, J.; Cheng, G.; Li, W.; Sha, Y.; Yang, Y. On the Variation of NDVI with the Principal Climatic Elements in the Tibetan Plateau. *Remote Sens.* **2013**, *5*, 1894–1911. [CrossRef]
19. Zhou, X.Y.; Shi, H.D.; Wang, X.R. Impact of Climate Change and Human Activities on Vegetation Coverage in the Mongolian Plateau. *Arid Zone Res.* **2014**, *31*, 604–610. (In Chinese)
20. Shi, S.; Yu, J.; Wang, F.; Wang, P.; Zhang, Y.; Jin, K. Quantitative Contributions of Climate Change and Human Activities to Vegetation Changes over Multiple Time Scales on the Loess Plateau. *Sci. Total Environ.* **2021**, *755*, 142419. [CrossRef]
21. Zhang, B.; Wang, D.; Wang, G.G.; Qiong, M.A.; Zhang, G.B.; Ding-Min, J.I. Vegetation cover change over the southwest China and its relation to climatic factors. *Resour. Environ. Yangtze Basin.* **2015**, *24*, 956–964. (In Chinese)
22. Du, C.C.; Bai, X.Y.; Li, Y.B.; Tan, Q.; Zhao, C.W.; Luo, G.J.; Wu, L.H.; Chen, F.; Li, C.J.; Ran, C.; et al. Inventory of China's Net Biome Productivity since the 21st Century. *Land* **2022**, *11*, 1244. [CrossRef]
23. Yao, J.Q.; Li, M.Y.; Zheng, J.H. Assessing the Spatiotemporal Evolution of Anthropogenic Impacts on Remotely Sensed Vegetation Dynamics in Xinjiang, China. *Remote Sens.* **2021**, *13*, 4651. [CrossRef]
24. Guo, E.L.; Wang, Y.F.; Wang, C.L.; Sun, Z.Y.; Bao, Y.L.; Mandula, N.; Jirigala, B.; Bao, Y.H.; Li, H. NDVI Indicates Long-Term Dynamics of Vegetation and Its Driving Forces from Climatic and Anthropogenic Factors in Mongolian Plateau. *Remote Sens.* **2021**, *13*, 688. [CrossRef]
25. McDowell, N.G.; Allen, C.D. Darcy's law predicts widespread forest mortality under climate warming. *Nat. Clim. Chang.* **2015**, *5*, 669–672. [CrossRef]
26. Williams, A.P.; Allen, C.D.; Macalady, A.K. Temperature as a potent driver of regional forest drought stress and tree mortality. *Nat. Clim. Chang.* **2013**, *3*, 292–297. [CrossRef]
27. Feng, K.; Wang, T.; Liu, S.; Yan, C.; Kang, W.; Chen, X.; Guo, Z. Path analysis model to identify and analyze the causes of aeolian desertification in Mu Us Sandy Land. *China. Ecol. Indic.* **2021**, *124*, 107386. [CrossRef]
28. Liu, X.; Pei, F.; Wen, Y.; Li, X.; Wang, S.; Wu, C.; Cai, Y.L.; Wu, J.; Chen, J.; Feng, K.; et al. Global urban expansion offsets climate-driven increases in terrestrial net primary productivity. *Nat. Commun.* **2019**, *10*, 5558. [CrossRef]
29. Liu, Z.; He, C.; Zhang, Q.; Huang, Q.; Yang, Y. Extracting the dynamics of urban expansion in China using DMSP-OLS nighttime light data from 1992 to 2008. *Landse Urban Plan.* **2012**, *106*, 62–72. [CrossRef]
30. Ma, T.; Zhou, C.; Pei, T.; Haynie, S.; Fan, J. Quantitative estimation of urbanization dynamics using time series of DMSP/OLS nighttime light data: A comparative case study from China's cities. *Remote Sens. Environ.* **2012**, *124*, 99–107. [CrossRef]
31. Fu, H.; Shao, Z.; Fu, P.; Cheng, Q. The Dynamic Analysis between Urban Nighttime Economy and Urbanization Using the DMSP/OLS Nighttime Light Data in China from 1992 to 2012. *Remote Sens.* **2017**, *9*, 416. [CrossRef]
32. Zhu, S.Q.; Wei, L.M.; Chen, Z.R.; Zhang, C.G. A Preliminary Study on Biomass Components of Karst Forest in Maolan of Guizhou Province, China. *ACTA Phytoecol. Sin.* **1995**, *19*, 358. (In Chinese)

33. Liu, Y.; Liu, C.; Wang, S.; Guo, K.; Yang, J.; Zhang, X.; Li, G. Organic Carbon Storage in Four Ecosystem Types in the Karst Region of Southwestern China. *PLoS ONE* **2013**, *8*, e56443. [CrossRef]
34. Ni, J.; Luo, D.H.; Xia, J.; Zhang, Z.H.; Hu, G. Vegetation in Karst Terrain of Southwestern China Allocates More Biomass to Roots. *Solid Earth Discuss.* **2015**, *7*, 1209–1235. [CrossRef]
35. Tu, Y.L.; Yang, J. Study on biomass of the karst scrub community in central region of Guizhou province. *Carsol. Sin.* **1995**, *3*, 199–208. (In Chinese)
36. Huang, Q.H.; Cai, Y.L. Spatial Pattern of Karst Rock Desertification in the Middle of Guizhou Province, Southwestern China. *Environ. Geol.* **2007**, *52*, 1325–1330. [CrossRef]
37. Wu, Y.Y.; Liu, L.B.; Guo, C.Z.; Zhang, Z.H.; Hu, G.; Ni, J. Low carbon storage of woody debris in a karst forest in southwestern China. *Acta Geochim.* **2019**, *38*, 576–586. [CrossRef]
38. Chen, F.; Bai, X.Y.; Liu, F.; Luo, G.J.; Tian, Y.C.; Qin, L.Y.; Li, Y.; Xu, Y.; Wang, J.F.; Wu, L.H.; et al. Analysis Long-Term and Spatial Changes of Forest Cover in Typical Karst Areas of China. *Land* **2022**, *11*, 1349. [CrossRef]
39. Yang, J.L.; Dong, J.W.; Xiao, X.M.; Dai, J.H.; Wu, C.Y.; Xia, J.Y.; Zhao, G.S.; Zhao, M.M.; Li, Z.L.; Zhang, Y.; et al. Divergent Shifts in Peak Photosynthesis Timing of Temperate and Alpine Grasslands in China. *Remote Sens. Environ.* **2019**, *233*, 111395. [CrossRef]
40. Yang, J.; Huang, X. The 30 m annual land cover datasets and its dynamics in China from 1990 to 2021 [Data set]. *Earth Syst. Sci. Data* **2022**, *13*, 3907–3925. [CrossRef]
41. Peng, S.; Ding, Y.; Liu, W.; Zhi, L. 1 Km Monthly Temperature and Precipitation Dataset for China from 1901 to 2017. *Earth Syst. Sci. Data* **2019**, *11*, 1931–1946. [CrossRef]
42. Qu, L.S.; Zhu, Q.; Zhu, C.F. Monthly Precipitation Data Set with 1 km Resolution in China from 1960 to 2020. Available online: https://www.scidb.cn/en/detail?dataSetId=ff7ee051d2d44ab4a221cd810bf37251 (accessed on 18 August 2022).
43. Chen, Y.Z.; Feng, X.M.; Fu, B.J. An improved global remote-sensing-based surface soil moisture (RSSSM) dataset covering 2003–2018. *Earth Syst. Sci. Data* **2021**, *13*, 1–31. [CrossRef]
44. Fick, S.E.; Hijmans, R.J. WorldClim 2: New 1 km Spatial Resolution Climate Surfaces for Global Land Areas. *Int. J. Climatol.* **2017**, *37*, 4302–4315. [CrossRef]
45. Jiang, W.G.; Yuan, L.H.; Wang, W.J.; Cao, R.; Zhang, Y.F.; Shen, W.M. Spatio-temporal analysis of vegetation variation in the Yellow River Basin. *Eco. Indic.* **2015**, *51*, 117–126. [CrossRef]
46. Tosic, I. Spatial and Temporal Variability of Winter and Summer Precipitation over Serbia and Montenegro. *Theor. Appl. Climatol.* **2004**, *77*, 47–56. [CrossRef]
47. Correa-Díaz, A.; Romero-Sánchez, M.E.; Villanueva-Díaz, J. The greening effect characterized by the Normalized Difference Vegetation Index was not coupled with phenological trends and tree growth rates in eight protected mountains of central Mexico. *For. Ecol. Manag.* **2021**, *496*, 119402. [CrossRef]
48. Hurst, H.E. Long-Term Storage Capacity of Reservoirs. *Trans. Am. Soc. Civ. Eng.* **1951**, *116*, 770–799. [CrossRef]
49. Wang, J.F.; Zhang, T.L.; Fu, B.J. A measure of spatial stratified heterogeneity. *Ecol. Indic.* **2016**, *67*, 250–256. [CrossRef]
50. Deng, X.J.; Hu, S.; Zhan, C.H. Attribution of vegetation coverage change to climate change and human activities based on the geographic detectors in the Yellow. *Environ. Sci. Pollut. Res.* **2022**, *29*, 44693–44708. [CrossRef]
51. Yang, Q.Q.; Wang, K.L.; Zhang, C.; Yue, Y.M.; Tian, R.C.; Fan, F.D. Spatio-Temporal Evolution of Rocky Desertification and Its Driving Forces in Karst Areas of Northwestern Guangxi, China. *Environ. Geol.* **2011**, *64*, 383–393. [CrossRef]
52. Wu, L.H.; Wang, S.J.; Bai, X.Y.; Chen, F.; Li, C.J.; Ran, C.; Zhang, S.R. Identifying the Multi-Scale Influences of Climate Factors on Runoff Changes in a Typical Karst Watershed Using Wavelet Analysis. *Land* **2022**, *11*, 1284. [CrossRef]
53. Hou, W.J.; Gao, J.B.; Wu, S.H.; Dai, E.F. Interannual Variations in Growing-Season NDVI and Its Correlation with Climate Variables in the Southwestern Karst Region of China. *Remote Sens.* **2015**, *7*, 11105–11124. [CrossRef]
54. Gao, J.B.; Li, S.C.; Zhao, Z.Q.; Cai, Y.L. Investigating Spatial Variation in the Relationships between NDVI and Environmental Factors at Multi-Scales: A Case Study of Guizhou Karst Plateau, China. *Int. J. Remote Sens.* **2012**, *33*, 2112–2129. [CrossRef]
55. Zhang, W.; Zhou, L.; Zhang, Y.; Chen, Z.J.; Hu, F.N. Impacts of Ecological Migration on Land Use and Vegetation Restoration in Arid Zones. *Land* **2022**, *11*, 891. [CrossRef]
56. Muradyan, V.; Tepanosyan, G.; Asmaryan, S.; Saghatelyan, A.; Dell'Acqua, F. Relationships between NDVI and climatic factors in mountain ecosystems: A case study of Armenia. *Remote Sens. Appl. Soc. Environ.* **2019**, *14*, 158–169. [CrossRef]
57. Snyder, K.A.; Tartowski, S.L. Multi-Scale Temporal Variation in Water Availability: Implications for Vegetation Dynamics in Arid and Semi-Arid Ecosystems. *J. Arid Environ.* **2006**, *65*, 219–234. [CrossRef]
58. Fu, B.J.; Wang, S.; Liu, Y.B.; Liu, J.; Liang, W.; Miao, C.Y. Hydrogeomorphic Ecosystem Responses to Natural and Anthropogenic-Changes in the Loess Plateau of China. *Annu. Rev. Earth Planet. Sci.* **2016**, *45*, 223–243. [CrossRef]
59. Cairns, J.J. Ecosystem Services: An Essential Component of Sustainable Use. *Environ. Health Perspect.* **1995**, *103*, 534. [CrossRef]
60. Ma, S.B.; An, Y.L.; Yang, G.; Zhang, Y. The Analysis of the Difference Vegetation Variation and Driver Factors on NDVI Change in Karst Region: A Case on Guizhou. *Ecol. Environ. Sci.* **2016**, *25*, 1106–1114. [CrossRef]
61. Hua, W.J.; Chen, H.S.; Zhou, L.M.; Xie, Z.H.; Qin, M.H.; Li, X.; Ma, H.D.; Huang, Q.H.; Sun, S.L. Observational quantification of climatic and human influences on vegetation greening in China. *Remote Sens.* **2017**, *9*, 425. [CrossRef]
62. He, B.; Chen, C.; Lin, S.G.; Yuan, H.W.; Chen, D.L.; Zhang, Y.F.; Guo, L.L.; Zhao, X.; Liu, X.B.; Piao, S.L.; et al. Worldwide impacts of atmospheric vapor pressure deficit on the interannual variability of terrestrial carbon sinks. *Natl. Sci. Rev.* **2022**, *9*, nwab150. [CrossRef] [PubMed]

63. Du, X.L.; Wang, S.J. Space-time distribution of soil water in a karst area: A case south of the Wangjiazhai catchment, Qingzhen, Guizhou province. *Earth Environ.* **2008**, *3*, 193–201. (In Chinese)
64. Bai, X.Y.; Wang, S.J.; Xiong, K.N. Assessing spatial temporal evolution processes of karst rocky desertification land: Indications for restoration strategies. *Land Degrad. Dev.* **2013**, *24*, 47–56. [CrossRef]
65. Yan, X.; Cai, Y.L. Multi-scale anthropogenic driving forces of karst rocky desertification in southwest China. *Land Degrad. Dev.* **2015**, *26*, 193–200. [CrossRef]
66. Deng, Y.H.; Wang, S.J.; Bai, X.Y.; Luo, G.J.; Tian, S.Q. Characteristics of Soil Moisture Storage from 1979 to 2017 in the Karst Area of China. *Geocarto Int.* **2019**, *36*, 903–917. [CrossRef]
67. Yang, X.D.; Ali, A.; Xu, Y.L.; Jiang, L.M.; Lv, G.H. Soil Moisture and Salinity as Main Drivers of Soil Respiration across Natural Xeromorphic Vegetation and Agricultural Lands in an Arid Desert Region. *Catena* **2019**, *177*, 126–133. [CrossRef]
68. Li, S.X.; Wu, Q. Effects of China's ecological restoration on economic development based on Night-Time Light and NDVI data. *Environ. Sci. Pollut. Res.* **2021**, *28*, 65716–65730. [CrossRef]
69. Xu, B.; Lin, B. How industrialization and urbanization process impact on CO_2 emissions in China: Evidence from nonparametric additive regression models. *Energy Econ.* **2015**, *48*, 188–202. [CrossRef]
70. Massmann, A.; Gentine, P.; Lin, C. When does vapor pressure deficit drive or reduce evapotranspiration? *J. Adv. Model Earth Syst.* **2019**, *11*, 3305–3320. [CrossRef]
71. Fu, Z.; Ciais, P.; Prentice, I.C.; Gentine, P.; Makowski, D.; Bastos, A.; Luo, X.Z.; Green, J.K.; Stoy, P.C.; Yang, H.; et al. Atmospheric dryness reduces photosynthesis along a large range of soil water deficits. *Nat. Commun.* **2022**, *13*, 989. [CrossRef] [PubMed]
72. Humphrey, V.; Berg, A.; Ciais, P.; Gentine, P.; Jung, M.; Reichstein, M.; Seneviratne, S.I.; Frankenberg, C. Soil moisture–atmosphere feedback dominates land carbon uptake variability. *Nature* **2021**, *592*, 65–69. [CrossRef] [PubMed]

MDPI
St. Alban-Anlage 66
4052 Basel
Switzerland
www.mdpi.com

Land Editorial Office
E-mail: land@mdpi.com
www.mdpi.com/journal/land

www.ingramcontent.com/pod-product-compliance
Lightning Source LLC
Chambersburg PA
CBHW061600220926
51850CB00005B/380
* 9 7 8 3 0 3 6 5 9 2 2 0 6 *